or a listing of recent titles in the *Artech House Mobile Communications Library*,
turn to the back of this book.

Multiantenna Wireless Communication Systems

Multiantenna Wireless Communication Systems

Sergio Barbarossa

ARTECH HOUSE

BOSTON | LONDON

artechhouse.com

Library of Congress Cataloging-in-Publication Data

A catalog record for this book is available from the Library of Congress.

British Library Cataloguing in Publication Data
Barbarossa, Sergio
Multiantenna wireless communication systems.—(Artech House mobile communications library).
1. Mobile communication systems 2. Wireless communication systems
3. Antennas (Electronics) 4. Antenna arrays
I. Title
621.3'84135

ISBN 1-58053-634-4

Cover design by Igor Valdman

685 Canton Street
Norwood, MA 02062

International Standard Book Number: 1-58053-634-4
A Library of Congress Catalog Card number is available for this book.

10 9 8 7 6 5 4 3 2 1

To my parents,
Lillo Barbarossa and Maria Morettini,
the two people from whom I learned the most.
Even though they are not physically here anymore,
I am sure that their joyful spirit is alive and smiling somewhere.

"To
The inhabitants of Space in General
And H. C. in particular
This Work is Dedicated
By a Humble Native of Flatland
In the Hope that
Even as he was initiated into the Mysteries
Of THREE Dimensions
Having been previously conversant
With ONLY TWO
So the Citizens of that Celestial Region
May aspire yet higher and higher
To the Secrets of FOUR FIVE OR EVEN SIX Dimensions
Thereby contributing
To the Enlargement of the IMAGINATION
And the possible Development
Of that most rare and excellent Gift of MODESTY
Among the Superior Races
of SOLID HUMANITY"

—Edwin A. Abbott, from the Preface to Flatland

"Quelli che s'innamoan di pratica sanza scienza son come il nocchiere, ch'entra in navilio sanza timone o bussola, che mai ha certezza dove si vada."

"He who loves practice without theory is like the sailor who boards ship without a rudder and compass and never knows where he may cast."

—Leonardo da Vinci

Contents

Preface

The most relevant contradiction to be faced in the current development of modern telecommunication systems is the contrast between, on one side, the increasing demand for higher bit rates for a higher number of users and, on the other side, the scarceness of the available resources, typically bandwidth. This conflict pushes the system designers towards the development of systems capable of making the best use of the available resources or, in more technical terms, capable of maximizing the spectral efficiency, measured as the number of bit/s/Hz that can be transmitted and retrieved reliably. But coding theory has now arrived at developing coding strategies whose performance is so close to the fundamental Shannon's limits that there is no much further gain to be expected.

Nevertheless, there is a way to provide a *significant* improvement with respect to current systems. The idea comes from realizing that most current systems implicitly assume that the information is carried by a physical support (information signal) that is defined over a one-dimensional independent variable: time. However, in wireless communications, the information flows, naturally and inevitably, through space. Adding space to time creates a high number of additional degrees of freedom that can be exploited to increase the spectral efficiency. In other words, the way to go around the scarceness of resources is to add dimensionality to the problem, without increasing neither the bandwidth nor the power.

The question is then how to create additional degrees of freedom and how to use them to achieve the potential benefits. To incorporate space as an additional domain where the information flows, besides time, it is necessary to transmit from different points in space and to gather information from different points. This is achieved through *multiantenna* systems. Furthermore, since electromagnetic waves are polarized, additional degrees of freedom may be introduced by spreading information across different polarizations axes. All these different domains create the possibility for *multichannel* communications. It is this multiplicity that allows for a potential dramatic increase of spectral efficiency with respect to conventional systems.

The scope of this book is to provide the basic theoretical tools for designing and analyzing multiantenna communication systems. The goal is to supply the reader with the methodologies useful to design communication systems that exploit the multi-dimensionality made available by multiantenna systems. The number of publications in this area has grown so large in the last few years that it is impossible to cover all aspects of multiantenna systems in a single book with sufficient detail. Many topics which are historically part of multiantenna systems have been left out, like direction finding or adaptive beamforming. The choice made in writing this book has been to concentrate on the fundamental aspects of multiantenna communications, emphasizing their capability to increase the bit rate and/or increase the reliability in transmitting information over fading channels. Wherever possible, there is the attempt to provide a physical interpretation of the solutions.

Space-time coding plays a fundamental role in this book, as the way to achieve most of the promised benefits. Three chapters are entirely devoted to space-time coding and one more chapter addresses the relatively novel strategy of *distributed space-time coding*, as a novel paradigm to design cooperative communication systems. However, this is not a book on space-time coding. There are many other tools which are not, typically, part of a book on multiantenna systems. These include: *game theory*, as a tool to devise the optimal transmission strategies in a multiuser scenario, where different user terminals (players) compete with each other for the use of the available resources; *convex optimization* and *majorization theory*, as the basic tool for optimizing the transmission scheme when the transmitter has some knowledge about the channel; *random geometric graphs*, as the basic tool for studying the connectivity of wireless networks; *eigendecomposition of weakly inhomogeneous operators*, as a way to analyze the optimal transmission strategies over time-varying channels. The hope is to provide the reader with a series of tools, borrowed from different disciplines but presented in a single context, whose combination may give rise to a strong synergism and cross-fertilization.

It is my pleasure to thank a series of people with whom I enjoyed discussing problems and sharing ideas. I began my research activity on synthetic arrays long time ago, but in a context, imaging radars, completely different from the one described in this book. I started working on space-time coding with Prof. Georgios Giannakis years ago and I enjoyed the many animated discussions about the intricacies and challenges of this field. I am grateful towards the European Commission for funding two projects on multiantenna and multihop systems, namely SATURN and ROMANTIK, that supported the work of my research group. I have enjoyed working with these projects partners and, in particular, I wish to mention the many friends at the Polytechnic University of Barcelona (too many to be mentioned one by one without running the risk of forgetting someone), the colleagues at the University of Bristol, for having provided the real MIMO data that have been important to test the proposed algorithms, and the colleagues from Dune Systems.

I wish to send heartfelt tranks to Dr. Ananthram Swami for his encouragement, for reading the manuscript and giving me invaluable feedback. I also wish to thank the anonymous reviewer who gave me important feedback about the really basic physical motivations underlying the potentials of MIMO systems. I am especially thankful to Dr. Antonio Fasano, for his incredibly careful reading of the book, for having discussed together line by line and having contributed substantially in several chapters, in particular on the trace-orthogonal space-time coding design. I am also deeply grateful to Daniele Ludovici, untiring in his availability to help me in all the aspects related to the writing of this book, from the development of new algorithms, to simulations and book editing. I am also deeply thankful to the many people who contributed on different subjects covered throughout this book, and, in particular to Gesualdo Scutari, who introduced me to the field of game theory and for having contributed to the development of distributed space-time coding, to Francesca Romano, for her early works on the connectivity of random graphs, to Francesco Bucciarelli, for his contributions on the connectivity of wireless networks and cooperative communication systems, to Loreto Pescosolido for his work on distributed space-time coding. I have also a great memory of the invaluable discussions with Daniel Palomar, whose approach to transceiver optimization forms the bulk of the chapter on optimization. I also wish to acknowledge the support from the staff of Artech House, who encouraged me throughout this project: from Julie Lancashire, who proposed to write this book, to Tiina Ruonamaa, for her patient and careful help in handling all the phases necessary to make this project to become a real book.

My final thoughts are for my beloved Francesca, whose love and support have been precious through the years. My only regret in writing this book is for all the time that we have not spent together.

Chapter 1

Introduction

One of the key features of our time is the dramatic increase in the flow of information with respect to the previous centuries. The availability of technologies enabling the exchange of huge amounts of information has induced new needs and opened perspectives unforeseeable until not too long ago. Telesurgery, where the patient is in a small hospital whereas the surgeon is, possibly, on another continent is not a futuristic scenario anymore. A student living in a small village can have access, in principle, to the same information as a colleague living in a big city, provided that he or she is properly connected to the right network and has sufficiently fast access. These examples suggest that *connectivity* through the proper telecommunications infrastructure is the first basic concept enabling the share of information.

Adopting a very basic classification procedure, connectivity can be insured using *wired* or *wireless* technologies. From a historical point of view, even though most of us have experienced the first connection through the wired telephone system, the most primitive forms of telecommunications were wireless and used a very high frequency band, namely the electromagnetic spectrum between, approximately, 4×10^{14} and 8×10^{14} Hz. As weird as it might appear, this was the first band allocated for telecommunications at a very low (indeed null) cost. This is, in fact, the region of the electromagnetic spectrum where one of our receivers (our eyes) is tuned. And the reason our eyes are tuned in such a region is because this is the band where the natural transmitter available since the very beginning of our presence in this planet, namely our Sun, emits most of its energy. Therefore, humankind indeed started communicating in the optical region of the electromagnetic spectrum, close to the band where modern, highly sophisticated fiber optics technology enables transmission of huge amounts of information and constitutes the basic technology for the Internet backbones. It might appear quite singular that we rediscovered wired optical communications, through fiber optics, after wireless optical communications had taken place for millions of years, using rather inexpensive equipment. Indeed, although wired communications, through fiber optics, can support transmission rates unbeatable by any wireless system, there is one key feature of wireless communications that assigns them a fundamental role: the possibility of establishing a communication link from virtually any place towards any other place.

Wireless connectivity is not only the basis for the success of current cellular telephony, but it represents in many cases the only form of connectivity that we can establish. Communications between people, machines, or sensors, placed in cars, ships, airplanes, or even submarines, is obviously possible only through wireless systems. Deployment of sensor networks for gathering information in places hard, or even impossible for human beings to reach, is an example of a new perspective unforeseeable within the framework of wired communications. We can think of ad-hoc wireless networks, between cars circulating in high density roads, that emerge spontaneously to manage traffic or even prevent car accidents, or sensors thrown in disaster sites to collect data useful to guide rescue operations. Self-organizing wireless networks can create network infrastructures whenever and wherever they are needed, provided that there are radio nodes available. Coupling these immense potentials, from the application point of view, with the technology advances, especially in terms of miniaturization, we can easily foresee an immense potential for wireless systems, both for telecommunications as well as for remote sensing[1].

Having talked about their potentials, it is now necessary to pay attention to the limitations of wireless systems. The major limitations come from the wireless medium that is typically shared among many users. Sharing is necessary to reduce the cost per user, but it induces interference, if the access is not managed properly. On the other hand, the frequency spectrum is already full of existing services. Given the scarcity of this precious resource, it is then necessary to look for methods capable of increasing the system *efficiency*, defined as the number of bit per second per Hertz (bit/s/Hz) that can be transmitted *reliably*. Reliability here means the possibility for the receiver to recover the transmitted messages with negligible errors. The efficiency can be improved by proper coding. However, the coding algorithms developed through the last decades have already brought the efficiency very close to its fundamental, unsurpassable, limits. We have then from one side great potentials, and from the other side, strong limitations. What to do?

In 1884, Edwin A. Abbott wrote a novel describing a fantasy world, *Flatland*, whose inhabitants were two-dimensional geometric shapes, segments, triangles,

1 Of course, in spite of all the potentials of wireless communications, we cannot forget the huge transmission rates made possible by wired systems. The best way of looking at wireless or wired systems is to look at them not as competitors but as systems complementing each other: Wireless systems have a degree of connectivity impossible for wired systems, whereas wired (fiber optics) systems offer transmission rates still unthinkable with today wireless technology.

squares, etc., moving in a two-dimensional world [1]. One of the major problems for Flatlanders (Flatland's inhabitants) was how to recognize each other. One day, A. Square, a Flatlander, was brought by an alien visitor to *Spaceland*, a three-dimensional world. From there, at a certain height above Flatland, it became immediately clear to Square that Flatland was populated by polygons and how easy it was to distinguish between different polygons. This novel was used as a metaphor to show how introducing additional dimensions one could simplify the solution of otherwise extremely difficult problems. Sixty years later, the further mathematical fantasy, *Sphereland*, revisited the world of Flatland in the four-dimensional space-time domain, to explore the revolutionary theories of Albert Einstein about relativity.

Using the same metaphor here, the way for overcoming the limitations about wireless communications mentioned before is to add dimensionality to the problem. Coding theory has been historically thought as a function of a single independent variable: time. But in wireless communications, the temporal flow of information spreads, inevitably, also over space. Furthermore, since electromagnetic waves are polarized, there are also extra dimensions we can play with: the polarization axes. All these extra domains make available a high number of additional degrees of freedom that can be exploited to increase the system efficiency, without requiring neither more bandwidth nor more power. Multiantenna systems can be thought as systems that, thanks to spatial sampling of the information waves, are capable of achieving the advantages made available by these additional degrees of freedom. The major goal of this book is to describe basic algorithms capable of providing these advantages. In the following section, we will first recall the basic limitations of single-input/single-output (SISO) systems to motivate the need for adopting multiple-input/multiple-output (MIMO) systems.

1.1 SISO SYSTEMS

The first basic concept that a telecommunications engineer learns is that there exist fundamental limits that prevent the reliable transmission of data at any arbitrary rate, for a given transmission bandwidth and power. These limits were discovered in 1948, when Claude Shannon published a series of papers where he established the principles of information and communication theory [2]. Since the pioneering work of Shannon, the word *information* started assuming a new meaning, scientifically motivated and justified. Given any phenomenon, whose outcomes are characterizable in probabilistic sense, the amount of uncertainty removed through the observation of these outcomes is precisely the information that we acquire about the phenomenon. Starting from this basic idea, Shannon introduced the concept of information, as a well defined parameter, quantifiable in a unique sense, that can be associated to any random phenomenon. Besides quantifying the information carried by random events, Shannon established also the conditions that enable the transmission of information over a noisy channel, at a given rate, for a given power of the useful signal and noise. This leads to the concept of channel capacity.

1.1.1 Capacity of a Deterministic Channel

Given a channel, characterized by a double-side bandwidth of $2B$ Hertz, with additive noise described by a Gaussian random process with zero mean and variance σ_n^2 and a transmitted sequence, independent of the noise, with average power σ_x^2, Shannon proved that it is possible to transmit information with a rate of r bits per second (bps), with error probability arbitrarily small, provided that r is less than a parameter, the channel *capacity*, that uniquely characterizes the channel. In the previous example, the capacity is [3]

$$C = B \log_2 \left(1 + \frac{\sigma_x^2}{\sigma_n^2}\right) \quad \text{bps}, \tag{1.1}$$

where σ_x^2/σ_n^2 is the signal-to-noise ratio (SNR). Shannon proved that, if $r < C$, there exists an error correction code that allows us to achieve arbitrarily low (possibly zero) error probabilities, provided that the code is sufficiently (possibly infinitely) long. This theorem is known as the *channel coding* theorem. There is also a converse form of the channel coding theorem, stating that, if r is greater than C, the error probability is lower bounded by a positive nonzero quantity. Hence, in such a case, there is no coding scheme that can yield an arbitrarily low error probability.

The channel coding theorem establishes a rigorous link between three fundamental quantities: power, bandwidth, and complexity. As we will see throughout this book, the design of even the most sophisticated communication system has to do, ultimately, with a trade-off between these fundamental parameters. At a first glance, since the channel coding theorem establishes an unsurpassable limit on reliable information transmission, it might appear as a negative, although important, result. However, looking at the theorem only from this perspective would be too restrictive. In fact, the theorem states something that, conversely, is not trivial at all, and it has an incredibly positive feature: Even if a channel is affected by noise, whose power is not necessarily lower than the useful signal power, the transmitted information can be, potentially, recovered without any error, provided that we use a sufficiently long code and we limit the transmission rate!

This fundamental property allowed people to envisage communication links between very distant points. The view of the amazingly beautiful pictures of the most distant planets of our solar system, sent back to Earth by the Voyager, from a distance of billions of kilometers, using a very low power transmitter, is probably one of the most striking experimental confirmations of Shannon theorems.

The capacity formula tells us also how to trade power, bandwidth, and complexity. Let us assume that we have a channel with additive noise with variance σ_n^2 and that we have some freedom in choosing the bandwidth B or the average transmission power σ_x^2, to set up a reliable link to send r bps. From the channel coding theorem, we know that C, as given in (1.1), must be greater than r, but we still have one degree of freedom in the choice of bandwidth (B) and power (σ_x^2). From (1.1), we realize that if, for a given SNR, we wish to double C, we have to double the bandwidth B. Conversely, if we want to double C, for a given B, we have to elevate the SNR to the square (assuming, realistically, that SNR $\gg 1$). This simple example shows that the trade-off between bandwidth and average transmission power necessary to achieve a desired capacity implies that a linear increase of the bandwidth by a factor of, let us say n, can only be traded by an elevation of the SNR to its nth power. Hence, bandwidth is clearly a precious resource.

1.1.2 Capacity of Flat-Fading SISO Channels

Equation (1.1) refers to the simplest possible scenario of an ideal channel, where the only negative effect of the channel is the introduction of additive noise. However, a wireless channel is never so simple. One of the distinguishing features of wireless channels is indeed their strong variability. Thinking, for example, of communications in a urban environment, we have to take into account the simultaneous presence of (possible) line-of-sight (LOS) propagation, superimposed to other waves generated by reflections, scattering, and diffraction from the environment surrounding transmitters and receivers. Typically, the superposition of all these effects is so complicated that it is impossible to characterize the wireless channel in a deterministic sense. In such a case, a statistical model is more appropriate, as it can catch some general, yet meaningful, properties. The simplest model consists of describing the channel as a multiplicative coefficient, modelled as a random variable. Thus, denoting by $x(n)$ the nth transmitted symbol and by $y(n)$ the corresponding received symbol, the input/output relationship can be expressed very simply as

$$y(n) = h\,x(n) + w(n) \tag{1.2}$$

where h is a random variable, whereas $w(n)$ is additive noise. This is the so-called *flat-fading* model, where the "flat" attribute comes from modeling the channel transfer function with a constant (flat) modulus.

The capacity of the channel, conditioned to the value of h, when h is known at the receiver side but unknown to the transmitter, and the noise $w(n)$ is Gaussian with zero mean and variance σ_n^2, independent of $y(n)$, is

$$C(h) = B\log_2\left(1 + |h|^2\frac{\sigma_x^2}{\sigma_n^2}\right) \text{ bps.} \tag{1.3}$$

For values of $|h|^2$ and of the SNR such that $|h|^2\sigma_x^2 \gg \sigma_n^2$, the capacity can be approximated as follows

$$C(h) \approx B\log_2\left(\frac{\sigma_x^2}{\sigma_n^2}\right) + B\log_2\left(|h|^2\right) \text{ bps.} \tag{1.4}$$

If the transmission rate r is fixed, depending on the value assumed by h, the capacity $C(h)$ may be less than r. When this happens, the converse of the channel coding theorem tells us that there is no coding capable of making the error probability arbitrarily small. Let us dig a little further into this problem.

In the case of random channels, it is better to characterize the system performance by computing the outage probability i.e., the probability that the rate r is greater than $C(h)$. As an example, let us consider the widely used Rayleigh fading model, where h is a zero mean, circularly symmetric, complex Gaussian random variable (rv). In such a case, the outage probability P_{out} is

$$P_{\text{out}} := Pr\{r > C(h)\} = Pr\{B\log_2(1 + \alpha\,\text{SNR}) < r\}, \tag{1.5}$$

where $\alpha := |h|^2$. Since h is a zero-mean, circularly symmetric, complex Gaussian random variable rv, α is exponential i.e., it has probability density function (pdf)

$$p_A(\alpha) = \frac{1}{\alpha_0}e^{-\alpha/\alpha_0}u(\alpha), \tag{1.6}$$

where α_0 is the mean value of α and $u(x)$ is the unit step function. Combining (1.5) and (1.6), we get the relationship between the rate r and the outage probability P_{out}. In particular, we can express the rate r as a function of P_{out}, as follows

$$r = B\log_2\left[1 - \alpha_0\,\text{SNR}\,\ln(1 - P_{\text{out}})\right]. \tag{1.7}$$

From (1.7) it turns out that, if we wish to have a null outage probability, we obtain $r = 0$. Hence, there cannot be any reliable transmission, at any rate, guaranteeing a null outage probability, regardless of the value of the bandwidth B and of the transmit power! This is indeed a rather annoying result: The Rayleigh-fading model is probably the most widely used model, largely supported by experimental measurement campaigns, and we have just verified that, for such a channel, it is impossible to achieve zero outage probability, for any rate greater than zero!

1.1.3 Bit Error Rate

Let us now turn our attention to bit error rate. The channel coding theorem states that, if we transmit at an information rate smaller than channel capacity, we can make the bit error rate arbitrarily low, provided that we use sufficiently long coding. But in all applications where decoding delay is one of the key performance parameters, as for example in voice links, we cannot use very long codes. It thus makes sense to analyze the bit error rate of an uncoded system. For example, using antipodal binary signaling, the error probability over an additive white Gaussian noise (AWGN) channel is

$$P_e = \frac{1}{2}\,\mathrm{erfc}\left(\sqrt{\frac{\alpha\mathcal{E}}{N_0}}\right) := \frac{1}{2}\,\mathrm{erfc}\left(\sqrt{\gamma_b}\right), \tag{1.8}$$

where $\mathcal{E}$ is the energy per bit, N_0 is the one-sided noise spectral density, and $\gamma_b := \alpha\mathcal{E}/N_0$.

In case of wireless channels, α is a random variable. For example, in case of Rayleigh-fading, α has the exponential pdf given in (1.6). The variability of α affects then the variability of P_e in (1.8). To have a global parameter describing the performance in a wireless channel, we can take, for example, the average of (1.8) with respect to α. In this simple example, the behavior of the average BER, at high SNR, is[2]:

$$\overline{P}_e \simeq \frac{1}{4\,\mathrm{SNR}}, \tag{1.9}$$

where SNR:= $\alpha_0\mathcal{E}/N_0$. This result is, again, rather annoying, as it shows that, passing from a deterministic channel to a Rayleigh-fading channel, the average BER degrades considerably: In the former case, the BER decays exponentially, whereas in the second case, the average BER decreases only as $1/\mathrm{SNR}$. This implies a huge loss of power of the Rayleigh-fading case, if we try to enforce the same average BER obtained with a deterministic channel.

The intuitive reason explaining such a behavior is that the most likely outcomes of random variable α, in case of Rayleigh-fading, are small values. These values give rise to high BER values that give the dominant contribution to the average BER.

Put it in more mathematical terms, let us consider a flat-fading channel with power channel gain α, having pdf $p_A(\alpha)$. The average BER goes asymptotically, i.e. at high SNR, as

$$\overline{P}_e \simeq \frac{G_k}{(G_c\,\mathrm{SNR})^{G_d}}. \tag{1.10}$$

2 The details of these derivations are in Chapter 4.

The parameter G_d is the so-called *diversity gain*, whereas G_c is the *coding gain*. Among the three parameters G_k, G_c, and G_d, G_d plays the most important role as it determines the (asymptotic) decaying rate of the average BER. The diversity gain depends on the smoothness of $p_A(\alpha)$ around the origin [4]. More precisely, let us indicate with p the minimum order of the derivative of $p_A(\alpha)$, evaluated in $\alpha = 0$, such that the derivative is different from zero, that is

$$\left.\frac{d^k p_A(\alpha)}{d\alpha^k}\right|_{\alpha=0} = 0,\ 0 \le k < p,\ \text{and}\ \left.\frac{d^p p_A(\alpha)}{d\alpha^p}\right|_{\alpha=0} \neq 0. \tag{1.11}$$

The diversity gain is related to p by the following relationship [4]

$$G_d = p + 1. \tag{1.12}$$

As a pictorial example, let us consider a family of flat-fading channels whose channel gain α has a gamma pdf, with parameters L and σ_h^2, i.e.

$$p_A(\alpha) = \frac{1}{(L-1)!\sigma_h^2}\,\alpha^{L-1}\,e^{-\alpha/\sigma_h^2}. \tag{1.13}$$

These pdfs are reported in Figure 1.1, for different values of L. Intuitively speaking, the channels with increasing L should have better average BER, because as L increases, small values of α are less and less likely to occur. Indeed, applying (1.12) to (1.13), is is straightforward to verify that the diversity gain G_d is precisely L.

These arguments suggest, as expected, that the channel pdf has a deep impact on the performance. But the main problem is that the channel statistics are given by the propagation environment and they are not under control of the communication system designer. The main question is then: How can we obtain a diversity gain for transmissions over a given random channel?

1.2 MIMO SYSTEMS

Let us increase the dimensionality of our observations [3]. Let us consider a multiple-input/multiple-output (MIMO) system, with n_T transmit and n_R receive antennas. Assuming, as in previous section, that all channels are flat-fading, the input/output

3 In this section, we will anticipate some concepts that justify the use of MIMO systems. In this introduction, we just want to give a flavor of the most important aspects of MIMO systems, simply to motivate the writing of this book. As a consequence, we will follow a more intuitive approach, as the rigorous treatment of the topics touched in this section will be the subject of the following chapters.

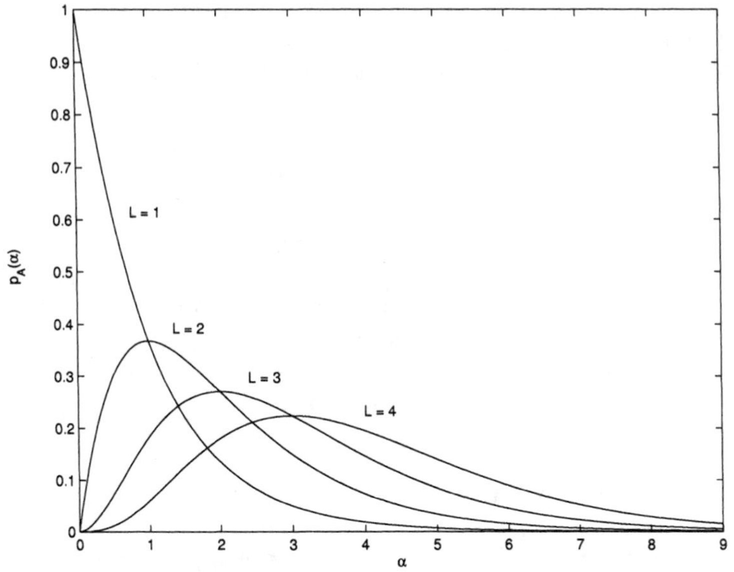

Figure 1.1 Probability density functions with gamma-distribution, for different values of L.

relationship (1.2) generalizes into

$$\boldsymbol{y}(n) = \boldsymbol{H}\boldsymbol{x}(n) + \boldsymbol{w}(n), \tag{1.14}$$

where $\boldsymbol{H}(n)$ is an $n_R \times n_T$ matrix, whose (k, l) entry is the complex flat-fading coefficient between the lth transmit antenna and the kth receive antenna, $\boldsymbol{x}(n)$ is the nth transmitted vector, composed of n_T symbols, $\boldsymbol{y}(n)$ contains n_R receive symbols, and $\boldsymbol{w}(n)$ is additive noise, assumed to be white Gaussian, with zero mean and variance σ_n^2.

1.2.1 Channel Eigenmodes

From basic algebra properties, we know that every matrix can be expanded through its singular value decomposition (SVD). Thus, for example, we can write $\boldsymbol{H}$ as

$$\boldsymbol{H} = \boldsymbol{U}\boldsymbol{\Sigma}\boldsymbol{V}^H = \sum_{i=1}^{q} \sigma_i \boldsymbol{u}_i \boldsymbol{v}_i^H \tag{1.15}$$

where $\boldsymbol{U}$ is an $n_R \times n_R$ matrix, $\boldsymbol{V}$ is $n_T \times n_T$, and $\boldsymbol{\Sigma}$ is an $n_R \times n_T$ matrix whose only nonzero elements are given by $\Sigma(i, i) = \sigma_i$, $i = 1, ..., q$; the columns $\boldsymbol{u}_i$ $(\boldsymbol{v}_i)$

of $\boldsymbol{U}$ ($\boldsymbol{V}$) are orthonormal, so that $\boldsymbol{U}^H\boldsymbol{U} = \boldsymbol{I}_{n_R}$ and $\boldsymbol{V}^H\boldsymbol{V} = \boldsymbol{I}_{n_T}$; q denotes the rank of $\boldsymbol{H}$. Equivalently, we can introduce the matrix $\boldsymbol{H}\boldsymbol{H}^H$ which, being a square Hermitian matrix, admits the eigenvalue decomposition (EVD):

$$\boldsymbol{H}\boldsymbol{H}^H = \boldsymbol{U}\boldsymbol{\Sigma}\boldsymbol{\Sigma}^H\boldsymbol{U}^H = \sum_{i=1}^{q} \lambda_i \boldsymbol{u}_i \boldsymbol{u}_i^H, \tag{1.16}$$

where $\lambda_i = \sigma_i^2$. When we transmit a vector $\boldsymbol{x}$ through a MIMO channel, we excite the so-called *eigenmodes* of the channel. Each eigenmode (or simply mode) is directly related to one of the channel singular vectors. Each mode is then received with a gain proportional to the corresponding singular value σ_i or, in a quadratic sense, with a power gain equal to the corresponding eigenvalue λ_i. Thus, the eigenvalues of $\boldsymbol{H}\boldsymbol{H}^H$ play a fundamental role in characterizing the performance of a MIMO communication system.

1.2.2 Capacity of a MIMO Channel

The presence of multiple, simultaneous, modes makes possible to foresee a rate gain, with respect to a SISO system. However, to make this concept more precise, it is necessary to look at the capacity of a MIMO system, to compare it with the capacity of a SISO system, transmitting with the same total power. This implies that, if the power radiated by a SISO system is σ_x^2, the power radiated from each antenna of a MIMO system having n_T transmit antennas, must be σ_x^2/n_T. In such a case, the capacity corresponding to the situation where the channel is deterministic, the channel matrix is unknown to the transmitter, but known to the receiver, can be written as [3]

$$C(\boldsymbol{H}) = B \log_2 \left| \boldsymbol{I}_{n_R} + \frac{\text{SNR}}{n_T} \boldsymbol{H}\boldsymbol{H}^H \right| \text{ bps}, \tag{1.17}$$

where $|\boldsymbol{A}|$ denotes the determinant of $\boldsymbol{A}$, SNR $:= \sigma_x^2/\sigma_n^2$ is the same as in (1.3), and $\boldsymbol{I}_{n_R}$ indicates the identity matrix of dimension $n_R \times n_R$. Equation (1.3) is indeed a special case of (1.17), corresponding to $n_T = n_R = 1$.

Using the channel EVD (1.16), we can rewrite (1.17) as

$$C(\boldsymbol{H}) = B \log_2 \left| \boldsymbol{I}_{n_R} + \frac{\text{SNR}}{n_T} \boldsymbol{U}\boldsymbol{\Sigma}\boldsymbol{\Sigma}^H\boldsymbol{U}^H \right| \text{ bps}. \tag{1.18}$$

Since $\boldsymbol{U}$ is unitary, the capacity can be rewritten as

$$C(\boldsymbol{H}) = B \log_2 \left| \boldsymbol{I}_q + \frac{\text{SNR}}{n_T} \boldsymbol{\Sigma}\boldsymbol{\Sigma}^H \right| = B \sum_{i=1}^{q} \log_2 \left(1 + \frac{\text{SNR}}{n_T} \lambda_i \right) \text{ bps}, \tag{1.19}$$

where q is the rank of $\boldsymbol{H}$. At high SNR, that is for SNR $\lambda_i \gg n_T$, $\forall\, i$, we can approximate the previous relationship as follows[4]

$$C(\boldsymbol{H}) \simeq q\,B \log_2(\text{SNR}) + B \sum_{i=1}^{q} \log_2\left(\frac{\lambda_i}{n_T}\right) \text{ bps.} \tag{1.20}$$

If the channel is random and ergodic, the ergodic capacity is[5]

$$C = E\{C(\boldsymbol{H})\} \simeq \min(n_T, n_R)\, B\, \log_2(\text{SNR}) + B \sum_{i=1}^{q} E\left\{\log_2\left(\frac{\lambda_i}{n_T}\right)\right\} \text{ bps,} \tag{1.21}$$

where $E_H\{C(\boldsymbol{H})\}$ denotes the expected value of $C(\boldsymbol{H})$ with respect to the channel statistics. Comparing (1.20) or (1.21) with (1.4), if we neglect for the moment the second term on the right-hand side of (1.20) or (1.21), we see that, at high SNR, *a MIMO channel makes possible a potential increase of channel capacity, with respect to the SISO case, by a factor equal to the rank of the channel matrix* $\boldsymbol{H}$. The maximum value for the rank is $\min(n_T, n_R)$. This is indeed a very strong result, because this rate increase can be reached without increasing neither the bandwidth nor the overall transmitted power, which are both precious resources. Furthermore, this improvement is achievable without assuming any knowledge of the channel at the transmitter side.

The previous considerations show that, as far as the rate is concerned, the rank of $\boldsymbol{H}$ plays a fundamental role. In practice, if transmitter and receiver are in line of sight and there is no scattering from the environment, the rank is one and thus there is no rate gain. However, line of sight propagation is not a really critical situation, as the channel attenuation in such a case is limited. The most critical situation occurs indeed when there is no line of sight propagation. This happens, typically, in urban environments, where the mobile users do not see the access point (or base station), because of the surrounding buildings. In such a case, the received signal is the result of the rich backscattering from the medium surrounding transmitter and receiver. Under such circumstances, the attenuation can be quite elevated and the maximum reliable rate is then limited. However, at the same time, the angular diffusion of the electromagnetic waves produced by the propagation medium induces also an increase of the rank of $\boldsymbol{H}$ that, if properly exploited, yields the rate gain mentioned above. Hence, interestingly, the very same criticality of no line of sight propagation

4 Indeed, the approximation in (1.20) is critical, because, for any given SNR, there is always a nonnull probability that the smallest eigenvalue does not satisfy the approximation used in deriving (1.20). In this introduction, rigor is somehow sacrificed to simplify the intuitive explanation of some basic concepts. But these arguments will be reconsidered with more attention in Chapter 5.

5 The derivation of these equations will be given in Chapter 5.

that determines the strong channel attenuation and randomness, if occurring in a strongly diffusing environment, makes available a considerable rate gain that can more than counteract the channel attenuation effects. Of course, the gain is achievable only if all potentials of the MIMO system are exploited appropriately.

1.2.3 Rate Gain Versus Diversity Gain

In the previous section, we have seen how the channel eigenvalues are ultimately responsible for the system performance. Since a wireless channel is random, its eigenvalues are themselves random variables. It is then important to study the statistical properties of the channel eigenvalues to evaluate the performance of MIMO wireless systems. Let us consider a MIMO channel where the channels are statistically independent and each channel is Rayleigh-fading, exactly as in the previous section. In the MIMO case, instead of only one parameter α, we need to consider a set of parameters, namely the eigenvalues λ_i of the channel matrix $\boldsymbol{H}\boldsymbol{H}^H$, which we will simply call channel eigenvalues in the following. Let us consider for simplicity, a square MIMO system, where $n_T = n_R := N$. In case of independent Rayleigh fading MIMO channels, the histograms of the channel eigenvalues are reported, as an example, in Figure 1.2, for different values of N. The eigenvalues are assumed to be ordered in increasing sense, so that $\lambda_1 \leq \lambda_2 \leq \ldots \leq \lambda_N$. The symbol $p_k(\lambda)$ in Figure 1.2 denotes the pdf of λ_k. Going from the upper left to the lower right corner, we have the cases: $N = 1, 2, 3$, and 4. Each plot reports N histograms [6]. From Figure 1.2, we notice that the smallest eigenvalue has an exponential pdf. This means that the channel mode associated to the smallest eigenvalue has diversity one, as in the SISO case. However, we also see that higher order eigenvalues have smoother pdfs, around the origin. This implies that if different modes carry different symbols, we have different diversity gains over different modes. More precisely, the diversity gain increases with the eigenvalue order. This property was shown in [5], in case of MIMO systems with at most two antennas on one side (and three or four antennas at the other side). More generally, we will show in Chapter 5 that, in case of independent fading, the mode associated to the largest eigenvalue has diversity gain

$$G_d = n_T n_R,$$

using the definition of diversity gain given in (1.10). This is indeed the maximum achievable diversity gain for transmissions over flat-fading channels.

6 The histograms of $p_1(\lambda)$ in Figures 1.2 c) and d) are cut on the top, simply to better compare the different pdfs. We will showin Chapter 5 that, in the case analyzed in Figure 1.2, $p_1(\lambda)$ always follows an exponential law.

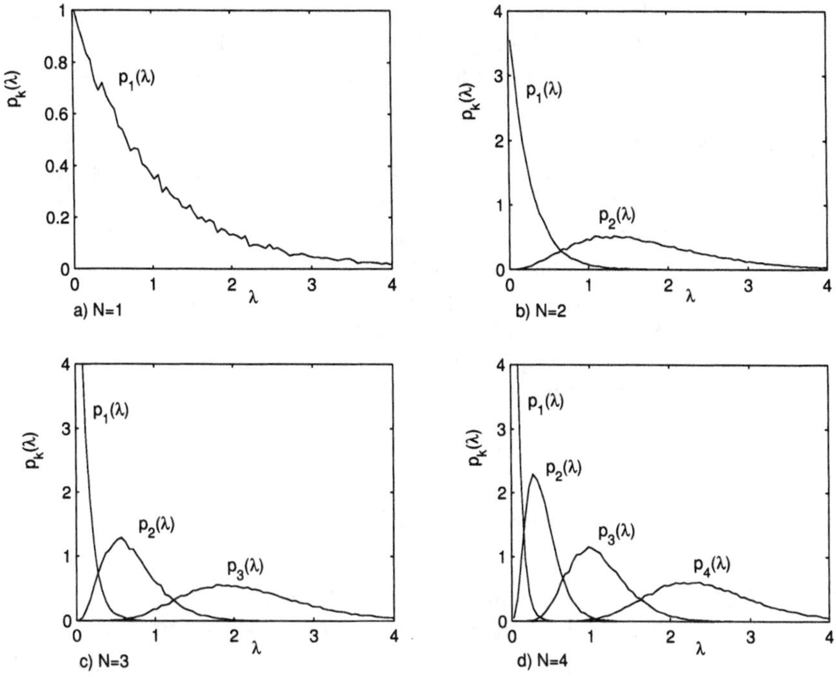

Figure 1.2 Histograms of the channel eigenvalues, for different MIMO sizes.

The previous considerations show one of the great potentials of MIMO systems: *Even though each channel follows the same statistical laws as the reference SISO channel (that has no diversity gain by itself), the overall MIMO channel exhibits modes that have diversity gain.*

From the analysis of the histograms of the ordered eigenvalues, it is also possible to make one more fundamental remark. If we use all channel modes, we may increase the transmission rate, as we have q parallel (spatial) channels, where q is the rank of $\boldsymbol{H}$. However, the mode associated to the smallest eigenvalue will strongly affect the final average BER, as it is the most faded mode. We may thus expect that a maximum rate system might have minimum diversity gain. Conversely, using only the mode associated to the largest eigenvalue, we would get the maximum diversity gain, but with low rate gain, as we would be using only one of the available modes. Clearly, deciding which modes to use requires channel knowledge at the transmitter side. But this information is not always available. The previous intuitive arguments suggest that, in the absence of channel knowledge at the transmit side, there should be some kind of trade-off between reliable transmission rate gain and diversity gain. In fact, Zheng and Tse proved analytically that there indeed exists a fundamental limit on the best trade-off between information rate gain and diversity gain [6].

Looking again at the pdf of the channel modes, it may also be guessed that the channel knowledge at the transmitter side can provide considerable benefits to a MIMO system. In fact, knowing the channel, we can load power differently over different modes, to optimize some performance criterion. More precisely, loading power as a function of the channel itself, we can modify the pdf of each mode and then its diversity gain. Thus, in principle, we can get the maximum diversity gain over all the modes. In such a case, we could use all the modes and still have the maximum diversity gain. The price to be paid for this potential advantage is the need to know the channel at the transmit side.

1.3 COOPERATIVE COMMUNICATIONS

As a final general concept that will be studied in this book, there is the case of systems with single antenna transceivers, where the advantages of MIMO systems are made possible through the *cooperation* among radio nodes. As a simple example, let us consider the scenario sketched in Figure 1.3, where there are a few mobile users (MU) wishing to send data to an access point (AP), or base station (BS). The grey

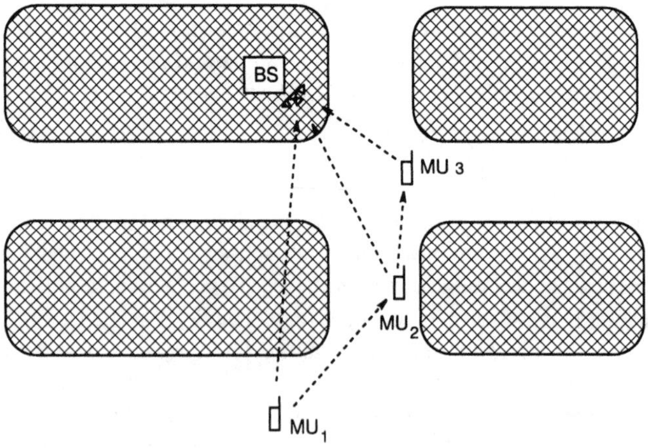

Figure 1.3 Multihop wireless network.

blocks are buildings that attenuate the transmitted signal and prevent direct line-of-sight propagation. A mobile terminal can experience a deep fade, depending on its position with respect to the base station. For example, the link between MU_1 and BS could be highly attenuated. It would be then advantageous for MU_1 to send its data to another user, instead of sending them directly to BS. This idea leads to the so-called *multihop* wireless networks. In current third generation systems, MU_1 would have to increase its transmit power to allow the BS to recover its data reliably. However, power control is not simple to manage because of the interference among users. Increasing the power of one user inevitably causes more interference to the other users and so on. However, adopting the multihop idea, MU_1 can initially send its data to MU_2, which can act as a relay. Under common propagation conditions, the sum of the powers necessary to both MU_1 and MU_2 to convey the information of MU_1 to the BS can be much lower than the power necessary to MU_1 alone.

Building on this initial idea, the situation can improve further. In fact, if MU_2 is able to decode the data sent by MU_1 without errors, after the initial information exchange, MU_1 can transmit its data to BS using a *virtual* double antenna, composed of the antennas of *both* MU_1 and MU_2. Hence, if the BS has two (real) receive antennas, a *virtual* MIMO system can be established, with all the possible advantages in terms of rate or diversity gains. Clearly, this requires a good synchronization between MU_1 and MU_2. However, the possible advantage makes this scenario worth of being investigated.

The principle of sending information packets through multiple paths is not novel, as it is employed, for example, in the routing mechanisms of wired networks. However, there is a fundamental difference between wired and wireless networks. In wired networks, each link is necessarily, and inevitably, point-to-point. Conversely, in wireless networks, each transmitter broadcasts its data towards many directions. This means that the information sent by any node can be received and decoded by, in principle, many receivers, at the same time. If the information is decoded without errors, all the nodes that have received the information correctly can retransmit it, acting all together like a huge array. Each radio node can act, in principle, in a cooperative manner to convey the received information to the final destination. Hence, overall, cooperation makes possible the creation of *virtual* arrays that are potentially capable of the MIMO advantages mentioned before. This, although difficult to implement, would be impossible to envisage over a wired network.

1.4 A BRIEF HISTORICAL PERSPECTIVE

Although the advantages of using multiantenna receivers have been known for a long time, the rate and diversity gains achievable using multiple antennas at *both* transmit and receive sides have been fully understood only relatively recently. Winters [7] was among the first to show that MIMO systems provide a substantial capacity increase. Then Telatar [8] and Foschini [9] proved the fundamental results on the capacity of flat-fading MIMO channels. Independently of [8, 9], Paulraj and Kailath had proven that the capacity of cellular CDMA systems equipped with multiple antennas at both the transmit and receive sides can increase considerably with respect to SISO systems [10]. The first experimental testbed proving the potentials of MIMO communications was the Bell Laboratories layered space-time system (BLAST), developed at Bell Labs under Foschini's guidance [9]. A strong impulse to the design of space-time coding strategies that provide some of the advantages foreseen by the theoretical limits came from Alamouti [11], who devised a very simple scheme for a system with two transmit and any number of receive antennas. The beauty of Alamouti design is that, in case of two transmit and one receive antenna, it has all the good qualities, as it is optimal in terms of rate, diversity, and decoder simplicity. The Alamouti approach was then generalized into orthogonal design [12], which is optimal in terms of diversity gain and receiver simplicity, but suboptimal in terms of rate. Since then, a huge amount of works have been published on the design of space-time coding techniques capable of striking the best compromise between the fundamental quantities: rate gain, diversity, and receiver complexity [13, 14, 15, 16]. Today, space-time coding is mature enough to be considered for implementation over third (and beyond) generation wireless systems.

1.5 A QUICK DETOUR THROUGH THE WORLD OF NUMBERS

Before illustrating the structure of the book, it is interesting to approach the problem of communicating through multiantenna transceivers from a purely numerical perspective. The most basic building block in any communication system is ultimately given by information bits, which are *binary digits*. Bits can be combined and mapped into *integer numbers* (symbols), belonging to a finite alphabet. In a baseband digital transmission system, these numbers can be transmitted using, for example, pulse amplitude modulation (PAM) that converts a sequence of numbers (symbols) into a *real*, continuous time, waveform. In a radio system, it is necessary to transmit over a frequency spectrum centered around a nonnull carrier frequency. To employ the antennas efficiently, it is indeed necessary to transmit narrowband passband signals i.e. signals whose bandwidth is much smaller than the carrier frequency. Passband signals have an equivalent baseband signal, which is composed of a pair of real signals, namely the in-phase and quadrature components. These two signals can be seen as the real and imaginary part of a *complex* signal. Of course, signals traveling through the air are real. However, the complex model, albeit an abstract model, is very convenient to use for mathematical derivations. A typical example of digital communication is quadrature amplitude modulation (QAM), used, for example, in third generation systems, where the transmitted signal can be represented as the linear modulation of complex symbols, resulting from mapping the information bits into the complex field.

Why are complex numbers interesting in digital communications? Because they provide a simple tool to analyze and design systems where, instead of transmitting one (real) symbol at the time, we transmit a *pair* of symbols *at the same time, without requiring an increase the bandwidth.* This way of looking at the problem induces then a question: Is it possible to envisage other number structures that allow the transmission of more symbols, at the same time, without requiring an increase of time or bandwidth? Indeed, several generalizations are possible. To grasp this possibility, it is necessary to dig a little further into the world of numbers.

Complex numbers are typically represented as linear combinations of two basic units: the real unit 1 and the imaginary unit $j = \sqrt{-1}$. A complex number is then written as $a \cdot 1 + b \cdot j$ or, simply, $a + jb$. It is known that the field of complex numbers is isomorphic to the set of 2×2 real matrices having a specific structure. In particular, given any *complex number* $a + jb$, with a and b real, we can put $a + jb$ in a one-to-one correspondence with the 2×2 *real matrix*

$$\begin{pmatrix} a & -b \\ b & a \end{pmatrix}. \tag{1.22}$$

The correspondence is such that common operations between complex numbers, like addition, multiplication, and inversion, have a direct counterpart among the members of the set of 2×2 real matrices having the structure (1.22), with the operations of addition, multiplication, and inversion defined in the matrix algebra.

Let us now see how this concept can be generalized to higher order numeric structures. As a first step, let us consider the so-called *quaternions*. Quaternions are obtained by adding the units i, j, and k to the real numbers, with i, j, and k satisfying the following relations:

$$i^2 = j^2 = k^2 = ijk = -1. \tag{1.23}$$

Every quaternion is a real linear combination of the *unit quaternions* $1, i, j$, and k, i.e., every quaternion is uniquely expressible in the form $a + bi + cj + dk$, where a, b, c, and d are real numbers. Hence, each quaternion carries, equivalently, four real numbers. Addition and multiplication between quaternions are carried out taking into accounts the rules (1.23). Without going into the details of the operations between quaternions, which will be explored in more detail in Chapter 9, it is useful to mention that the algebra of quaternions is isomorphic to the set of 2×2 *complex* matrices. In particular, every quaternion $a + bi + cj + dk = (a + bi) + (c + di)j$ is in a unique correspondence with the complex matrix[7]

$$\mathcal{X} \equiv \begin{pmatrix} a+bi & c+di \\ -c+di & a-bi \end{pmatrix} = \begin{pmatrix} x_1 & x_2 \\ -x_2^* & x_1^* \end{pmatrix}, \tag{1.24}$$

having introduced the complex numbers $x_1 := a + bi$ and $x_2 := c + di$.

Interestingly, *the matrix structure in (1.24) is exactly the same structure used in the two-transmit antenna systems employing Alamouti space-time coding* [11], where x_1 and x_2 are the two transmitted (complex) symbols. More specifically, Alamouti coding, which will be studied in full detail in Chapter 7, means that, in the first time slot, the first antenna transmits the symbol x_1 whereas the second antenna transmits $-x_2^*$. In the second slot, the first antenna transmits x_2 whereas the second antenna transmits x_1^*. Let us consider, for simplicity, a 2×1 MIMO system, where h_1 and h_2 denote the two (flat-fading) channel coefficients between the two transmit antennas and the receive one. The symbols y_1 and y_2 received in two successive slots are

$$\begin{aligned} y_1 &= h_1 x_1 - h_2 x_2^* + v_1 \\ y_2 &= h_1 x_2 + h_2 x_1^* + v_2, \end{aligned} \tag{1.25}$$

where v_1 and v_2 are the noise samples. Arranging each pair of complex symbols into one quaternion, as follows

$$\begin{aligned} \mathcal{Y} &\equiv \begin{pmatrix} y_1 & y_2 \\ -y_2^* & y_1^* \end{pmatrix}, \ \mathcal{H} \equiv \begin{pmatrix} h_1 & h_2 \\ -h_2^* & h_1^* \end{pmatrix}, \\ \mathcal{X} &\equiv \begin{pmatrix} x_1 & x_2 \\ -x_2^* & x_1^*, \end{pmatrix}, \ \mathcal{V} \equiv \begin{pmatrix} v_1 & v_2 \\ -v_2^* & v_1^*, \end{pmatrix} \end{aligned} \tag{1.26}$$

it is easy to see that the input/outut relationship (1.25) can be rewritten as

$$\mathcal{Y} = \mathcal{H}\mathcal{X} + \mathcal{V}.$$

7 Recall that $i \cdot j = k$.

Interestingly, a 2×1 MIMO system, with Alamouti space-time coding, is equivalent to a SISO system, in the algebra of quaternions! In other words, Alamouti coding, which is one of the most well known space-time coding techniques, can be seen as a possible way to transmit quaternions. Even more interestingly, the optimal maximum likelihood estimator consists in simply multiplying the received quaternion $\mathcal{Y}$ by the inverse of the channel quaternion $\mathcal{H}$.

This example shows that, to transmit one quaternion, it is necessary to use two domains, time and space, *jointly*. This means that, to transmit quaternions we must add one more physical domain to the conventional time domain: space.

Generalizing upon this digression about numbers, it is then natural to ask ourselves whether it is possible to devise transmission systems that allow the transmission of more complicated numeric structures. Indeed, the mathematical tools are available through, for example, the so called Cayley-Dickson construction [17] that leads to *hyper-complex numbers*, like octonions (composed of eight real numbers), sedenions (sixteen real numbers), and so on. However, the design of practical systems capable of exploiting such structures is still an open problem[8]. Hopefully, this digression will trigger the brain of some bright reader who will invent such systems[9].

1.6 SCOPE OF THIS BOOK

As anticipated before, this book will be primarily driven by two basic conceptual thrusts: multichannel propagation and diversity. These concepts apply to a large variety of cases, not necessarily multiantenna systems. We can in fact build multichannel, or multi-modal, transmissions and/or achieve diversity gain not only in the space domain, but in the time, frequency, and polarization domains, possibly in a joint manner. The basic intuitive idea is to convert the channel fluctuations, through possibly many domains e.g., time, space, and polarization, into a useful source of diversity. The fundamental goal of this book is to illustrate the methods that are capable of achieving the best compromise between the rate increase, offered by multichannel propagation, the diversity gain, and the receiver complexity. We will consider both the optimal theoretical limits, as well as the ways to achieve or approach the theoretical limits.

One key aspect of the book that we wish to outline is its methodological nature. Our basic goal is to provide the reader with some fundamental theoretical tools useful to

8 An interesting recent approach in this sense is [18].

9 We only wish to warn the reader that the solution is probably not going to be searched within the algebra of matrices. In fact, the division ring of octonions or higher order hyper-complex numbers does not satisfy the associative property, whereas the matrix algebra is associative.

tackle challenging problems posed by modern communication systems. Throughout the book, the reader will find basic principles of optimization theory, game theory, and random graphs, which are not typically covered in books on smart antennas. Our hope is that the study and acquisition of these theoretical tools, although presented in the specific context of multiantenna systems, will be a good investment for readers wishing to work in communications fields, not necessarily limited to multiantenna systems.

Before illustrating the book structure, it is necessary to mention the topics that will not be covered in this book, or that will only be mentioned very quickly. Some of the major topics regarding multiantenna systems that will not be covered are adaptive beamforming and direction finding. This choice has nothing to do with the importance of such topics. Both topics are undoubtedly important in the applications. However, they are nowadays classical topics in multiantenna systems and thus they are covered by already available texts (see, for example, [19] and the references therein).

1.7 BOOK OVERVIEW

The book is roughly organized in five parts. The first part is composed of Chapter 2, which is devoted to channel models. In Chapter 2, we start describing SISO systems, as the first step to introducing the basic channel parameters, using both deterministic as well as stochastic modeling. Then we consider truly MIMO channels. Special attention in this chapter is devoted to the analysis of the modes of SISO and MIMO channels, as this is fundamental for the ensuing sections.

In the second part, composed of Chapters 3 and 4, we review some basic principles of SISO systems, both single user and multiuser. In particular, Chapter 3 is devoted to multicarrier systems. Special attention is given to the so-called orthogonal frequency division multiplexing (OFDM) systems, for their importance in current wireless local area network (LAN) standards. In Chapter 4, we study the multiple access problem, considering time, frequency, code, and space division strategies. In particular, we give special attention to code division multiple access (CDMA) systems, as the standard radio access scheme of third generation (3G) cellular systems.

The third part, composed of Chapters 5 and 6, studies the fundamental limits of MIMO systems and possible optimizations of a MIMO system. In particular, in Chapter 5, we study the so-called *open-loop* systems, where the transmitter has no knowledge of the channel. We show, in such a case, which are the fundamental limits in terms of average bit error rate (BER) and capacity. Chapter 6 is then devoted to *closed-loop* systems, where the transmitter has some knowledge of the channel. In such a case, we show how to design the coding strategy in order to optimize alternative performance criteria. Convex optimization theory, together with majorization theory, will play a central role in Chapter 6 to derive the optimal coding structures. Then, we formulate the multiple access problem as a multiobjective optimization problem and show how to find out the best multiple access strategy. This leads to a centralized solution that would be difficult to implement in a wireless network. A decentralized, yet sub-optimal solution is then achieved adopting a game-theoretic approach, where players compete with each other over the available resources, with the aim to optimize their own performance. Within this broader perspective, we show that common multiple access schemes, like TDMA, FDMA and CDMA, are only specific game equilibria.

The fourth part is specifically devoted to space-time coding systems. This part is composed of Chapters 7, 8 and 9. In particular, in Chapter 7 we study the space-time coding techniques valid for transmissions over flat-fading channels. We illustrate the basic compromise between rate, BER, and complexity and the ways to achieve the desired trade-off. We illustrate in detail the trace-orthogonal design, recently proposed in [20], as a very flexible tool to strike a good trade-off between rate, diversity, and complexity. In Chapter 8, we consider the extension of space-time coding to frequency-selective channels. In particular, we show how to achieve multipath diversity. Whereas chapters 7 and 8 are primarily concerned with single user systems, Chapter 9 is entirely devoted to multiuser systems.

Finally, the goal of the fifth part of this book, composed of Chapter 10, is to illustrate some novel concepts concerning cooperative communications. The chapter shows how to improve the connectivity of wireless networks through cooperation among radio nodes. The connectivity is studied on the basis of random geometric graphs. Finally, the novel concept of distributed space-time coding is illustrated, as a means for coordinating the transmission in multihop relay networks.

1.8 PRELIMINARIES

Before starting our study of the channel, it is useful to recall a few fundamental concepts. The transmitter of a multiantenna digital communication system is usually composed of the blocks sketched in Figure 1.4. The corresponding receiver is reported in Figure 1.5. The successive blocks sketched in Figure 1.4 perform the following basic operations: 1) coding of information bits; 2) mapping of the coded bits onto a symbols constellation; 3) user-coding; 4) space-time (frequency) encoder; 5) pulse shaping; 6) frequency up-conversion; 7) (high) power amplifier;

and 8) antenna subsystem. We briefly review the basic operations of each block,

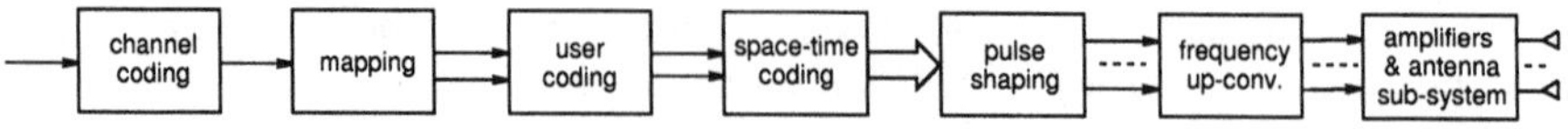

Figure 1.4 Block diagram of the transmitter.

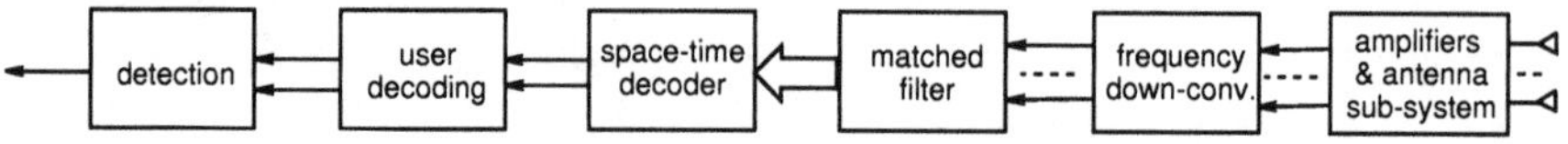

Figure 1.5 Block diagram of the receiver.

even though we assume a certain familiarity of the reader with the basic principles of digital communications.

- *Coding:* The input information bits may come from a discrete source, as a keyboard, for example, or from an analog source. In the latter case, an analog-to-digital converter is necessary to sample the analog waveform in time and then to convert the samples into a stream of bits. The coding operation performed by the initial block in Figure 1.4 refers to channel coding and its aim is to introduce error detection or correction coding, in order to increase the robustness of the system to noise.

- *Mapping:* The encoded bits pass through a mapper that associates to each set of let us say L bits, a symbol belonging to an alphabet of 2^L symbols. Typically, in wireless communications the symbols are complex, as in phase-shift keying (PSK) or quadrature amplitude modulation (QAM) systems. The two operations of channel coding and mapping into a symbol constellation may also be unified into a single transformation, as in trellis-coded modulation systems.

- *User-coding:* In multiuser systems, it is necessary to apply some form of user-coding whose purpose is to make possible (or at least facilitate) the discrimination of the different user signals at the receiver side.

- *Space-time encoding:* This kind of encoding, evidently present only when we have a multiantenna transmitter, maps the information symbols, which is a one-dimensional (1D) stream flowing in time, onto a two-dimensional (2D) stream flowing in both space and time.

- *Pulse shaping:* This operation is necessary to convert the sequence to be transmitted into a continuous time waveform. If we denote by T_s the duration of each transmitted symbol, or, equivalently by $1/T_s$ the symbol transmission rate, and by $g_T(t)$ the (real) impulse response of the shaping filter, the continuous-time waveform is related to the symbols sequence $x[k]$ as follows

$$x(t) = \sum_{k=-\infty}^{\infty} x[k] g_T(t - kT_s). \tag{1.27}$$

The signal $x(t)$ is our *baseband* transmitted signal. Typically, using PSK or QAM constellations, the symbols $x[k]$ are complex and thus we need to form the in-phase component $x_R(t) := \Re\{x(t)\} = \sum_{k=-\infty}^{\infty} \Re\{x[k]\} g_T(t - kT_s)$ and the quadrature component $x_I(t) := \Im\{x(t)\} = \sum_{k=-\infty}^{\infty} \Im\{x[k]\} g_T(t - kT_s)$. Using a multiple antenna system, we will have a set of continuous-time waveforms $x_i(t)$, one for each incoming symbol stream.

- *Frequency up-conversion:* The baseband waveform $x(t)$ has to be properly modulated before transmission through an antenna. The aim of the frequency up-converter is to shift the spectrum of the baseband signal $x(t)$ around the carrier allocated for our system. The frequency conversion is performed through mixers, which are basically nonlinear devices, whose final operation, after proper filtering, can be well described by the following relationship

$$\bar{x}(t) = x_R(t)\cos(2\pi f_0 t) - x_I(t)\sin(2\pi f_0 t), \tag{1.28}$$

  where f_0 is the carrier frequency.

- *High-power amplification:* Next, the high frequency signal $\bar{x}(t)$ has to be properly amplified. This operation is usually performed by a high-power amplifier. To fully exploit all the potential amplification of these devices, one can make them work in a amplitude range close to their saturation region. However, in such a case, the amplifier is nonlinear. The nonlinearity can be really detrimental when the transmitted signal has an amplitude modulation, which could result in severe distortion. Therefore, how close the power amplifier works to the saturation region depends on the kind of modulation that we are using.

- *Antenna subsystem:* The task of the final antenna subsystem is to send the set of signals $\bar{x}_i(t)$ to each antenna.

The blocks composing the receiver, shown in Figure 1.5, perform the inverse operations: 1) The front-end amplifier needs to have a low noise figure and sufficient gain to provide a global low system noise figure; 2) the frequency down-conversion brings the received signal into baseband; 3) the matched filter has the aim to maximize the SNR, 4) after matched filtering, the continuous-time signal is sampled and converted in a string of numbers. The successive stages have the task of recovering the information bits. We may distinguish, in principle, space-time decoding and multi-user detection, but these operations could also be performed jointly.

References

[1] Abbott, E., A., *Flatland : A Romance of Many Dimensions*, Dover Pubblications, New York, 1992.

[2] Shannon, C., "A mathematical theory of communications," *Bell Labs Tech. J.*, July and October 1948, pp. 379–423 and 623–656.

[3] Cover, T. M., Thomas, J. A., *Elements of Information Theory*, New York, John Wiley & Sons, 1991.

[4] Wang, Z., Giannakis, G.B., "A simple and general parameterization quantifying performance in fading channels," *IEEE Transactions on Communications*, Aug. 2003, pp. 1389–1398.

[5] Getu, B.N., Andersen, J.B., "BER and spectral efficiency of a MIMO system," *Proc. of the* 5th *Int. Symp. on Wireless Personal Multimedia Communications*, Vol. 2, Honolulu, Oct. 2002, pp.397–401.

[6] Zheng, L., Tse, D.N.C., "Diversity and multiplexing: A fundamental tradeoff in multiple antenna channels," *IEEE Trans. on Information Theory*, Vol. 49, May 2003, pp. 1073–1096.

[7] Winters, J., H., "On the capacity of radio communication systems with diversity in a Rayleigh fading environment," *IEEE Journal on Selected Areas in Communications*, June 1987, pp. 871–878.

[8] Telatar, I., E., "Capacity of multi-antenna Gaussian channels," *AT&T Technical Memo*, 1995; see also *European Trans. on Telecommun.*, Vol. 10, no. 6, 1999, pp. 586–595.

[9] Foschini, G., J., "Layered space-time architecture for wireless communication in a fading environment when using mutiple antennas," *Bell Lab. Tech. J.*, Vol. 1, 1996, pp. 41–59.

[10] Paulraj, A., Kailath, T., "Increasing capacity in wireless broadcast systems using distributed transmission/directional reception," *US Patent*, 5 345 599, 1994.

[11] Alamouti, S.,M., "A simple transmitter diversity scheme for wireless communications," *IEEE Journal on Selected Areas in Communications*, Vol. 16, Oct. 1998, pp. 1451–1458.

[12] Tarokh, V., Jafarkhani, H., Calderbank, A.R., "Space-time block codes from orthogonal designs," *IEEE Transactions on Information Theory*, Vol. 45, July 1999, pp. 1456–1467.

[13] Hottinen, A., Tirkkonen, O., Wichman, R., *Multi-antenna - Transceiver techniques for 3G and beyond*, West Sussex, UK: John Wiley & Sons, 2003.

[14] Larsson, E., G., Stoica, P., *Space-time block coding for wireless communications*, Cambridge, UK: Cambridge University Press, 2003.

[15] Paulraj, A., Nabar, R., Gore, D., *Introduction to space-time wireless communications*, Cambridge, UK: Cambridge University Press, 2003.

[16] Vucetic, B., Yuan, J., *Space-time coding*, New York, John Wiley & Sons, 2003.

[17] Dixon, G., M., *Division Algebras: Octonions, Quaternions, Complex Numbers and the Algebraic Design of Physics*, Dordrecht, The Neherlands: Kluwer, 1994.

[18] Sethuraman, B.A., Sundar Rajan, B., Shashidar, V., "Full-diversity, high-rate space-time block codes from division algenras," *IEEE Trans. on Information Theory*, Vol. 49, Oct. 2003, pp. 2596–2616.

[19] Martone, M. , *Multiantenna Digital Radio Transmission*, Norwood MA: Artech House, 2002.

[20] Barbarossa, S., "Trace-orthogonal design of MIMO systems with simple scalar detectors, full diversity and (almost) full rate," *Proc. of the V IEEE Signal Proc. Workshop on Signal Proc. Advances in Wireless Commun., SPAWC '2004*, Lisbon, Portugal, July 11–14, 2004.

Chapter 2

Channel Models and Modes

2.1 INTRODUCTION

Appropriate modeling of multiple-input/multiple-output (MIMO) channels is naturally the first step in the study of multiantenna systems. Before starting the analysis, it is important to clarify that, even though MIMO channel systems are usually intended to be multiantenna systems, this is not always the case. Multiple antennas are necessary only to create multiple *spatial channels*, but other dimensions can be exploited as well, such as time, frequency, and polarization. For example, multiple *temporal channels* can be implemented sending the information bits through different time slots, as in time division multiple access (TDMA) systems; *frequency channels* can be induced by sending bits through different (sub-) carriers, as in frequency division multiple access (FDMA) systems. Moreover, since in wireless communications the transmitted signals are carried by electromagnetic fields, which are vectors characterized by a certain polarization, we can also associate different bits to different polarization states, so as to create multiple *polarization channels*. One further possibility arises when planar antennas are used. Planar antennas (like drums) are in fact characterized by multiple modes that could be excited or sounded independently of each other. Hence, space, time, frequency, polarization, and even vibration modes, are all examples of physical channels that can be used to transmit information. We can thus see a MIMO system as the instrument to convey information through multiple channels. The nature of the channels can be very diverse. What is important is the multiplicity of the channels. This is indeed a possibility not far from practical implementations, as it is under consideration for the incoming third generation (3G) systems, where spatial and multi-polarization channels are considered, besides the conventional temporal channels.

Indeed, time and frequency variables are not independent of each other, as they are related by a Fourier transform (FT), so that temporal and frequency channels are different ways of managing the same physical domain: time. More generally, one can also associate different bits to different codes, as happens in code division multiple access (CDMA) systems, so that TDMA, FDMA, and CDMA can all be seen as different ways to handle the time domain.

In general, the choice of the domains to be used for communications depends on several aspects, including complexity, portability, and cost. For example, incorporating multiple antennas on a portable cellular phone poses some implementation problems, because of the phone dimensions. In fact, most of the potential advantages "promised" by multiple antenna systems are really achieved only if the channels are statistically independent. This requires the antenna elements to be spatially separated by more than half a wavelength. Working, for example, with a carrier frequency of 1 GHz, the wavelength is 30 cm, so that placing elements more distant than half a wavelength on an handset cellular phone would be troublesome. In such a case, one could place the elements at less than half a wavelength, thus sacrificing part of the benefits, or use polarization instead of space. The price to be paid in this case is the need to use a receive antenna able to catch all polarization components of the electric field, but the system can achieve a significant improvement without increasing the phone size. The idea of exploiting all polarization components to increase the capacity was proposed, for example, in [1].

One distinguishing feature of the wireless channel is its time-variability. Moreover, especially in an urban environment for example, the propagation occurs through multiple paths and undergoes to a series of different phenomena, like refractions, reflections, diffusions, scattering. The transmitted signal arrives then at the receiver through a number of paths. Intuitively, one could look at these phenomena as a source of troubles for establishing reliable communication links. However, a proper design can convert these shortcomings in sources of diversity that can improve the performance. Clearly, the first step in the design of such systems consists of a proper channel modeling. We start with single-input/single-output (SISO) channel, to introduce the basic functions characterizing the channels in the time-frequency plane, examining both deterministic as well as statistical channel models. Then we will extend the modeling to the MIMO case. We will introduce the important concepts of coherence time, frequency, and space. After having illustrated the main channel models for both SISO and MIMO structures, we will introduce the concept of channel *mode* and show that the modes are the fundamental vehicles to increase transmission rate and/or diversity.

2.2 WAVEFIELDS

In this chapter we review a few basic concepts about electromagnetic fields and propagation. The reader interested in a more in-depth treatment of this subject is invited to check, for example [2] or, for a direct link to multiantenna systems, [3].

The electromagnetic field is a vector field, and it is described, in each point of space, by two vectors, the *electric* and *magnetic* fields. Each field is described by a vector which is a function of both space and time. Indicating with $\vec{x}_0$, $\vec{y}_0$, and $\vec{z}_0$, the unit vectors of a cartesian reference system, the electric field in a point of position $\boldsymbol{r} := \vec{x}_0 x + \vec{y}_0 y + \vec{z}_0 z$ can be written as

$$\vec{e}(\boldsymbol{r};t) = \vec{x}_0 e_x(\boldsymbol{r};t) + \vec{y}_0 e_y(\boldsymbol{r};t) + \vec{z}_0 e_z(\boldsymbol{r};t) \tag{2.1}$$

where all the components are function of time. If we consider a monochromatic plane wave, characterized by the wavelength λ, with propagation vector $\vec{k}$, the electric field varies as a function of space and time as follows

$$\vec{e}(\boldsymbol{r};t) = \vec{E}_0 e^{j(\omega t - \vec{k}\cdot\vec{\boldsymbol{r}})}, \tag{2.2}$$

where $\vec{E}_0$ is the electric field in $\boldsymbol{r} = \boldsymbol{0}$, at $t = 0$. The propagation vector, in such a case, is $\vec{k} = (2\pi/\lambda)\vec{n}$, where $\vec{n}$ is the unit vector indicating the direction of propagation.

In particular, if we consider wave propagation in a medium with no charges or currents, and solve Maxwell's equations, it turns out that the electric field vector $\vec{e}(\boldsymbol{r};t)$ and the associated magnetic field $\vec{h}(\boldsymbol{r};t)$ are orthogonal to each other, and they are both orthogonal to the direction of propagation, established by the vector $\vec{k}$. More specifically,

$$\vec{k}\cdot\vec{e}(\boldsymbol{r};t) = 0, \tag{2.3}$$

and

$$\vec{h}(\boldsymbol{r};t) = \frac{1}{Z_0}\vec{n}\times\vec{e}(\boldsymbol{r};t), \tag{2.4}$$

where $\cdot$ denotes scalar product, $\times$ denotes vector product, and Z_0 is the characteristic impedance of free-space, equal, approximately, to 377 Ω.

We express the properties established by (2.3) and (2.4) by saying that the propagating wave is a *transverse* electromagnetic (TEM) field. In a TEM field, the electric and magnetic vectors lie on a plane perpendicular to the direction of propagation.

Let us consider a radiating antenna of size a. We choose the origin of the reference system as the center of this radiation source. The field in the immediate neighborhood of the source is composed of several different components. Each component is characterized by a specific attenuation, as the component travels away from the source. Beyond a certain distance from the antenna, in a region defined as the antenna *far-field*, many components can be neglected, and the only significant field components attenuate as $1/r$, where r is the distance from the antenna. The antenna *far-field* is conventionally defined as the region of points more distant than

$$r_f = \frac{2a^2}{\lambda} \tag{2.5}$$

from the antenna.

Introducing a polar reference system (r, θ, ϕ) centered at the radiating source and denoting by $E_r(\boldsymbol{r};t)$, $E_\theta(\boldsymbol{r};t)$, and $E_\phi(\boldsymbol{r};t)$ the three components of the electric field in the far-field region, from (2.3) we have $E_r = 0$ and [2]

$$E_\theta = E_0\frac{e^{-jkr}}{4\pi r}, \quad E_\phi = \alpha e^{j\beta} E_0 \frac{e^{-jkr}}{4\pi r}, \tag{2.6}$$

where α and β are real. Assuming, without any loss of generality, that E_0 is also real, we can also write the relationship between E_θ and $E\phi$ as follows [2]:

$$\left(\frac{E_\phi}{\alpha}\right)^2 + E_\theta^2 - \frac{2\cos\beta}{\alpha}E_\theta E_\phi = \frac{E_0^2\sin^2\beta}{(4\pi r)^2}. \tag{2.7}$$

This is the equation of an ellipse and it means that, in any given point in space, the electric field traces out an ellipse, as time passes. A pictorial description of the electromagnetic field is sketched in Figure 2.1, where we report the electric and magnetic fields at different points of space. In each point, the vector rotates as a function of time. The meaning of (2.7) can be explained by thinking of the action of the electric field on a charged particle. If there is an electrically charged particle at $\boldsymbol{r}$, with charge q, immersed in an electric field $\vec{e}(\boldsymbol{r};t)$, the particle is subject to a force $\vec{f}(\boldsymbol{r};t) = q\vec{e}(\boldsymbol{r};t)$. Therefore, the particle tends to oscillate in a direction specified by the direction of the electric field vector, with an intensity given by the product of the electric field intensity [the modulus of $\vec{e}(\boldsymbol{r};t)$] times the particle charge. Equation (2.7) can degenerate into a circle or a straight line, in which cases we say that we have *circular* or *linear* polarization. In particular, we have a circularly polarized wave when $\alpha = 1$ and $\beta = \pm\pi/2$. Conversely, we have *linear* polarization when $\beta = k\pi$, with k integer, which implies that $E_\phi = \alpha E_\theta$.

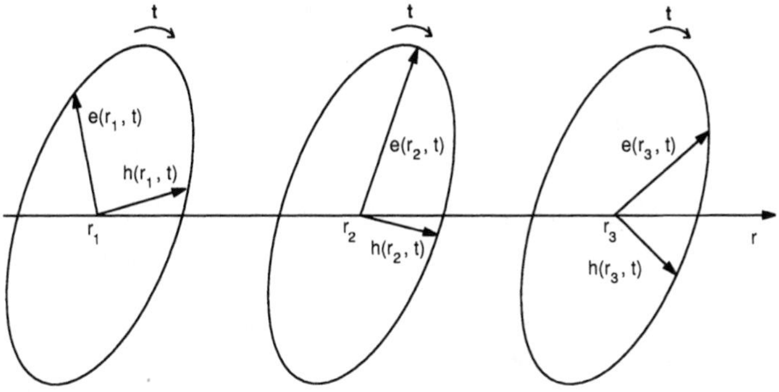

Figure 2.1 Field vectors as a function of space and time.

As already mentioned in the introduction of this chapter, polarization can be very important because it may be exploited to obtain multimode propagation even using a single antenna at both the transmitter and receiver, as suggested in [1]. We can in fact associate different information bits with the two polarization components of the transmitted electric field. At the receiver, we need then to be able to measure at least two polarization components. In general because of the complicated interaction of the transmitted field with the propagation environment, the electric filed impinging on the receive antenna does not lie, necessarily, on the same plane as the transmitted field.

We introduce now a completely general relationship between the transmitted and the received wave. Let us indicate with $\vec{e}_t(\boldsymbol{r}';t)$ the electric field transmitted from an antenna placed at $\boldsymbol{r}'$ and with $\vec{e}_r(\boldsymbol{r};t)$ the electric field impinging on the receive antenna, located at $\boldsymbol{r}$. Assuming linear propagation, the relationship between the transmitted and received fields is

$$\vec{e}_r(\boldsymbol{r};t) = \int_S \int_{t'=-\infty}^{t} \underline{K}(\boldsymbol{r},\boldsymbol{r}';t,t')\vec{e}_t(\boldsymbol{r}';t')d\boldsymbol{r}'dt', \tag{2.8}$$

where the first (triple or volume) integral is over space, and more specifically over the domain S indicating the region comprising all sources of radiation, whereas the second integral is over time. The integration limits go from $-\infty$ to t reflect the causality property, as the field $\vec{e}_r(\boldsymbol{r};t)$ at any instant t depends only on the past (i.e., on fields radiating up to the instant t itself) but not on the future.

The matrix $\underline{K}(\boldsymbol{r}, \boldsymbol{r}'; t, t')$ is a matrix kernel with components

$$\underline{K}(\boldsymbol{r}, \boldsymbol{r}'; t, t') = \begin{pmatrix} K_{xx}(\boldsymbol{r}, \boldsymbol{r}'; t, t') & K_{xy}(\boldsymbol{r}, \boldsymbol{r}'; t, t') & K_{xz}(\boldsymbol{r}, \boldsymbol{r}'; t, t') \\ K_{yx}(\boldsymbol{r}, \boldsymbol{r}'; t, t') & K_{yy}(\boldsymbol{r}, \boldsymbol{r}'; t, t') & K_{yz}(\boldsymbol{r}, \boldsymbol{r}'; t, t') \\ K_{zx}(\boldsymbol{r}, \boldsymbol{r}'; t, t') & K_{zy}(\boldsymbol{r}, \boldsymbol{r}'; t, t') & K_{zz}(\boldsymbol{r}, \boldsymbol{r}'; t, t') \end{pmatrix}. \tag{2.9}$$

If $\underline{K}(\boldsymbol{r}, \boldsymbol{r}'; t, t')$ is diagonal, the received field is parallel to the transmitted field, otherwise there is a *cross-polarization* effect. In a dense scattering environment, for example, we observe cross-polarization due to the superposition of several reflections, scattering and diffraction effects. The interactions of the waves with the environment are so complicated that it is not meaningful to try to write a deterministic relationship between the transmitted and received fields. For this reason, we will proceed, in Section 2.3.2, to provide a statistical characterization of the channel.

2.3 SISO CHANNELS

Let us consider first a SISO channel. This constitutes a particular case of (2.8), occurring when the source of radiation and the receiving antenna can be both assimilated to (dimensionless) points in space, and we consider only one polarization. In such a case, we may concentrate on the time coordinate only, because there is no integration over the space dimension. Denoting by $x(t)$ the (scalar) signal sent through the channel and with $y(t)$ the corresponding received signal, any linear channel can be described by the following input/output relationship:

$$y(t) = \int_{-\infty}^{\infty} h(t, \tau) x(t - \tau) d\tau \tag{2.10}$$

where $h(t, \tau)$ is the time-varying channel impulse response. With respect to the more general relationship (2.8), (2.10) does not have an explicit dependence over the space dimension. In this case, in fact, since both source and destination are point-like, and we are using only one polarization, there is no integration over space. The impulse response $h(t, \tau)$ depends on the coordinates of the transmit and receive antennas, but since there is no integration over space, to keep the notation as simple as possible, we neglect the dependence of $h(t, \tau)$ on the spatial coordinates.

In general, given the high complexity of the physical interactions characterizing the transmission through a real channel, including reflections, scattering, refraction, or diffraction (see, for example, [2]), the most appropriate modeling of the impulse response is probabilistic, so that $h(t, \tau)$ is a 2D random process. In several cases of practical interest, the realizations of this random process can be described very accurately by a parametric model whose parameters represent physically meaningful quantities, such as delays, Doppler frequencies, and reflection coefficients. Modeling the parameters of a multipath channel as random variables, for example, provides a simple yet important random channel model. Parametric modeling is especially important in devising channel estimation and tracking algorithms (see, for example, [4]). In general, both deterministic and stochastic approaches are equally useful in describing a time-varying channel, as they embrace different aspects: The stochastic model is better suited for describing global behaviors, whereas the deterministic one is more useful for studying the transmission through a specific channel realization. For all these reasons, in the next section we start with a deterministic characterization; the random model counterpart will be studied in the ensuing section.

2.3.1 Deterministic Models

2.3.1.1 Continuous-Time Model

A SISO channel can be described equivalently by one out of four interchangeable functions. Following the same notation introduced in the pioneering work of Bello [5], any linear SISO channel can be fully described by its impulse response $h(t, \tau)$, which was introduced in (2.10), or by any of the following characteristic system functions:

1) Time-varying transfer function:

$$H(t, f) := \int_{-\infty}^{\infty} h(t, \tau) e^{-j2\pi f\tau} d\tau; \tag{2.11}$$

2) Delay-Doppler spread function:

$$S(\nu, \tau) := \int_{-\infty}^{\infty} h(t, \tau) e^{-j2\pi\nu t} dt; \tag{2.12}$$

3) Output-Doppler spread function:

$$Q(\nu, f) := \int_{-\infty}^{\infty} \int_{-\infty}^{\infty} h(t, \tau) e^{-j2\pi(\nu t + f\tau)} dt d\tau. \tag{2.13}$$

The relationship between these functions is depicted in Figure 2.2. In the figure, the symbol $\mathcal{F}_{1D}$ indicates one-dimensional Fourier transform (FT), whereas $\mathcal{F}_{2D}$ stands for a two-dimensional FT (the variables within parentheses indicate the pair of dual variables related by a FT, as established by (2.11,2.12), and (2.13)).

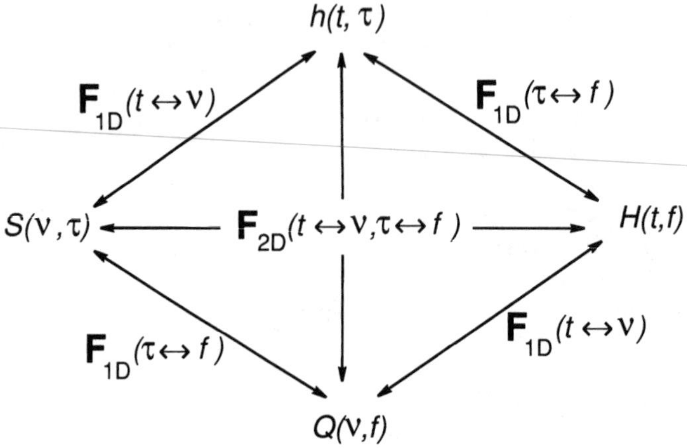

Figure 2.2 Relationship between system functions.

Example: Multipath channel

Let us consider the multipath model depicted in Figure 2.3, where the the receiver gets a superposition of replicas of the transmitted signal. We consider the situation where the transmitter is fixed and the receiver is moving with a constant speed v along the direction specified by the angle γ. We assume, for simplicity, that the transmitted signal is monochromatic, with carrier frequency f_0. Each path in Figure 2.3 is characterized by a triplet of values: the amplitude h_k, the distance r_k, and the angle θ_k. We further assume that all these values are constant within the observation interval. Since the electromagnetic waves propagate at the speed of light $c = 3 \cdot 10^8$ m/s, the delays τ_k and the Doppler frequency shifts f_k are related to the distances r_k and the angles θ_k through the following relationships

$$\tau_k = \frac{r_k}{c}, \quad f_k = \frac{v}{\lambda} cos(\theta_k - \gamma), \tag{2.14}$$

where $\lambda = c/f_0$. Hence, denoting with $x(t)$ the complex envelope of the transmitted signal, the corresponding received signal $y(t)$ is

$$y(t) = \sum_{k=0}^{K-1} h_k x(\tau - \tau_k) e^{j2\pi f_k t}. \tag{2.15}$$

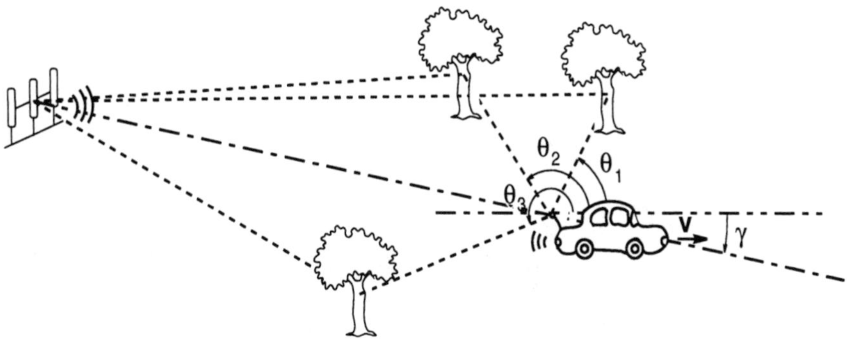

Figure 2.3 Multipath channel.

Such a system is characterized by the impulse response

$$h(t,\tau) = \sum_{k=0}^{K-1} h_k \delta(\tau - \tau_k) e^{j2\pi f_k t} \tag{2.16}$$

or, exploiting the relationship between the different system functions, by the delay-Doppler spread function

$$S(\nu,\tau) = \sum_{k=0}^{K-1} h_k \delta(\tau - \tau_k)\delta(\nu - f_k), \tag{2.17}$$

the time-varying transfer function

$$H(t,f) = \sum_{k=0}^{K-1} h_k e^{j2\pi(f_k t - \tau_k f)} \tag{2.18}$$

or the Doppler spread function

$$Q(\nu,f) = \sum_{k=0}^{K-1} h_k e^{-j2\pi(\tau_k f)} \delta(\nu - f_k). \tag{2.19}$$

As we can see from (2.17), the delay-Doppler spread of the multipath channels function assumes then a very specific form, as it is composed only of Dirac pulses. An example of spread function of a multipath channel is reported in Figure 2.4.

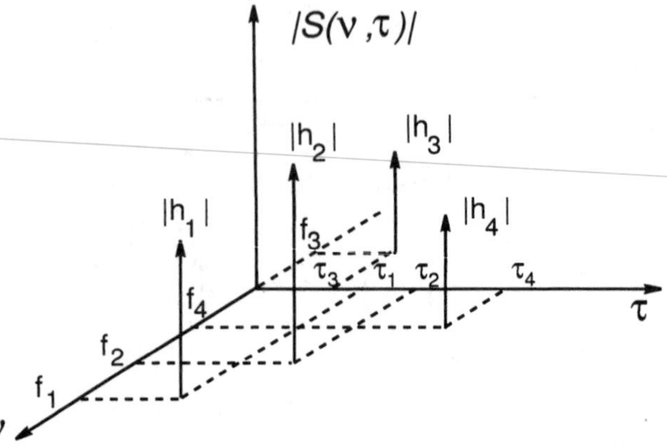

Figure 2.4 Modulus of a multipath channel spread function.

Delay-Doppler spread function

Each channel function has its own specific meaning. For example, applying an inverse FT to (2.12), we can express $h(t, \tau)$ in terms of $S(\nu, \tau)$

$$h(t, \tau) := \int_{-\infty}^{\infty} S(\nu, \tau) e^{j2\pi\nu t} d\nu. \tag{2.20}$$

Then, plugging (2.20) into (2.10), we can write the I/O relationship (2.10) as

$$y(t) = \int_{-\infty}^{\infty} \int_{-\infty}^{\infty} S(\nu, \tau) x(t - \tau) e^{j2\pi\nu t} d\nu d\tau. \tag{2.21}$$

Expressing the integral as the limit value of the series

$$y(t) = \lim_{\Delta\tau, \Delta\nu \to 0} \sum_{k=-\infty}^{\infty} \sum_{l=-\infty}^{\infty} S(k\Delta\nu, l\Delta\tau) x(t - k\Delta\tau) e^{j2\pi l \Delta\nu t} \Delta\tau \Delta\nu, \tag{2.22}$$

we can state that the output of *every* linear system can always be written as the superposition of replicas of the input signal shifted in time by τ, in frequency by ν, and scaled in amplitude by the factor $S(\nu, \tau)\Delta\tau\Delta\nu$. The function $S(\nu, \tau)$ assumes thus the meaning of a (complex) amplitude density, quantifying the contribution on the received signal corresponding to Doppler shift ν and delay τ. In case of a multipath channel, the function $S(\nu, \tau)$ is composed only of Dirac pulses and the output is the sum of a finite number of replicas, whereas in the most general case, $S(\nu, \tau)$ might be a continuous plus possible Dirac pulses.

Since the range of values for Doppler shift and delay is limited, the spread function is necessarily concentrated in a certain region of the (ν, τ) plane. We can quantify the concentration of $S(\nu, \tau)$ by introducing the moments of its modulus. In particular, we define the following absolute (uncentered) moments:

$$m_S^{(k,l)} := \frac{\int_{-\infty}^{\infty} \int_{-\infty}^{\infty} |S(\nu, \tau)||\tau|^k |\nu|^l d\nu d\tau}{\int_{-\infty}^{\infty} \int_{-\infty}^{\infty} |S(\nu, \tau)| d\nu d\tau} \tag{2.23}$$

and the centered absolute moments:

$$\mu_S^{(k,l)} := \frac{\int_{-\infty}^{\infty} \int_{-\infty}^{\infty} |S(\nu, \tau)||\tau - \tau_0|^k |\nu - \nu_0|^l d\nu d\tau}{\int_{-\infty}^{\infty} \int_{-\infty}^{\infty} |S(\nu, \tau)| d\nu d\tau}, \tag{2.24}$$

where τ_0 and ν_0 can be chosen to minimize, for example, $\mu_S^{(1,0)}$ and $\mu_S^{(0,1)}$, separately. In such a case, τ_0 is equal to the median value of $S(\tau) := \int_{-\infty}^{\infty} |S(\nu, \tau)| d\nu$, and ν_0 is the median value of $S(\nu) := \int_{-\infty}^{\infty} |S(\nu, \tau)| d\tau$. Alternatively, if $S(\nu, \tau)$ has exactly finite support, we can measure the extent of its support through its area, as

$$\mathcal{A} := \max(|\nu|) \cdot \max(|\tau|), \text{ such that } S(\nu, \tau) \neq 0. \tag{2.25}$$

A channel is said to be *underspread* if $\mathcal{A} \ll 1$ or, equivalently, the normalized moments (2.24) are all much smaller than one. This property holds true when the spread function is very concentrated around the origin of the delay-Doppler domain. Indeed, this is the situation of interest, as most communication channels are underspread.

Let us consider, for example, a narrowband system, transmitting around a frequency $f_0 = 3$ GHz, with wavelength $\lambda = c/f_0 = 0.1$ m. Let us assume that the maximum delay among the received signal components is $\tau_{\max} = 1\ \mu$s. This corresponds to a difference in the distance travelled by the fastest and slowest received components equal to $d = c\tau_{\max} = 3 \cdot 10^8 \cdot 10^{-6}$ m= 300m. Let us suppose also that the receiver is moving at a speed of $v = 120$ km/h. The maximum Doppler frequency is thus $\nu_{\max} \approx 666$ Hz. Hence, we have $\mathcal{A} \approx 6.66 \cdot 10^{-4}$, which is indeed much less than 1.

Time-varying transfer function

The time-varying transfer function $H(t, f)$ is particularly useful to describe time-invariant systems. In such a case, in fact, the spectrum of the system output $Y(f)$ is related to the spectrum of the input by the simple relationship

$$Y(f) = H(f)X(f). \tag{2.26}$$

This equation explains the filtering operation operated by a time-invariant channel. In fact, (2.26) makes it immediately evident that the frequency components of the input are more or less attenuated by the transit through the channel, depending on the corresponding values of the channel transfer function. If the channel is time-varying, (2.26) does not hold anymore. However, if we consider a block transmission system, where each block has duration T and we assume that the channel transfer function is constant within each block, we can generalize (2.26) as follows. If we denote by $X(n; f)$ and $Y(n; f)$ the spectrum of the transmit and receive signal, respectively, during the nth time slot, and if the channel within each time slot is time-invariant, we can write

$$Y(n; f) = H(nT, f)X(n; f). \tag{2.27}$$

How fast $H(t, f)$ varies with t and f? The rapidity of variation is dictated by the support of the delay-Doppler spread function $S(\nu, \tau)$. In fact, $H(t, f)$ is related to $S(\nu, \tau)$ by a two-dimensional FT:

$$H(t, f) = \int_{-\infty}^{\infty} \int_{-\infty}^{\infty} S(\nu, \tau) e^{j2\pi(\nu t - f\tau)} d\tau \, d\nu. \tag{2.28}$$

Therefore, $H(t, f)$ is smooth along t (or f) if the support of $S(\nu, \tau)$ along ν (or τ) is small and vice versa. More specifically, the smoothness of $H(t, f)$ can be quantified through the moments of $S(\nu, \tau)$. More specifically, differentiating (2.28) with respect to both t (k times) and f (l times), we have

$$\frac{\partial^{k+l} H(t, f)}{\partial t^k \partial f^l} = (j2\pi)^{k+l} \int_{-\infty}^{\infty} \int_{-\infty}^{\infty} \nu^k (-\tau)^l S(\nu, \tau) e^{j2\pi(\nu t - f\tau)} d\tau \, d\nu. \tag{2.29}$$

Therefore,

$$\begin{aligned} \left| \frac{\partial^{k+l} H(t, f)}{\partial t^k \partial f^l} \right| &\leq (2\pi)^{k+l} \int_{-\infty}^{\infty} \int_{-\infty}^{\infty} |\nu|^k |\tau|^l |S(\nu, \tau)| d\tau \, d\nu \\ &= (2\pi)^{k+l} m_S^{(k,l)} \int_{-\infty}^{\infty} \int_{-\infty}^{\infty} |S(\nu, \tau)| d\nu d\tau, \end{aligned} \tag{2.30}$$

where $m_S^{(k,l)}$ is given by (2.23). Hence, the time-varying transfer function is a smooth function if the absolute moments of $S(\nu, \tau)$ are small (i.e., if the channel is underspread).

Output Doppler spread function

The output Doppler spread function $Q(\nu, f)$ is useful to derive the dual input/output relationship of (2.10) in the frequency domain:

$$Y(f) = \int_{-\infty}^{\infty} Q(f-\nu,\nu)X(\nu)d\nu = \int_{-\infty}^{\infty} Q(\nu, f-\nu)X(f-\nu)d\nu, \quad (2.31)$$

where $X(f)$ and $Y(f)$ denote the FT of the input and output signals, respectively.

2.3.1.2 Discrete-Time Model

The continuous-time characterization is useful for grasping some channel properties, but, as we are basically interested in digital communications, it is important to model the equivalent discrete-time channel existing between the transmit and receive sequences. We consider the system depicted in Figure 1.4 and we denote with $x[k]$ the (generally complex) k-th transmitted symbol and with $g_T(t)$ the transmit lowpass filter whose bandwidth is directly proportional to the symbol rate $1/T_s$. The impulse response $g_T(t)$ has a Nyquist characteristic, and it is usually a root raised cosine filter [6]. Using linear modulation, the baseband transmitted signal can be expressed as

$$x(t) = \sum_{k=-\infty}^{\infty} x[k]g_T(t-kT_s). \quad (2.32)$$

The channel output can then be written as

$$z(t) = \sum_{k=-\infty}^{\infty} x[k]\int_{-\infty}^{\infty} h(t,\tau)g_T(t-\tau-kT_s)d\tau. \quad (2.33)$$

The received signal is demodulated, low-pass filtered, and sampled. Denoting by $g_R(t)$ the impulse response of the receive low-pass filter, the baseband received signal is

$$y(t) = \sum_{k=-\infty}^{\infty} x[k]\int_{-\infty}^{\infty}\int_{-\infty}^{\infty} g_R(t-\theta)g_T(\theta-\tau-kT_s)h(\theta,\tau)d\tau d\theta. \quad (2.34)$$

Hence, sampling $y(t)$ at symbol period T_s, we obtain the sequence

$$y[n] := y(nT_s) = \sum_{k=-\infty}^{\infty} h[n, n-k]x[k], \quad (2.35)$$

where we have introduced the equivalent discrete-time impulse response

$$h[n, n-k] := \int_{-\infty}^{\infty}\int_{-\infty}^{\infty} g_R(nT_s - \theta)g_T(\theta - \tau - kT_s)h(\theta, \tau)d\tau d\theta. \quad (2.36)$$

Equation (2.35) is the discrete-time counterpart of (2.10).

To gain better insight into the transmission through LTV channels, it is useful to express (2.36) in the frequency domain. Specifically, introducing the transfer functions $G_T(f)$ and $G_R(f)$ of the transmit and receive filters and using the output Doppler-spread function (2.13), we may rewrite (2.36) as

$$h[n, n-k] := \int_{-\infty}^{\infty}\int_{-\infty}^{\infty} G_R(f)G_T(\nu)Q(f-\nu, \nu)e^{j2\pi(nf-\nu k)T_s}d\nu df. \quad (2.37)$$

Using now the multipath channel model (2.16), we get

$$h[n, n-k] = \sum_{q=0}^{L} h_q e^{j2\pi f_q nT_s}\int_{-\infty}^{\infty} G_R(\nu + f_q)G_T(\nu)e^{j2\pi\nu((n-k)T_s+\tau_q)}d\nu. \quad (2.38)$$

Substituting the transmit and receive transfer functions $G_T(f)$ and $G_R(f)$ in (2.38), we obtain the equivalent discrete-time impulse response in the most general case. To comprehend some of the basic features of the discrete-time equivalent channels, it is useful to analyze the simple case where both transmit and receive shaping filters are ideal lowpass filters. In particular, setting $G_T(f) = G_R(f) = \sqrt{T_s}\text{rect}(fT_s)$, where the rectangular function $\text{rect}(f)$ is equal to one for $|f| < 1/2$ and is null otherwise, (2.38) gives rise to the following DT impulse response

$$h[n, k] = \sum_{q=0}^{L}[1 - |f_q|T_s]^+ h_q e^{j\pi\nu_q(2n-k+\theta_q)}\text{sinc}[\pi(1-|\nu_q|)(k-\theta_q)], \quad (2.39)$$

where $[x]^+ \equiv \max(x, 0)$ and $\text{sinc}(x) := \sin(x)/x$. We have also introduced the normalized delay $\theta_q := \tau/T_s$ and Doppler $\nu_q := f_qT_s$. Since $\max_q|\nu_q| \ll 1$, (2.39) makes clear that the components corresponding to higher Doppler shifts ν_q are more attenuated. This happens because part of their energy falls outside of the receive filter bandwidth. In fact, since the transit through an LTV channel increases the bandwidth of the signal, the receive filter should have a bandwidth greater than $1/T_s$ to keep all the useful energy and, consequently, the sampling rate should also be higher than $1/T_s$, to avoid any loss of information. However, in practice $f_qT_s \ll 1$, so that usually the receive filter bandwidth and sampling rate can be maintained equal to $1/T_s$, without any appreciable loss. As a simple numerical example, using a carrier frequency of 2 GHz in a link of 1 Mbps between terminals in relative motion at a velocity of $v = 150$ km/h, the maximum normalized Doppler shift is approximately $2.8 \cdot 10^{-4}$ Hz. This explains why in most practical systems the receiver bandwidth is not greater than $1/T_s$, for the gain obtainable otherwise is not worth of the extra complications related to resampling. Therefore, (2.39) can be approximated with negligible error as

$$h[n,k] \simeq \sum_{q=0}^{L} h_q e^{j\pi\nu_q(2n-k+\theta_q)} sinc[\pi(k-\theta_q)]. \tag{2.40}$$

Ideal channels

We say that a continuous time (CT) channel is ideal if the channel response $y(t)$ to any input $x(t)$ is a delayed replica of the input, possibly multiplied by a coefficient A, so that $y(t) = Ax(t-t_0), \forall x(t)$. An ideal channel is then characterized by an impulse response $h(t) = A\delta(t-t_0)$ or, equivalently, by a transfer function $H(f) = Ae^{-j2\pi f t_0}$. Similarly, we say that a discrete-time (DT) channel is ideal if, for any input sequence $x[n]$, the corresponding output sequence $y[n]$ is a delayed version of $x[n]$, possibly scaled in amplitude, that is $y[n] = Ax[n-n_0]$.

Note that an ideal CT channel does not necessarily imply that the corresponding DT channel is ideal. This happens only if the delay t_0 is an integer multiple of the sampling time T_s.

2.3.2 Stochastic Models

The deterministic description of a wireless channel introduced in the previous sections is useful to relate the channel parameters to physically meaningful quantities, such as delays, Doppler frequencies, and amplitudes of the received signal components. However, such a characterization is specific to a given channel scenario and, as such, it cannot be used to describe a wireless channel, in general. In practice, the received signal is the result of the interaction of the radiated electromagnetic wave with the environment between transmitter and receiver. Such an interaction is too complicated to be described in a deterministic way. Hence, the only way to describe the *general* properties of a wireless channel is in probabilistic terms. More specifically, the impulse response of a wireless channel can be modeled as a stochastic (random) process, whose parameters (rather than the impulse responses) are related to physical quantities. In this more general setup, the deterministic impulse responses studied in the previous section, can be seen as realizations of this random process.

The first step in the characterization of the channel functions is establishing the link between the statistics of the channel functions and the parameters of the physical channel. In general, this relationship is rather complicated since it depends on too many parameters. To simplify the problem, it is useful to look at the problem at different scales, in terms of both space and time, considering first order, and then second order, statistics, as tools able to capture different meaningful features.

First order statistics

The first order statistics describe the properties of the single random variable extracted from the random process describing the channel impulse response. Since these properties depend on several propagation aspects, it is useful (and common practice) to distinguish between small, medium, and large scale models.

Small-scale propagation model

Small-scale or short-term fading is caused by the superposition of several received signals backscattered from the environment[1] A typical example is the urban environment, where the received signal is the superposition of paths induced by reflection, diffraction, and scattering from buildings, cars, and so on. If several paths arrive with delays and Doppler shifts which differ by less than the system resolution in time and frequency, we receive, for each pair of delay and Doppler, a superposition of, possibly, several contributions. Since these contributions arrive from different reflecting (diffracting) objects, we may well assume that they are statistically independent.

The scale of this kind of (spatial) fading is of the order of the wavelength. In fact, moving by a fraction of the wavelength, different field components might arrive in phase or out of phase. This is the phenomenon underlying small-scale fading. If the number of independent contributions is sufficiently high and there is no contribution clearly dominant with respect to the others, we may invoke the central limit theorem and state that each path is characterized by a complex amplitude modeled as a complex Gaussian random variable $x_r + jx_i$. This random variable (rv) may have a deterministic component, when there is, for example, a line-of-sight (LOS) path. In such a case, the rv has a non-null mean, otherwise the mean is zero. In case of no-LOS, the received complex variable has an amplitude $r = \sqrt{x_r^2 + x_i^2}$, which has

1 Assuming a (fairly) constant relative velocity between transmitter and receiver (and no moving scatterers), the spatial variation is directly proportional to the temporal variation. This is the basic hypothesis underlying the common approach that treats small-scale and short-term variabilities as different aspects of the same physical phenomenon.

a Rayleigh probability density function (pdf)

$$p_R(r) = \frac{r}{\sigma^2} e^{-r^2/2\sigma^2} u(r), \tag{2.41}$$

where $2\sigma^2$ is the variance of the complex Gaussian rv $x_r + jx_i$; $u(r)$ is the unit step function. This model is known as the *Rayleigh-fading* model.

Conversely, if there is a LOS, with amplitude A, the pdf of r is

$$p_R(r) = \frac{r}{\sigma^2} e^{-(r^2+A^2)/2\sigma^2} I_0\left(\frac{Ar}{2\sigma^2}\right) u(r) \tag{2.42}$$

where $I_0(x)$ is the modified Bessel function of the first kind of order zero. The model (2.42) is known as the *Rice-fading* model. To analyze the behavior of the Rice pdf, it is useful to introduce the so-called Rice factor, defined as

$$K := \frac{A^2}{2\sigma^2}. \tag{2.43}$$

In Figure 2.5 we report a few examples of the Rice pdf, for different values of K. As we can see from Figure 2.5, for $K = 0$, the Rice pdf is equal to the Rayleigh pdf, whereas, as K increases, the Rice pdf tends to follow a nearly Gaussian behavior (for positive r). Another pdf that is sufficiently versatile to include several situations of interest is the Nakagami pdf

$$p_R(r) = \frac{2m^m r^{2m-1}}{\Gamma(m)\sigma^{2m}} e^{-mr^2/(2\sigma^2)} u(r), \quad m \geq \frac{1}{2}, \tag{2.44}$$

where $\Gamma(x)$ is the Gamma function. The parameter m models different situations: $m = 1/2$ gives rise to a one-sided Gaussian pdf; with $m = 1$, we have the Rayleigh pdf; if m approaches infinity, (2.44) tends to a Dirac pulse, which indicates no fading at all. Hence, m models somehow the randomness of the channel: The higher is m, the less random is the channel, and vice versa. Some examples of Nakagami pdf's are reported in Figure 2.6, for different values of m. We can see that, indeed, as m increases, $p_R(r)$ tends to be more and more concentrated, thus revealing less randomness.

Medium-scale propagation model

The parameters m, A, or σ appearing in the pdf models examined in the previous section are also clearly related to physical properties of the channel as well as to the transmission parameters (e.g., wavelength, attenuations, and reflections). In

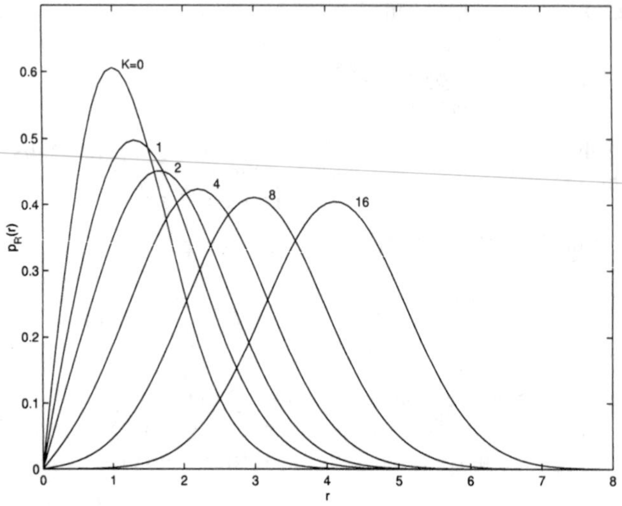

Figure 2.5 Rice and Rayleigh probability density functions.

general, these parameters are random variables themselves, as they reflect different propagation phenomena. For example, even staying at the same distance from the transmitter, the received power might change significantly because of shadowing effects. A typical pdf commonly used to characterize the medium-scale fluctuations is the *log-normal* pdf, where the received mean power p_0 is itself a random variable, with pdf

$$p_{P_0}(p_0) = \frac{1}{\sqrt{2\pi}\sigma_p p_0} e^{-(ln(p_0)-m_p)^2/2\sigma_p^2} u(r), \tag{2.45}$$

where m_p and σ_p^2 are the mean value and variance of $\ln(p_0)$. The scale of variability is now larger than small-scale models, as the receiver must move from, for example, a non-shadowed zone to a shadowed one. This scale is then dependent on the environment where the communications occurs. In an urban scenario, the variability is higher than in a rural environment.

Large-scale propagation model

The average power received over a large-scale model depends primarily on the distance. In general, indicating with R the distance between transmitter and receiver,

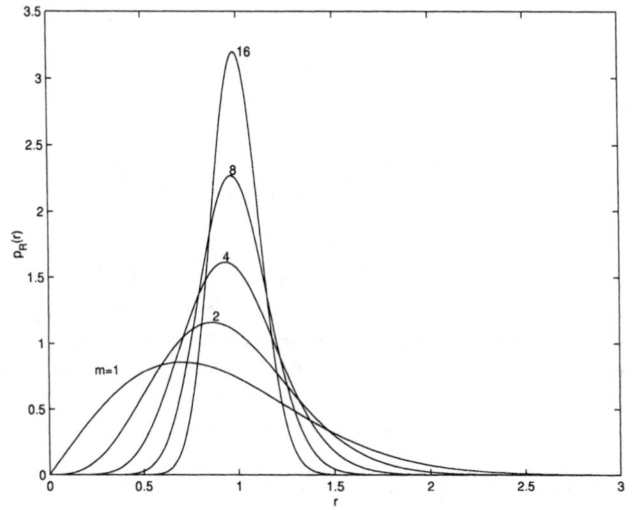

Figure 2.6 Nakagami probability density functions.

the average power p_a, received at a distance R from the transmitter, follows a power-law like

$$p_a(R) = \alpha \left(\frac{R}{R_0} \right)^{-\gamma} \tag{2.46}$$

where R_0 is a reference distance, α depends on the transmit wavelength and on the transmit and receive antenna radiation patterns; γ is the path-loss exponent, which is equal to 2 for free space propagation, but it can range from 2 to 6 in indoor or outdoor environments.

In summary, small, medium and large-scale models capture different aspects: Small scale variations are due to the fluctuation of the sum of (possibly many) contributions arriving at the receiver with different phase shifts; medium scale models reflect the possibility for different propagation phenomena, i.e., shadowing versus no shadowing, rich scattering versus LOS propagation, etc.; large-scale variations incorporate the dependence of the received power with the distance between transmitter and receiver. The parameters of the different scale models are related to each other. For example, the average value of p_0, in (2.45), at a distance R, is $p_a(R)$.

Second order statistics

To describe how the channel parameters fluctuate with time (or space), we need to introduce the second order statistics. We define the mean value of the impulse response

$$m_h(t,\tau) = E\{h(t,\tau)\} \tag{2.47}$$

and its correlation function

$$R_h(t_1,t_2;\tau_1,\tau_2) := E\{h^*(t_1,\tau_1)h(t_2,\tau_2)\}. \tag{2.48}$$

The symbol $E\{\cdot\}$ denotes expected value. We can also introduce the correlation functions of all the other functions describing the random channel. Clearly, because of the link between every pair of functions, there is a relationship between the corresponding correlation functions. For example, the correlation of the delay-Doppler spread function $R_S(\nu_1,\nu_2;\tau_1,\tau_2) := E\{S^*(\nu_1,\tau_1)S(\nu_2,\tau_2)\}$ is related to the correlation of the impulse response by

$$R_h(t_1,t_2;\tau_1,\tau_2) = \int_{-\infty}^{\infty}\int_{-\infty}^{\infty} R_S(\nu_1,\nu_2;\tau_1,\tau_2)e^{j2\pi(\nu_2 t_2-\nu_1 t_1)}d\nu_1 d\nu_2. \tag{2.49}$$

Similarly, we can introduce the correlation of the time-varying transfer function, which is related to the impulse response correlation by the following transformation

$$R_H(t_1,t_2;f_1,f_2) = \int_{-\infty}^{\infty}\int_{-\infty}^{\infty} R_h(t_1,t_2;\tau_1,\tau_2)e^{-j2\pi(f_2 t_2-f_1 t_1)}d\tau_1 d\tau_2. \tag{2.50}$$

As we can see from the previous expressions, all channel correlation functions are four-dimensional functions, and thus it is not simple to represent them and extract meaningful channel parameters. However, in many practical cases, the correlation functions can be described by a two-dimensional function, as we will show shortly.

Let us consider a channel where the contributions received from different angles are uncorrelated. This assumption is well justified in practice as arrivals from different angles refer to different scattering objects. In such a case, the function $R_S(\nu_1,\nu_2;\tau_1,\tau_2)$ is equal to zero for $\nu_2 \neq \nu_1$. We can thus describe this correlation function as

$$R_S(\nu_1,\nu_2;\tau_1,\tau_2) = R_S(\nu_1;\tau_1,\tau_2)\delta(\nu_2-\nu_1). \tag{2.51}$$

Inserting such an expression in (2.49), we get

$$\begin{aligned} R_h(t_1,t_2;\tau_1,\tau_2) &= \int_{-\infty}^{\infty}\int_{-\infty}^{\infty} R_S(\nu_1;\tau_1,\tau_2)\delta(\nu_2-\nu_1)e^{j2\pi(\nu_2 t_2-\nu_1 t_1)}d\nu_1 d\nu_2 \\ &= \int_{-\infty}^{\infty} R_S(\nu_1;\tau_1,\tau_2)e^{j2\pi\nu_1(t_2-t_1)}d\nu_1. \end{aligned} \tag{2.52}$$

The last term reveals that $R_h(t_1, t_2; \tau_1, \tau_2)$ depends on t_1 and t_2 only through their difference Δt, so that we can write

$$R_h(t_1, t_2; \tau_1, \tau_2) = R_h(\Delta t; \tau_1, \tau_2). \tag{2.53}$$

A channel satisfying (2.53) is a *wide-sense stationary* (WSS) channel. Therefore, a WSS channel is simply a channel where the contributions coming from different angles (Doppler frequencies) are uncorrelated.

Proceeding in a similar manner, if the contributions from different distances (e.g., delays) are uncorrelated; that is,

$$R_S(\nu_1, \nu_2; \tau_1, \tau_2) = R_S(\nu_1, \nu_2; \tau_1)\delta(\tau_2 - \tau_1), \tag{2.54}$$

the correlation of the impulse response assumes the form

$$R_h(t_1, t_2; \tau_1, \tau_2) = R_h(t_1, t_2; \tau_1)\delta(\tau_2 - \tau_1). \tag{2.55}$$

A channel satisfying (2.55) is denoted as an *uncorrelated scattering* (US) channel. This terminology is largely widespread, and this is the only reason we use it here. Nevertheless, it is worth pointing out that in this case, uncorrelated scattering refers only to different delays (and not necessarily Doppler shifts).

Combining the WSS and US properties, a channel where contributions with different delays *and* Doppler frequencies are uncorrelated is a WSS-US channel, and its impulse response is characterized by a correlation function

$$R_h(t_1, t_2; \tau_1, \tau_2) = R_h(\Delta t, \tau_1)\delta(\tau_2 - \tau_1), \tag{2.56}$$

where $\Delta t = t_2 - t_1$. Therefore, we can state that the correlation properties of a WSS-US channel are fully described by a two-dimensional, rather than four-dimensional, function, which we denote now as $R_h(\Delta t, \tau)$, where the meaning of the two independent variables t and τ is as specified in the previous derivations. To single out a few parameters, it is useful to analyze the behavior of $R_h(\Delta t, \tau)$ along the axes $\Delta t = 0$ and $\tau = 0$. More specifically, the function $P_h(\tau) := R_h(0, \tau)$ is the *power delay profile*, as it describes the power behavior as a function of the delay, considering contributions arriving at the same instant, (i.e., $t_2 = t_1$ or $\Delta t = 0$). We assume that there are no contributions with negative delays, so that $P_h(\tau) = 0$ for $\tau < 0$. Two parameters are useful to grasp some of the main features of $P_h(\tau)$, namely the *average delay*

$$\bar{\tau} := \frac{\int_0^\infty \tau P_h(\tau) d\tau}{\int_0^\infty P_h(\tau) d\tau} \tag{2.57}$$

and the *delay spread*

$$\sigma_\tau := \left[\frac{\int_0^\infty (\tau - \bar{\tau})^2 P_h(\tau) d\tau}{\int_0^\infty P_h(\tau) d\tau} \right]^{1/2}. \tag{2.58}$$

Example

A typical behavior of the power delay is exponential; that is,

$$P_h(\tau) = P_0 e^{-\tau/\tau_0} u(\tau), \tag{2.59}$$

where $u(\tau)$ is the unit step function. In such a case, the average delay is $\bar{\tau} = \tau_0$ and the delay spread is also $\sigma_\tau = \tau_0$.

So far, we have only considered the correlation properties of the channel impulse response. However, clearly, if the correlation of the channel impulse response has a certain structure, this induces a corresponding structure on the correlation of the other system functions. For example, if the channel is WSS-US, and thus its impulse response correlation is as in (2.56), the correlation of the delay-Doppler spread function is

$$R_S(\nu_1, \nu_2; \tau_1, \tau_2) = R_S(\nu_1, \tau_1)\delta(\tau_2 - \tau_1)\delta(\nu_2 - \nu_1), \tag{2.60}$$

and the correlation of the transfer function is

$$R_H(t_1, t_2; f_1, f_2) = R_H(t_2 - t_1, f_2 - f_1). \tag{2.61}$$

The function $R_S(\nu_1, \tau_1)$, for a WSS-US channel, has a direct physical meaning, as it represents the power spectral density gain of the channel at Doppler frequency ν_1 and delay τ_1. In fact, from (2.21), we can state that the contribution to the received signal $y(t)$, coming from the transmitted signal $x(t)$, within the delay-Doppler interval $[\nu - \Delta\nu/2, \nu + \Delta\nu/2; [\tau - \Delta\tau/2, \tau + \Delta\tau/2]$ is

$$y_{\nu,\tau}(t) \approx S(\nu, \tau) x(t - \tau) e^{j2\pi\nu t} \Delta\nu\Delta\tau. \tag{2.62}$$

Thus, the power spectral density of $y_{\nu,\tau}(t)$, as a function of both delay and Doppler shift is $E\{|S(\nu, \tau)|^2\} P_x = R_S(\nu, \tau)$.

Exploiting the relationship between spread function and time-varying transfer function, we can also evaluate $R_H(t_2 - t_1, f_2 - f_1)$ as

$$R_H(\Delta t, \Delta f) = \int_{-\infty}^{\infty} \int_{-\infty}^{\infty} R_S(\nu, \tau) e^{j2\pi(\Delta t\nu - \Delta f\tau)} d\nu d\tau, \tag{2.63}$$

where $\Delta t := t_2 - t_1$ and $\Delta f := f_2 - f_1$.

Physical channel models

So far, we have described the main channel correlation properties, but without making any explicit reference to the physics of the underlying phenomena. In this section, we recall some physically meaningful models. We start with narrowband channels and then we will extend the analysis to wideband channels.

Narrowband models

We start with a flat fading channel. In this case, all the contributions arrive with the same delay. More precisely, all contributions arrive with delays that differ from each other by less than the system time resolution. Since time resolution is inversely proportional to the system bandwidth, this is why flat fading typically occurs with narrowband systems. The channel impulse response assumes, in such a case, the form

$$h(t, \tau) = h(t)\delta(\tau - \tau_0), \tag{2.64}$$

where τ_0 represents the delay. We recall now the so-called Jakes' model, which is commonly employed to characterize flat fading narrowband channels [7]. The basic assumptions underlying such a model are that: 1) the receiver has an omnidirectional antenna, so that the power spectral density $P_\Theta(\theta)$ arriving at the receiver as a function of the angle θ is distributed uniformly; 2) the scatterers present in the communication channel are fixed so that they do not introduce further Doppler effects. Assumption 1) means that, since in general there is no reason to expect more power from a certain angle with respect to other angles, it is perfectly justifiable to assume that $P_\Theta(\theta)$ does not vary with θ. With reference to Figure 2.3, where $\boldsymbol{v}$ denotes the velocity vector of the receiver and γ is the angle between the direction of the velocity vector and the line-of-sight (LOS) between transmitter and receiver, the relationship between the Doppler shift ν and the angle θ is, as in (2.14),

$$\nu(\theta) = \frac{v}{\lambda}\cos(\theta - \gamma), \tag{2.65}$$

where v is the modulus of the vector $\boldsymbol{v}$. This relationship allows us to rewrite the received power spectral distribution as a function of the Doppler shift. We have

$$P_\nu(\nu) = \sum_i P_\Theta(\theta_i(\nu)) \left| \frac{\partial \theta(\nu)}{\partial \nu} \right|_{\theta = \theta_i(\nu)}, \tag{2.66}$$

where $\theta_i(\nu) = \gamma \pm \arccos(\lambda\nu/v)$ denotes the ith inverse solution of (2.65). In the interval $(-\pi, \pi]$, there are two inverse solutions. Thus, the power spectral density, as a function of ν, is

$$P_\nu(\nu) = \frac{P_0}{\pi}\frac{1}{\sqrt{f_{\max}^2 - \nu^2}}, \quad \nu \in [-f_{\max}, f_{\max}], \tag{2.67}$$

where $f_{\max} := v/\lambda$. Starting from $P_\nu(\nu)$, we can compute the correlation of the transfer function as the inverse Fourier transform of $P_\nu(\nu)$ and the result is known in closed form

$$R_H(\Delta t, 0) = P_0 J_0(2\pi f_{\max}\Delta t). \tag{2.68}$$

The corresponding correlation coefficient $\rho_H(\Delta t) := R_H(\Delta t)/R_H(0)$ is depicted in Figure 2.7. Assuming that two contributions are (nearly) uncorrelated when their correlation coefficient is smaller than one half, from Figure 2.7, we may infer that the contributions are (nearly) uncorrelated when $\Delta t \approx 1/(4f_{\max})$.

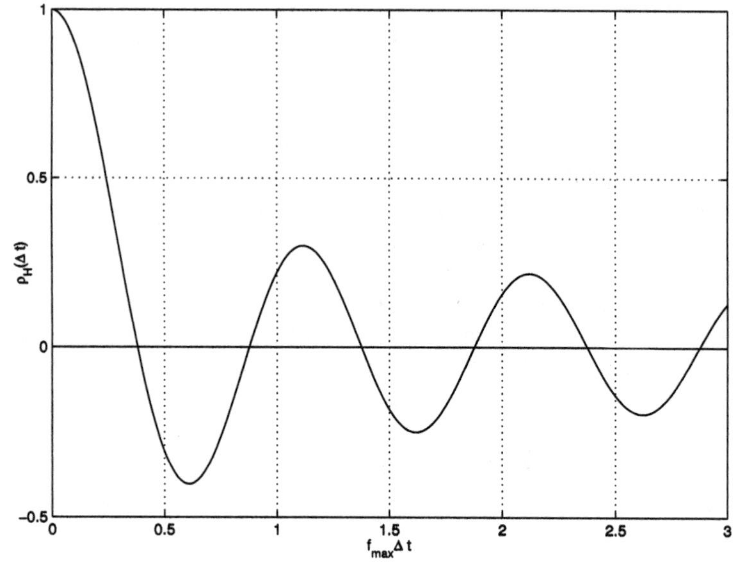

Figure 2.7 Correlation coefficient $\rho_H(\Delta t)$.

Wideband models

Wideband systems are capable of discriminating multipath arrivals in time. In particular, denoting by B the system bandwidth, the system is capable of discriminating arrivals whose delays differ by more than $\delta\tau = 1/B$. By duality arguments, a receiver is able to discriminate arrivals with different Doppler shift if the observation

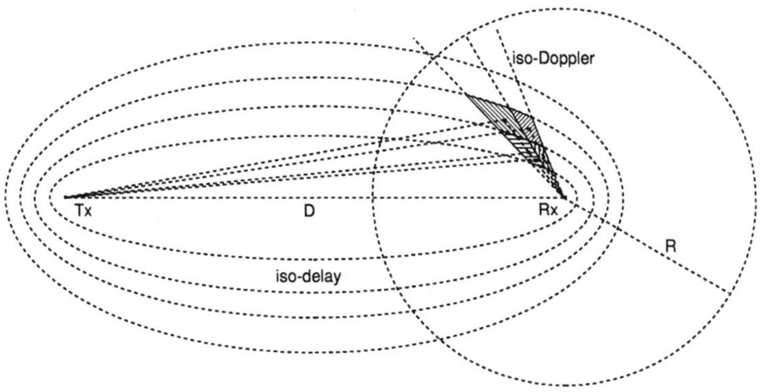

Figure 2.8 Space tessellation induced by the receiver discrimination capability in both time and frequency.

interval has a duration T greater than $1/\delta\nu$, where $\delta\nu$ is the Doppler difference. This discrimination capability implies that all the space around transmitter and receiver can be tasselled according to a grid such that each tassel contains points that are not distinguishable, at the receiver, neither in terms of delay nor in terms of Doppler frequency. The geometry is sketched in Figure 2.8. All the points located over the ellipses having as foci the transmitter and receiver locations give rise to the same delay, whereas all the points located over a straight line departing from the receiver give rise to the same Doppler frequency. Each point on the plane has then its own delay and Doppler coordinates. In the practical case where both time and frequency resolution are nonnull, there is a region of finite nonnull area (tassel) that gives rise to returns that are indistinguishable at the receiver. With reference to Figure 2.8, all points located within the dashed tassels around the receiver are indistinguishable in terms of delay or Doppler. The tessellation sketched in Figure 2.8 refers to the scenario where the following assumptions hold true: 1) the only moving terminal is the receiver; 2) there are no moving objects between transmitter and receiver; and 3) each path arrives with a single bounce (multiple-bounce arrivals are neglected).

The geometry depicted in Figure 2.8 establishes a relationship between the scatterers' distribution and the received power density, as a function of delay and Doppler. That geometry was first proposed in [8, 9] to find out a relationship between the received power and the Doppler frequency or angle of arrival. This same geometry can be used to derive a closed-form expression for the received power density. In particular, assuming that 1) the scatterers are uniformly distributed around the receiver, within a circle of radius R, and 2) the backscattering coefficient is constant within the circle of radius R, the power contribution coming from each tassel is proportional to the tassel area, with a proportionality coefficient equal for all the tassels. As an example, in Figure 2.9 we report the received power, as a function of the normalized Doppler frequency $\nu/f_{\max}$, where $f_{\max} = v/\lambda$, for two different values of τ: $\tau_0 = 0$ and $\tau_1 = T$. We set in this case $\gamma = \pi/2$ and we assumed a radius R of 150 m. We compare the theoretical behavior (solid line) and the results of a simulation program (dashed line) that generates the presence of a high number of scatterers, distributed uniformly within the circle around the receiver and computes the received power. Interestingly, we notice that when $\tau = 0$, we have contributions from all the Doppler shifts, whereas for $\tau > 0$, there are no returns for very small Doppler frequencies. The reason for this behavior is evident from Figure 2.8, where we can observe that, for τ greater than a threshold, there are directions (around the line of sight line) where there are no points inside the circle. Therefore, for Doppler shifts corresponding to those directions there is no received power.

Another model, easier to handle from the mathematical point of view, assumes that the power density, as a function of delay and Doppler, has the following behavior

$$R_s(\nu, \tau) = \frac{P_0}{2\pi\tau_0\sqrt{f_{\max}^2 - \nu^2}} e^{-|\tau|/\tau_0}, \tag{2.69}$$

where P_0 is the overall received power. The two-dimensional Fourier transform of this function is known in closed form, so that we can say that in this case, the correlation of the time-varying transfer function is

$$R_H(\Delta t, \Delta f) = P_0 \frac{J_0(2\pi f_{\max}\Delta t)}{1 + (2\pi\Delta f \tau_0)^2}. \tag{2.70}$$

2.3.3 Channel Classification

The correlation of the time-varying transfer function is an important function, as it allows for the introduction two important parameters: the channel coherence time and bandwidth. The channel *coherence time* is defined as the duration of the time interval in which the channel may be assumed to be approximately constant. In formulas, the channel coherence time T_c is the value such that $R_H(T_c, 0) = R_H(0,0)/2$. For example, assuming (2.70) to be valid, we have

$$T_c \approx \frac{1}{4f_{\max}}. \tag{2.71}$$

Similarly, the channel *coherence bandwidth* is defined as the value B_c such that $R_H(0, B_c) = R_H(0,0)/2$. Again, assuming (2.70), the coherence bandwidth is

$$B_c = \frac{1}{2\pi\tau_0}. \tag{2.72}$$

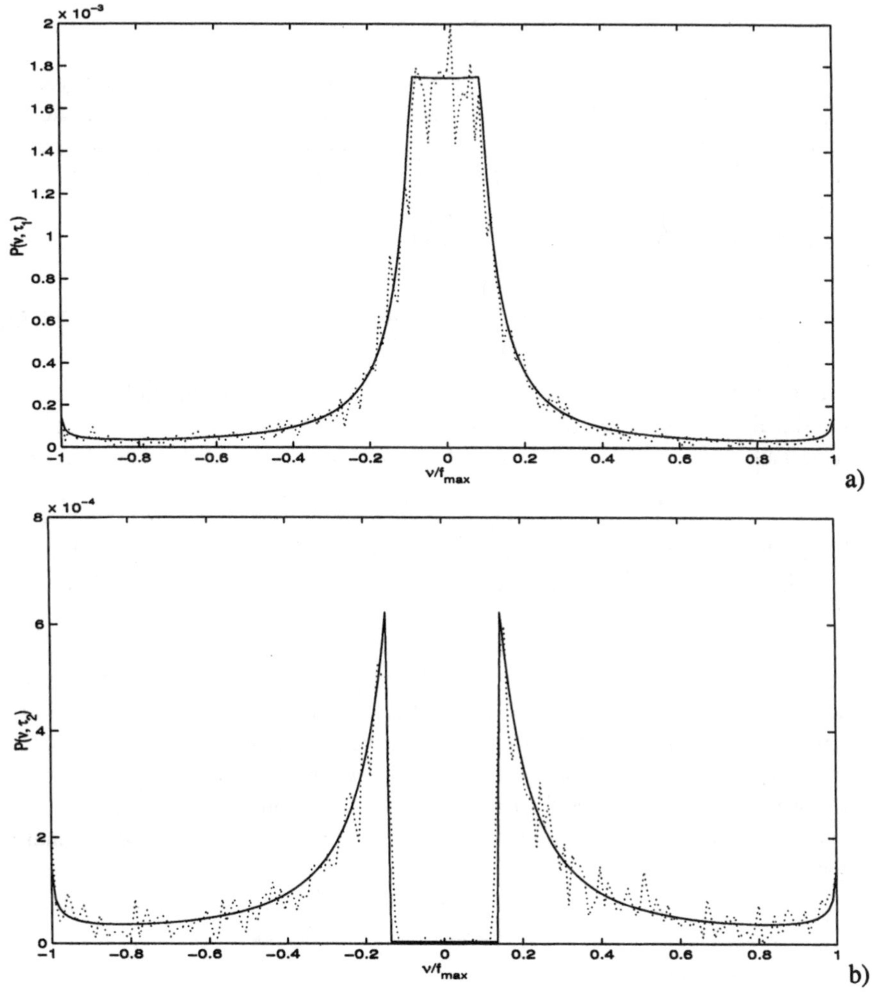

Figure 2.9 Power density as a function of ν, for two different values of τ: a) $\tau = 0$, and b) $\tau = T_s$; $\gamma = \pi/2$; theoretical values (solid line) and simulations results (dashed line).

As we will see in the next sections, the concepts of channel coherence time and bandwidth play a fundamental role in the design of proper transmission schemes. Depending on their value and, more specifically, on the relationship between these values and the time and bandwidth of the signals transiting through the channel, it is possible to classify the channels as follows.

Let us denote by B the bandwidth of the transmitted signal and by T its duration. Even though from a strict mathematical point of view a finite bandwidth implies an infinite duration, in practice we may assume both T and B to be finite. Actually, if we define B and T as the effective bandwidth and duration of the transmitted signals, the Heisenberg's uncertainty principle states that the time-bandwidth product is lower bounded by the following inequality:

$$BT \geq \frac{1}{2}. \tag{2.73}$$

We say that a channel is *frequency-selective* if its coherence bandwidth is less than the signal bandwidth (i.e., $B_c < B$). Equivalently, since B_c is inversely proportional to τ_0 and B is inversely proportional to T, a frequency-selective channel is *time-dispersive*, as $\tau_0 > T$.

Using duality arguments, we say that a channel is *time-selective* if its coherence time is less than the signal duration (i.e., $T_c < T$). This implies, in the frequency domain, that $f_{\max} > B$.

We emphasize that the property of a channel of being time, and/or frequency-selective, is not a property of the channel *per se*, but it is meaningful only in relationship with the properties of the signal passing through the channel. We anticipate that the property of being time, and/or frequency-selective is not necessarily a negative feature because, if properly exploited, time and/or frequency selectivity can become the source of time and frequency diversity, as we will see in detail in the next chapters.

2.4 MIMO CHANNELS

Let us consider now multiantenna systems. A MIMO system with n_T transmit and n_R receive antennas is characterized by an input/output relationship

$$y_m(t) = \sum_{n=1}^{n_T} \int_{-\infty}^{\infty} h_{mn}(t,\tau)x_n(t-\tau)d\tau, \quad m = 1, 2, \ldots, n_R, \tag{2.74}$$

where $x_n(t)$ is the signal transmitted from the nth transmit antenna, $y_m(t)$ is the signal received from the mth receive antenna, and $h_{mn}(t, \tau)$ is the impulse response of the channel between the nth transmit and the mth receive antennas. The system of equation in (2.74) can also be rewritten in a more compact form using a matrix notation, as

$$\boldsymbol{y}(t) = \int_{-\infty}^{\infty} \boldsymbol{H}(t, \tau)\boldsymbol{x}(t - \tau)d\tau, \tag{2.75}$$

where $\boldsymbol{H}(t, \tau)$ is the $n_R \times n_T$ channel matrix

$$\boldsymbol{H}(t, \tau) = \begin{pmatrix} h_{11}(t, \tau) & h_{12}(t, \tau) & \dots & h_{1,n_T}(t, \tau) \\ h_{21}(t, \tau) & h_{22}(t, \tau) & \dots & h_{2n_T}(t, \tau) \\ \vdots & \vdots & \vdots & \vdots \\ h_{n_R1}(t, \tau) & h_{n_R1}(t, \tau) & \dots & h_{n_Rn_T}(t, \tau) \end{pmatrix}, \tag{2.76}$$

$\boldsymbol{y}(t) := [y_1(t), \dots, y_{n_R}(t)]$ is the n_R-size vector containing the signals received from the n_R antennas, whereas $\boldsymbol{x}(t) := [x_1(t), \dots, x_{n_T}(t)]$ is the n_T-size vector containing the signals transmitted from the n_T antennas. The properties of each impulse response $h_{mn}(t, \tau)$ can be described by the same tools introduced in the previous sections on SISO channels. In practice, the entries of $\boldsymbol{H}(t, \tau)$ have all the same statistical properties. Hence, the coherence time and bandwidth of a MIMO system can be evaluated as in the SISO case. The really distinguishing features of MIMO systems is the spatial correlation among the impulse responses composing $\boldsymbol{H}(t, \tau)$.

First order models

In case of flat-fading channels, a MIMO system is described by the I/O relationship

$$\boldsymbol{y} = \boldsymbol{H}\boldsymbol{x}, \tag{2.77}$$

where $\boldsymbol{y}$ contains the n_R samples received by the n_R antennas, at a given time instant, $\boldsymbol{x}$ contains the n_T transmitted samples, and $\boldsymbol{H}$ is an $n_R \times n_T$ (complex) matrix, whose (k, l) entry is the (flat-fading) channel between the lth transmit and the kth receive antenna. In case of no-LOS, each entry has a zero mean value. The MIMO Rayleigh-fading model, with independent fading, consists of a matrix $\boldsymbol{H}$ composed of identically distributed (iid), circularly symmetric, zero mean complex Gaussian random variables. If there is a LOS, the elements of $\boldsymbol{H}$ have nonzero mean values that are related to each other. The relationship can be better understood with reference to Figure 2.10, where we see a transmit array, composed of n_T elements, uniformly spaced by d_T, and a receive array, composed of n_R elements, uniformly spaced by d_R. If each antenna is in the far field of the other, the channel matrix has

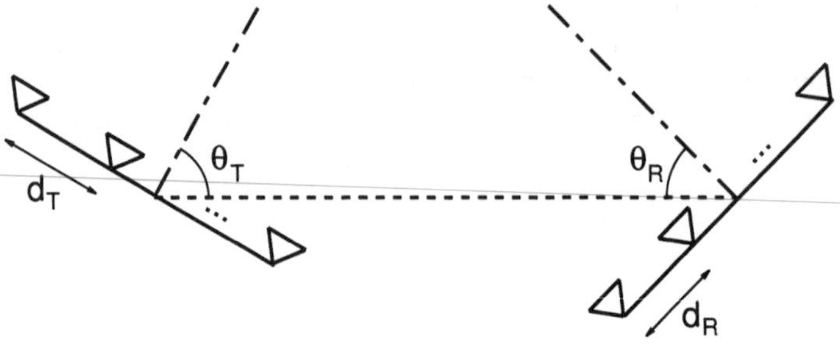

Figure 2.10 MIMO channel model.

mean value

$$\boldsymbol{H}_m = \alpha\, \boldsymbol{a}_R\, \boldsymbol{a}_T^T, \tag{2.78}$$

where alpha is a (complex) coefficient, whereas $\boldsymbol{a}_T$ and $\boldsymbol{a}_R$ are the transmit and receive steering vectors:

$$\begin{aligned} \boldsymbol{a}_R^T &= \left(1, e^{-j2\pi d_R \sin(\theta_R)/\lambda}, \ldots, e^{-j2\pi(n_R-1)d_R \sin(\theta_R)/\lambda}\right), \\ \boldsymbol{a}_T^T &= \left(1, e^{-j2\pi d_T \sin(\theta_T)/\lambda}, \ldots, e^{-j2\pi(n_T-1)d_T \sin(\theta_T)/\lambda}\right). \end{aligned} \tag{2.79}$$

The MIMO Rice-fading model is

$$\boldsymbol{H} = \sqrt{\frac{K}{K+1}}\boldsymbol{H}_m + \sqrt{\frac{1}{K+1}}\boldsymbol{H}_w, \tag{2.80}$$

where K is the Rice factor, $\boldsymbol{H}_m$ is the mean values matrix, given by (2.78), and $\boldsymbol{H}_w$ is composed of iid complex circularly symmetric, zero mean complex Gaussian random variables. If there is no LOS, $K = 0$, and (2.80) reduces to the MIMO Rayleigh-fading model.

Given the vicinity of the antenna elements, the statistical properties of the different channels are the same.

Second order models

For a MIMO system, it is important to derive the correlation between the signals received from different antenna elements. Narrowband models based on the physical

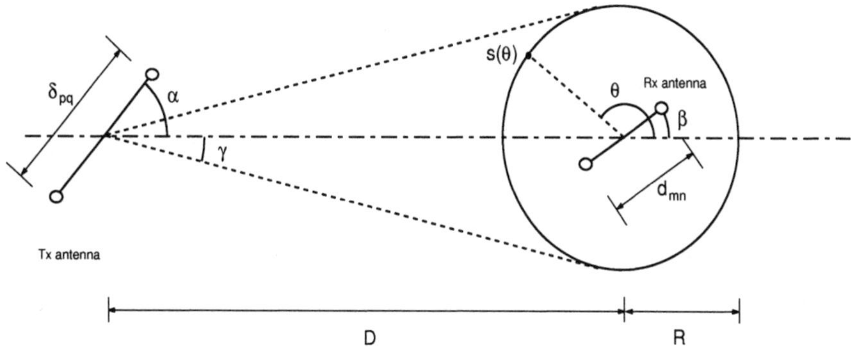

Figure 2.11 Geometry of the one-ring scattering model.

channel geometry, have been proposed, for example, in [10, 11, 12]. Here, we recall briefly the space-time statistical model derived in [12]. The geometry underlying the model derivation is sketched in Figure 2.11. The model assumes that the scatterers are located along the circumference of a circle of radius R, centered around the receiver, whereas there are no scatterers around the transmitter. This model refers to the scenario in which the transmitter is the base station, located typically at a certain height from the ground, whereas the receiver is a mobile terminal, surrounded by many scattering or reflecting buildings. The pdf of the scatterers is modeled as a Von Mises pdf

$$p_\Theta(\theta) = \frac{1}{2\pi I_0(\kappa)} e^{\kappa\cos(\theta-\mu)}, \quad \theta \in [-\pi, \pi), \tag{2.81}$$

where $I_0(\cdot)$ is the zero order modified Bessel function of the first kind and μ is the mean angle of arrival. In this model, the receiver is assumed to be moving along a direction specified by the angle θ. The parameter κ controls the concentration of the scatterers around the mean angle of arrival: $\kappa = 0$ indicates isotropic scattering, whereas $\kappa = \infty$ indicates that the scattering is concentrated only on one direction. The transmit array is composed of uniformly spaced elements; δ_{pq} is the distance between the p-th and q-th elements. The receive array is also uniformly spaced with inter-element distance d_{mn}. Some examples of $p_\Theta(\theta)$, parameterized according to κ, are reported in Figure 2.12. Denoting with $h_{p,n}(t)$ the multiplicative channel between the p-th transmit and the n-th receive antenna element, the space-time correlation $R^{q,n}_{p,m}(\tau) := E\{h^*_{p,n}(t)h_{q,m}(t+\tau)\}$ has been evaluated in [12].

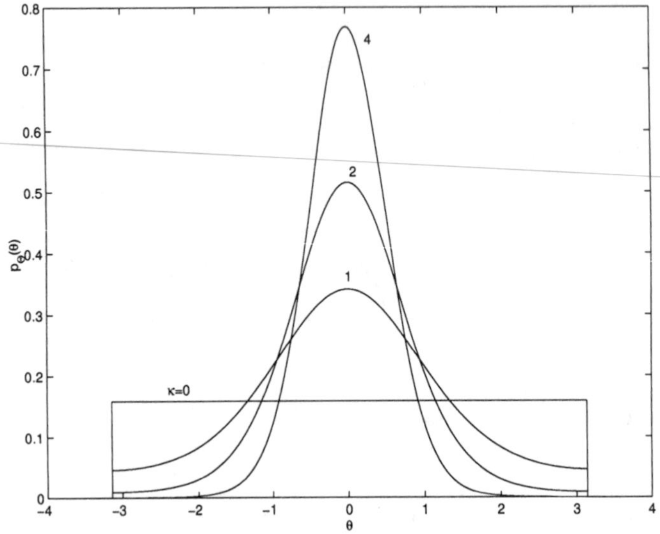

Figure 2.12 Angular distribution of scatterers according to the Von Mises pdf, for $\kappa = 0, 1, 2$, and 4.

In Figure 2.13 we report the spatial correlation coefficient

$$\rho_{p,m}^{q,n} = \frac{E\{h_{p,n}^*(t)h_{q,m}(t)\}}{\sqrt{E\{|h_{p,n}(t)|^2\} \cdot E\{|h_{q,m}(t)|^2\}}} \tag{2.82}$$

of a 2×2 MIMO system, between different channels, observed at the same time instant, as a function of the spacing between the transmit elements δ_{12} and of the receive elements d_{12}. In particular, Figure 2.13(a) refers to channels with no line-of-sight (LOS), with two different parameters of the Von Mises pdf: a) refers to isotropic scattering ($\kappa = 0$) and b) refers to nonisotropic scattering ($\kappa = 3$). From Figure 2.13 we observe that the mainlobe of the correlation is larger in the case of nonisotropic scattering, whereas the sidelobes are lower.

Possible LOS components play also an important role on the value assumed by the correlation coefficient. As an example, Figure 2.14 refers to the channel of Figure 2.13(a), except for the presence of a LOS component, with Rice factor $K = 4$. We see from Figure 2.14 that the correlation coefficient is higher in case of LOS.

The behavior of the spatial correlation allows us to introduce the concept of *coherence space* (or distance). Similarly to the coherence time and bandwidth of a SISO channel, the coherence space is the distance D_c between two points beyond

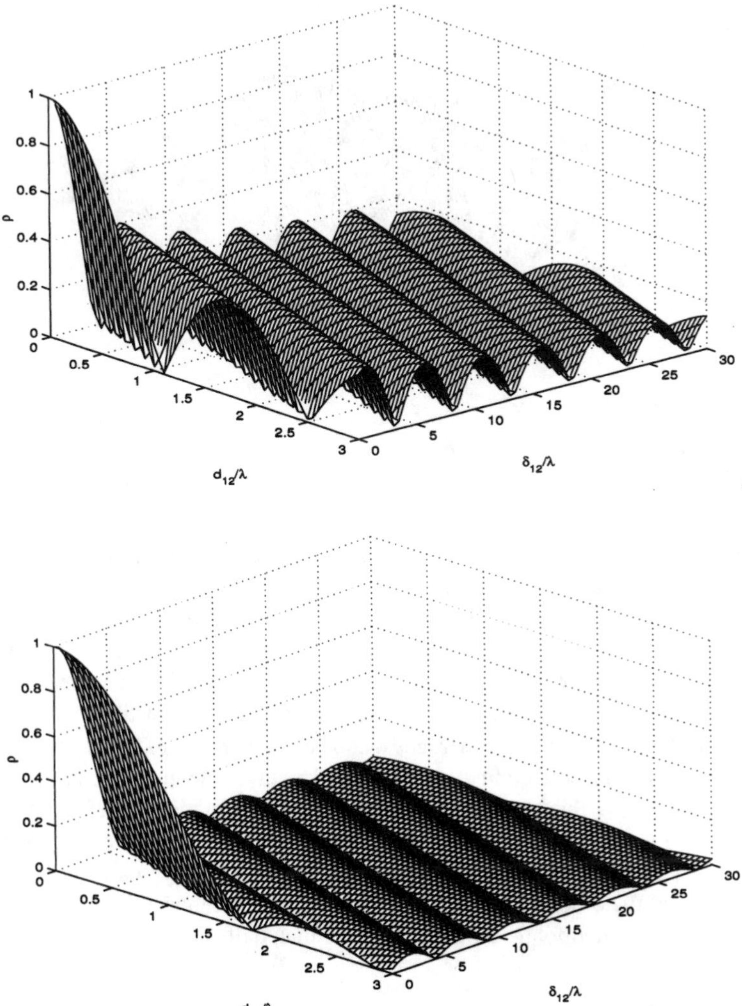

a)

b)

Figure 2.13 Spatial channel correlation coefficient for channels with no LOS, in case of: a) isotropic scattering ($\kappa = 0$), and b) nonisotropic scattering ($\kappa = 3$).

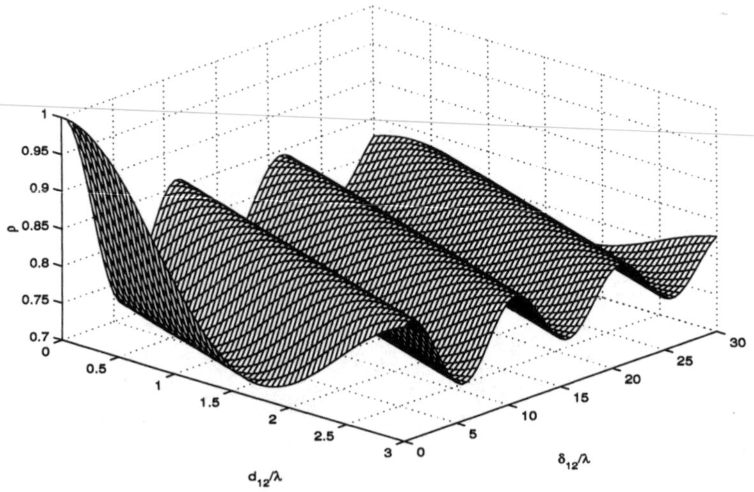

Figure 2.14 Spatial channel correlation coefficient for channels with LOS, with isotropic scattering ($\kappa = 0$), and Rice factor $K = 4$.

which the correlation coefficient between signals received from those points, at the same time, is smaller than one half. Dealing with systems with multielement arrays at both transmit and receive side, we need to introduce indeed two parameters: the transmit and receive coherence distances. Looking at Figure 2.12, for example, we see that the *transmit coherence space* is equal to a few times the wavelength, whereas the *receive coherence space* is equal to a fraction of the wavelength. The difference between the two coherence distances is the result of the asymmetry of the scatterers distribution around the transmit and receive antennas. Recalling that in the model underlying this result the transmitter is a base station, whereas the receiver is a mobile terminal, the previous results state that the elements of the base station must be spaced at a distance greater than the wavelength (typically, one should use a distance of the order of ten times the wavelength), whereas it is sufficient to space the antennas of the mobile terminal by more than half a wavelength.

2.5 MODES

One of the major advantages in a wireless communications system based on multiple antenna transceivers is the possibility of *multimodal* propagation. Different modes can be used to transmit, simultaneously, more information bits. We use the term *mode* in a very broad sense, to signify a physical channel that may be defined over the space, time, or polarization domains. In this section, after introducing the concept of mode, we will provide some physical insight about the meaning of modes in case of time-varying systems.

2.5.1 Discrete-Time Channel

As observed in the previous sections, both block transmission through SISO systems and MIMO systems are characterized by a discrete-time I/O relationship that assumes the general form[2]

$$\boldsymbol{y} = \boldsymbol{H}\boldsymbol{x} + \boldsymbol{w}, \tag{2.83}$$

where $\boldsymbol{x}$ and $\boldsymbol{y}$ are vectors of size M and N, respectively, $\boldsymbol{H}$ is necessarily $N \times M$, and $\boldsymbol{w}$ is a noise vector, of size N. We do not make any assumption about M or N, so as to leave (2.83) as general as possible. Also, for the same reason, we do not impose any structure on the matrix $\boldsymbol{H}$, so that (2.83) can describe a SISO or a MIMO system.

We know, from basic properties of linear algebra, that *every* matrix $\boldsymbol{H}$ can be factorized by its singular value decomposition (SVD):

$$\boldsymbol{H} = \boldsymbol{U}\boldsymbol{\Sigma}\boldsymbol{V}^H, \tag{2.84}$$

where $\boldsymbol{U}$ and $\boldsymbol{V}$ are unitary matrices. The matrix $\boldsymbol{\Sigma}$ may assume one of the following forms, depending on the relationship between N and M:

$$\boldsymbol{\Sigma} = \boldsymbol{D}, \text{if } N = M; \boldsymbol{\Sigma} = \begin{pmatrix} \boldsymbol{D} \\ \boldsymbol{0} \end{pmatrix}, \text{if } N > M; \boldsymbol{\Sigma} = \begin{pmatrix} \boldsymbol{D} & \boldsymbol{0} \end{pmatrix}, \text{if } N < M, \tag{2.85}$$

where $\boldsymbol{D} = \text{diag}\{\sigma_1, \sigma_2, \ldots, \sigma_q\}$, with $\sigma_1 \geq \sigma_2 \geq \cdots \geq \sigma_q \geq 0$, and $q = \min(M, N)$. The diagonal entries of $\boldsymbol{D}$ are the singular values of $\boldsymbol{H}$, and the columns of the matrices $\boldsymbol{U}$ and $\boldsymbol{V}$ are, respectively, the left and right singular vectors of $\boldsymbol{H}$. Denoting these last ones by $\boldsymbol{u}_k$ and $\boldsymbol{v}_k$, and introducing the matrices $\boldsymbol{U}_r = (\boldsymbol{u}_1 \cdots \boldsymbol{u}_r)$, $\boldsymbol{V}_r = (\boldsymbol{v}_1 \cdots \boldsymbol{v}_r)$, and $\boldsymbol{D}_r = \text{diag}\{\sigma_1, \ldots, \sigma_r\}$, where r is the rank of $\boldsymbol{H}$, the SVD in (2.84) can also be written as

$$\boldsymbol{H} = \sum_{k=1}^{r} \sigma_k \boldsymbol{u}_k \boldsymbol{v}_k^H = \boldsymbol{U}_r \boldsymbol{D}_r \boldsymbol{V}_r^H, \tag{2.86}$$

since r is also the number of non-zero singular values of $\boldsymbol{H}$. Note that $\boldsymbol{U}_r$ and $\boldsymbol{V}_r$ are paraunitary matrices, i.e. $\boldsymbol{U}_r^H \boldsymbol{U}_r = \boldsymbol{V}_r^H \boldsymbol{V}_r = \boldsymbol{I}_r$.

2 We drop the block index for simplicity of notation.

The SVD tool is sufficient to introduce the concept of multimode propagation. Let us suppose that we want to transmit r symbols, $s_1, s_2, \ldots, s_r$, simultaneously through the channel. If the transmitter knows the channel right singular vectors $\boldsymbol{v}_i$, it can transmit the M-size vector

$$\boldsymbol{x} = \sum_{i=1}^{r} c_i s_i \boldsymbol{v}_i, \tag{2.87}$$

where the coefficients c_i may be varied in order to optimize the transmission strategy according to some criterion of interest. Combining (2.83), (2.86), and (2.87), it is straightforward to check that, transmitting (2.87) produces, at the receiver, the vector

$$\boldsymbol{y} = \boldsymbol{H}\boldsymbol{x} + \boldsymbol{w} = \sum_{k=1}^{r}\sum_{i=1}^{r} c_i s_i \sigma_k \boldsymbol{u}_k \boldsymbol{v}_k^H \boldsymbol{v}_i + \boldsymbol{w}. \tag{2.88}$$

Exploiting the orthogonality of the vectors $\boldsymbol{v}_i$, (2.88) becomes

$$\boldsymbol{y} = \sum_{k=1}^{r} c_k s_k \sigma_k \boldsymbol{u}_k + \boldsymbol{w}. \tag{2.89}$$

Hence, we can recover every symbol s_l by projecting $\boldsymbol{y}$ along the vector $\boldsymbol{u}_l$, and scaling the result, as follows

$$\hat{s}_l = \frac{1}{\sigma_l c_l}\boldsymbol{u}_l^H \boldsymbol{y} = s_l + w_l, \quad l = 1, \ldots, r, \tag{2.90}$$

where $w_l := \boldsymbol{u}_l^H \boldsymbol{w}/(\sigma_l c_l)$. All previous equations can be collected in a more compact matrix form, as follows. We transmit the vector

$$\boldsymbol{x} = \boldsymbol{F}\boldsymbol{s} = \boldsymbol{V}_r \boldsymbol{\Phi} \boldsymbol{s}, \tag{2.91}$$

where $\boldsymbol{F} := \boldsymbol{V}_r\boldsymbol{\Phi}$ is an $M \times r$ linear encoding matrix, with $\boldsymbol{\Phi} = \text{diag}\{c_1, \ldots, c_r\}$. At the receiver, we multiply the vector $\boldsymbol{y}$ by the decoding matrix $\boldsymbol{G} := \boldsymbol{\Gamma}\boldsymbol{U}_r^H$, where $\boldsymbol{\Gamma}$ is diagonal. The result is

$$\boldsymbol{z} = \boldsymbol{G}\boldsymbol{H}\boldsymbol{x} + \boldsymbol{G}\boldsymbol{w} = \boldsymbol{\Gamma}\boldsymbol{U}_r^H \boldsymbol{U}_r \boldsymbol{D}_r \boldsymbol{V}_r^H \boldsymbol{V}_r \boldsymbol{\Phi}\boldsymbol{s} + \boldsymbol{\Gamma}\boldsymbol{U}_r^H \boldsymbol{w} = \boldsymbol{\Gamma}\boldsymbol{D}_r\boldsymbol{\Phi}\boldsymbol{s} + \boldsymbol{w}' \tag{2.92}$$

where $\boldsymbol{w}' := \boldsymbol{\Gamma}\boldsymbol{U}_r^H\boldsymbol{w}$. It is important to observe the two following properties about (2.92): 1) Thanks to the use of channel right singular vectors at the transmitter and left singular vectors at the receiver, the channel matrix is diagonalized, so that there is no intersymbol interference (ISI); and 2) if the received noise is white Gaussian, (i.e., the covariance matrix of $\boldsymbol{w}$ is diagonal), the final noise vector $\boldsymbol{w}'$ is also white and Gaussian, so that symbol-by-symbol decision is optimal. This property can be easily proved. In fact, if we denote by $\boldsymbol{C_w} = \sigma_n^s \boldsymbol{I}$ the covariance matrix of $\boldsymbol{w}$, the covariance matrix of $\boldsymbol{w}'$ is

$$\boldsymbol{C_{w'}} := E\{\boldsymbol{w}'\boldsymbol{w}'^H\} = \sigma_n^2 \boldsymbol{\Gamma} \boldsymbol{U}_r^H \boldsymbol{U}_r \boldsymbol{\Gamma}^H = \sigma_n^2 \boldsymbol{\Gamma}\boldsymbol{\Gamma}^H \tag{2.93}$$

and this matrix is certainly diagonal.

We say that every pair of vectors $(\boldsymbol{u}_k, \boldsymbol{v}_k)$ constitutes a channel *mode*. Transmitting using the scheme of (2.87) is equivalent to transmitting every symbol through one channel mode. From this general setup, it is clear that the maximum number of symbols that can be transmitted simultaneously is equal to the rank of the channel matrix $\boldsymbol{H}$. Depending on the structure of $\boldsymbol{H}$, we may talk of modes in the time, space, or polarization domain.

If the channel (or its right singular vectors) is unknown at the transmit side, the transmission strategy of (2.87) is not possible. In such a case, the transmitted vector will carry symbols that will excite a combination of modes. In such a case, the optimization and the symbol retrieval are more difficult. In Chapters 5 and 6, we will show the fundamental limits and performance of systems, withour or with channel knowledge at the transmit side.

2.5.2 Continuous-Time Channel

The discrete-time formulation presented in the previous section has been useful to introduce the concept of mode. However, in the discrete-time formulation given before it is not immediate to associate a physical meaning to the concept of mode. Finding such a interpretative model is indeed not an easy task, in general. Nevertheless, in some cases of practical interest, it is possible to provide a useful insight. This happens, for example, for underspread SISO channels. It is the scope of this section to give an analytic, albeit approximate, expression for the eigenfunctions of underspread SISO channels, useful to shed light on the physical meaning of mode.

We consider the case where the channel impulse response is square integrable; that is,

$$\int_{-\infty}^{\infty}\int_{-\infty}^{\infty} |h(t,\tau)|^2 dt d\tau < \infty. \tag{2.94}$$

In practice, there are at least three important cases where $h(t,\tau)$ does not respect (2.94): 1) LTI channels, where $h(t,\tau)$ is constant along t; 2) multiplicative channels, where $h(t,\tau) = m(t)\delta(\tau-\tau_0)$; and 3) multipath channels with specular reflections, where $h(t,\tau)$ contains Dirac pulses. We consider such situations as limiting cases, and we will show that the results for such cases can be described as limiting values of the expressions derived for the finite norm case. Nevertheless, it is worth pointing out that, in case of LTI, multiplicative, or two-ray multipath channels, the system eigenfunctions are known in closed form, exactly, so that in such a case, there is no real problem.

If $h(t,\tau)$ satisfies (2.94), there exists a countable series of real numbers λ_i and two sets of orthonormal functions $u_i(t)$ and $v_i(t)$ such that the following pair of integral equations hold true [3]:

$$\lambda_i u_i(t) = \int_{-\infty}^{\infty} h(t, t-\tau) v_i(\tau) d\tau \tag{2.95}$$

and

$$\lambda_i v_i(\tau) = \int_{-\infty}^{\infty} h^*(t, t-\tau) u_i(t) dt. \tag{2.96}$$

The scalar values λ_i are the so called *singular values* of the system, and the functions $u_i(t)$ and $v_i(t)$ are known as the *left* and *right* singular functions, associated with λ_i. Equations (2.95) and (2.96) are the continuous-time counterpart of the SVD (2.84).

Inserting (2.95) in (2.96), we get

$$\lambda_i^2 v_i(\tau) = \int_{-\infty}^{\infty} \int_{-\infty}^{\infty} h^*(t, t-\tau) h(t, t-\theta) v_i(\theta) d\theta dt \tag{2.97}$$

so that $v_i(\tau)$ is the eigenfunction of the system

$$\lambda_i^2 v_i(\tau) = \int_{-\infty}^{\infty} \tilde{h}(\tau,\theta) v_i(\theta) d\theta, \tag{2.98}$$

associated to the eigenvalue λ_i^2. The kernel of the integral equation (2.98) is

$$\tilde{h}(\tau,\theta) := \int_{-\infty}^{\infty} h^*(t, t-\tau) h(t, t-\theta) dt. \tag{2.99}$$

Similarly, substituting (2.96) in (2.95), $u_i(t)$ is an eigenfunction of the system with kernel

$$\bar{h}(t,\theta) := \int_{-\infty}^{\infty} h(t, t-\tau) h^*(\theta, \theta-\tau) d\tau. \tag{2.100}$$

3 There is no loss of generality in assuming that λ is real as every phase of λ could be incorporated n the channel singular functions.

In Appendix 1A.1, we review the basic theory of slowly varying linear Hermitian operators, which is useful for finding an approximate expression for the eigenfunctions of underspread channels. However, before recalling the major results of such a theory, it is useful to anticipate a few simple derivations that provide a first order approximate solution.

The best known example where the eigenfunctions are known exactly is the case of linear time-invariant channels. In such a case, $h(t,\tau) = h(\tau)$ and the channel eigenfunctions are $u_i(t) \equiv v_i(t) = e^{j2\pi f_i t}$; the eigenvalues are $\lambda_i = H(f_i)$, where $H(f)$ is the channel transfer function. This property can be checked immediately by simply substituting these expressions for $u_i(t)$ and $v_i(t)$ in (2.95) or (2.96). Let us now see how to extend the solution to the case of underspread channels. From what we have seen in Section 2.3.1.1, a channel is underspread if the support of its spread function is very limited in delay, Doppler, or both. Let us consider the case where the support along τ is very limited. We search for a solution of the system (2.95) within the class of functions having the following form

$$v_i(t) = A_i e^{j\phi_i(t)}, \quad t \in I_i, \tag{2.101}$$

where the function $\phi_i(t)$ and the interval I_i have to be determined. The amplitude A_i must insure that $v_i(t)$ has unit norm.

Inserting (2.101) in (2.95), we get

$$\lambda_i u_i(t) = \int_{-\infty}^{\infty} h(t,\tau) e^{j\phi_i(t-\tau)} d\tau. \tag{2.102}$$

Since the channel is assumed to have a very limited support along τ, we can expand $\phi_i(t-\tau)$ around t and keep only the first two terms[4]

$$\phi_i(t-\tau) \approx \phi_i(t) - \tau\dot{\phi}_i(t), \tag{2.103}$$

where $\dot{\phi}_i(t)$ denotes the first order time derivative of $\phi_i(t)$. Substituting (2.103) in (2.102), we get

$$\lambda_i u_i(t) \approx A_i e^{j\phi(t)} \int_{-\infty}^{\infty} h(t,\tau) e^{-j\tau\dot{\phi}_i(t)} d\tau. \tag{2.104}$$

Recalling (2.11), we may rewrite (2.104) as

$$\lambda_i u_i(t) \approx A_i e^{j\phi(t)} H\left(t, \frac{\dot{\phi}_i(t)}{2\pi}\right) = v_i(t) H\left(t, \frac{\dot{\phi}_i(t)}{2\pi}\right). \tag{2.105}$$

4 This approach for finding approximate solutions of integrals of the form (2.102) is known as the stationary phase method [13].

This relationship implies that, if $\phi(t)$ is such that $H(t,f)$ is constant along the curve $f(t) = \dot{\phi}_i(t)/2\pi$, then $v_i(t)$, as expressed in (2.101), is a possible solution of (2.95), with

$$\lambda_i = H(t, \dot{\phi}_i(t)/2\pi), \tag{2.106}$$

where I_i is the time interval where (2.106) admits a solution. In this case, $u_i(t) = v_i(t)$. In general, (2.106) may admit more than one solution, so that the most general form for the solution is

$$v_i(t) = \sum_{k=1}^{n_i(t)} A_{i,k} e^{j\phi_{i,k}(t)}, \quad t \in I_i, \tag{2.107}$$

where all the instantaneous phases $\phi_{i,k}(t)$ are such that the corresponding instantaneous frequencies $f_{i,k}(t) := \dot{\phi}_{i,k}(t)/2\pi$ are solutions of the implicit equation (2.106); $n_i(t)$ is the number of solutions of (2.106), as a function of t.

In summary, we can state that a first order solution for the eigenfunction problem is given by functions $v_i(t)$ composed of components having constant amplitude and instantaneous frequencies given by the curves, in the time-frequency domain, where $H(t,f)$ is constant and equal to λ_i, the singular value associated with $v_i(t)$.

The previous derivations started by assuming that $S(\nu,\tau)$ or, equivalently, $h(t,\tau)$, had a very limited support along τ. If we consider the case where $S(\nu,\tau)$ has very limited support along ν, instead of τ, using the duality property of the Fourier transform, we may follow similar arguments as before, but in the frequency domain, to show that an approximate model for $v_i(t)$ is given by those functions whose Fourier transform $V_i(f)$ assumes the form

$$V_i(f) = A_i e^{j\Phi_i(f)}, \quad t \in I_i, \tag{2.108}$$

where $\Phi_i(f)$ is such that

$$\lambda_i = H\left(\frac{\dot{\phi}_i(f)}{2\pi}, f\right) \tag{2.109}$$

is constant. The above derivations are first order approximations. A better approximation can be obtained by allowing the instantaneous amplitude to be time-varying. Exploiting the general results reported in Appendix 1A.2, based on the theory of slowly varying operators [14], $v_i(t)$ may be approximated as

$$v_i(t) \simeq \sum_m a_{i,m}(t) e^{j\phi_{i,m}(t)}, \tag{2.110}$$

where the instantaneous phase $\phi_{i,m}(t)$ of each component is such that the corresponding instantaneous frequency $f_{i,m}(t) := \dot{\phi}_{i,m}(t)/2\pi$ is a solution of

$$|H(t, f_{i,m}(t))|^2 = \lambda_i^2, \tag{2.111}$$

whereas the instantaneous envelope is

$$a_{i,m}(t) = \left[\frac{1}{\sqrt{\frac{\partial |H(t,f)|^2}{\partial f}}} \right]_{f=\dot{\phi}_{i,m}(t)/2\pi} . \tag{2.112}$$

In words, $v_i(t)$ is a multicomponent signal whose components have an instantaneous frequency given by the curve, in the time-frequency domain (t, f), where the modulus of the channel time-varying transfer function is constant and equal to λ_i^2. The relationship (2.111) between the channel eigenvalues and the time-varying transfer function was noticed for the first time in [15].

Invoking duality arguments, we can also state that if the channel spread along the Doppler axis is small, the function $v_i(t)$ is such that its spectrum $V_i(f)$ assumes the form

$$V_i(f) = \sum_m A_{i,m}(f) e^{j\Phi_{i,m}(f)}, \tag{2.113}$$

where the instantaneous phase $\Phi_{i,m}(f)$ of each spectrum component is such that the corresponding group delay $t_{i,m}(t) := -\dot{\Phi}_{i,m}(f)/2\pi$ is a solution of

$$|H(t_{i,m}(f), f)|^2 = \lambda_i^2, \tag{2.114}$$

and the instantaneous spectrum envelope of each component is

$$A_{i,m}(f) = \left[\frac{1}{\sqrt{\frac{\partial |H(t,f)|^2}{\partial f}}} \right]_{t=-\dot{\Phi}_{i,m}(f)/2\pi} . \tag{2.115}$$

The previous derivations show what is the (approximate) expression for the eigenfunctions associated to a given eigenvalue, but there are no hints about the values of the eigenvalues. Indeed, not all values of λ_i are admissible. First of all, from (2.111), evidently λ_i must belong to the range

$$\min_{t,f} |H(t, f)| \leq \lambda_i \leq \max_{t,f} |H(t, f)|. \tag{2.116}$$

But even within this interval, not all values are admissible. It is shown in Appendix 1A.1 that the only admissible values must satisfy the so called *area rule*, which can

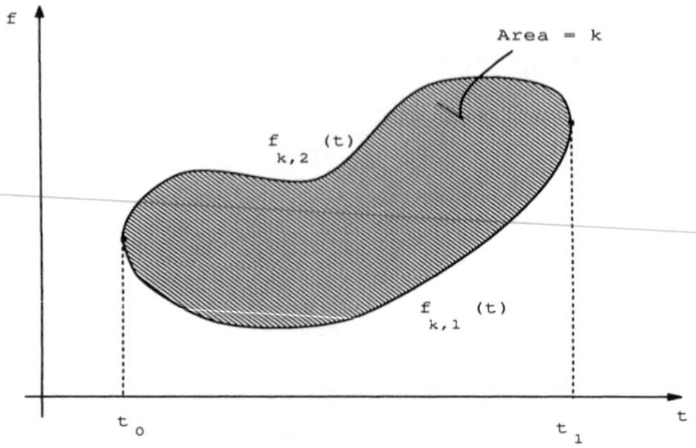

Figure 2.15 Contour plot of one amplitude level of $|H(t, f)|^2$.

be explained graphically, with the help of Figure 2.15 showing one possible contour plot of the square modulus of the channel transfer function, relative to a single level of $|H(t, f)|^2$. Given the relationship between instantaneous phase and frequency, if we consider the lower curve $f_{k,1}(t)$, we can write

$$\phi_k(t_2) = \phi_k(t_1) + 2\pi \int_{t_1}^{t_2} f_{k,1}(\theta) d\theta. \tag{2.117}$$

On the other hand, if we consider the upper curve, instead, we have

$$\phi_k(t_1) = \phi_k(t_2) + 2\pi \int_{t_2}^{t_1} f_{k,2}(\theta) d\theta. \tag{2.118}$$

Combining these two equations and considering that any phase value gives rise to the same signal value if an integer multiple of 2π is added to the phase, we can write

$$\int_{t_1}^{t_2} [f_{k,2}(\theta) - f_{k,1}(\theta)] d\theta = k, \tag{2.119}$$

where k is an integer number. Equation (2.119) states that the only levels of $|H(t, f)|$ that may be associated to an eigenvalue λ_i are the ones such that the area within the two instantaneous frequency curves (the dashed area in Figure 2.15) is a positive integer number. This statement is the so called area rule.

After having derived the approximate analytic model (2.110), with (2.111), we also need to check that the class of functions $v_i(t)$ and $u_i(t)$ constitute two classes of orthonormal functions. To prove this property, it is useful to resort to time-frequency distributions (TFDs). A few properties about TFD are recalled in Appendix 1A.2. In particular, we recall the Wigner-Ville distribution (WVD), as it plays a central role in the theory of TFDs. The WVD $W_s(t, f)$ of a signal $s(t)$ is defined as

$$W_s(t, f) = \int_{-\infty}^{\infty} s^*(t - \tau/2)s(t + \tau/2)e^{-j2\pi f\tau} d\tau. \tag{2.120}$$

One of the main properties of the $W_s(t, f)$, useful for our analysis here, is that it is maximally concentrated, in the time-frequency domain, along the instantaneous frequency of the signal. The degree of concentration depends on the amplitude modulation of the signal: The smoother is the amplitude modulation, the more concentrated is the WVD around the instantaneous frequency curve. This property is particularly useful in the analysis of underspread channels, because the amplitude modulation given by (2.112) is indeed a smooth function. The other important property to be used is the so-called Moyal's formula, see, for example (2.181), stating that the scalar product between the WVD's of two signals $x(t)$ and $y(t)$ is equal to the square modulus of the scalar product between the two signals. Therefore, we can check the orthogonality between two functions, looking at the scalar product of their WVD's. Any two singular functions $v_i(t)$ and $v_j(t)$, $i \neq j$, relative to different singular values, have instantaneous frequencies located, by construction, over nonoverlapping curves of the time-frequency plane. This happens because those instantaneous frequencies refer to different contour levels of $|H(t, f)|^2$. Therefore, if the WVDs are also maximally concentrated along those instantaneous frequencies, the scalar product between the WVDs of $v_i(t)$ and $v_j(t)$ is null. These arguments suggest that, even if two functions $v_i(t)$ and $v_j(t)$ may be partially overlapping when observed in the time or in the frequency domain, separately, they may be non-overlapping when observed in the *joint* time-frequency domain. Indeed, this is precisely what happens with the eigenfunctions of underspread systems.

2.5.3 Examples

A few examples are useful to understand the meaning of the derivations illustrated in the previous section.

2.5.3.1 LTI Channels

A linear time-invariant (LTI) channel is a channel whose impulse response $h(t, \tau)$ does not depend on t, that is,

$$h(t, \tau) = h(\tau).$$

Equivalently, its transfer function depends only on f, that is $H(t, f) = H(f)$. As a consequence, the curve, in the time-frequency domain where $|H(t, f)|$ is constant is given by lines of equation $f = f_i$. This behavior of the instantaneous frequency characterizes a signal having only one frequency, constant for all times. This signal is clearly a complex exponential of the form $\exp(j2\pi f_i t)$. In fact, this is a well known property that can be immediately verified by plugging $x(t) = \exp(j2\pi f_i t)$ in (2.10) to check that the corresponding output is $y(t) = H(f_i)x(t)$.

2.5.3.2 Multiplicative Channels

A multiplicative channel is characterized by the following input-output relationship

$$y(t) = m(t)x(t) \tag{2.121}$$

or, incorporating a possible delay, $y(t) = m(t)x(t - \tau_0)$. The impulse response of such a channel is

$$h(t, \tau) = \delta(\tau - \tau_0)m(t).$$

As an example, the multipath model (2.16) degenerates into a multiplicative channel when the delays are all equal to each other (e.g., $\tau_k = \tau_0, \forall k$). In such a case,

$$h(t, \tau) = \delta(\tau - \tau_0) \sum_{q=0}^{Q-1} h_q e^{j2\pi f_q t}. \tag{2.122}$$

The time-varying transfer function of a multiplicative channel is $H(t, f) = e^{-j2\pi\tau_0 f} m(t)$ and its modulus $|H(t, f)| = |m(t)|$ is constant over lines of equation $t = t_i$. Since $|H(t, f)|$ is constant along f, according to (2.113) and (2.114), the group delay is constant, and thus the spectrum of the channel eigenfunctions has constant amplitude and linear phase. Hence, the eigenfunctions are Dirac pulses. In fact, if the input is $x(t) = \delta(t - t_i)$, the corresponding output $y(t)$ is proportional to the input and it is $y(t) = m(t_i)\delta(t - t_i) = m(t_i)x(t)$.

Before concluding this section, we wish to remark that even though the LTI and multiplicative channels are characterized by impulse responses that are not square-integrable, the models (2.110) and (2.113) in these two cases hold *exactly*.

2.5.3.3 Two-Ray Multipath Channel

Let us consider now a multipath channel composed of two rays (see, e.g., (2.16) with $Q = 2$). We assume that the delays τ_k are not equal to each other (as this would be a particular case of the multiplicative channel analyzed before). In such a case, the channel singular function must be a solution of

$$\lambda_i e^{j\psi_0} v_i(t - t_d) e^{j2\pi f_d t} = h_0 v_i(t - \tau_0) e^{j2\pi f_0 t} + h_1 v_i(t - \tau_1) e^{j2\pi f_1 t}. \tag{2.123}$$

Setting $t_d = \tau_0$ and $f_d = f_0$, we may rewrite (2.123) as

$$\lambda_i v_i(t-\tau_0) = h_0 v_i(t-\tau_0) + h_1 v_i(t-\tau_1) e^{j2\pi(f_1-f_0)t}. \tag{2.124}$$

Setting $\theta = t - \tau_0$, we have

$$\lambda_i v_i(\theta) = h_0 v_i(\theta) + h_1 v_i(\theta - \Delta\tau) e^{j2\pi\Delta f(\theta+\tau_0)}, \tag{2.125}$$

where $\Delta f := f_1 - f_0$ and $\Delta\tau := \tau_1 - \tau_0$. It is straightforward to verify, by direct substitution, that the solution of (2.125) is given by the functions

$$v_i(t) = e^{j2\pi\frac{\alpha}{\Delta\tau}t} e^{j\pi\frac{\Delta f}{\Delta\tau}t^2} \tag{2.126}$$

parameterized with respect to the variable α, which is related to λ_i by

$$\lambda_i e^{j\psi_0} = h_0 + h_1 e^{j2\pi\alpha} e^{-j\pi\Delta f(\tau_1+\tau_0)}, \tag{2.127}$$

where the phase ψ_0 is chosen in order to have a real non-negative value for λ_i. Therefore, the singular functions of two-ray channels are *chirp* signals (i.e., signals whose instantaneous frequency is a linear function of time and, in particular, whose sweep rate is $\Delta f/\Delta\tau$). Furthermore, the eigenvalues are related to the channel parameters by (2.127). We can now verify the validity of (2.110) and (2.111) in this case. In fact, the channel transfer function is

$$\begin{aligned} H(t,f) &= h_0 e^{j2\pi(f_0 t - f\tau_0)} + h_1 e^{j2\pi(f_1 t - f\tau_1)} \\ &= e^{j2\pi(f_0 t - f\tau_0)}(h_0 + h_1 e^{j2\pi(\Delta f t - \Delta\tau f)}) \end{aligned} \tag{2.128}$$

and the contour lines of $|H(t,f)|$ relative to the level λ_i are described by the equation

$$\Delta\tau \cdot f - \Delta f \cdot \tau = \alpha, \tag{2.129}$$

where the parameter α is related to λ_i by

$$\lambda_i e^{j\varphi_0} = h_0 + h_1 e^{j2\pi\alpha}; \tag{2.130}$$

φ_0 is constant phase. Solving (2.129) for f and using (2.110), the instantaneous frequency of the channel eigenfunctions components is

$$f = f_i(t) = \frac{\Delta f}{\Delta\tau}t + \frac{\alpha}{\Delta\tau} \tag{2.131}$$

so that, according to (2.110), the channel eigenfunctions are

$$v_i(t) = e^{j2\pi\frac{\alpha_i}{\Delta\tau}t} e^{j\pi\frac{\Delta f}{\Delta\tau}t^2}. \tag{2.132}$$

Notice that this expression coincides with (2.126). Furthermore, if the channel is underspread [i.e., $(\tau_1 + \tau_0)\Delta f \ll 1$] the channel eigenvalues obtained through (2.127) or (2.130) coincide by simply setting $\varphi_0 = \psi_0$.

2.5.3.4 General Multipath Channel

To check the validity of the analytic model in the general case, it is useful to analyze the channel eigenfunctions in the *joint* time-frequency domain, rather than simply in the time or in the frequency domain. The basic tool for verifying the analytic model is given by time-frequency distributions (TFDs).

We checked the validity of (2.110) or (2.113) using the following scheme. Given the impulse response $h(t,\tau)$ of the continuous-time system, we 1) build the channel matrix $\boldsymbol{H}$ of the equivalent DT system; 2) compute the SVD of $\boldsymbol{H}$; 3) compute the WVD of the right and left singular vectors associated to the generic singular value λ_i; and 4) compare the energy distribution of these TFDs with the contour plot of $|H(t,f)|$ corresponding to level λ_i. We used as a basic tool to analyze the signals in the time-frequency domain the smoothed pseudo-Wigner-Ville distribution (SPWVD) with reassignment, introduced in [16], for its property of having low cross terms without degrading the resolution. We considered as a test system a communication channel affected by multipath propagation, thus described by the CT impulse response

$$h(t,\tau) = \sum_{q=0}^{Q-1} h_q e^{j2\pi f_q t} \delta(\tau - \tau_q),$$

where each path is characterized by the triplet of amplitude h_q, delay τ_q, and Doppler shift f_q. We generated the amplitudes h_q as independent identically distributed (iid) complex Gaussian random variables with zero mean and unit variance (the Rayleigh-fading model), and the variables τ_q and f_q as iid random variables with uniform distribution within the intervals $[0, \Delta\tau]$ and $[-\Delta f/2, \Delta f/2]$, respectively. An example, relative to a multipath channel composed of $Q = 12$ paths, with $\Delta\tau = 4T_s$ and $\Delta f = 4/NT_s$, $N = 128$, is reported in Figure 2.16, where we show: a) the mesh plot of $|H(t,f)|$, b) two contour plots of $|H(t,f)|$ corresponding to the levels λ_{16} (dashed line) and λ_{32} (solid line); c) the contour plot of the SPWVD of $\boldsymbol{v}_{16}$; and d) the contour plot of the SPWVD of $\boldsymbol{v}_{32}$.

It is worth noticing how, in spite of the rather peculiar structure of the contour plots of $|H(t,f)|$, the SPWVDs of the two singular functions are strongly concentrated along curves coinciding with the contour lines of $|H(t,f)|$ corresponding to the associated singular values, as predicted by the theory.

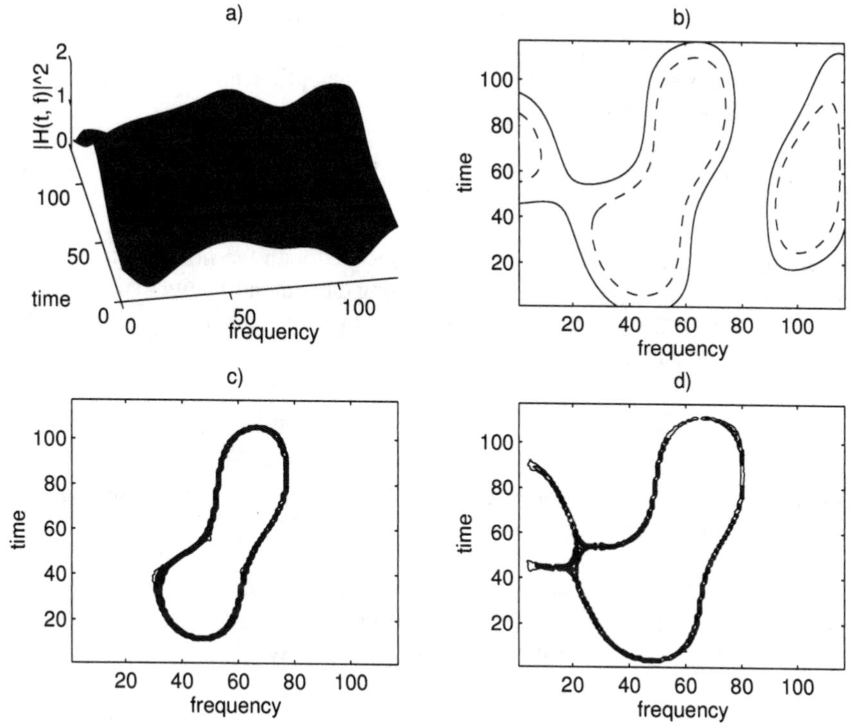

Figure 2.16 Comparison between contour lines of $|H(t,f)|$ and TFDs of channel singular vectors: a) $|H(t,f)|$; b) contour lines of $|H(t,f)|$ corresponding to levels λ_{16} (dashed line) and λ_{32} (solid line); and c) SPWVD of $\boldsymbol{v}_{16}$; and d) SPWVD of $\boldsymbol{v}_{32}$.

It is also interesting to observe the *bubble*-like structure of the two SPWVDs. Indeed this behavior is quite common, because in general the contour lines of the time-varying transfer function are typically closed curves.

The singular functions $u_i(t)$ and $v_i(t)$ are the continuous time channel modes. They play, in fact, a similar role to the left and right singular vectors of the channel matrix, in the discrete-time domain. In fact, the optimal strategy for transmitting a set of symbols $s[k], k = 0, \ldots, K-1$, in the presence of additive white Gaussian noise (AWGN), is to send the signal

$$x(t) = \sum_{k=0}^{K-1} c_k s[k] v_k(t) \tag{2.133}$$

where $v_k(t)$ is the right singular function associated with the kth eigenvalue of the channel response $h(t, \tau)$ and c_k are coefficients used to allocate the available power among the transmitted symbols according to some optimization criterion [5]. Using (2.95), the received signal is thus

$$y(t) = \int_{-\infty}^{\infty} h(t, \tau)x(t - \tau)d\tau + w(t) = \sum_k c_k \lambda_k s[k] u_k(t) + w(t), \qquad (2.134)$$

where $u_k(t)$ is the left singular function associated with the kth singular value of $h(t, \tau)$ and $w(t)$ is AWGN. Hence, by exploiting the orthonormality of the functions $u_k(t)$, the transmitted symbols can be estimated by simply taking the scalar products of $y(t)$ with the left singular functions; that is,

$$\hat{s}[m] = \frac{1}{\lambda_m c_m} \int_{-\infty}^{\infty} y(t) u_m^*(t) dt = s[m] + w[m], \qquad (2.135)$$

where the noise samples sequence $w[m] := \int_{-\infty}^{\infty} w(t) u_m^*(t) dt$ constitutes a sequence of iid Gaussian random variables.

In summary, also in the case of continuous-time channels, the knowledge of the channel eigenfunctions allows an easy recovery of the transmitted symbols, with the possibility of optimizing the performance of interest.

2.6 ANALYSIS OF REAL DATA

In this section, we show some real MIMO channel measurements gathered at the University of Bristol, from the group coordinated by Professor M. Beach, within the project denominated Smart Antenna Technology in Universal Broadband Networks (SATURN), supported by the European Union, within the fifth Information Society Technology (IST) program framework. The experimental setup is as follows. The transmitter has four transmit antennas, each antenna with dual polarization. The transmit antennas were spaced apart by 20λ. The transmitted signal occupies a bandwidth of 20 MHz, from $1,920$ to $1,930$ MHz. The element spacing at the receiver is 0.5λ. Several sets of data have been gathered, in different conditions. We show some results concerning two scenarios: (a) time-varying channel, where the receiver is moving, and (b) stationary channel, where the receiver is fixed.

In Figure 2.17 we show the modulus of the time-varying transfer function and of the corresponding delay-Doppler spread function of a channel under scenario (A). The data reported in Figure 2.17 refers to a time observation window of 0.8 sec. It

5 The optimization criteria will be described in Chapter 6.

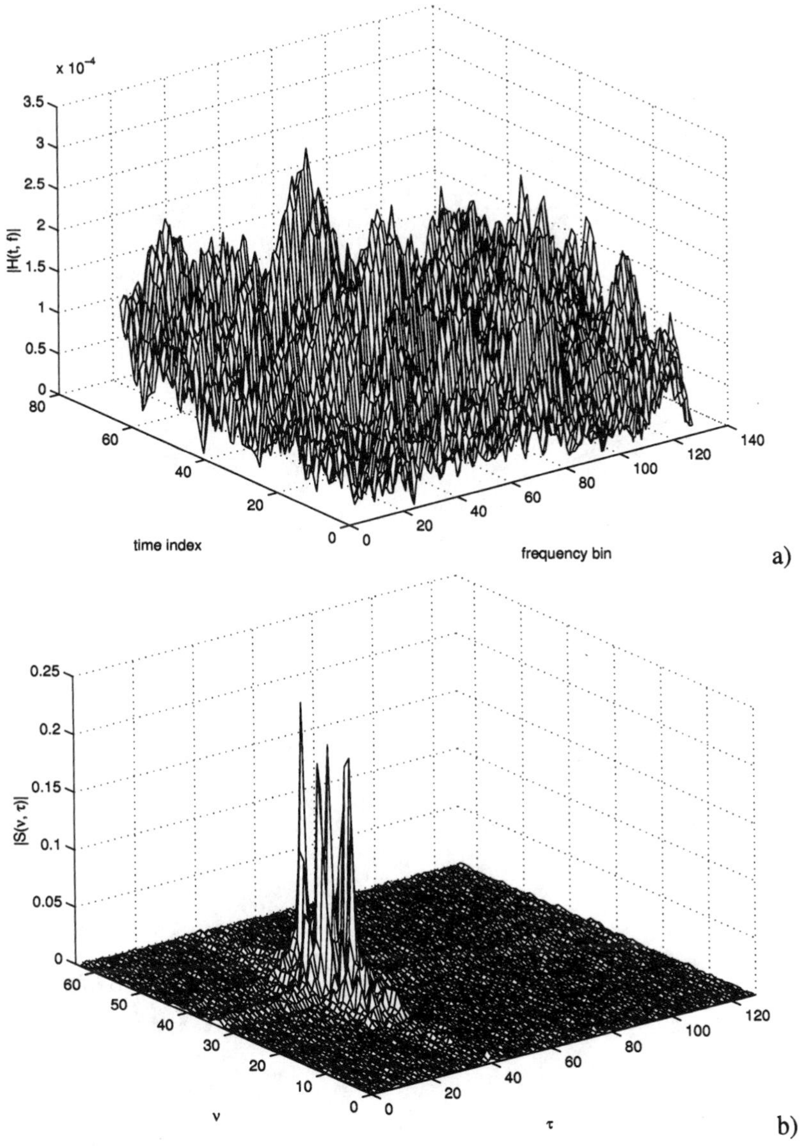

Figure 2.17 Analysis of a real channel; co-polarized channels; mobile receiver: a) Modulus of $H(t, f)$ and b) modulus of $S(\nu, \tau)$.

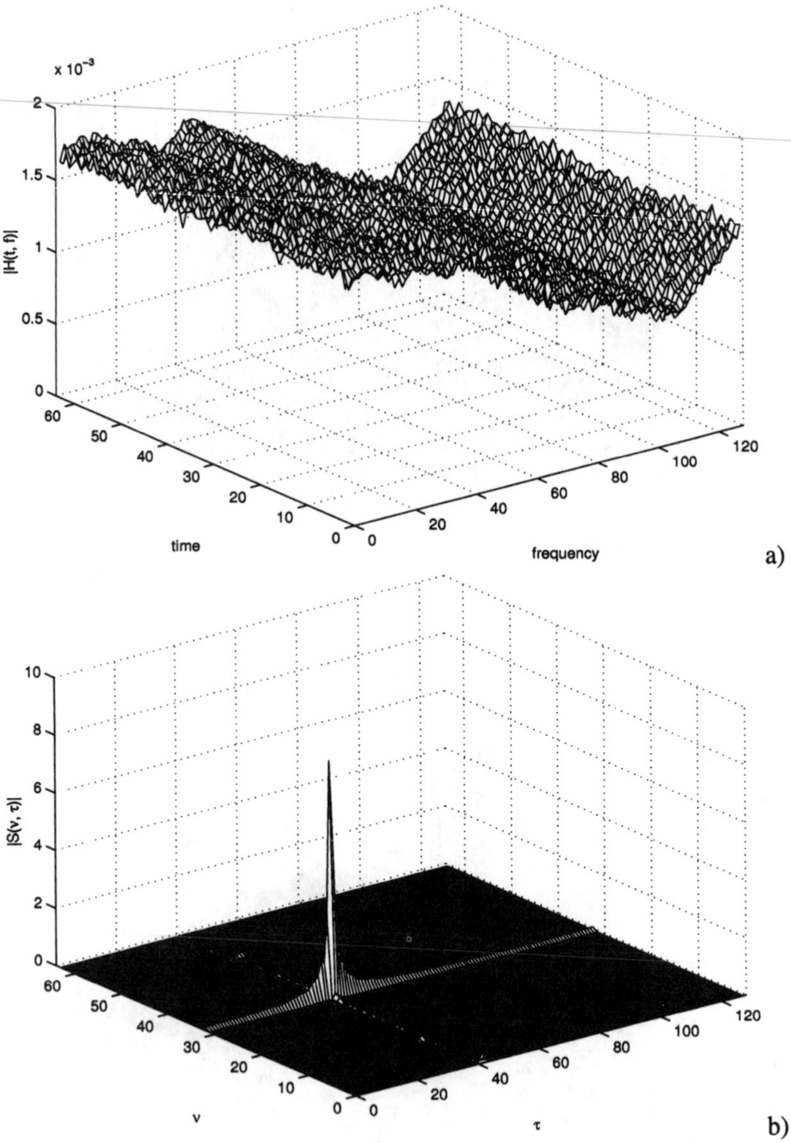

Figure 2.18 Analysis of a real channel; co-polarized channels; mobile receiver: a) Modulus of $H(t, f)$ and b) modulus of $S(\nu, \tau)$.

is evident from Figure 2.17 that the spread function has a nonnull support and thus that the channel is really time-variant.

As an example of time-invariant channel, in Figure 2.18 we report the behaviors of $|H(t, f)|$ and $|S(\nu, \tau)|$, taken as in Figure 2.17, except that now the receiver is fixed and there are no moving objects around the receiver. We can see from Figure 2.18 that now the spread function is highly concentrated around the origin, and in fact the transfer function is practically constant over time, whereas it varies along frequency, because of multipath propagation.

2.7 SUMMARY

In this chapter, we have introduced the main parameters characterizing a wireless channel. We have illustrated the important concepts of coherence time, bandwidth, and space, which will play a fundamental role in devising the proper transmission strategy. Basic references on time-varying channel models are, for example, [7, 17, 18]. A more in-depth analysis of MIMO channels is given in the recent books [3, 19] and in journal papers [9, 10, 11, 12, 20, 21]. We have provided also an approximate model for the eigenfunctions of slowly varying SISO channels. This model will be useful, in Chapter 6, to get a physical insight into the optimal way of allocating the transmit power as a function of time and frequency.

APPENDIX 1A.1: EIGENFUNCTIONS OF SLOWLY VARYING LINEAR OPERATORS

A linear operator is a rule for mapping an "input" signal $a(x)$ onto an "output" signal $b(x)$, which respects the superposition principle. We denote this mapping through the symbol $b(x) = \mathcal{L}[a(x)]$. The superposition principle states that if we apply the operator to a linear combination of functions, namely to $a(x) = \sum_{m=1}^{M} c_m a_m(x)$, the result is the superposition, or linear combination, of the partial results $b_k(x) = \mathcal{L}[a_k(x)]$ through the same combination coefficients. In formulas, if $a(x) = \sum_{m=1}^{M} c_m a_m(x)$ and $b_m(x) = \mathcal{L}[a_m(x)]$, then $b(x) = \mathcal{L}[a(x)] = \sum_{m=1}^{M} c_m b_m(x)$.

Every linear operator is characterized by an input/output relationship that can be expressed through the following integral equation

$$b(x) = \int_{-\infty}^{\infty} \mathcal{K}(x, y) a(y) dy, \tag{2.136}$$

where $\mathcal{K}(x, y)$ is known as the system *kernel* and describes the linear operator completely. The variables x and y that we use here may have different physical meanings, as they may denote spatial coordinates or temporal variables, for example. In (2.136) we have assumed the variables x and y to be scalar (or monodimensional) variables. However, (2.136) can be generalized to the multidimensional case, as follows

$$b(\boldsymbol{x}) = \int_{-\infty}^{\infty} \cdots \int_{-\infty}^{\infty} \mathcal{K}(\boldsymbol{x}, \boldsymbol{y}) a(\boldsymbol{y}) d\boldsymbol{y}, \tag{2.137}$$

where now $\boldsymbol{x}$ and $\boldsymbol{y}$ are multidimensional vectors. In such a case, $\mathcal{K}(\boldsymbol{x}, \boldsymbol{y})$ is often referred to as the Green's function.

In this section we do not specify the physical meaning of the variables $\boldsymbol{x}$ and $\boldsymbol{y}$ so as to derive general results that can be applied to a variety of situations. In practice, $\boldsymbol{x}$ and $\boldsymbol{y}$ may indicate temporal quantities (and in that case they would be scalar variables), or spatial variables, or even joint space-time coordinates. In the following, we will use a terminology related to spatial coordinates, but the same considerations apply equally well to temporal coordinates. We will also consider only the scalar case, because in such a case it is possible to derive some important results in closed form, and this is useful to better understand the main properties of linear operators.

Specifically, we will concentrate on some functions that play a special role in the study of linear operators, namely the *eigenfunctions* or *singular functions*. If the operator has finite norm, or

$$\int_{-\infty}^{\infty} \int_{-\infty}^{\infty} |\mathcal{K}(x, y)|^2 \, dt \, d\tau < \infty, \tag{2.138}$$

then there exists a countable set of *singular values* λ_i and two classes of orthonormal functions $v_i(x)$ and $u_i(x)$, named *right* and *left singular functions*, such that the following system of integral equations holds true

$$\lambda_i u_i(y) = \int_{-\infty}^{\infty} K(x, y) v_i(x) dx, \tag{2.139}$$

$$\lambda_i v_i(x) = \int_{-\infty}^{\infty} K^*(x, y) u_i(y) dy. \tag{2.140}$$

In particular, if the kernel is Hermitian, that is if $K(y, x) = K^*(x, y)$, the singular functions $u_i(x)$ and $v_i(x)$ coincide and we may talk, properly, of eigenfunctions.

Hence, a function $\psi_i(x)$ is an eigenfunction of a Hermitian operator having kernel $\mathcal{K}(x, y)$, associated to the eigenvalue λ_i, if $\psi_i(x)$ is solution of the following integral equation

$$\lambda_i \psi_i(x) = \int_{-\infty}^{\infty} \mathcal{K}(x, y)\psi_i(y)dy. \tag{2.141}$$

This means that the effect of applying an operator on one of its eigenfunctions consists simply in the multiplication of the eigenfunction by a constant. In general, since the eigenfunctions are solution of (2.141), they have a structure that depends on the kernel itself. The only exceptions of this statement are given by two special classes of systems: the space-invariant systems and the multiplicative systems.

HOMOGENEOUS SYSTEMS

A homogeneous, or space-invariant, system is a system whose kernel satisfies the following property

$$\mathcal{K}(x, y) = \mathcal{K}(x - y). \tag{2.142}$$

The structure of the eigenfunctions of this kind of operators is well known. It is straightforward to verify, in fact, that the eigenfunctions of such operators are the exponential functions $\psi(x) = e^{jpx}$. In fact, substituting such an expression into (2.142), we get

$$\int_{-\infty}^{\infty} \mathcal{K}(x-y)e^{jpy}dy = \int_{-\infty}^{\infty} K(u)e^{jp(x-u)}du = e^{jpx}\int_{-\infty}^{\infty} \mathcal{K}(u)e^{-jpu}du = \lambda\psi(x) \tag{2.143}$$

where

$$\lambda := \int_{-\infty}^{\infty} \mathcal{K}(u)e^{-jpu}du. \tag{2.144}$$

It is important to point out that exponential functions of the form e^{jpx} are eigenfunctions of *every* linear homogeneous system, irrespective of the specific structure of the system or, in other words, of the system kernel function.

MULTIPLICATIVE SYSTEMS

A multiplicative system is characterized by a kernel having the form $K(x, y) = K(x)\delta(y - x)$, where $\delta(x)$ denotes the Dirac pulse. The input/output relationship of such systems is

$$b(x) = \int_{-\infty}^{\infty} K(x)\delta(y - x)a(y)dy = K(x)a(x). \tag{2.145}$$

This relationship shows that the action of such a system on the input signals is simply to multiply the input by a function $K(x)$. It is easy to verify that the eigenfunctions of such systems are the Delta functions. In fact, if $a(x) = \delta(x - x_0)$, we have

$$b(x) = \int_{-\infty}^{\infty} K(x)\delta(y-x)\delta(y-x_0)dy = K(x_0)\delta(x-x_0) = K(x_0)a(x). \quad (2.146)$$

Once again, the Dirac functions are eigenfunctions of *every* multiplicative system.

WEAKLY HOMOGENEOUS SYSTEMS

If the operator is neither homogeneous nor multiplicative, it is much more difficult to derive an analytic expression of its eigenfunctions, and, in general, there is no such closed form expression. Nonetheless, since in many practical applications, we have operators that depart only slightly from the homogeneous condition, it is useful to derive a closed-form expression, although approximate, for the eigenfunctions of weakly inhomogeneous (or slowly time-varying) operators. We start by using a change of variable that enables us to write the kernel $\mathcal{K}(x, y)$ as follows

$$\mathcal{K}(x, y) = K\left[x - y, \frac{\epsilon}{2}(x + y)\right], \quad (2.147)$$

where ϵ is an instrumental parameter that measures the departure from homogeneity. When $\epsilon = 0$, the system is perfectly homogeneous, whereas a small value of ϵ denotes weak inhomogeneity. In the following, we will review the perturbation method illustrated in [14] to derive an approximate expression for the eigenfunctions of weakly inhomogeneous operators, valid for small values of ϵ.

We start considering the Taylor's series expansion of the kernel around the point $\epsilon = 0$:

$$K\left[x - y, \frac{\epsilon}{2}(x + y)\right] = K_0(x - y) + \frac{\epsilon}{2}(x + y)K_1(x - y) + \cdots \quad (2.148)$$

where $K_n(x - y) = \left.\frac{\partial^n K(x-y,\xi)}{\partial \xi^n}\right|_{\xi=0}$.
Similarly, we make explicit the dependence of the eigenfunctions on the parameter ϵ and, for small ϵ, we consider the Taylor's series development

$$\psi(x, \epsilon) = \psi_0(x) + \epsilon\psi_1(x) + \cdots, \quad (2.149)$$

where $\psi_k(x) = \left.\frac{\partial^k \psi(x,\xi)}{\partial \xi^k}\right|_{\xi=0}$.
Hence, we wish to solve the following system

$$\lambda\psi(x,\epsilon) = \int_{-\infty}^{\infty} K\left[x-y, \frac{\epsilon}{2}(x+y)\right]\psi(y,\epsilon)dy. \tag{2.150}$$

In particular, according to [14], we search for a solution expressed in the analytic form

$$\psi(x,\epsilon) = A(x\epsilon)e^{j\frac{1}{\epsilon}\phi(x\epsilon)},$$

where $A(x\epsilon)$ and $\phi(x\epsilon)$ are functions that depend on x and ϵ only through the product $x\epsilon$.

Following the same line of reasoning as in [14], we introduce the parameter $q = \epsilon x$ and consider the Taylor's series expansion of $K[x-y, \frac{\epsilon}{2}(x+y)]$ along its second variable, in the neighborhood of $y = x$; that is

$$\begin{aligned} K\left[x-y, \frac{\epsilon}{2}(x+y)\right] &= K(x-y,q) - \frac{\epsilon}{2}(x-y)K_{01}(x-y,q) \\ &+ \frac{\epsilon^2}{8}(x-y)^2 K_{02}(x-y,q) + \cdots \end{aligned} \tag{2.151}$$

where we have introduced the symbol

$$K_{lm}(p,q) := \frac{\partial^{l+m}K(p,q)}{\partial p^l q^m}.$$

We expand, similarly, $\psi(x,\epsilon)$. In this case, however, we have to be careful because, in general, small variations of the phase $\frac{1}{\epsilon}\phi(x\epsilon)$ have a greater impact on the integral (2.150) than corresponding variations of the amplitude $A(x\epsilon)$. To take into account this different sensitivity, we expand $\psi(y,\epsilon)$ in the neighborhood of $y = x$ as follows

$$\psi(y,\epsilon) = e^{\frac{j}{\epsilon}\left[\sum_{n=0}^{\infty}(-1)^n\epsilon^n(x-y)^n\phi^{(n)}(q)/n!\right]}\sum_{n=0}^{\infty}(-1)^n\epsilon^n\frac{A_n(q)(x-y)^n}{n!} \tag{2.152}$$

where $A_n(q) = \left.\frac{\partial^n A(\xi)}{\partial \xi^n}\right|_{\xi=q}$.
In this way, the first order expansion is

$$\psi(y,\epsilon) \approx e^{j\frac{\phi(q)}{\epsilon}}e^{-j(x-y)\phi'(q)}A_0(q), \tag{2.153}$$

(i.e., it retains two terms of the phase and one term of the amplitude). Similarly, the second order expansion retains three terms of the phase and two terms for the amplitude; that is,

$$\begin{aligned}\psi(y,\epsilon) &\approx e^{j\frac{\phi(q)}{\epsilon}} e^{-j\left[(x-y)\phi'(q)-\epsilon(x-y)^2\phi''(q)/2\right]}\left[A_0(q)-\epsilon(x-y)A_1(q)\right] \\ &\approx e^{j\frac{\phi(q)}{\epsilon}} e^{-j(x-y)\phi'(q)}\left[1+j\epsilon(x-y)^2\frac{\phi''(q)}{2}\right]\left[A_0(q)-\epsilon(x-y)A_1(q)\right] \\ &\approx e^{j\frac{\phi(q)}{\epsilon}} e^{-j(x-y)\phi'(q)}A_0(q)+e^{j\frac{\phi(q)}{\epsilon}} e^{-j(x-y)\phi'(q)}(x-y) \\ &\quad \cdot\left[j(x-y)\frac{\phi''(q)}{2}A_0(q)-A_1(q)\right]\epsilon.\end{aligned} \tag{2.154}$$

First Order Solution

We derive now the first order solution of (2.150). We consider the following expansions

$$\begin{aligned}&K(x-y,\tfrac{\epsilon}{2}(x+y))\approx K(x-y,q); \\ &\psi(y,\epsilon)\approx e^{j\frac{\phi(q)}{\epsilon}} e^{-j(x-y)\phi'(q)}A_0(q).\end{aligned} \tag{2.155}$$

Introducing (2.155) in (2.150) and setting $x-y=u$, we get

$$A_0(q)e^{j\frac{\phi(q)}{\epsilon}}\int_{-\infty}^{\infty}K(u,q)e^{-ju\phi'(q)}du=\lambda A_0(q)e^{j\frac{\phi(q)}{\epsilon}}. \tag{2.156}$$

This equation is satisfied if $\phi(q)$ is solution of the following implicit equation

$$\int_{-\infty}^{\infty}K(u,q)e^{-j\phi'(q)u}du=\lambda. \tag{2.157}$$

Hence, introducing the function

$$\tilde{K}(p,q):=\int_{-\infty}^{\infty}K(u,q)e^{-jup}du, \tag{2.158}$$

$\phi(q)$ is, equivalently, solution of

$$\tilde{K}\left[\phi'(q),q\right]=\lambda. \tag{2.159}$$

This means that $\phi'(q)$ describes the curve, in the (u,q) plane, where the function $\tilde{K}(u,q)$ is constant and equal to λ. In other words, $\phi'(q)$ is the *contour line* of $\tilde{K}(u,q)$ corresponding to the level λ. In general, (2.157) admits more than one real

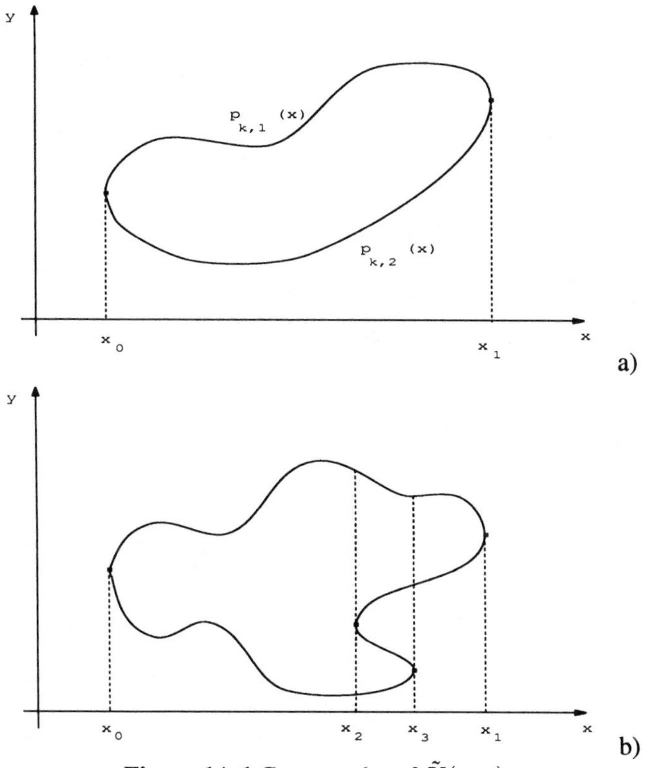

Figure 1A.1 Contour plot of $\tilde{K}(x, y)$.

solution. Let us consider, for example, a function whose contour plot is reported in Figure 1A.1. We see that in the example reported in Figure 1A.1(a) there are two solutions for $x \in [x_0, x_1]$. In Figure 1A.1(b), we observe that there are two solutions for $x \in [x_0, x_2]$ and $x \in [x_3, x_1]$ and four solutions for $x \in [x_2, x_3]$.

If we reincorporate the eigenvalue index i in our expressions, and we indicate with $p_{i,m}(q)$ one of the solutions of

$$\tilde{K}\left[p_{i,m}(q), q\right] = \lambda_i, \tag{2.160}$$

the first order expression for the eigenvalues of an inhomogeneous system assumes the form

$$\psi_i(x) = \sum_m A_{i,m} e^{j\phi_{i,m}(x)}, \tag{2.161}$$

where m runs over all the solutions, for the given x, and

$$\phi_{i,m}(x) = \int_{x_0}^{x} p_{i,m}(y)dy + \phi_{i,m}(x_0). \tag{2.162}$$

The initial phases $\phi_{i,m}(x_0)$ can be neglected without any loss of generality, as they may be incorporated in the amplitudes $A_{i,m}$. The only constraint on the amplitudes $A_{i,m}$ is that the eigenfunctions have unit norm.

It is important to remark that not all values of λ_i are eigenvalues of the system. In fact, if we denote by $\mathcal{A}_i$ the area enclosed within the contour line associated to the ith eigenvalue, the ith eigenvalue must respect the following rule

$$\mathcal{A}_i = (2i + 1)\pi. \tag{2.163}$$

Second Order Solution

Let us consider now the second order expansion (2.154) and see how to get a refined solution of (2.150). Substituting (2.154) (the second approximation) and the expansion (2.151), up to the linear term in ϵ, in (2.150) and setting $u = x - y$, we get the following integral

$$\begin{aligned}
&\int_{-\infty}^{\infty}\left[K(u,q)-\frac{\epsilon}{2}uK_{01}(u,q)\right]\left[A_0(q)-\epsilon uA_1(q)\right]e^{j\frac{\phi(q)}{\epsilon}}e^{-ju\phi'(q)}\\
&\cdot\left[1+j\epsilon u^2\frac{\phi''(q)}{2}\right]du\\
\approx\;& A_0(q)e^{j\frac{\phi(q)}{\epsilon}}\int_{-\infty}^{\infty}K(u,q)e^{-ju\phi'(q)}du+\epsilon e^{j\frac{\phi(q)}{\epsilon}}\int_{-\infty}^{\infty}e^{-ju\phi'(q)}\\
&\cdot\left\{K(u,q)\left[j\phi''(q)\frac{u^2}{2}A_0(q)-uA_1(q)\right]-\frac{u}{2}K_{01}(u,q)A_0(q)\right\}du.
\end{aligned} \tag{2.164}$$

We may seek for a solution of the previous equation using the same phase found in the first order solution and finding the amplitude $A_0(q)$ that nulls the term proportional to ϵ in the previous equation. To this end, it is useful to introduce the following function

$$\tilde{K}_p(p,q):=\frac{\partial\tilde{K}(p,q)}{\partial p}=-j\int_{-\infty}^{\infty}uK(u,q)e^{-jup}du, \tag{2.165}$$

where we have used (2.158). Using (2.165), we may write

$$\int_{-\infty}^{\infty}K(u,q)ue^{-ju\phi'(q)}du=j\tilde{K}_p(\phi'(q),q). \tag{2.166}$$

Furthermore, deriving $j\tilde{K}_p(\phi'(q),q)$ with respect to q, we obtain

$$\begin{aligned}
j\frac{\partial\tilde{K}_p(\phi'(q),q)}{\partial q} &= \int_{-\infty}^{\infty}uK_{01}(u,q)e^{-ju\phi'(q)}du\\
&- j\phi''(q)\int_{-\infty}^{\infty}u^2K(u,q)e^{-ju\phi'(q)}du.
\end{aligned} \tag{2.167}$$

Therefore, substituting (2.167) and (2.166) in the last line of (2.164), we can write the last integral as

$$-j\epsilon e^{j\frac{\phi(q)}{\epsilon}}\left\{A_1(q)\tilde{K}_p[\phi'(q),q]+\frac{A_0(q)}{2}\frac{\partial\tilde{K}_p[\phi'(q),q]}{\partial q}\right\}. \tag{2.168}$$

The function $A_0(q)$ that nulls the previous expression is a possible solution of our system, as it nulls the last integral of (2.164). Equation (2.168) can be rewritten, equivalently (omitting $-j\epsilon e^{j\frac{\phi(q)}{\epsilon}}$), as

$$\tilde{K}_p^{1/2}[\phi'(q),q]\left\{A_1(q)\tilde{K}_p^{1/2}[\phi'(q),q]+\frac{A_0(q)}{2}\frac{\partial\tilde{K}_p[\phi'(q),q]}{\partial q}\frac{1}{\tilde{K}_p^{1/2}[\phi'(q),q]}\right\}. \tag{2.169}$$

Assuming $\tilde{K}_p^{1/2}(\phi'(q),q)\neq 0$, $A_0(q)$ must be solution of the following differential equation

$$\frac{\partial}{\partial q}\left\{A_0(q)\tilde{K}_p^{1/2}[\phi'(q),q]\right\}=0. \tag{2.170}$$

The solution of this equation is, up to a multiplicative factor,

$$A_0(q)=\frac{1}{\tilde{K}_p^{1/2}[\phi'(q),q]} \tag{2.171}$$

Combining (2.171) with (2.153), we can state that the second order expression for the eigenfunctions of a weakly inhomogeneous system is

$$\psi_i(q;\epsilon)=\sum_m\frac{e^{j\phi_{i,m}(q)}}{\sqrt{\tilde{K}_p[\phi'_{i,m}(q),q]}}, \tag{2.172}$$

where the instantaneous phase $\phi_{i,m}(q)$ of each component is such that the corresponding derivative $\phi'_{i,m}(s)$ is a solution of (2.157).

The previous approximation breaks down for values of q where $\tilde{K}_p[\phi'(q),q]=0$. These points are turning points and are indicated by the dots in Figure 1A.1. Around such points, higher order expansions are needed.

TIME-VARYING SYSTEMS

We specialize now some of the results achieved in the previous section to time-varying systems. In particular, we wish to give a physical interpretation of the results shown above. For time-varying channels, the kernel is related to the channel impulse response through the following equation

$$\mathcal{K}(t,\theta)=\int_{-\infty}^{\infty}h(t,t-\tau)h^*(\theta,\theta-\tau)d\tau. \tag{2.173}$$

Hence, proceeding as in (2.147), we can introduce the function $K(p,q)$ such that

$$\mathcal{K}(t,\theta)=K\left[t-\theta,\frac{\epsilon}{2}(t+\theta)\right], \tag{2.174}$$

or, equivalently, setting $p = t - \theta$ and $q = \epsilon(t+\theta)/2$,

$$K(p,q) = \mathcal{K}\left(\frac{p}{2}+\frac{q}{\epsilon}, -\frac{p}{2}+\frac{q}{\epsilon}\right). \tag{2.175}$$

It is also useful to introduce the function

$$\begin{aligned}\tilde{K}(u,q) &:= \int_{-\infty}^{\infty} K(p,q)e^{-j2\pi up}dp \\ &= \int_{-\infty}^{\infty} \mathcal{K}\left(\frac{p}{2}+\frac{q}{\epsilon}, -\frac{p}{2}+\frac{q}{\epsilon}\right) e^{-j2\pi up}dp \\ &= \int_{-\infty}^{\infty}\int_{-\infty}^{\infty} h\left(\frac{p}{2}+\frac{q}{\epsilon}, \frac{p}{2}+\frac{q}{\epsilon}-\tau\right) h^*\left(-\frac{p}{2}+\frac{q}{\epsilon}, -\frac{p}{2}+\frac{q}{\epsilon}-\tau\right) \\ &\cdot \quad e^{-j2\pi up}dpd\tau.\end{aligned}$$

It is interesting to express $\tilde{K}(u,q)$ as a function of the channel transfer function $H(t,f)$. Substituting in the previous expression $h(t,\tau)$ with $\int_{-\infty}^{\infty} H(t,f)e^{j2\pi f\tau}df$, we get

$$\tilde{K}(f,t) = \int_{-\infty}^{\infty}\int_{-\infty}^{\infty} H\left(\frac{p}{2}+\frac{t}{\epsilon}, \nu\right) H^*\left(-\frac{p}{2}+\frac{t}{\epsilon}, \nu\right) e^{-j2\pi(f-\nu)p}dpd\nu. \tag{2.176}$$

We wish to find out the equation of the line where $\tilde{K}(f,t)$ is constant, as a function of $H(t,f)$. Setting $f = f(t) := \phi'(t)/2\pi$ and considering that, for $\epsilon \to 0$ we have $H\left(\frac{p}{2}+\frac{t}{\epsilon}, \nu\right) \approx H(\frac{t}{\epsilon}, \nu)$, we can rewrite $\tilde{K}(f,t)$ as

$$\begin{aligned}\tilde{K}(f(t),t) &= \int_{-\infty}^{\infty}\int_{-\infty}^{\infty} H\left(\frac{p}{2}+\frac{t}{\epsilon}, \nu\right) H^*\left(-\frac{p}{2}+\frac{t}{\epsilon}, \nu\right) e^{-j2\pi(f(t)-\nu)p}dpd\nu \\ &\approx \int_{-\infty}^{\infty}\int_{-\infty}^{\infty} H\left(\frac{t}{\epsilon}, \nu\right) H^*\left(\frac{t}{\epsilon}, \nu\right) e^{-j2\pi(f(t)-\nu)p}dpd\nu \\ &\approx \int_{-\infty}^{\infty} \left|H\left(\frac{t}{\epsilon}, \nu\right)\right|^2 \int_{-\infty}^{\infty} e^{-j2\pi(f(t)-\nu)p}dpd\nu = \lambda^2 \\ &\approx \int_{-\infty}^{\infty} \left|H\left(\frac{t}{\epsilon}, \nu\right)\right|^2 \delta(f(t)-\nu)d\nu = \left|H\left(\frac{t}{\epsilon}, f(t)\right)\right|^2.\end{aligned} \tag{2.177}$$

Hence, the condition $\tilde{K}(f,t) = \lambda^2$ becomes

$$\left|H\left(\frac{t}{\epsilon}, f(t)\right)\right|^2 = \lambda^2. \tag{2.178}$$

In words, the eigenfunctions of a slowly varying channel are composed of the superposition of components whose instantaneous frequency is given by the curve, in the time-frequency domain, where the modulus of the channel transfer function is constant and equal to the eigenvalue associated to the eigenfunction.

Proceeding similarly with the instantaneous envelope $a(t)$, we find out that

$$a(t) = \left[\frac{1}{\sqrt{\frac{\partial |H(t,f)|^2}{\partial f}}} \right]_{f=\phi'(t)/2\pi} . \tag{2.179}$$

APPENDIX 1A.2: TIME-FREQUENCY REPRESENTATIONS

We recall here a few basic properties of time frequency distributions (the interested reader may refer to [22, 23], for example). Within the generalized Cohen's class of time-frequency distributions (TFDs), the so called Wigner-Ville distribution (WVD) plays a prominent role because all other TFDs can be derived from the WVD through a convolution in the time-frequency domain with the smoothing function characterizing the desired distribution. Given a signal $s(t)$, its WVD is defined as

$$W_s(t,f) = \int_{-\infty}^{\infty} s^*(t-\tau/2)s(t+\tau/2)e^{-j2\pi f\tau}d\tau. \tag{2.180}$$

The WVD satisfies the following properties:

1. Moyal's formula: Given two signals $x(t)$ and $y(t)$, their scalar product is preserved in the time-frequency domain, in the sense that the following property, known as Moyal's formula, holds true:

$$\int_{-\infty}^{\infty}\int_{-\infty}^{\infty} W_x(t,f)W_y(t,f)dtdf = \left| \int_{-\infty}^{\infty} x^*(t)y(t)dt \right|^2 . \tag{2.181}$$

2. Inversion formula: Given the WVD $W_s(t,f)$ of a signal $s(t)$, it is possible to recover $s(t)$ from $W_s(t,f)$, up to a scalar factor, using the following inversion formula:

$$s(t) = \frac{1}{s^*(0)} \int_{-\infty}^{\infty} W_s\left(\frac{t}{2}, f\right) e^{j2\pi ft}df. \tag{2.182}$$

3. Moments: The instantaneous frequency $f_s(t)$ of a complex signal $s(t)$ is the center of gravity of its WVD along f:

$$f_s(t) = \frac{\int_{-\infty}^{\infty} fW_s(t,f)df}{\int_{-\infty}^{\infty} W_s(t,f)df} . \tag{2.183}$$

By duality, the group delay $t_s(f)$ is the center of gravity of the WVD along t:

$$t_s(f) = \frac{\int_{-\infty}^{\infty} tW_s(t,f)dt}{\int_{-\infty}^{\infty} W_s(t,f)dt}. \tag{2.184}$$

References

[1] Andrews, M. R., Mitra, P. P. , de Carvalho, R., "Tripling the capacity of wireless communications using electromagnetic polarization," *Nature*, vol. 409, Jan. 2001, pp. 316–318.

[2] Collin, R.E. , *Antennas and Radiowave Propagation*, New York: McGraw-Hill International Editions, 1985.

[3] Martone, M. , *Multiantenna Digital Radio Transmission*, Norwood MA: Artech House, 2002.

[4] Giannakis, G. B., Tepedelenlioglu, "Basis expansion models and diversity techniques for blind identification and equalization of time-varying channels," *Proc. of the IEEE*, Vol. 86, No. 10, Oct. 1998, pp. 1969–1986.

[5] Bello, P. A., "Characterization of randomly time-variant linear channels," *IEEE Trans. on Circuits and Systems*, vol. CS-11, No. 4, Dec. 1963, pp. 360–393.

[6] Proakis, J., *Digital Communications*, (4th edition), New York: McGraw Hills, 2000.

[7] Jakes, W. C., *Microwave Mobile Communications*, Section 1.5, Piscataway, NJ: IEEE Press, 1974.

[8] Liberti, J. C., Rappaport, T. S., "A geometrically based model for line-of-sight multipath radio channels," *IEEE 46th Vehicular Technology Conference*, Vol. 2, Atlanta (GA), 28 Apr–1 May 1996, pp. 844–848.

[9] Ertel, R. B., et al., "Overview of spatial channel models for antenna array communication systems," *IEEE Personal Communications*, Feb. 1998, pp. 10–22.

[10] Da-Shan Shiu, Foschini, G. J., Gans, M. J., Kahn, J. M., "Fading correlation and its effect on the capacity of multielement antennas systems," *IEEE Trans. on Communications*, Vol. 48, No. 3, March 2000, pp. 502–513.

[11] Gesbert, D., Bölcskei, H., Gore, D., Paulraj, A., "MIMO wireless channels: Capacity and performance," *Proc. of Globecom 2000*, Vol. 2, Nov. 2000, pp. 1083–1088.

[12] Abdi, A., Kaveh, M., "A space-time correlation modeling of multielement antenna systems in mobile fading channels," *IEEE Journal on Selected Areas in Communications*, Vol. 20, No. 3, April 2002, pp. 550–560.

[13] Papoulis, A., *Signal Analysis*, New York: Mc Graw Hill, 1977.

[14] Sirovich, L., Knight, B. W., "On the eigentheory of operators which exhibit a slow variation," *Quart. Appl. Math.*, Vol. 38, 1980, pp. 469–488.

[15] Barbarossa, S., Scaglione, A., "Optimal precoding for transmissions over linear time-varying channels," *Proc. of GLOBECOM '99*, Rio de Janeiro, 1999, Vol. 5, pp. 2545–2549.

[16] Auger, F., Flandrin, P., "Improving the Readability of Time-Frequency and Time-Scale Representations by the Reassignment Method," *IEEE Trans. on Signal Proc.*, Vol. 43, No. 5, May 1995, pp. 1068-1089.

[17] Grenwood, D. J., Hanzo, L., "Characterization of Mobile Radio Channels," in *Mobile Radio Communications*, Steele, R., (ed), Piscataway, NJ: IEEE Press, 1992.

[18] Lee, W. C. Y., *Mobile Communications Design Fundamentals*, (second ed.), New York: Wiley Series in Telecommunications, 1993.

[19] Da-shan Shiu, *Wireless Communication Using Dual Antenna Arrays*, Boston, MA: Kluwer Academic Publisher, 1999.

[20] Svantesson, T., "A physical MIMO radio channel model for multi-element multi-polarized antenna systems," *Proc. of VTC Fall*, Vol. 2, 2001, pp. 1083–1087.

[21] Yu, K., et al., "A wideband statistical model for NLOS inddor MIMO channels," *Proc. of VTC Spring*, Vol. 1, 2002, pp. 370–374.

[22] Cohen, L., *Time-Frequency Analysis*, Englewood Cliffs, NJ: Prentice-Hall, 1995.

[23] Flandrin, P., *Time-Frequency/Time-Scale Analysis (Wavelet Analysis and Its Applications)*, San Diego, CA: Academic Press, 1999.

Chapter 3

Multicarrier Systems

3.1 INTRODUCTION

In this chapter we review the basic principles of single antenna multicarrier (MC) systems, considering both single user as well as multiuser systems. This chapter will be propaedeutic for the ensuing chapters, where we will address the multiple antenna MC case. We will devote special attention to orthogonal frequency division multiplexing (OFDM), whose success is testified by its being chosen as the standard radio interface for digital audio broadcasting (DAB), terrestrial digital video broadcasting (DVB), wireless local area networks (LANs), and wireless metropolitan area networks (MANs). OFDM is also under consideration for the fourth generation cellular systems. The primary reason why OFDM and, more generally, MC systems are becoming so popular is that they lead to a very simple receiver structure for broadband transmissions, with good performance. An excellent introductory tutorial on MC modulation is [1]. A more detailed description of systems employing MC modulation is given in [2], [3].

In this chapter, we review first, in Section 3.2, the basic principles underlying block transmission systems, as opposed to serial schemes. Then, we will concentrate on OFDM systems in Section 3.3, where we will show how the combination of OFDM with forward error correction coding creates indeed a simple, yet powerful tool to tackle, at the same time, channel equalization and error correction. We will then provide the criteria to design the main parameters of an OFDM system in Section 3.4. We will consider the channel estimation and the synchronization problems in Sections 3.5 and 3.6, respectively. Multiuser systems, based on MC interfaces, will be reviewed in Section 3.7. Finally, we will illustrate a case study of a broadband wireless access system, where OFDM is the basic building block: the standard IEEE 802.11a.

3.2 BLOCK TRANSMISSION SYSTEMS

The main principle underlying a block transmission system is that the stream of bits to be transmitted, instead of being sent in a serial manner, is parsed into consecutive blocks, which are coded and decoded, at the receiver, independently from each other. The reason why a block strategy may be preferred to a serial one can be explained through the following example. Let us consider a linear time-invariant channel described by a finite impulse response (FIR) $h(n)$. Transmitting a sequence $x(n)$ through this channel generates, at the receiver, a sequence $y(n)$ equal to the convolution between $x(n)$ and $h(n)$, that is,

$$y(n) = \sum_{k=-\infty}^{\infty} h(k)x(n-k). \tag{3.1}$$

The convolution shows that, transmitting over a time dispersive channel entails the insurgence of intersymbol interference (ISI). One of the main purposes of the receiver is to equalize the channel. If we decide to use a linear equalizer, we need to design a filter $g(n)$, such that the convolution between $h(n)$ and $g(n)$ is proportional to a unit pulse. Introducing the transfer functions $H(z)$ and $G(z)$, defined as the $\mathcal{Z}$-transform of $h(n)$ and $g(n)$, perfect equalization requires that

$$H(z)G(z) = z^{-d}, \tag{3.2}$$

where d incorporates a possible delay. Clearly, if $H(z)$ is a polynomial of degree L, $G(z)$ cannot be a polynomial of finite order, so that the equalization filter is necessarily an infinite impulse response (IIR) filter. This poses a practical problem, as an IIR filter can be unstable, depending on the zeros of $H(z)$, which are clearly not under control.

Let us consider now a block system. In block transmission systems, besides ISI, there is also, in general, interblock interference (IBI). However, assuming that the discrete-time equivalent channel is FIR of order L, inserting *guard intervals* between consecutive transmitted blocks of length at least equal to L, it is possible to get rid of IBI simply by discarding the first L samples of each received block. In this way, after discarding the guard interval at the receiver, we end up with a block input/output relationship that can always be written in a matrix form

$$\boldsymbol{y} = \boldsymbol{H}\boldsymbol{x} + \boldsymbol{v}, \tag{3.3}$$

where $\boldsymbol{H}$ is the channel matrix and $\boldsymbol{v}$ is the noise vector (details about the structure of $\boldsymbol{H}$ will be given in the following, since it depends on the transmission strategy).

Equation (3.3) means that, in a block system, the symbol vector can be recovered by inverting the system (3.3), that is without resorting to any IIR filter. In this way, we avoid the problem of a potentially unstable equalization filter. What we pay for this advantage is that we require now the inversion of a matrix, which can be troublesome, especially for real-time implementations. Therefore, many efforts are devoted, in block systems, to devise a transmission strategy that simplifies symbol recovery as much as possible. We will now review the basic strategies in detail.

In a block transmission system, given a sequence of information bits $b(n)$ to be transmitted, the sequence is first parsed into blocks. Each block of bits is encoded using some kind of forward error correction (FEC) code. The encoded bits are then mapped onto (typically complex) symbols belonging to a finite order constellation. The result is a set of, let us say M symbols, which form a vector $\boldsymbol{s}(n)$ (the index n indicates the block). The vector $\boldsymbol{s}(n)$ is then further encoded using a linear precoder whose operation can be described as the multiplication of the vector $\boldsymbol{s}(n)$ by the precoding matrix $\boldsymbol{F}$ of size $P \times M$. The result is an P-size vector $\boldsymbol{x}(n) = \boldsymbol{F}\boldsymbol{s}(n)$. Each block is then sent, one after the other.

At the receiver, we get a sequence of samples $\boldsymbol{y}(n)$. If we parse this sequence in blocks of size P, we obtain the block input/output relationship:

$$\bar{\boldsymbol{y}}(n) = \sum_{l=-\infty}^{0} \boldsymbol{H}_l \boldsymbol{x}(n-l) + \bar{\boldsymbol{v}}(n), \tag{3.4}$$

where $\bar{\boldsymbol{v}}(n)$ is the receiver noise vector and the $P \times P$ matrices $\boldsymbol{H}_l$ are defined as:

$$\boldsymbol{H}_l := \begin{pmatrix} h(lP) & h(lP-1) & \cdots & h(lP-P+1) \\ h(lP+1) & h(lP) & \ddots & h(lP-P+2) \\ \vdots & \ddots & \ddots & \vdots \\ h(lP+P-1) & h(lP+P-2) & \cdots & h(lP) \end{pmatrix}. \tag{3.5}$$

The summation in (3.4) reveals the presence of IBI. In general, in fact, if the channel is dispersive in time, every received block depends on all previously transmitted blocks. The relationship (3.4) is fully general. However, in most practical block transmission systems, the block length P is greater than the channel order L. In such a case, in each received block there is IBI only from the previous block. Therefore, under the assumption $P > L$, (3.4) simplifies into

$$\bar{\boldsymbol{y}}(n) = \boldsymbol{H}_0 \boldsymbol{x}(n) + \boldsymbol{H}_1 \boldsymbol{x}(n-1) + \bar{\boldsymbol{v}}(n), \tag{3.6}$$

where the matrices $\boldsymbol{H}_0$ and $\boldsymbol{H}_1$ are $P \times P$ and have the following structure:

$$\boldsymbol{H}_0 := \begin{pmatrix} h(0) & 0 & 0 & \cdots & 0 \\ \vdots & h(0) & 0 & \cdots & 0 \\ h(L) & \ddots & \ddots & \ddots & \vdots \\ \vdots & \ddots & \ddots & \ddots & 0 \\ 0 & \cdots & h(L) & \cdots & h(0) \end{pmatrix},$$

$$\boldsymbol{H}_1 := \begin{pmatrix} 0 & \cdots & h(L) & \cdots & h(1) \\ \vdots & \ddots & 0 & \ddots & \vdots \\ 0 & \ddots & \ddots & \ddots & h(L) \\ \vdots & \ddots & \ddots & \ddots & \vdots \\ 0 & \cdots & 0 & \cdots & 0 \end{pmatrix}. \tag{3.7}$$

The term $\boldsymbol{H}_1\boldsymbol{x}(n-1)$ in (3.6) represents the interference to the nth block from the $(n-1)$th block. The simplest way to get rid of IBI is to make the last L entries of each transmitted block equal to zero. In this way, $\boldsymbol{H}_1\boldsymbol{x}(n-1) = \mathbf{0}$ by construction. This implies the insertion of null guard intervals between consecutive blocks. Then, setting $N = P - L$ and discarding the first L samples of each received block, we get the following I/O relationship

$$\boldsymbol{y}(n) = \boldsymbol{H}\boldsymbol{F}\boldsymbol{s}(n) + \boldsymbol{v}(n), \tag{3.8}$$

where $\boldsymbol{H}$ is now $N \times N$ and all the vectors have size N. The structure of $\boldsymbol{H}$ is as $\boldsymbol{H}_0$ in (3.7), except that its size is now $N \times N$.

This insertion of the null guard intervals is paid in terms of transmission rate. In fact, the insertion implies that to transmit N useful symbols, we need to transmit $P = N + L$ samples. Hence, the rate is reduced by a factor

$$\epsilon := \frac{N}{N+L}. \tag{3.9}$$

3.2.1 Symbol Detection

Given the block $\boldsymbol{y}(n)$ in (3.8), we can recover the symbols $\boldsymbol{s}(n)$ using for example the zero-forcing (ZF) or the minimum mean square error (MMSE) decoder.

ZF detector

The ZF decoder derives first a ZF estimate and then it takes decisions on the basis of the ZF estimate. The ZF estimate yields a vector equal to the true vector $\boldsymbol{s}(n)$ in the absence of noise:

$$\hat{\boldsymbol{s}}_{\mathrm{ZF}}(n) = (\boldsymbol{HF})^{\dagger}\boldsymbol{y}(n), \tag{3.10}$$

where † denotes pseudo-inverse, that is $(\boldsymbol{HF})^{\dagger} = (\boldsymbol{F}^H\boldsymbol{H}^H\boldsymbol{HF})^{-1}\boldsymbol{F}^H\boldsymbol{H}^H$ when $\boldsymbol{HF}$ is full column rank. The decision about which symbols have been transmitted are then taken using a scalar detector that associates to each entry of $\hat{\boldsymbol{s}}_{ZF}(n)$ the nearest symbol belonging to the constellation of the transmitted symbols.

MMSE detector

As with the ZF decoder, the MMSE decoder provides the MMSE estimate first and then it applies a scalar detector to each entry of the MMSE estimate. Denoting with $\hat{s}_k(n) = \boldsymbol{g}_k^H\boldsymbol{y}(n)$ the linear MMSE estimate of the kth symbol $s_k(n)$, belonging to the nth block, the vector $\boldsymbol{g}_k$ is found as the vector that minimizes the MSE

$$\mathrm{MSE}_k = E\{|\boldsymbol{g}_k^H\boldsymbol{y}(n) - s_k(n)|^2\}. \tag{3.11}$$

Introducing the vector $\boldsymbol{e}_k$, with all null entries except the k-th one, so that $s_k(n) = \boldsymbol{e}_k^H\boldsymbol{s}(n)$, indicating with $\boldsymbol{C}_{ss}$ and $\boldsymbol{C}_{nn}$ the covariance matrices of the transmitted symbols and noise, respectively, and assuming that the transmitted symbols are uncorrelated from the noise, the MSE is equal to

$$\mathrm{MSE}_k = (\boldsymbol{g}_k^H\boldsymbol{HF} - \boldsymbol{e}_k)\boldsymbol{C}_{ss}(\boldsymbol{F}^H\boldsymbol{H}^H\boldsymbol{g}_k - \boldsymbol{e}_k) + \boldsymbol{g}_k^H\boldsymbol{C}_{nn}\boldsymbol{g}_k. \tag{3.12}$$

Equating the gradient of MSE_k with respect to $\boldsymbol{g}_k^H$ to zero, we obtain the vector

$$\boldsymbol{g}_k = (\boldsymbol{HFC}_{ss}\boldsymbol{F}^H\boldsymbol{H}^H + \boldsymbol{C}_{nn})^{-1}\boldsymbol{HFC}_{ss}\boldsymbol{e}_k. \tag{3.13}$$

Stacking all column vectors $\boldsymbol{g}_k$ one after each other, we get the MMSE matrix

$$\boldsymbol{G} = (\boldsymbol{HFC}_{ss}\boldsymbol{F}^H\boldsymbol{H}^H + \boldsymbol{C}_{nn})^{-1}\boldsymbol{HFC}_{ss}. \tag{3.14}$$

In the most common case where the symbols are uncorrelated, with variance σ_s^2 and the noise samples are also uncorrelated, with variance σ_n^2, the MMSE decoder is

$$\hat{\boldsymbol{s}}_{\mathrm{MMSE}}(n) = \boldsymbol{G}^H\boldsymbol{y}(n) = \boldsymbol{F}^H\boldsymbol{H}^H\left(\boldsymbol{HFF}^H\boldsymbol{H}^H + \frac{\sigma_n^2}{\sigma_s^2}\boldsymbol{I}\right)^{-1}\boldsymbol{y}(n). \tag{3.15}$$

Applying the matrix inversion lemma (6.14), the estimate can be rewritten as

$$\hat{\boldsymbol{s}}_{\mathrm{MMSE}}(n) = \left(\boldsymbol{F}^H\boldsymbol{H}^H\boldsymbol{HF} + \frac{\sigma_n^2}{\sigma_s^2}\boldsymbol{I}\right)^{-1}\boldsymbol{F}^H\boldsymbol{H}^H\boldsymbol{y}(n). \tag{3.16}$$

This expression is more suitable to implement than (3.15) when $P > M$, i.e., when the number of encoded symbols M is strictly less than the blocklength, because in such a case the size of the matrix to be inverted is smaller than in (3.15). It is useful to verify that (3.16) reduces to (3.15), when the noise contribution becomes negligible (i.e., $\sigma_n^2 \ll \sigma_s^2$). Hence, at high SNR, the ZF and the MMSE estimators tend to coincide.

3.2.2 Guard Intervals

The main problem related to the ZF and MMSE decoders shown above is complexity. In fact, the computation of a pseudo-inverse or of an inverse is troublesome to implement in real-time. The number of complex multiplication needed to compute $\boldsymbol{H}^\dagger$ is in fact in the order of N^3. Even though the computation of $\boldsymbol{H}^\dagger$ needs to be done only once within the channel coherence time, in broadband applications such computation would represent a real bottleneck. Hence, all methods able to avoid the matrix inversion are clearly welcome. We show now two methods that make the channel inversion very simple (see [1] for more details).

Zero padding and overlap

The first method consists of introducing null guard intervals between consecutive blocks, of length L, at least equal to the channel order. At the receiver, instead of discarding the initial L samples of the received block, we add the last L samples to the first L ones. If we take the first N samples of the resulting block, we obtain a vector $\boldsymbol{y}(n)$ of size N related to the transmitted block by the same relationship as in (3.8), except that now the channel matrix has the following structure

$$\boldsymbol{H} := \begin{pmatrix} h(0) & 0 & \cdots & 0 & h(L) & h(L-1) & \cdots & h(1) \\ h(1) & h(0) & 0 & \ddots & 0 & h(L) & \cdots & h(2) \\ \vdots & \ddots & \ddots & \ddots & \ddots & \ddots & \ddots & \vdots \\ h(L-1) & \ddots & \ddots & \ddots & \ddots & \ddots & \ddots & h(L) \\ h(L) & \ddots & \ddots & \ddots & \ddots & \ddots & \ddots & 0 \\ 0 & \ddots & \ddots & \ddots & \ddots & \ddots & \ddots & \vdots \\ \vdots & \ddots & \ddots & \ddots & \ddots & \ddots & \ddots & 0 \\ 0 & \cdots & 0 & h(L) & \cdots & \cdots & \cdots & h(0) \end{pmatrix}. \tag{3.17}$$

This matrix is Toeplitz and circulant. Such a structure simplifies the inversion of $\boldsymbol{H}$ drastically. In fact, a Toeplitz circulant $\boldsymbol{H}$ can be always diagonalized as follows

$$\boldsymbol{H} = \boldsymbol{W}\boldsymbol{\Lambda}\boldsymbol{W}^H, \tag{3.18}$$

where $\boldsymbol{W}$ has entries $W(k,l) = \exp(j2\pi kl/N)/\sqrt{N}$ and $\boldsymbol{\Lambda}$ is a diagonal matrix whose diagonal entries are the values of the channel transfer function:

$$\Lambda(i,i) = H(i) = \sum_{l=0}^{L} h(l)e^{-j2\pi il/N}. \tag{3.19}$$

The matrix $\boldsymbol{W}$ is unitary, so that $\boldsymbol{W}^{-1} = \boldsymbol{W}^H$. From (3.18), the inverse of $\boldsymbol{H}$ in (3.17), if it exists, is simply $\boldsymbol{H}^{-1} = \boldsymbol{W}\boldsymbol{\Lambda}^{-1}\boldsymbol{W}^H$. This implies that the only inversion necessary to compute the inverse of $\boldsymbol{H}$ is the inversion of $\boldsymbol{\Lambda}$, but since $\boldsymbol{\Lambda}$ is a diagonal matrix, the inversion is obtained with only N complex divisions. As we will show later on, the multiplication by $\boldsymbol{W}$ (or $\boldsymbol{W}^H$) can also be implemented in an efficient manner, using the fast Fourier transform (FFT) algorithm.

Cyclic prefix

The second method consists in inserting, between any two successive blocks, instead of a null guard interval, a cyclic prefix (CP). Inserting a CP means to copy the last L samples of each block in front of the block. At the receiver, if we discard the first L samples, we get an I/O relationship still equal to (3.8), with $\boldsymbol{H}$ as in (3.17). Hence, once again, the channel matrix is Toeplitz and circulant, and thus its inversion is very simple.

Therefore, in general in a block transmission where we insert either a null guard interval or a CP, the I/O relationship can always be written as

$$\boldsymbol{y}(n) = \boldsymbol{H}\boldsymbol{F}\boldsymbol{s}(n) + \boldsymbol{w}(n), \tag{3.20}$$

where $\boldsymbol{H}$ is an $N \times N$ Toeplitz circulant matrix and $\boldsymbol{F}$ is an $N \times M$ encoding matrix. This expression is valid for any linear coding strategy, as there are no restrictions on the choice of $\boldsymbol{F}$. We will exploit these degrees of freedom to optimize the system performance later on.

From (3.20), using (3.18), if we multiply the received vector $\boldsymbol{y}(n)$ by $\boldsymbol{W}^H$, from the left side, we get

$$\boldsymbol{z}(n) := \boldsymbol{W}^H\boldsymbol{y}(n) = \boldsymbol{\Lambda}\boldsymbol{W}^H\boldsymbol{F}\boldsymbol{s}(n) + \boldsymbol{w}'(n) = \boldsymbol{\Lambda}\tilde{\boldsymbol{F}}\boldsymbol{s}(n) + \boldsymbol{w}'(n), \tag{3.21}$$

where $\tilde{\boldsymbol{F}} := \boldsymbol{W}^H \boldsymbol{F}$ and $\boldsymbol{w}'(n) = \boldsymbol{W}^H \boldsymbol{w}(n)$. Given the structure of $\boldsymbol{W}$, the product $\boldsymbol{W}^H \boldsymbol{x}$ is equivalent to computing the discrete Fourier transform (DFT) of the vector $\boldsymbol{x}$. If N is a power of two, the DFT can be implemented very efficiently using the FFT algorithm, which is now available also on a chip. Hence, the matrix $\tilde{\boldsymbol{F}}$ has, as columns, the DFT of each column of $\boldsymbol{F}$.

We consider now the choice of $\boldsymbol{F}$ that provides the simplest equalization scheme.

3.3 OFDM

OFDM is characterized by having a precoding matrix $\boldsymbol{F} = \boldsymbol{W}$. This is the best choice as far as complexity is concerned, as $\tilde{\boldsymbol{F}} = \boldsymbol{I}$. As a consequence, from (3.21) we have

$$\boldsymbol{z}(n) = \boldsymbol{\Lambda s}(n) + \boldsymbol{w}'(n), \tag{3.22}$$

where $\boldsymbol{w}'(n)$ has exactly the same statistical properties as $\boldsymbol{w}(n)$. Since $\boldsymbol{\Lambda}$ is diagonal, each sample $z_k(n)$ contains only one symbol (i.e., there is no ISI). In particular,

$$z_k(n) = \Lambda(k,k)s_k(n) + w'_k(n) = H(k)s_k(n) + w'_k(n). \tag{3.23}$$

Therefore, OFDM converts a time-dispersive channel into a set of N flat-fading channels, easy to equalize. The equalization can in fact be carried out symbol-by-symbol. In particular, we have the zero-forcing (ZF) equalizer

$$\hat{s}_k(n) = \frac{1}{H(k)} z_k(n) \tag{3.24}$$

or the minimum mean square error (MMSE) equalizer

$$\hat{s}_k(n) = \frac{\sigma_s^2 H^*(k)}{|H(k)|^2\sigma_s^2 + \sigma_n^2} z_k(n). \tag{3.25}$$

The ZF equalizer guarantees that, in the absence of noise, $z_k(n) = s_k(n)$. However, in the presence of noise, the division by $H(k)$ in (3.24) can enhance the noise excessively. This happens when the channel transfer function, in the z-domain, $H(z)$ has nulls close to the unit circle. In particular, the division is not even possible when $H(k) = 0$, that is, when $H(z)$ has a null exactly on the grid $\exp(j2\pi n/N)$, with n integer. Conversely, the MMSE equalizer does not suffer from the same noise enhancement effects of the ZF equalizer.

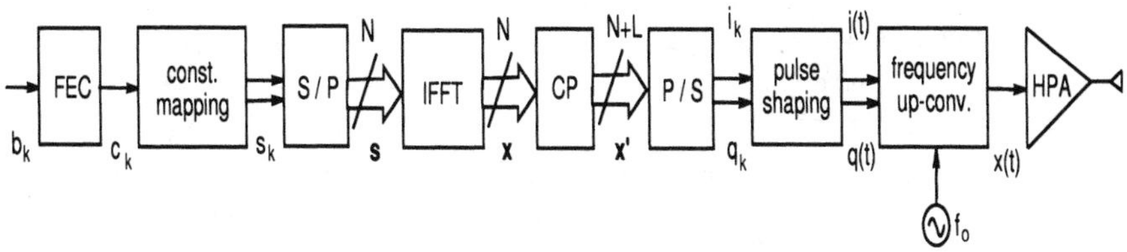

Figure 3.1 OFDM transmitter block diagram.

In summary, the transmitter block diagram of an OFDM system is depicted in Figure 3.1. The sequence of information bits b_k passes through a forward error correction (FEC) block that produces the encoded bit stream c_k. These bits are then mapped onto constellation symbols s_k. In general, we assume a complex constellation (i.e., QPSK or QAM) so that at the output of the constellation mapper, we have two (real) output streams. The sequence of symbols s_k is then parsed into consecutive blocks of N symbols, through the serial-to-parallel (S/P) converter. Each block is a vector $\boldsymbol{s}$ of size N. The main distinguishing feature of an OFDM transmitter is that the vector $\boldsymbol{s}$ is sent to a processor that computes the inverse fast Fourier transform (IFFT) of its input[1]. Specifically, if we consider the generic nth block, the vector $\boldsymbol{x}(n)$ at the output of the IFFT block is related to the input vector $\boldsymbol{s}(n)$ by the relationship

$$\boldsymbol{x}(n) = \boldsymbol{W}\boldsymbol{s}(n), \tag{3.26}$$

where $\boldsymbol{W}$ is the DFT matrix, whose entries are $W_{kl} = e^{j2\pi kl/N}/\sqrt{N}$. We include the normalization term $\sqrt{N}$, so that the energy of $\boldsymbol{x}(n)$ is equal to the energy of $\boldsymbol{s}(n)$. More specifically, the entries of the vector $\boldsymbol{x}$ are related to the entries of $\boldsymbol{s}$ by the inverse DFT

$$x_l(n) = \frac{1}{\sqrt{N}} \sum_{k=0}^{N-1} s_k(n)\, e^{j2\pi \frac{kl}{N}}, \quad l = 0, 1, \ldots, N-1. \tag{3.27}$$

The successive block appends a cyclic prefix (CP) of length L to each block. This corresponds to copying the last L samples of the vector $\boldsymbol{x}(n)$ in front of the block. In formulas, after insertion of the CP, each block has a duration $P = N + L$ and its samples are

$$x_l(n) = \frac{1}{\sqrt{N}} \sum_{k=0}^{N-1} s_k(n)\, e^{j2\pi \frac{kl}{N}}, \quad l = -L, \ldots, -1, 0, 1, \ldots, N-1. \tag{3.28}$$

1 The block size N is indeed always chosen so that the DFT can be efficiently computed using the FFT.

The block is then put in serial form and transmitted. We assume in general a complex constellation, so that the sequence is composed of the in-phase (I) $i_l(n)$ and quadrature (Q) $q_l(n)$ sequences, with $x_l(n) = i_l(n) + jq_l(n)$. Each sequence passes through the shaping filter, to form the continuous time signals $i_n(t) = \sum_{l=-L}^{N-1} i_l(n)g_T(t - nT_s)$ and $q_n(t) = \sum_{l=-L}^{N-1} q_l(n)g_T(t - nT_s)$. Then both I and Q signals are modulated so as to generate the transmitted signal (relative to the nth block) $x_n(t) = i_n(t)\cos(2\pi f_0 t) - q_n(t)\sin(2\pi f_0 t)$.

To get a physical insight about OFDM, it is useful to consider the baseband continuous-time (CT) signal. The CT baseband signal transmitted by an OFDM system, within each block, has the following approximate form

$$x_n(t) = \frac{1}{\sqrt{N}} \sum_{k=0}^{N-1} s_k(n)\, e^{j2\pi f_k t}, \quad t \in [-LT, NT], \tag{3.29}$$

where $f_k = k/NT$. Considering that, at the receiver, L samples are discarded, the effective duration is, equivalently, NT. Therefore, each symbol is carried by a complex sinusoid, whose spectrum is $\text{sinc}[\pi NT(f - f_k)]$. The spectra corresponding to different symbols are depicted in Figure 3.2. Each curve is the modulus of the spectrum carrying the symbol labeled near the curve. The spectra are clearly overlapping, which means that the system is not exactly FDMA. However, the waveforms carrying different symbols are orthogonal within a time window of duration NT. The orthogonality is indeed what is really important to separate the symbols at the receiver and in fact it is this basic property that gives the name to OFDM. The solid arrows in Figure 3.2 represent the points where we sample the spectrum at the receiver, in case of perfect frequency synchronization. In such a case, we see that each sample contains information only about one symbol. Figure 3.2 anticipates also one of the weaknesses of OFDM: If the receiver is not perfectly synchronous with the transmitter (i.e., we sample at the points represented by the dashed arrows), each sample contains intersymbol interference (ISI). Hence, OFDM pays some of its advantages with an increased sensitivity with respect to synchronization errors. We will study the OFDM synchronization problem in more detail in Section 3.6.

3.3.1 Performance of Uncoded OFDM

The performance is usually evaluated using the bit error rate (BER) or symbol error rate (SER). From (3.23), since there is no ISI, the SER can be computed in closed form. In particular, we assume that the noise is a complex Gaussian random variable with zero mean and variance σ_n^2. If we transmit a PAM symbol of dimension M on each subcarrier and the symbols are equally probable, the SER on each subcarrier

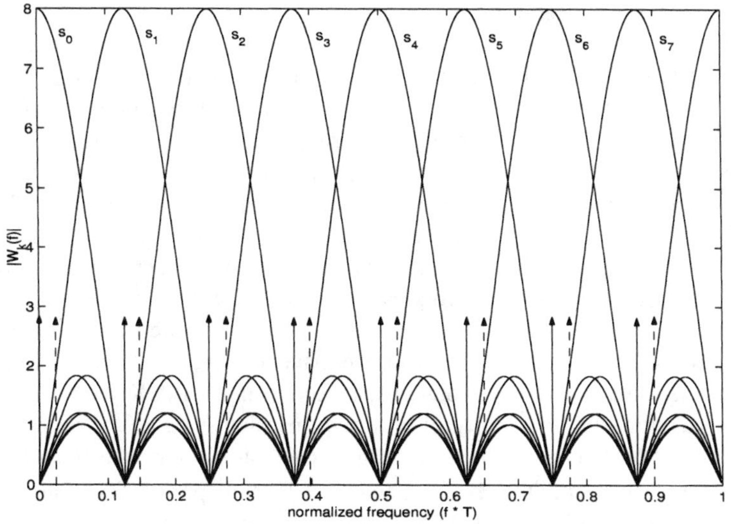

Figure 3.2 Spectra of the waveforms carrying OFDM symbols.

is [4]:

$$\mathrm{SER}_{M-\mathrm{PAM}}(k) = \frac{M}{M-1}\mathrm{erfc}\left(\sqrt{\frac{3\sigma_s^2|H(k)|^2}{(M^2-1)\sigma_n^2}}\right), \tag{3.30}$$

where σ_s^2 is the variance (power) of the transmitted symbols. Transmitting QAM symbols belonging to a square constellation of size M^2, the SER on each subcarrier is $\mathrm{SER}_{M^2-\mathrm{QAM}}(k) = 2\mathrm{SER}_{M-\mathrm{PAM}}(k) - \mathrm{SER}_{M-\mathrm{PAM}}(k)^2$, with $\mathrm{SER}_{M-\mathrm{PAM}}(k)$ given by (3.30).

From the previous expression of the symbol-error rate, it is evident that an uncoded OFDM system is very sensitive to channel frequency selectivity. From (3.30) in fact, it is clear that over the subchannels where the transmitted signal experiences the strongest attenuation, the SER is high, and then it conditions the performance of the overall block. To prevent this circumstance and then add robustness against channel frequency selectivity, we must avoid to send any information bit on one subcarrier only. The idea is to spread the information bits over as many subcarriers as possible, so that even if one subcarrier is strongly attenuated, there is a chance of recovering the information bit from the other subcarriers. This can be achieved using a redundant or a nonredundant encoder. We analyze the two possibilities in the following sections.

3.3.2 OFDM with Redundant Coding

The most typical OFDM system is, as depicted in Figure 3.1, a coded-OFDM (COFDM) system, where the information bits pass first through a FEC encoder. In OFDM systems, the combination of FEC and equalization is particularly interesting because, besides the obvious correction capabilities of the FEC encoder, FEC helps also against channel selectivity. To explain this concept, we consider, for simplicity, a BPSK transmission scheme. In COFDM, every set of, let us say k, bits is encoded and gives rise to a codeword of n bits. Let us assume a block error correction code capable of correcting up to t errors, with $n - k = 2t + 1$. In every transmitted OFDM block, we transmit a codeword. The encoded bits are transmitted through the parallel subcarriers. At the receiver, we make an error on the coded word only if there are more than t erroneous bits. Hence, using a BPSK modulation, after error correction, the probability of a correct decision is equal to the probability of making up to t errors on the transmitted codeword. In formulas, denoting with $p_e(l)$ the BER over the lth subcarrier and with $P_e^{(c)}$ the probability of error on the transmitted codeword, the probability of a correct decision is

$$\begin{aligned} 1 - P_e^{(c)} &= \prod_{l=1}^{n}[1 - p_e(l)] \left[1 + \sum_{k_1=1}^{n} \frac{p_e(k_1)}{1 - p_e(k_1)} \right. \\ &+ \sum_{k_1 \neq k_2}^{n} \frac{p_e(k_1)p_e(k_2)}{[1 - p_e(k_1)][1 - p_e(k_2)]} \\ &+ \left. \cdots + \sum_{k_1 \neq k_2 \neq ..k_t}^{n} \frac{p_e(k_1)\cdots p_e(k_t)}{[1 - p_e(k_1)]\cdots[1 - p_e(k_t)]} \right], \end{aligned} \tag{3.31}$$

where the first term on the left-hand side is the probability of making no errors, the second term is the probability of making one error, and so on; the last term is the probability of making t errors (the last summation is extended to all indexes $k_1, \ldots k_t$ different among them).

This combination of FEC and OFDM is particularly appealing because it allows us to get two of the most distinguishing advantages of OFDM and FEC coding, namely simplicity of the equalizer and error correction coding. What is most interesting is that the combination here introduces robustness against frequency selective fading. In fact, with this combination, if the number of subchannels with high attenuation is less than t, the number of bits that the decoder is able to correct, the system is robust against frequency selectivity even without knowing which are the most attenuated subchannels.

3.3.3 OFDM with Nonredundant Linear Precoding

The COFDM system described above leads to a simple equalization and decoding scheme, robust against frequency selectivity. Its only price is the redundancy introduced by the FEC. We show now that, with a slight increase of complexity, we can increase the robustness of an OFDM system, even without introducing any redundancy.

If we look back at (3.20), we recall that to induce a Toeplitz and circulant structure of the channel matrix, it is sufficient to insert a CP. Since there are no restrictions on the choice of the matrix $\boldsymbol{F}$, we can choose $\boldsymbol{F}$ in order to optimize some system performance parameter. The optimal design depends, of course, on the optimality criterion and on the amount of channel knowledge available at the transmitter side. In some systems, it is possible to acquire such a knowledge and exploit it advantageously. We will study in detail when this is possible and the corresponding optimal $\boldsymbol{F}$ in Chapter 6. However, in some cases, it makes no sense to optimize the transmission with respect to a specific channel. This is the case, for example, of TV broadcasting, where the transmitter broadcasts the same data to all the listeners. Each receiver gets its data through its own channel and thus it is not meaningful to optimize the transmission strategy with respect to a specific channel. Nevertheless, even when the transmitter does not know the channel, it is possible to design the precoding matrix $\boldsymbol{F}$ in order to optimize some performance parameter. Clearly the design depends on the optimality criterion. Since BER (or SER) is one of the ultimate targets in the design of a digital communication system, we show how to choose $\boldsymbol{F}$ in order to minimize the average SER.

Before deriving the optimal $\boldsymbol{F}$, it is useful to state the main assumptions. We assume that the transmitted symbols are uncorrelated and have the same variance (i.e., the covariance matrix of the vector $\boldsymbol{s}$ is $\boldsymbol{R}_s = \sigma_s^2\boldsymbol{I}$). We require that $\boldsymbol{F}$ must allow for perfect symbol recovery and, moreover, that its application does not alter the covariance matrix of the symbol vector $\boldsymbol{s}$. This last property, in particular, implies the following constraint on $\boldsymbol{F}$

$$\boldsymbol{F}\boldsymbol{R}_s\boldsymbol{F}^H = \sigma_s^H\boldsymbol{F}\boldsymbol{F}^H = \sigma_s^H\boldsymbol{I}, \tag{3.32}$$

that is satisfied if and only if $\boldsymbol{F}\boldsymbol{F}^H = \boldsymbol{I}$. Note that, in general, a matrix $\boldsymbol{F}$ satisfying (3.32) may be rectangular. However, the additional requirement on lossless symbols recovery implies that $\boldsymbol{F}$ must be a *unitary* matrix. Under this assumption, the MMSE decoder applied to (3.21) assumes the following expression

$$\hat{\boldsymbol{s}}(n) = \tilde{\boldsymbol{F}}^H\boldsymbol{\Lambda}^*\left(|\boldsymbol{\Lambda}|^2 + \frac{1}{\gamma}\boldsymbol{I}\right)^{-1}\boldsymbol{z}(n), \tag{3.33}$$

where $|\boldsymbol{\Lambda}|^2 = \boldsymbol{\Lambda}\boldsymbol{\Lambda}^*$, and $\gamma = \sigma_s^2/\sigma_n^2$ and $\tilde{\boldsymbol{F}} = \boldsymbol{W}^H\boldsymbol{F}$. The corresponding mean square error MSE_k on the kth symbols is[2]

$$\text{MSE}_k = \sigma_s^2\left[1 - \tilde{\boldsymbol{f}}_k^{\ H}\boldsymbol{W}|\boldsymbol{\Lambda}|^2\left(|\boldsymbol{\Lambda}|^2 + \frac{1}{\gamma}\boldsymbol{I}\right)^{-1}\tilde{\boldsymbol{f}}_k\right], \tag{3.34}$$

where $\tilde{\boldsymbol{f}}_k$ is the kth column of $\tilde{\boldsymbol{F}}$. From (3.34), thanks to the unitarity of $\tilde{\boldsymbol{F}}$, the sum of all the MSEs is

$$\sum_{k=1}^{N}\text{MSE}_k = \sum_{k=1}^{N}\frac{\sigma_s^2}{\gamma|\Lambda(k,k)|^2 + 1}, \tag{3.35}$$

where $|\Lambda(k,k)|^2$ is the kth diagonal element of $|\boldsymbol{\Lambda}|^2$. Note that the right-hand side term in (3.35) does not depend on the precoding matrix $\boldsymbol{F}$. This property will be crucial for the subsequent optimization.

The MMSE estimate is not immune from intersymbol interference (ISI) and this complicates the derivation of the SER. However, invoking the central limit theorem, when N is sufficiently large one can get a fairly good approximation of the final SER by modeling ISI as additive complex Gaussian noise. Within the validity limit of such an approximation, the error probability for the kth symbol can be expressed as

$$\text{SER}_k = \alpha\,\text{erfc}\left(\sqrt{\beta\,\text{SINR}_k}\right), \tag{3.36}$$

where α and β are coefficients that depend on the constellation (see [4]), and SINR_k is the signal to interference plus noise (SINR) on the kth symbol, defined as

$$\text{SINR}_k = \frac{\sigma^2_{sig_k}}{\sigma^2_{int_k+noise_k}} \tag{3.37}$$

where $\sigma^2_{sig_k}$ is the variance of the useful component in $\hat{s}_k$, and $\sigma^2_{int_k+noise_k}$ is the variance of the ISI and noise contained in $\hat{s}_k$.

It is possible to prove that MSE_k is related to SINR_k through the following relation

$$\text{SINR}_k = \frac{\sigma_s^2}{\text{MSE}_k} - 1. \tag{3.38}$$

The average SER, taking into account (3.38), is then

$$\overline{\text{SER}} = \frac{1}{N}\sum_{k=1}^{N}\text{SER}_k = \frac{1}{N}\sum_{k=1}^{N}\alpha\,\text{erfc}\left[\sqrt{\beta\left(\frac{\sigma_s^2}{\text{MSE}_k} - 1\right)}\right]. \tag{3.39}$$

2 Note that the product of diagonal matrices is commutative.

Let us consider now the $\mathrm{erfc}(\cdot)$ function inside the summation in (3.39). This is in general a convex function for values of MSE_k below a threshold that depends on β, which, in turn, depends on the constellation. However, for BPSK and QPSK constellations that function is convex for all values of MSE_k. Assuming that all MSE_k are below that threshold, or either BPSK or QPSK is used, an application of Jensen's inequality [5] to (3.39) leads to

$$\overline{\mathrm{SER}} \geq \alpha\,\mathrm{erfc}\left[\sqrt{\beta\left(\frac{\sigma_s^2}{\frac{1}{N}\sum_{k=1}^{N}\mathrm{MSE}_k}-1\right)}\right], \tag{3.40}$$

with equality if and only if all the MSE_k are equal. The term in the denominator does not depend on the precoding matrix $\boldsymbol{F}$, because of (3.35). This means that, for each channel realization, (3.40) is the minimum achievable average SER. The minimum of (3.40) is reached if and only if $\mathrm{MSE}_1 = \cdots = \mathrm{MSE}_N$, i.e., the MSE is the same over all estimated symbols. From (3.34), this condition is achieved if the vectors $\tilde{\boldsymbol{f}}_k = \boldsymbol{W}^H \boldsymbol{f}_k$ satisfy the following constraints

$$|\tilde{f}_k(n)| = |\tilde{f}_j(n)|\,, \quad n = 1,\ldots,N,\ \forall\, k,j \in \{1,\ldots,N\} \tag{3.41}$$

where $\tilde{f}_k(n)$ is the nth entry of $\tilde{\boldsymbol{f}}_k$. A special case is when all the entries of the vectors $\tilde{\boldsymbol{f}}_k$ have the same modulus. Note that, according to (3.38) the condition for equal MSE_k is equivalent to the condition for equal SINR_k.

In summary, to reach the minimum average SER (when either BPSK or QPSK is used, or when all MSE_k are below a threshold that depends on the constellation) a sufficient condition is to use a unitary coding matrix $\boldsymbol{F}$ such that $\tilde{\boldsymbol{F}} = \boldsymbol{W}^H \boldsymbol{F}$ has all constant modulus entries. Denoting with $\boldsymbol{\mathcal{H}}$ the Walsh-Hadamard matrix, two possible choices are:

$$\tilde{\boldsymbol{F}} = \boldsymbol{\mathcal{H}} \quad \Rightarrow \boldsymbol{F} = \boldsymbol{W}\boldsymbol{\mathcal{H}}; \tag{3.42}$$

$$\tilde{\boldsymbol{F}} = \boldsymbol{W}^H \quad \Rightarrow \boldsymbol{F} = \boldsymbol{I}. \tag{3.43}$$

The first choice is possible only if N is an integer multiple of 4. Interestingly, we notice that with the second choice the optimal $\boldsymbol{F}$ is simply the identity matrix. This means that the corresponding transmitted block is composed of a set of N successive symbols, as in TDMA. The only difference with respect to TDMA is that, to each block we have to append a CP of length L (i.e., we must copy the last L symbols in front of the block). In spite of this very small change with respect to conventional TDMA, we obtain a very simple equalizer, as in (3.33) with also minimum average SER. An example of comparison of different precoding strategies is reported in Figure 3.3, that shows the average SER obtained using OFDM

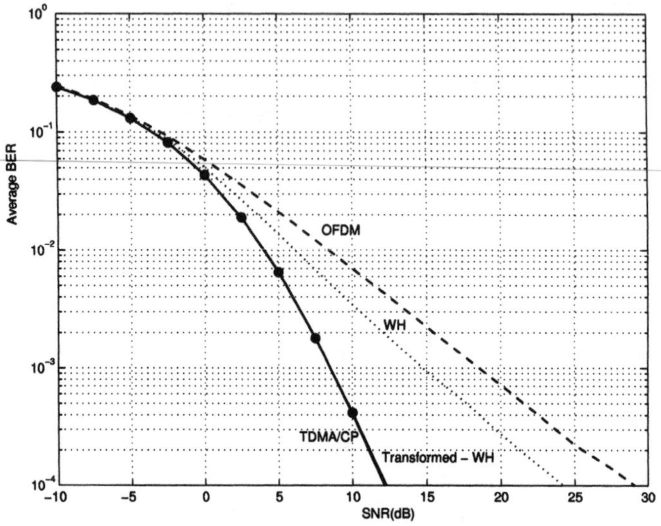

Figure 3.3 Average BER versus SNR=σ_s^2/σ_n^2, for different precoding matrices.

(dashed line), Walsh-Hadamard with CP (dotted line), identity precoder, i.e., $\boldsymbol{F} = \boldsymbol{I}$ or TDMA/CP, (solid line with $*$), transformed Walsh-Hadamard, i.e., $\boldsymbol{F} = \boldsymbol{W}\boldsymbol{\mathcal{H}}$ (solid line with o). The other parameters are $L = 6$ and $N = M = 32$. We can see that, at moderate SNR, all precoders give rise to the same slope (approximately one decade per 10 dB of SNR). We notice that, at high SNR, the identity precoders and the transformed WH codes perform in the same way, as predicted by the above theory and, most important, they perform much better than uncoded OFDM or WH.

What is also interesting to remark is that the slope of the average BER is much higher in case of TDMA/CP or transformed WH. Even though the diversity gain is a parameter that will be introduced later, in Chapter 5, we can anticipate here that OFDM and WH do not have any diversity gain, whereas TDMA/CP and the transformed WH have a multipath diversity gain.

3.4 SYSTEM DESIGN ISSUES

In this section, we provide the main criteria underlying the choice of the critical parameters of an OFDM system. The first decision concerns the block length N or, equivalently, the block duration NT. In general, the main requirements of a digital communication system are given in terms of error probability. If we refer to 3G systems, for example, it is necessary to deliver voice with 10^{-3} of BER and data with BER $= 10^{-6}$. Furthermore, given the randomness of the radio channel, it is also important to evaluate the probability of the so-called out-of-service event (i.e., the event that the error probability is greater than a specified target BER). We may tolerate an out-of-service event for no more than one percent of time, for example. One of the major constraints in current wireless communication links is given by the bandwidth allocated for the service. The bandwidth is typically assigned by regulatory bodies. Given a bandwidth B, the transmission rate, as measured in number of bits per second (bps), using a constellation of cardinality M, is

$$R = \alpha B \log_2(M), \tag{3.44}$$

where $\log_2(M)$ is the number of bits per symbol, and αB is the number of symbols per second. The coefficient α is less than one, and it incorporates a series of rate losses, as, for example, the roll-off factor or the rate loss due to the insertion of the cyclic prefix.

Given a channel with coherence time T_c, the overall duration of the OFDM block must be a fraction of T_c. The smaller the symbol duration, the more valid the assumption, underlying any OFDM system, that the channel is time-invariant within one block. On the other hand, though, the shorter the block length, the greater the rate reduction induced by the insertion of the CP. Since the duration of the CP is LT_s, the rate is reduced by a factor

$$\epsilon = \frac{N}{N+L}. \tag{3.45}$$

Typically, the loss should never exceed 20%, which means $\epsilon = 0.8$. This implies that the prefix length should not exceed one-quarter of the OFDM symbol, without CP (i.e., $L \leq N/4$). In general, given ϵ, N is related to L by

$$N = \frac{\epsilon L}{1-\epsilon}. \tag{3.46}$$

In practice, the duration of the CP must be greater than the channel delay spread, which implies $LT_s > \sigma_\tau$. Since σ_τ is inversely proportional to the channel coherence bandwidth, we can also write $LT_s > 1/B_c$. Requiring that the OFDM symbol duration be smaller than the channel coherence time and that $LT_s > 1/B_c$, implies that N must belong to the following interval

$$\frac{\epsilon}{1-\epsilon}\frac{1}{B_c} < NT_s < T_c. \tag{3.47}$$

The existence of a nonempty interval requires that the coefficient ϵ must be smaller than a function of the time-bandwidth product

$$\epsilon < \frac{B_c T_c}{1 + B_c T_c}. \quad (3.48)$$

3.5 CHANNEL ESTIMATION

Clearly, equalization requires the knowledge, and thus the estimate, of the channel. The easiest way to estimate the channel in an OFDM system consists of inserting pilot tones in the transmitted block. Inserting pilot tones means that the transmitter sends known (training) symbols on some subcarriers. The receiver knows these symbols and on which subcarriers they are transmitted. We assume that the pilot tones have unit amplitude and we indicate with I_{PT} the set of indexes corresponding to the pilot tones. At the receiver we have, over the pilot tones

$$y(i) = H(i) + w(i), \quad i \in I_{PT}, \quad (3.49)$$

The main questions are then: 1) how many pilot tones do we need? and 2) where should they be put? The number of pilot tones depends on the number of unknowns. Since $H(i) = \sum_{l=0}^{L} h(i) e^{-j2\pi il/N}$, the number of unknowns is $L+1$. Therefore, we need at least $L+1$ pilot tones. Having $L+1$ pilot tones, we can write a set of $L+1$ linear equations

$$y(i) = \sum_{l=0}^{L} h(i) e^{-j2\pi il/N} + w(i) \quad (3.50)$$

with the $L+1$ unknowns $h(l), l = 0, \ldots, L$. Introducing the matrix $\boldsymbol{V}_{L+1}$ with entries $V_{L+1}(k,l) := e^{-j2\pi kl/N}$, $k, l \in I_{PT}$, we can rewrite (3.50) in matrix form

$$\boldsymbol{y} = \boldsymbol{V}_{L+1} \boldsymbol{h} + \boldsymbol{w}, \quad (3.51)$$

where $\boldsymbol{h}^T := [h(0), \ldots, h(L)]$ and $\boldsymbol{w}^T := [w(0), \ldots, w(L)]$. The ZF channel estimate is then

$$\hat{\boldsymbol{h}} = \boldsymbol{V}^{\dagger} \boldsymbol{y}. \quad (3.52)$$

Combining (3.51) and (3.52), it turns out that the channel estimation error is $\boldsymbol{e} = \boldsymbol{V}^{\dagger} \boldsymbol{w}$. Therefore, this error vector is a complex Gaussian vector with zero mean and covariance matrix $\boldsymbol{C}_e = \sigma_n^2 \boldsymbol{V}^{\dagger} \boldsymbol{V}^{\dagger^H}$. The location of the pilot tones conditions the error, as it alters the error covariance matrix. It may be verified that the location of the pilot tones that minimizes the sum of the variances of the channel estimation errors is the one where the pilot tones are at the maximum distance from each other. This implies that the pilot tones must be equispaced in the frequency domain. This is in fact the choice adopted in all practical implementations of OFDM.

We evaluate now the effect of channel estimation errors on the performance of an OFDM system. We adopt a small perturbation analysis, valid for small estimation errors. We denote with $\hat{H}(k) = H(k) + \epsilon(k)$ the estimate of the kth sample of the channel transfer function, and we assume that the estimation error $\epsilon(k)$ is much smaller than $H(k)$. The ZF decoder (3.24) produces then a symbol estimate

$$\hat{s}_k(n) = \frac{1}{\hat{H}(k)} z_k(n) = \frac{H(k)}{\hat{H}(k)} s_k(n) + \frac{w_k(n)}{\hat{H}(k)}. \tag{3.53}$$

Assuming $\epsilon(k) \ll H(k)$, we may adopt a first order approximation

$$\hat{s}_k(n) \approx s_k(n) - \frac{\epsilon(k)}{\hat{H}(k)} s_k(n) + \frac{w_k(n)}{\hat{H}(k)}. \tag{3.54}$$

Within the limit of validity of such an expression, we see that the channel estimation error introduces an additional noise. This term is proportional to the symbol through a multiplicative coefficient, the channel estimation error. Therefore, assuming this additional noise contribution independent of the symbol, and denoting with σ_e^2 the variance of the channel estimate, the SNR on the kth symbol is then

$$SNR(k) \approx \frac{\sigma_s^2}{\frac{\sigma_s^2 \sigma_e^2}{|H(k)|^2} + \frac{\sigma_n^2}{|H(k)|^2}} = \frac{|H(k)|^2 \sigma_s^2}{\sigma_s^2 \sigma_e^2 + \sigma_n^2}. \tag{3.55}$$

Hence, as the noise variance vanishes (or the signal power increases with respect to the receiver noise), the SNR tends to an asymptote dictated by the channel estimation error. If we further assume that the channel estimation error is Gaussian and we transmit BPSK symbols, the floor on the BER over the kth subchannel is obtained by computing the error probability corresponding to the SNR in (3.55) with $\sigma_n^2 = 0$; that is,

$$P_{\text{floor}}(k) = \frac{1}{2} \operatorname{erfc}\left(\sqrt{\frac{|H(k)|^2}{2\sigma_e^2}} \right). \tag{3.56}$$

This formula, albeit approximate, gives us a rule to derive the maximum variance of the channel estimation error necessary to guarantee that the error floor be less than the desired BER. Inverting (3.56), we have

$$\sigma_{e_{\max}}^2 = \frac{\max_k |H(k)|^2}{2\{\operatorname{erfc}^{-1}[2\, P_{\text{floor}}(k)]\}^2}, \tag{3.57}$$

where $\mathrm{erfc}^{-1}(\cdot)$ denotes the inverse of the function $\mathrm{erfc}(\cdot)$.

The insertion of pilot tones leads to a very simple channel estimator. The only price paid for the insertion of pilot tones is a rate reduction. If we wish to be able to perform the channel estimation using only one OFDM block, we need to insert at least $L+1$ pilot tones in one block. This implies that the number of information symbols is no longer N, but $N-L-1$. Therefore, the rate reduces by a factor

$$\epsilon = \frac{N-L-1}{N+L}, \tag{3.58}$$

where we have incorporated also the reduction due to the insertion of the cyclic prefix. To reduce this loss, it is possible to distribute the pilot tones over successive blocks. In this case, if we put l pilot tones in one block, with $l < L+1$, the rate reduction factor is

$$\epsilon = \frac{N-l}{N+L}, \tag{3.59}$$

but we need at least $\lfloor \frac{L+1}{l} \rfloor + 1$ blocks to perform the estimate (assuming that the channel will remain unaltered over these blocks).

3.6 SYNCHRONIZATION

One of the most critical aspects of an OFDM system is synchronization in both time and frequency. The synchronization in time can be recovered using standard techniques, as in many digital communication systems (i.e., by periodically sending a training sequence). However, the OFDM system is particularly sensitive to frequency errors, due to carrier offsets between transmitter and receiver and/or Doppler frequencies. The situation is represented, pictorially, in Figure 3.2, where the solid arrows represent the points where the received signal is sampled. We observe that, sampling at the right points, there is no ISI. However, if the sampling points are slightly shifted, as shown by the dashed arrows, each sample contains the contribution from many adjacent symbols. Therefore, a frequency error less than $1/NT$ can cause a serious performance degradation due to the insurgence of ISI. The problem can be studied analytically as follows.

We assume perfect synchronization in time and we indicate by f_D the carrier frequency offset between transmitter and receiver. As a consequence, the received sequence is

$$\hat{\boldsymbol{y}}(n) = \boldsymbol{D}(f_D)\boldsymbol{H}\boldsymbol{F}\boldsymbol{s}(n) + \boldsymbol{w}(n), \tag{3.60}$$

where $\boldsymbol{D}(f_D)$ is an $N \times N$ diagonal matrix

$$\boldsymbol{D}(f_D) = \begin{pmatrix} 1 & 0 & \cdots & \cdots & 0 \\ 0 & e^{j2\pi f_D} & \ddots & \ddots & \vdots \\ \vdots & \ddots & e^{j2\pi 2 f_D} & \ddots & \vdots \\ \vdots & \ddots & \ddots & \ddots & 0 \\ 0 & \cdots & \cdots & 0 & e^{j2\pi(N-1)f_D} \end{pmatrix}. \tag{3.61}$$

We perform now a perturbation analysis, valid for small frequency offset errors. We assume in our ensuing derivations that the error frequency offset is sufficiently small so that we can use the following approximation

$$e^{j2\pi f_D N} \approx 1 + j2\pi f_D N.$$

Introducing the diagonal matrix $\boldsymbol{Q} := diag\{0, 1, 2, \ldots, N-1\}$, we can write

$$\boldsymbol{D}(f_D) \approx \boldsymbol{I} + j2\pi f_D \boldsymbol{Q}. \tag{3.62}$$

Inserting (3.62) in (3.60), we get

$$\tilde{\boldsymbol{y}}(n) \approx \boldsymbol{HFs}(n) + j2\pi f_D \boldsymbol{QHFs}(n) + \boldsymbol{w}(n). \tag{3.63}$$

If we exploit the diagonalization $\boldsymbol{H} = \boldsymbol{W\Lambda W}^H$ and we multiply $\boldsymbol{y}(n)$ from the left side by $\boldsymbol{W}^H$, we get

$$\begin{aligned} \boldsymbol{W}^H \tilde{\boldsymbol{y}}(n) &= \boldsymbol{\Lambda}\tilde{\boldsymbol{F}}\boldsymbol{s}(n) + j2\pi f_D \boldsymbol{W}^H \boldsymbol{QW\Lambda}\tilde{\boldsymbol{F}}\boldsymbol{s}(n) + \boldsymbol{W}^H \boldsymbol{w}(n) \\ &:= \boldsymbol{\Lambda}\tilde{\boldsymbol{F}}\boldsymbol{s}(n) + j2\pi f_D \tilde{\boldsymbol{Q}}\boldsymbol{\Lambda}\tilde{\boldsymbol{F}}\boldsymbol{s}(n) + \boldsymbol{W}^H \boldsymbol{w}(n), \end{aligned} \tag{3.64}$$

having introduced the matrix $\tilde{\boldsymbol{Q}} := \boldsymbol{W}^H \boldsymbol{QW}$. Applying zero-forcing (ZF) equalization, the estimated symbol vector is

$$\hat{\boldsymbol{s}}(n) = (\boldsymbol{\Lambda}\tilde{\boldsymbol{F}})^\dagger \boldsymbol{W}^H \tilde{\boldsymbol{y}}(n) = \boldsymbol{s}(n) + j2\pi f_D (\boldsymbol{\Lambda}\tilde{\boldsymbol{F}})^\dagger \tilde{\boldsymbol{Q}}\boldsymbol{\Lambda}\tilde{\boldsymbol{F}}\boldsymbol{s}(n) + (\boldsymbol{\Lambda}\tilde{\boldsymbol{F}})^\dagger \boldsymbol{W}^H \boldsymbol{w}(n). \tag{3.65}$$

The second term on the right-hand side represents the ISI induced by the non-zero carrier frequency offset. In general, ISI is symbol-dependent, and thus the exact computation for the error probability is not an easy task. Here, we will perform a perturbation analysis, approximately valid for small frequency offset errors. We consider the case in which the symbols in each block are independent. Assuming that we sum a sufficiently large number of independent terms and invoking the central limit theorem, we can model, to a first order approximation, the ISI as a complex Gaussian random variable. Under such an approximation, which will be checked later on, we can consider the last two terms in (3.65) as a complex Gaussian random vector with zero mean and covariance matrix

$$\boldsymbol{C}_{\text{IN}} = 4\pi^2 f_D^2 \sigma_s^2 (\boldsymbol{\Lambda}\tilde{\boldsymbol{F}})^{\dagger} \tilde{\boldsymbol{Q}} (\boldsymbol{\Lambda}\tilde{\boldsymbol{F}})(\boldsymbol{\Lambda}\tilde{\boldsymbol{F}})^{H} \tilde{\boldsymbol{Q}} (\boldsymbol{\Lambda}\tilde{\boldsymbol{F}})^{\dagger^H} + \sigma_n^2 (\boldsymbol{\Lambda}\tilde{\boldsymbol{F}})^{\dagger} (\boldsymbol{\Lambda}\tilde{\boldsymbol{F}})^{\dagger^H}. \quad (3.66)$$

Modeling ISI and noise as a complex Gaussian vector independent of the symbols (even though this is not exactly true), it is possible to derive the BER in closed form. Using BPSK symbols, for example, the average BER (averaged over all the symbols) is

$$\overline{\text{BER}}(f_D) := \frac{1}{N}\sum_{k=1}^{N} \text{BER}(k) \approx \frac{1}{2N}\sum_{k=1}^{N} \text{erfc}\left(\sqrt{\frac{\sigma_s^2}{C_{\text{IN}}(k,k)}}\right), \quad (3.67)$$

where $C_{\text{IN}}(k,k)$ is the kth entry on the diagonal of $\boldsymbol{C}_{\text{IN}}$, as given in (3.66).

Clearly, since we have made many approximations in deriving this closed-form expression, it is necessary to check the validity of the approximations. As an example, in Figure 3.4 we report the average BER versus SNR $:= \sigma_s^2/\sigma_n^2$, for different values of the frequency offsets: $f_D NT = 1/10, 1/20$, and $f_D = 1/40$; the solid lines correspond to the (approximated) closed-form expression, whereas the circles refer to the simulation results. The other system parameters are the following: The block length is $N = 64$, the channel order is $L = 6$. The coding matrix $\boldsymbol{F}$ is a Walsh-Hadamard matrix. We can observe how, in spite of all the approximations, the closed-form expression is able to predict the true values quite well. There is a slight difference only when the offset is not too small, because in that case the small perturbation analysis is no longer valid. Therefore, (3.67) can be used to derive the maximum frequency offset error that can be tolerated to guarantee that the BER floor due to ISI be less than the value required by the quality of service.

Before concluding this section, it is interesting to compare how different coding matrices $\boldsymbol{F}$ affect the performance, in the presence of frequency synchronization errors. As an example, in Figure 3.5, we report the average BER obtained using three different kinds of encoders: $\boldsymbol{F} = \boldsymbol{I}$ (solid line); OFDM (dotted line), and Walsh-Hadamard coding (dashed line), obtained with different errors. We can observe that, for low-frequency offset, the three precoders tend to behave very similarly.

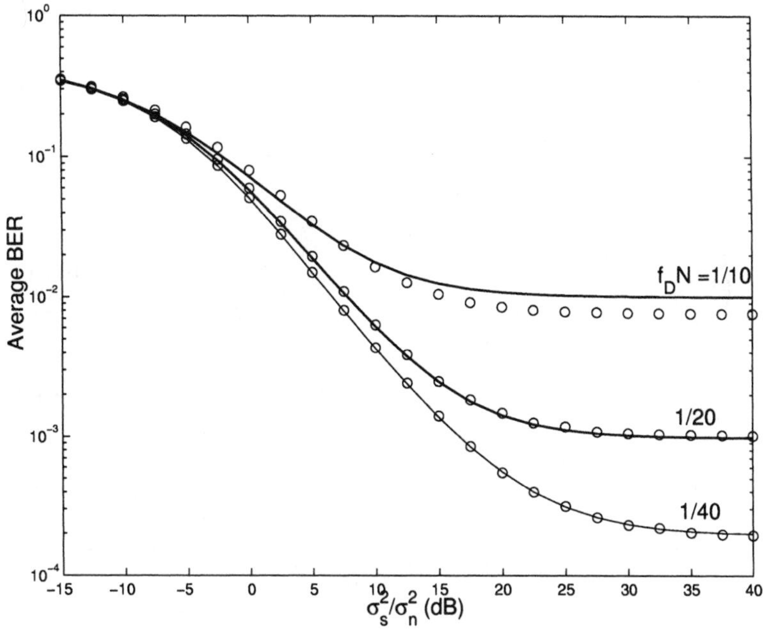

Figure 3.4 Average BER versus SNR=σ_s^2/σ_n^2, for different values of frequency offset f_D.

3.7 MULTIUSER SYSTEMS

So far, we have analyzed an OFDM system for a single-user system. It is now important to generalize the analysis to a multiuser case. The multiple access strategies will be studied in detail in Chapter 4. Here, we anticipate only some basic properties of multiple access strategies based on multicarrier modulation. There are basically three different ways to design a multiplexing scheme that allows for multiuser communication using an underlying OFDM scheme. The three schemes are detailed in the following sections.

3.7.1 OFDM-TDMA

The method known as OFDM-TDMA is a a combination of the TDMA idea with OFDM. In such a scheme, the frame is divided into consecutive slots. To each user accessing the network is assigned one (or more) time slot within the frame. Within each time slot, the user transmits data using an OFDM strategy. In this way, equalization is simple, as in OFDM system, and there is no multiuser interference

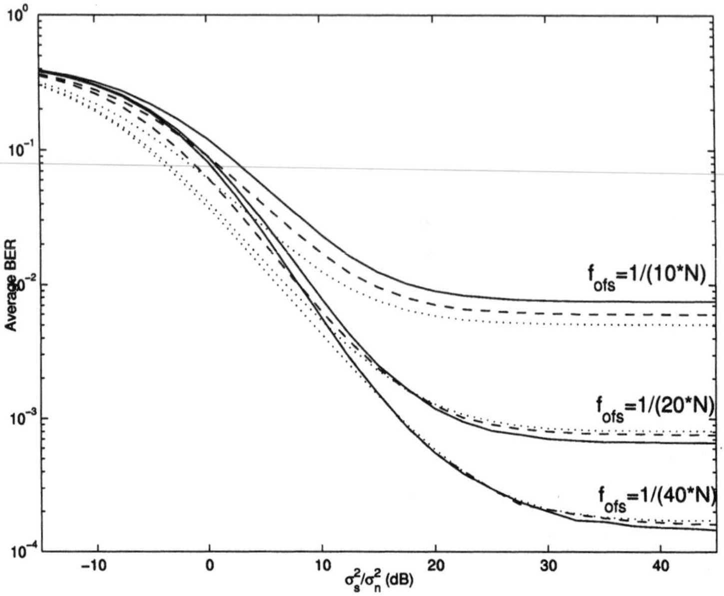

Figure 3.5 Average BER versus SNR=σ_s^2/σ_n^2, for different values of frequency offset f_D and different precoding matrices: $\boldsymbol{F} = \boldsymbol{I}$ (solid line); OFDM (dotted line), and Walsh-Hadamard (dashed line).

(MUI), as different users are separated on a time basis. The major advantage of this scheme is the absence of MUI. However, the price paid for this advantage is a rigid use of resources. In fact, for all the time a time slot is assigned to a user, until the user releases the slot, no other users can use that resource. As far as synchronization is concerned in an OFDM-TDMA system, the synchronization in frequency can be dealt with exactly as in single-user OFDM, as different users are separated in time. However, some guard intervals must be introduced between time slots associated with different users to prevent the spillover of some user data into other user time slots. Clearly, the longer the guard interval, the more robust the system against MUI and time synchronization, but the higher the rate loss, so that the guard interval has to result as a trade-off between these two needs.

3.7.2 OFDMA

Orthogonal frequency division multiple access (OFDMA) is a combination of the FDMA multiplexing scheme with OFDM. In particular, with OFDMA different users get a set of orthogonal subcarriers. The spectra associated to different users might be partially overlapping, as in the OFDM spectrum, but the important property is that the carriers associated to different users are orthogonal. The most critical aspect of an OFDMA system is synchronization in frequency, as a slight frequency offset can cause both ISI, as in OFDM, plus multiuser interference (MUI). Indeed, MUI is really critical, because it requires a very accurate synchronization among all the users. Some possible methods to compensate for the frequency synchronization problem in OFDMA systems were proposed, for example, in [6, 7] (see also the references therein).

3.7.3 MC-CDMA

Neither OFDMA nor OFDM/TDMA suffers from MUI. However, both systems pay for this immunity from MUI with a rigid resource allocation, such that once a time slot (or a subcarrier) has been allocated to a given user, that resource is not available to any other user until the resource itself is released from the user to whom it was assigned. A different approach is followed by multicarrier code division multiple access (MC-CDMA) systems, where in principle all time and all bandwidth is assigned to every user asking for the service. The idea is similar to CDMA, which will be studied in detail in the next chapter. Here, we consider a specific combination of CDMA with MC systems. There is no a unique MC-CDMA system, as explained in detail in [8]. Nevertheless, the basic idea is common to all MC-CDMA systems, and thus we concentrate here on the basic principles of MC-CDMA systems. We start with the downlink channel (i.e., the channel from the base station to the mobile users).

3.7.3.1 Downlink channel

The base station (BS) sends data to M users assigning to each user the whole available bandwidth and time, within each time slot. This can be done by assigning orthogonal codes to different users. Let us denote with $\boldsymbol{c}_k$ the code assigned to the kth user. This is a vector composed of N entries, which we call *chips*, as in the standard CDMA terminology. Denoting with $s_k(n)$ the symbol transmitted by the kth user within the nth time slot, the signal transmitted from the BS, in the nth time slot, is $\boldsymbol{u}(n) = \sum_{k=1}^{M} \boldsymbol{c}_k s_k(n)$. With MC-CDMA, the BS transmits this sequence using an OFDM strategy, so that the transmitted sequence is

$$\boldsymbol{x}(n) = \boldsymbol{W}\boldsymbol{u}(n) = \boldsymbol{W}\sum_{k=1}^{M} \boldsymbol{c}_k s_k(n) = \sum_{k=1}^{M} \boldsymbol{W}\boldsymbol{c}_k s_k(n). \tag{3.68}$$

The signal received by the lth user, characterized by a channel matrix $\boldsymbol{H}_l$ is then

$$\boldsymbol{y}_l(n) = \boldsymbol{H}_l \sum_{k=1}^{M} \boldsymbol{W}\boldsymbol{c}_k s_k(n) + \boldsymbol{v}_l(n), \tag{3.69}$$

where $\boldsymbol{v}_l(n)$ is the noise vector present in the lth user's receiver. Because of the OFDM structure, the matrix $\boldsymbol{H}$ is an $N \times N$ circulant and Toeplitz matrix. Therefore, exactly as in OFDM, it can be diagonalized as in (3.18). Hence, if we multiply (3.69) by $\boldsymbol{W}^H$ from the left side, we get

$$\boldsymbol{z}_l(n) := \boldsymbol{W}^H \boldsymbol{y}_l(n) = \boldsymbol{\Lambda}_l \boldsymbol{W}^H \sum_{k=1}^{M} \boldsymbol{W}\boldsymbol{c}_k s_k(n) + \boldsymbol{w}_l(n) = \boldsymbol{\Lambda}_l \sum_{k=1}^{M} \boldsymbol{c}_k s_k(n) + \boldsymbol{w}_l(n), \tag{3.70}$$

where $\boldsymbol{w}_l(n) := \boldsymbol{W}^H \boldsymbol{v}_l(n)$ is a noise having the same statistical properties as $\boldsymbol{v}_l(n)$. To recover its own symbol $s_l(n)$, the lth user multiplies $\boldsymbol{z}_l(n)$ by the row vector $\boldsymbol{c}_l^H \boldsymbol{\Lambda}_l^{-1}$. Exploiting the orthogonality among the code vectors $\boldsymbol{c}_k$, the result is

$$\hat{s}_l(n) := \boldsymbol{c}_l^H \boldsymbol{\Lambda}_l^{-1} \boldsymbol{z}_l(n) = s_k(n) + \boldsymbol{c}_l^H \boldsymbol{\Lambda}^{-1} \boldsymbol{w}_l(n). \tag{3.71}$$

This equation shows the basic property of an MC-CDMA system, capable of nulling ISI completely using a very simple receiver structure, in spite of having assigned the same spectrum and bandwidth to each user and having transmitted through a frequency-selective channel. This property is the result of the combination of CDMA and OFDM.

3.7.3.2 Uplink channel

In the uplink channel the situation is more complicated than in the downlink. The main disadvantage is that if the whole bandwidth is assigned to all the users and the channels are time-dispersive, the MC-CDMA system is not immune from MUI in the uplink channel, as we will show next.

In the uplink channel, the lth user transmits, for each symbol $s_l(n)$, a vector $\boldsymbol{c}_l s_l(n)$. As in the downlink case, each user terminal transmits its sequence using an OFDM scheme. The transmitted sequence is then $\boldsymbol{W}\boldsymbol{c}_l s_l(n)$. The BS receives the sum of the sequences transmitted from all users. The result is

$$\boldsymbol{y}(n) = \sum_{l=1}^{M} \boldsymbol{H}_l \boldsymbol{W}\boldsymbol{c}_l s_l(n) + \boldsymbol{v}(n), \tag{3.72}$$

where $\boldsymbol{v}(n)$ now is the noise at the BS. Again, all channel matrices can be diagonalized as in (3.18). Therefore, multiplying $\boldsymbol{y}(n)$ by $\boldsymbol{W}^H$ from the left side and exploiting (3.18), we get

$$z(n) := \mathbf{W}^H y(n) = \sum_{l=1}^{M} \mathbf{\Lambda}_l c_l s_l(n) + w(n). \tag{3.73}$$

If we multiply $z(n)$ by $c_k^H \mathbf{\Lambda}_k^{-1}$, to recover the symbol pertaining to the kth user, we obtain

$$\hat{s}_k(n) := c_k^H \mathbf{\Lambda}_k^{-1} z(n) = s_k(n) + \sum_{l=1, l \neq k}^{M} c_k^H \mathbf{\Lambda}_k^{-1} \mathbf{\Lambda}_l c_l s_l(n) + w_k'(n), \tag{3.74}$$

with $w_k'(n) = c_k^H \mathbf{\Lambda}_k^{-1} w(n)$. From (3.74) we notice that now we have the right symbol plus the interference from all other users. This happens because multiplication by $\mathbf{\Lambda}_k^{-1}$ equalizes *only* the kth channel, but not all other channels. As a consequence, after equalization, the vectors carrying the symbols from different users are not orthogonal (the orthogonality is destroyed by the channel), and this causes MUI. Hence, the performance of a MC-CDMA system, in the uplink, depends on the number of active users.

3.8 CASE STUDY: IEEE 802.11

In this section we review some of the basic parameters of the standard IEEE 802.11a, that is the system devised for broadband wireless local area network (LAN) [9]. The physical (PHY) layer of IEEE 802.11a is based on OFDM. The nominal carrier frequencies for IEEE 802.11a are allocated in two frequency bands: the lower frequency band, between 5.150 MHz and 5.350 MHz, and the upper frequency band, between 5.725 MHz and 5.825 MHz. To improve the radio link capability due to different interference scenarios and distances between mobile terminals (MT) and access point (AP), a multirate strategy is implemented, where the appropriate mode is selected by a link adaptation scheme. The data rate ranges from 6 to 54 Mbps, and it can be varied by combining various signal alphabets for modulating the OFDM subcarriers and different puncturing patterns applied to a mother convolutional code. The constellations BPSK, QPSK, 16-QAM are used as mandatory modulation formats, whereas 64-QAM is applied as an optional one for both access point and mobile terminal.

The information bits are coded using a convolutional code and then interleaved. The use of convolutional coding is particularly useful as it allows for puncturing (i.e., the possibility of eliminating some bits, according to a predefined pattern, as specified in [9]), without impairing the correction coding capabilities. In this way, playing with the convolutional code rate, the puncturing rate, and the dimension of the transmitted constellation, it is possible to enforce the multirate capability.

The stream of coded bits is parsed into groups composed of a number of bits per subcarrier (CBPSC) equal to 1, 2, 4, or 6 bits and converted into complex numbers representing BPSK, QPSK, 16-QAM, or 64-QAM constellation points. The mapping is performed according to a Gray-coded constellation mapping, so that any two adjacent points in the constellation differ by no more than one bit. The choice of all the parameters is performed in order to guarantee the desired rate. Some possible combinations of constellation, convolutional coding (CC) rate, nominal bit rate in Megabits per second (Mbps), number of coded bits per subcarrier (CBPSC), number of coded bits per OFDM symbol (CBPS), and number of data bits per symbol (DBPS) are listed in the following table. The stream of complex valued subcarrier

Table 3.1

Code rates and constellation settings for IEEE 802.11a

Modulation	CC rate	Mbps	CBPSC	CBPS	DBPS
BPSK	1/2	6	1	48	24
BPSK	3/4	9	1	48	36
QPSK	1/2	12	2	96	48
QPSK	3/4	18	2	96	72
16-QAM	1/2	24	4	192	96
16-QAM	3/4	36	4	192	144
64-QAM	2/3	48	6	288	192
64-QAM	3/4	54	6	288	216

modulation symbols at the output of mapper is parsed into groups of $N_{SD} = 48$ complex numbers. Each group is transmitted in one OFDM symbol. In each OFDM symbol there are $N_{SD} = 48$ data carriers and $N_{SP} = 4$ pilot carriers for channel estimation. Thus, each symbol is composed of a set of $N_{ST} = 52$ carriers and transmitted with a duration T_S. The pilot carriers for reference signal transmissions are: $l = -21, -7, 7, 21$. Each symbol interval has a useful symbol part with duration T_U and a cyclic prefix with duration T_{CP}. The length of the useful symbol part is equal to 64 samples, and its duration is $T_U = 3.2\ \mu s$. For the cyclic prefix length there are two possible values in the IEEE 802.11a system: 800ns (mandatory) and 400ns (optional). A possible configuration of the system has the numerical values listed in Table 3.2. It is worth noticing that the outmost subcarriers are not used, so as to create a null guard interval at the edges of the available spectrum (only 16.25 MHz are used out of the 20 MHz available). This is done to limit the spectrum spillover of each channel over adjacent channels.

Table 3.2

Typical Parameters Setting for IEEE 802.11a

Parameter	Value
Sampling rate $F_s = 1/T$	20 MHz
Useful duration $T_U = 64T$	$3.2\mu s$
Cyclic prefix duration $T_{CP} = 16T$	$0.8\ \mu s = 8T$
Number of data subcarriers N_{SD}	48
Number of pilot subcarriers N_{SP}	4
Subcarrier spacing Δf	$1/T_U = 0.3125$ MHz
Total number of subcarriers N_{ST}	$N_{SD} + N_{SP} = 52$
Spacing between the two outmost subcarriers	$N_{ST}\Delta f = 16.25$ MHz

More recently, a new version of the standard for wireless LANs has been issued: IEEE 802.11g. This release operates in the 2.4 GHz frequency band, at a rate of 54 Mbps.

3.9 SUMMARY

In this chapter we have shown that block transmission, with proper guard intervals between successive blocks, can simplify the equalization task considerably. A more in-depth analysis of block transmission systems is given , for example, in [10, 11, 12]. We have studied in detail OFDM systems, as a system that guarantees a very good trade-off between simplicity and performance. In particular, we have seen how a proper combination of precoding, either in the form of error correction coding or linear redundant coding, gives rise to a very effective, yet simple, way to combat the channel frequency-selectivity and noise. More specifically, we have shown how to design the linear precoder in order to minimize the average bit error rate, in the case where the receiver adopts an MMSE decoder. Then, we have provided the main criteria for choosing the critical parameters of an OFDM system. The major drawback of multicarrier systems is their strong sensitivity to synchronization errors. For this reason, we have analyzed the effect of channel estimation error and synchronization errors. Then, we have studied multiuser systems based on multicarrier modulation. We have considered as a case study one specific telecommunication standard, developed for broadband wireless LAN, namely IEEE 802.11a. In Chapter 8, we will show how to combine the milticarrier idea with space-time coding, in multiantenna systems. A detailed description of MC-CDMA systems is given in [8]. The interested reader may refer to books entirely devoted to multicarrier system, such as [2] or [3].

References

[1] Wang Z. and Giannakis G. B., "Wireless Multicarrier Communications: Where Fourier meets Shannon," *IEEE Signal Processing Magazine*, vol. 17, no. 3, May 2000, pp. 29–48.

[2] Hanzo L., et al. *OFDM and MC-CDMA for Broadband Multi-User Communications, WLANs and Broadcasting*, John Wiley & Sons, New York: August 2003.

[3] van Nee R., Prasad R., *OFDM for Wireless Multimedia Communications*, Artech House Publishers, Boston-London: 2000.

[4] Proakis J. G., *Digital Communications*, McGraw-Hill, (fourth edition), New York: August 2000.

[5] Cover, T. M., Thomas, J. A., *Elements of Information Theory*, John Wiley & Sons, New York: 1991.

[6] Barbarossa S., Pompili M., Giannakis G. B., "Channel-independent synchronization of orthogonal frequency division multiple access systems," *IEEE Journal on Selected Areas in Communications*, Feb. 2002, pp. 474–486.

[7] Morelli, M., "Timing and frequency synchronization for the uplink of an OFDMA system," *IEEE Transactions on Communications*, Vol. 52, Feb. 2004, pp. 296–306

[8] Fazel K., Fettweis G. P. (ed.), *MultiCarrier Spread-Spectrum*, Kluwer Academic Publishers: Sept. 1997.

[9] IEEE Std 802.11a/D7.0-1999 "Part 11: Wireless LAN Medium Access Control (MAC) and Physical Layer (PHY) Specifications: High-Speed Physical Layer in the 5 GHz Band", 1999.

[10] Scaglione A., Giannakis G. B., and Barbarossa S., "Redundant Filterbank Precoders and Equalizers - Part I: Unification and Optimal Designs," *IEEE Trans. on Signal Processing*, Vol. 47, No. 7, July 1999, pp. 1988–2006.

[11] Scaglione A., Giannakis G. B., and Barbarossa S., "Redundant Filterbank Precoders and Equalizers - Part II: Blind Channel Estimation, Synchronization, and Direct Equalization," *IEEE Trans. on Signal Processing*, Vol. 47, No. 7, July 1999, pp. 2007–2022.

[12] Scaglione A., Barbarossa S, and Giannakis G. B., "Filterbank Transceivers Optimizing Information Rate in Block Transmissions over Dispersive Channels," *IEEE Trans. on Info. Theory*, Vol. 45, No. 3, April 1999, pp. 1019–1032.

Chapter 4

Multiple Access Systems

4.1 INTRODUCTION

The increasing demand for communication links is making the radio spectrum a scarce resource that has to be managed as efficiently as possible. Different services, such as cellular systems, wireless LAN, radio and television broadcasting, and satellite communications are usually allocated over nonoverlapping frequency bands. However, within each kind of service, there are several users who share the same frequency spectrum. One of the most challenging tasks of modern communications is then the search for the best possible strategies for making many potential users share a communication medium and still guarantee the required quality of service (QoS).

The choice among the available alternatives is not unique and it is typically dictated by the QoS and by the traffic properties. The parameters assessing the QoS depend, on their turn, on the kind of message transmitted through the network. Typically, *voice* transmission is *delay-sensitive*, as an excessive delay is considered a very annoying factor. To give some figures of merit, interpacket delays in excess of 100 ms are perceived as rather annoying for the listener. Conversely, the delay is not as important in *data* transmissions. However, the bit error rate (BER) is a crucial parameter for data communications. As an example, in 3G cellular systems, the maximum tolerable BER for voice links is 10^{-3}, whereas the target BER for data links is 10^{-6}.

As a consequence of these different QoS requirements, it is customary to distinguish between two main classes of networks: (1) *voice-oriented* or *isochronous* networks and (2) *data-oriented* or *asynchronous* networks. Nowadays, the distinction between these two classes is becoming more and more blurred. We observe in fact the increasing deployment of data links over cellular networks, originally intended for voice communications. At the same time, there is an always increasing interest in transmitting voice over the Internet, using the so-called voice-over-Internet protocol (VoIP), even though the Internet protocol was originally designed only for data communications. Nevertheless, the conceptual distinction between isochronous and asynchronous networks is an important point to keep in mind.

In this chapter, we will concentrate on the physical layer strategies that allow the sharing of the available resources with acceptable performance. We will consider first the case where the transmitter has no knowledge about the channel and then we will analyze the situation where the transmitter knows the channel exactly. In the first case, the design of the transmission strategy must follow a robustness principle. In such a context, we will review the basic principles underlying the access methods that control the resource assignment to the users, considering both fixed and random access techniques.

Talking about physical channels, we distinguish between the situations where the transceivers have single or multiple antennas. In the first case, the channels are distributed over the *time* domain, or its dual counterpart, the frequency domain. When the transceivers have multiple antennas, one more domain comes into play: the space domain. In such a case, there are many more degrees of freedom to accommodate multiple users. In such a context, we review the basic principles underlying frequency, time, code, and space division multiple access. We pay special attention to code division multiple access, as it is the standard radio access technique for the third generation cellular system.

We consider then the case where the transmitters know the channel and then derive the optimal access strategies for the multiple access (multipoint-to-point) and the broadcast (point-to-multipoint) channel.

4.2 GENERAL PRINCIPLES

The strategies for allocating the available resources among users sharing a common communication channel depend on the traffic properties, on the network topologies, and on the channel characteristics. In this section, we will simply review some of the basic aspects of this complicated field, whose rigorous treatment goes beyond the scope of this book. We start with a general formulation of the problem, and then we will recall some of the most typical access methods. First of all, talking about access methods, we have to clarify which are the network topologies of interest.

4.2.1 Network Topologies

Making a taxonomy of network topologies is becoming more and more difficult, as the connection between heterogeneous networks tends to mix different topologies. Nevertheless, we may identify two fundamental topologies: 1) *centralized* or "hub-and-spoke" networks and 2) *peer-to-peer* networks [1]. An example of a centralized

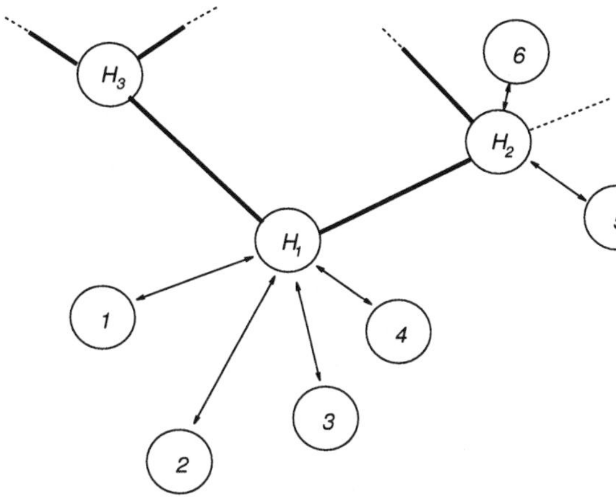

Figure 4.1 Centralized network topology.

network is illustrated in Figure 4.1, where we observe a number of users (circled numbers) who access the network passing through *access points* (APs) or *hubs* (labeled as $H_1, H_2, \cdots$). The basic principle underlying centralized topologies is that any communication from one user to the other must pass through a *hub*, or through multiple hubs, connected among them with high-capacity links (thick lines in Figure 4.1). In the single hub case, if user 1 needs to send data to user 2, the data is first sent from user 1 to the hub, which retransmits it to user 2. In the multiple-hub case, if user 1 wishes to send data to user 5, for example, the data goes from user 1 to hub H_1, then from hub H_1 to hub H_2 and, finally, it reaches user 5. The hub stations control and manage the access. There is no possibility for direct links from one user terminal to another. Typical examples of wireless centralized networks are wireless local area networks (WLANs) and cellular mobile systems, where the access point is called the *base station* (BS).

Looking at Figure 4.1, we may distinguish two typical channels: the *multiple access channel,* where multiple users send data to the hub or base station, and the *broadcast channel,* where the hub sends data to multiple users. The broadcast channel is also typically named, in cellular terminology, the *forward* or the *downlink* channel.

Conversely, the multiple access channel is typically identified as the *reverse* or the *uplink* channel.

Different from centralized networks, peer-to-peer networks allow for direct links

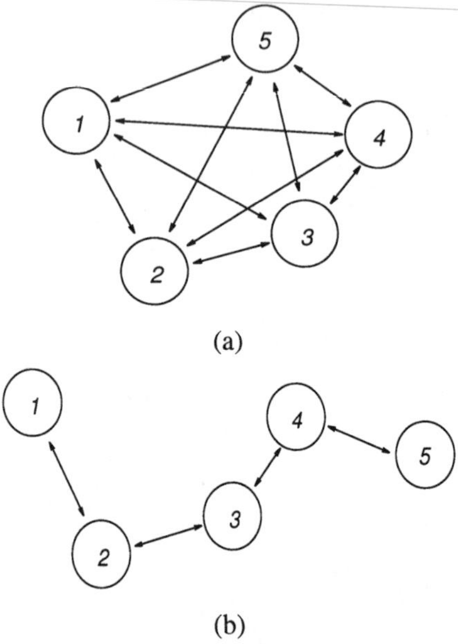

Figure 4.2 Peer-to-peer network topologies: (a) fully connected network and (b) multihop network.

among the users, without passing through any hub. In Figure 4.2(a), we see a *fully connected* network, where every node is (potentially) connected to every other node. This situation occurs, for example, during the deployment phase of an ad-hoc network, where there are no roles assigned to any nodes. Another typical situation is depicted in Figure 4.2 (b), where a *multihop* network is reported. In such a case, the information sent by a source may pass through many intermediate nodes, denoted as *relays,* before reaching the final destination.

4.2.2 A Signal Processing Perspective

Before going into the details of multiple access, it is useful to look at the access problem from a basic signal processing perspective. Let us consider the situation where there are M users transmitting toward a common base station. We assume, for simplicity, that each terminal has a single antenna. To leave the treatment as general as possible, we do not make any a priori choice about the transmission strategy, except that the modulation is linear and the channels are linear (but, possibly, time-varying). The signal transmitted by user k is then

$$x_k(t) = \sum_n s_k(n)\psi_{kn}(t), \quad t \in [0, T), \tag{4.1}$$

where $s_k(n)$ is the nth symbol transmitted by the kth user and $\psi_{kn}(t)$ is the waveform used by the kth user to carry the nth symbol; the summation index n runs over the symbols transmitted by user k. Denoting with $h_k(t, \tau)$ the time-varying channel impulse response between the kth user and the receiver, the signal observed at the access point is given by the superposition

$$y(t) = \sum_{k=1}^{M} \int_0^T h_k(t, t-\tau)x_k(\tau)d\tau, \tag{4.2}$$

where T denotes the duration of the transmitted signals. Substituting (4.1) in (4.2), the received signal can be written as

$$y(t) = \sum_{k=1}^{M} \sum_n s_k(n) \int_0^T h_k(t, t-\tau)\psi_{kn}(\tau)d\tau. \tag{4.3}$$

Given (4.3), we would like to find the set of waveforms $\{\psi_{kn}(t)\}$, to assign to each user, in order to satisfy some optimality criterion. For example, to simplify the receiver design, it would be important to use waveforms that simplify the extraction of the transmitted symbols from the received signal $y(t)$. Recalling the general analysis of time-varying systems carried out in Chapter 2, we know that each impulse response can be expanded in terms of its eigen-modes; that is,

$$h_k(t, t-\tau) = \sum_i \lambda_i^{(k)} u_i^{(k)}(t) v_i^{(k)*}(\tau), \tag{4.4}$$

where the sets of functions $\{u_i^{(k)}(t)\}$ and $\{v_i^{(k)}(t)\}$ are the left and right singular functions of $h_k(t, t-\tau)$. Generalizing the approach of Chapter 2, we could think of using as waveforms for user k the right singular functions $v_i^{(k)}(t)$ of $h_k(t, t-\tau)$. We know, from Chapter 2, that this choice is indeed the best choice, in terms of decoding simplicity. However, in the multiuser case, the situation is much more complicated. This happens because the sets $\{v_i^{(k)}(t)\}$ corresponding to different channels are, in general, distinct and nonorthogonal to each other. This means that it is not possible, in general, to devise an overall transmission strategy, for each user, such that the signals transmitted from each user remain orthogonal to each other after having passed through (different) communication channels.

Nevertheless, there are two important exceptions to the previous statement, that play an important role in the applications.

Time-invariant channels: $h_k(t,\tau) = h_k(\tau), \forall k$

If the channels are all time-invariant, we know from Chapter 2 that the eigenfunctions are complex exponentials of the form $v_i^{(k)}(t) = e^{j2\pi f_{i,k}t}$, *whichever is the behavior of* $h_k(t)$. In such a case, the k-th user can transmit the signal

$$x_k(t) = \sum_n s_k(n) e^{j2\pi f_{n,k}t}. \tag{4.5}$$

The received signal is then

$$y(t) = \sum_{k=1}^{M} \sum_n s_k(n) H_k(f_{n,k}) e^{j2\pi f_{k,n}t}, \tag{4.6}$$

where $H_k(f_{n,k})$ is the transfer function of the kth channel, evaluated at frequency $f_{n,k}$. From (4.6), we infer that a simple way to eliminate both intersymbol and multiuser interference, at the receiver, consists in assigning different sets of frequencies to different users; each user associates also different frequencies to different symbols. In fact, if the frequencies $f_{k,n}$ are all different from each other, a simple spectrum analysis is sufficient to recover all the symbols $s_k(n)$, without any interference. Hence, in case of time-invariant channels, complex exponentials (with linear phase) are waveforms that remain orthogonal to each other, even after passing through distinct channels, whatever the channels may be. This multiplexing strategy is indeed the (theoretical) principle underlying the so called frequency division multiple access (FDMA) system.

How many symbols can be transmitted in this way? The answer depends on the duration of the time interval over which the signals are transmitted and then received. Denoting with T the duration of the transmitted signal (and thus the duration of the complex exponentials) and with B the bandwidth allocated to the service, there are $N = BT$ orthogonal waveforms. Hence, in this way we can have an aggregate rate (the sum of the rates of all the users) equal to BT.

Multiplicative channels: $h_k(t,\tau) = m_k(t)\delta(\tau), \forall k$

If all the channels are multiplicative channels, we know from Chapter 2 that the eigenfunctions are delta pulses, *whatever is the behavior of* $m_k(t)$. Hence, in this case, the most suitable waveforms are, in the idealistic case of an infinite bandwidth system, Dirac pulses. The transmitted signals are then

$$x_k(t) = \sum_n s_k(n)\delta(t - t_{k,n}). \tag{4.7}$$

The received signal is, consequently,

$$y(t) = \sum_{k=1}^{M}\sum_n s_k(n)m_k(t_{k,n})\delta(t - t_{k,n}). \tag{4.8}$$

From (4.8), we see that to avoid ISI and MUI it is sufficient to assign short pulses, located over distinct time instants $t_{k,n}$ to different users and different symbols. This is indeed the basic theoretical principle underlying time division multiple access (TDMA): Different users get disjoint time slots to transmit their data. Within each slot, the symbols are transmitted using non-overlapping (in time) signals.

In practice, there is no infinite bandwidth available to any service. Thus, denoting by B the available bandwidth, instead of the Dirac pulses we use, in practice, pulses of duration approximately equal to $1/B$. Hence, in T seconds we can have an aggregate rate of BT symbols, which can be easily separated provided that all the channels are (approximately) multiplicative, even assuming a finite bandwidth.

Even though the previous considerations come from a purely theoretical signal processing point of view, they provide a simple justification of two of the most common multiple access techniques [i.e., TDMA and FDMA (or OFDMA)]. In summary, the above considerations suggest that, if the choice among alternative access techniques should rely only upon the channel characteristics, the best choices would be to use FDMA (or OFDMA) schemes for time-invariant, frequency-selective, channels or TDMA for time-varying, flat-fading, channels. The choice, in practice, is more complicated, as it depends on QoS requirements, as well as on traffic properties. There are two fundamentally distinct categories of assignment: fixed assignment and random access, which we will review in the next sections.

4.2.3 Fixed Assignment

With fixed assignment methods, some of the available resources are assigned to a user, whenever the user requests a link, depending of course on the availability of resources. The resources are kept by the user to whom they have been assigned until the user releases them. The choice of the assignment method depends primarily on the domains available for the multiple access. The first fundamental distinction concerns single-antenna and multiantenna systems. If all terminals have only one antenna, the only available domain is the time domain. In this case, there are three typical fixed assignment strategies: frequency division multiple access (FDMA), time division multiple access (TDMA), and code division multiple access (CDMA). We can see all these three methods as systems where *orthogonal* waveforms are assigned to distinct users. The orthogonality can be built, in fact, in several different ways. Having a time interval of duration NT_s to accommodate for N users, we can use, for example:

- *Time-division:* A set of N waveforms $\{g_k(t), k = 1, \ldots, N\}$ nonoverlapping in time, each one of duration T_s; that is,

$$g_k(t) = rect_{T_s}(t - T_s/2 - kT), k = 0, \ldots, N-1;$$

- *Frequency-division:* A set of N signals whose spectra do not overlap in frequency; this amounts to use as waveforms the set

$$g_k(t) = e^{j2\pi kt/(NT_s)}, t \in [0, NT_s], k = 0, \ldots, N-1;$$

- *Code-division:* A set of orthogonal waveforms $\{g_k(t)\}$ occupying the whole available bandwidth and time interval, but still satisfying the orthogonality conditions

$$\int_0^{NT_s} g_k^*(t) g_m(t) dt = \delta_{km}. \tag{4.9}$$

In all these cases, we have N degrees of freedom to play with for accommodating up to N users. These degrees of freedom are used differently in all previous systems, but all the waveforms satisfy the orthogonality condition (4.9).

In practice, the situation is complicated by the fact that we can never have an infinite time or infinite bandwidth to allocate our resources. Hence, having a finite bandwidth, we cannot design waveforms that have an exactly finite support in time. Similarly, by duality, using finite duration waveforms, we cannot have waveforms that have a rigorously finite support in frequency (i.e., finite bandwidth). This implies that, with both TDMA or FDMA, there is an unavoidable interference among adjacent channels. The situation is even more complicated by the transit of the signal through the channel that may introduce dispersion in time and/or frequency. To avoid the resulting interference, it is necessary to insert some guard intervals, either in time or frequency, as illustrated in Figures 4.3 and 4.4. The presence of these guard intervals constitutes a rate loss, as some of the physically available resources are wasted only to insure the users' separability.

Up to now, we have considered systems with only one antenna in each terminal. However, if the access point, or base station, has a multiantenna transceiver, we have more degrees of freedom available to design the access method. In principle, if the access point has n_R receive antennas, it can accommodate up to n_R users, even if they use the same temporal resources, provided that they are sufficiently separated in angle of arrival. This leads to what is known as *space division multiple access* (SDMA). Using block (as opposed to scalar) transmissions over N time intervals, we have $N \cdot n_R$ degrees of freedom to accommodate up to as many users. Hence, combining SDMA with any of the multiple access techniques described before, there is the possibility to accommodate up to $N \cdot n_R$ users. Thus, the availability of multiple antenna transceivers adds great flexibility in the design of the multiple access scheme.

We review now some of the basic properties of the different access techniques. The interested reader may check, for example, [2, 1] for a much more in-depth analysis.

4.2.3.1 Frequency Division Multiple Access

FDMA is the simplest and oldest form of multiplexing. In FDMA systems, the band available for the service is divided into several subchannels that are assigned to the users on a fixed basis. Once assigned, a subchannel is held by a user until it is released by the user terminal itself. The receiving terminal filters the incoming signal and retains only the signal passing through the band of interest. The FDMA principle of operation is sketched in Figure 4.3, where we see a series of nonoverlapping frequency subchannels. Different subchannels are assigned to different users. Between two adjacent subchannels there is typically a guard interval to facilitate the separation of different user signals at the receiver. The width of the guard intervals depends on the frequency synchronization properties of the transmit/receive systems, as well as on the selectivity of the receive filters. To limit the waste of resources due to the insertion of the guard bands, it is necessary to have a good frequency synchronization.

FDMA is efficient when the flow of information to be sent is steady. This happens, for example, with voice signals. However, the method is inefficient with data that are in nature more bursty or sporadic, like, for example, computer data.

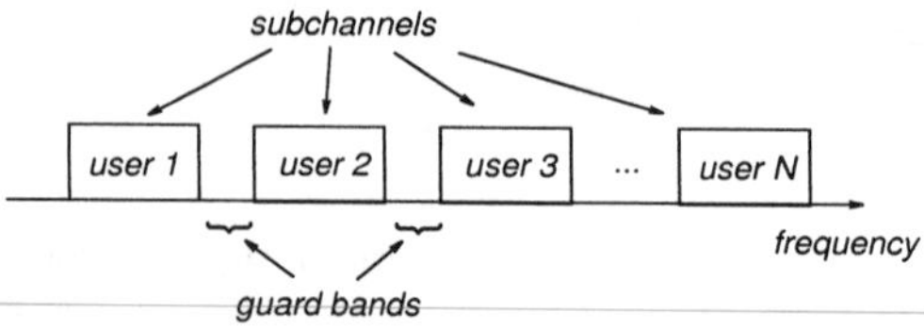

Figure 4.3 FDMA format.

4.2.3.2 Time Division Multiple Access

With TDMA, different users get nonoverlapping time slots. The situation is dual with respect to the FDMA case and it is pictorially represented in Figure 4.4. Between any two adjacent time slots there are now time guard intervals. The duration of the time guard intervals must be necessarily greater than the channel delay spread. To limit the rate losses due to the insertion of these guard intervals, the time slots must have a duration much longer than the delay spread. On the other hand, the duration cannot be increased too much, because otherwise there would be an excessive decoding delay. Clearly, from this point of view, TDMA systems are more appropriate for narrowband channels. The time slots are also organized in frames, where each frame contains the time slot for a certain number of users plus overhead bits carrying signaling information. As opposed to FDMA, where each user gets only a portion of the bandwidth, in TDMA systems all bandwidth allocated for a given service is assigned to each user, upon request.

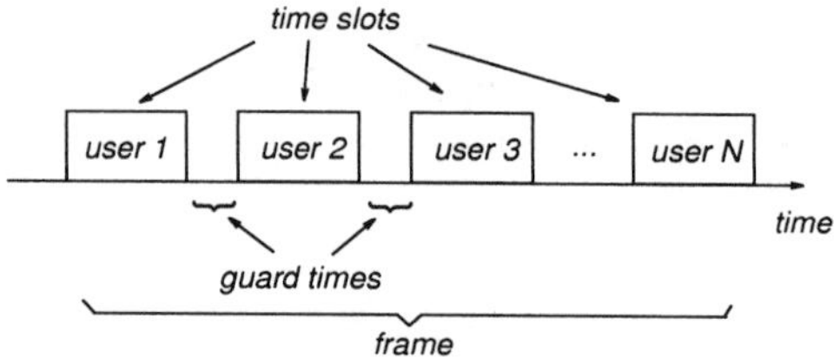

Figure 4.4 TDMA format.

The most typical example of TDMA is the so called *T1 carrier*, used in the public telephone network. A T1 frame is formed by multiplexing 24 PCM-encoded voice channels, each one carrying bits at a rate of 64 kbps. The rate of the T1 carrier is 1.544 Mbps, as given by the 24 channels plus signaling bits.

r hand, equalization or diversity schemes require channel knowledge a side. This, in turn, requires channel estimation. The simplest way to esti- annel consists of sending, periodically, some kind of training sequence, oth transmitter and receiver. To limit the rate loss due to the insertion sequences in the transmitted signal, we can use packet durations much n the duration of the training sequence. But the packet length should be gh to guarantee that the channel to be estimated is constant within each t long enough to get a sufficiently reliable estimate. How often the training should be transmitted depends on the channel coherence time. Hence, the channel variation, the greater is the rate insertion loss.

Code Division Multiple Access

nd TDMA strategies may be seen as methods that assign to different users tions of the available frequency or time. A different philosophy is followed *ode division multiple access* (CDMA) methods where, in principle, every get the whole bandwidth *and* the whole time. The waveforms assigned to t users (and different symbols) are orthogonal, as with FDMA or TDMA, are localized neither in the frequency nor in the time domain. Conversely, aveform tends to occupy the whole time and band available. CDMA is the d radio access technique for third generation (3G) systems. The CDMA rd for 3G systems is essentially based on the direct-sequence spread spec- DS-SS) technique, which we will review in Section 4.3.1. For this reason, the method used for 3G systems could be better identified as DS-CDMA.

ain advantages of CDMA are the following.

iming flexibility: The main advantage of CDMA with respect to TDMA is that t can operate without timing coordination among the users.

Robustness against selective fading: Since the CDMA waveforms occupy the whole spectrum and time available for transmission, the transmission is inherently robust against selective fading. In fact, as opposed to TDMA, where a deep fade in time can cause the loss of bits transmitted over that time interval, or FDMA, where a deep fade in frequency can preclude the detection of the symbols transmitted over the corresponding frequency, in CDMA, each symbol is transmitted over the whole time and band, so it can never be totally attenuated. If properly exploited, frequency and/or time selectivity can be turned into

The TDMA receiver must synchronize with th
extract the time slots of the user of interest. As
with steady data traffic, but it is inefficient with
the basic access method of second generation c

It is useful to compare FDMA and TDMA alon

- *Flexibility:* TDMA is more flexible than FD
atively simpler to assign time slots in order
transmission rates to different users. This is
supporting multimedia applications, where
rates.

- *Nonlinear amplifiers:* With FDMA, the transm
perposition of subchannels. This addition give
amplitude (as well as in phase). Hence, the am
should operate in a linear regime. This limits t
vents the amplifiers from operating close to thei
with TDMA systems, in principle, the transmitt
modulus so that it is less critical to make the pow
saturation point.

- *Immunity to interferences:* In a FDMA system, nar
only one subchannel, whereas in TDMA, the sa
the channels. At the same time, whereas all the
interference acts over one subchannel, in a TDM
power is split among all the subchannels and then ea
only with a portion of the interference power. Hen
to narrowband interference than FDMA. Conversely,
interference is impulsive, in nature, it is better to use F

- *Diversity:* In FDMA system, each subchannel is ty
might appear as an advantage as it implies that ea
need an equalization filter at the receiver. Converse
is typically frequency-selective, and thus the receive
form of equalization. However, it is also important to
clearer in Chapter 5, that frequency selectivity or, equ
dispersiveness can be turned into a useful form of divers
bandwidth large enough to discriminate different paths i

On the othe
the receive
mate the c
known to
of training
greater th
short eno
packet, bu
sequence
faster the

4.2.3.3

FDMA
only po
by the
user ca
differe
but the
each w
standa
standa
trum (
acces

The r

useful time and/or frequency diversity gains in DS-CDMA systems.

- *Interference resistance:* As will be reviewed in Section 4.3.1, one of key features of spread-spectrum systems is their robustness against both narrowband and wideband interference. Since DS-CDMA is built on SS systems, this property is shared by all DS-CDMA systems.

- *Soft handoff:* In cellular systems, when a mobile passes from one cell to the next, it has to switch the link from the old base station to the new one. This procedure is known as *handoff*. In 2G systems, where adjacent cells transmit over nonoverlapping frequency bands, the handoff is *hard*, meaning that the receiver has to switch from the frequency band to the other, as it crosses the boundary from one cell to the next. Conversely, adjacent cells in a DS-CDMA cellular network use the same frequency band. Hence, when a mobile moves from one cell to the next, it can communicate with both cells and even combine the two signals advantageously. Only when a reliable link has been established with the new station does the mobile user stop communicating with the previous station. This technique is called *soft handoff*.

- *Soft capacity limits:* With CDMA, as opposed to both TDMA and FDMA, there is no hard limit to the number of users that can be accommodated within a cell. Clearly, as the number of active users increases, the interference increases and this sets an upper bound on the maximum number of users that can get access. However, this bound is soft. Typically, different access points can give access to different numbers of users. Also the boundary of a cell is not as well defined as in TDMA or FDMA cellular systems. The boundary also changes through time, depending on the traffic within the cell and the position of the users that get access. This phenomenon is known as *cell breathing*.

The main disadvantages of CDMA systems are the following.

- *Complexity:* The immunity against interference and the diversity gain come at the price of increased complexity of the receiver. The receiver needs in fact to estimate the channel and set up appropriate countermeasures against the interference. In cellular systems, it is particularly important to keep weight and cost of the mobile handset as low as possible. Another crucial issue is the battery life of the mobile handset. All these factors prevent the use of too complicated decoding techniques at the receiver in the mobile handset. Hence, the design of the receiver has to result in a trade-off between performance and complexity.

The comparison of alternative decoding techniques will be the subject of Section 4.3.

- *Power control:* Since DS-CDMA systems are inherently affected by interference, it is particularly important to adapt the power used on each link in order to limit the detrimental effects of mutual interference. A typical problem is the so-called *near-far* problem, where users close to a base station can create a huge interference toward users accessing the same station from the boundary of the cell. To cope with this problem and try to maximize the number of users accessing the system with satisfying QoS, it is necessary to implement a power control (i.e., a feedback mechanism that forces the users to adapt their transmission power depending on the interference level). Power control will be analyzed in Section 4.3.4.

4.2.3.4 Space Division Multiple Access

Let us consider now a base station that has a multiantenna transceiver. In such a case, one more dimension can be exploited to separate the different users: the space domain. An example is shown in Figure 4.5, where the base station, located at the center of the figure, uses a set of six radiation patterns, indicated as beams 1 through 6 in the figure. The distinct radiation patterns are obtained by properly combining the signals received by each array element in reception or, by antenna reciprocity, by properly illuminating each array element, in transmission. The example sketched in Figure 4.5 is typically called a *sectorized* antenna, as the antenna patterns divide the space into a certain number of sectors. In principle, if the six radiation patterns are perfectly nonoverlapping, the space is divided into six distinct regions so that users belonging to distinct sectors are perfectly separated through the array. Statistically speaking, each sector, in the cell covered by the BS at the center of the figure, contains, on average, one-sixth of the users. Hence, the interference problem is mitigated. In practice, each radiation pattern has sidelobes through which enter the signals coming from users outside of the desired sector. Hence, we cannot expect perfect elimination of the users outside the sector. Nevertheless, there is certainly a reduction in multi-user interference thanks to the spatial filtering offered by the antenna array. An important point to emphasize with sectorized antennas is that we can reuse the same codes, time, or frequency slots for all users that pertain to different sectors, as far as they are well separated in angle. This increases the capacity of the system, without requiring any bandwidth expansion.

A more interesting possibility resulting from the availability of a multiantenna receiver at the base station, is that we may form different radiation patterns for each user. An example is sketched in Figure 4.6, where we have a base station with

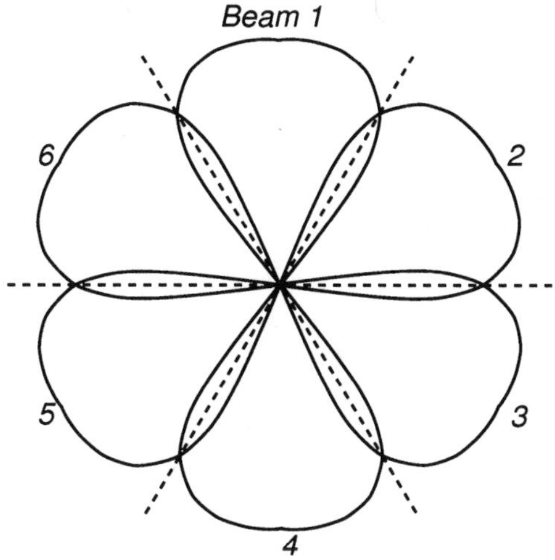

Figure 4.5 Space division multiple access.

an eigth-element antenna. Such an antenna is capable of enforcing seven zeros of its radiation pattern in as many directions. This means that if eigth users try to access the base station, we can form eigth distinct radiation patterns, one for each user. Each pattern maximizes the gain in the direction of the user and puts a null in the direction of all other users. Figure 4.6 depicts, as an example, the pattern for user 1. In this way, we have the possibility of a system with potentially no interference, provided that the number of antenna elements at the base station is equal to the number of users. Playing with both time and space, if each user transmits over blocks of N samples, with a receive array of n_R elements, we have the possibility of accommodating $N \cdot n_R$ users.

4.2.4 Random Assignment Strategies

The fixed assignment methods seen in the previous section make an efficient use of the radio resources when the flows of information exchanged among the users are relatively steady. In fact, with a fixed assignment, once a resource is assigned to a user, it cannot be given to any other user. Thus, if during the period where the resource is assigned, the user has no data to transmit, there is a clear waste of resources. This happens with TDMA and FDMA systems, when the data to transmit

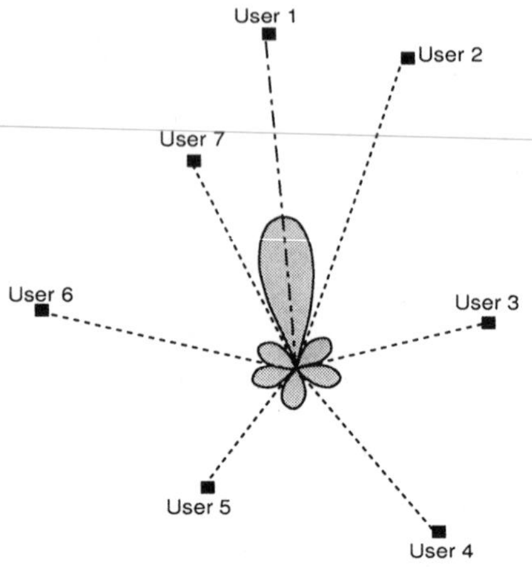

Figure 4.6 Example of receive beamforming.

are sporadic by nature. TDMA and FDMA are in fact designed to be interference-free: No time or frequency slots are shared by any pair of users. With CDMA systems, the situation is different, as they are not interference-free. In principle, CDMA systems assign all the radio resources, both time slots and band, to every user. As the number of users increases, the interference increases and the reception becomes more and more complicated. However, if a user with an assigned code does not have data to transmit, it does not create any interference to the other users and then it is as if that resource had not been assigned. Nevertheless, also with CDMA, once a code has been assigned to a user, it cannot be given to any other user.

A different philosophy is followed by *random access* methods, where each user may try to access the network whenever it has data to transmit, *without asking for a dedicated radio resource in advance*. In such a case, two, or more, users might try to access the network at the same time creating collisions. The system needs then a protocol to resolve the channel contention. One of the most known strategies for resolving the problem is the so-called *pure ALOHA* [3, 4][1].

In ALOHA, users send packets properly encoded with error-detection codes. After sending a packet, the user waits for an acknowledgment from the receiver that the packet has been received without errors. If no acknowledgment is received, the

1 ALOHA derives its name from the place where it was first applied, the University of Hawaii.

packet is assumed to be lost in a collision with other users. If a user realizes that its packet has collided with another user, it waits for a random time and then it tries again to retransmit. Since the waiting time is random and independent from user to user, it is less likely that, with the second attempt, the two users will collide again.

One of the key performance parameters of an access method is the *throughput* S, defined as the percentage of successful packet transmissions, per unit of time. A simple analysis is useful to note the basic characteristics of ALOHA (see, e.g., [2] for more details). Let us assume that all packets have a standard length T_p. We may compute S as the the average number of packets received successfully in the time interval T_p. Let us also assume that there is an infinite population of users and that the traffic generated by the users is stationary, with intensity G. The traffic intensity measures the average number of packet transmissions *attempted* per packet time T_p, including new packets as well as retransmissions. The standard unit of traffic flow is the *erlang*: A traffic flow of one packet per T_p seconds has a value of one erlang. By the previous definition of throughput, S cannot exceed 1. If more than one packet is sent within an interval T_p, there is a collision and then the packet cannot be decoded successfully. A typical assumption about the traffic, is that the probability that k packets are generated during a given packet time follows a Poisson distribution, with mean G, so that the probability of having k packets in a time T_p is

$$P(k) = \frac{G^k e^{-G}}{k!}. \tag{4.10}$$

Denoting with P_e the probability that the packet is received with an error, the throughput S is simply the traffic load G times the probability of correct decision $P_c := 1 - P_e$; that is,

$$S = (1 - P_e)G = P_c G. \tag{4.11}$$

Given a packet sent at time t_0, there is a collision if some other user attempted to send packets from time $t_0 - T_p$ to time $t_0 + T_p$. For a given traffic intensity G, the probability that there are no packets in a time window of duration $2T_p$ is e^{-2G}. Assuming that the only source of error in packet decoding is the presence of collisions, the probability of a correct decision is equal to

$$P_c = e^{-2G}. \tag{4.12}$$

As a consequence, the throughput of the pure ALOHA system is

$$S = Ge^{-2G}. \tag{4.13}$$

This curve is plotted in Figure 4.7. We can see, from Figure 4.7, that as the offered traffic increases -albeit is still low- the throughput increases. However, if the traffic increases excessively, the number of collisions increases consequently and this

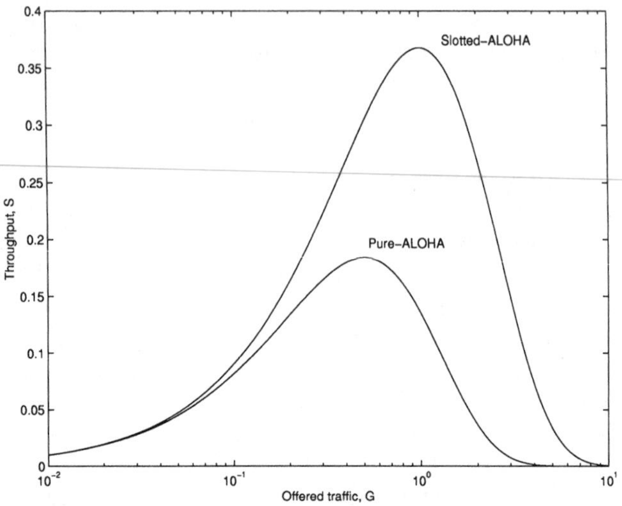

Figure 4.7 Throughput versus traffic offered by pure and slotted ALOHA systems.

induces a throughput reduction. In fact, we observe that when the traffic exceeds the value $G = 0.5$, the throughput starts to decrease. It is also important to remark that the maximum throughput is $S_{\max} = 1/(2e) \approx 0.184$. This shows that the best channel utilization achievable with pure ALOHA is quite small.

A better efficiency can be achieved using the so-called *slotted ALOHA*. With slotted ALOHA, the transmission is slotted into time slots of duration equal to the packet length T_p. Each user is then required to transmit its packet only within the available slots. This implies a time synchronization of the overall system. As a consequence of time slotting, the vulnerability window is now only T_p, instead of $2T_p$ of pure ALOHA. Hence, the probability of no collision is now

$$P_c = e^{-G} \tag{4.14}$$

and the throughput is then

$$S = Ge^{-G}. \tag{4.15}$$

The throughput S of slotted ALOHA is also plotted in Figure 4.7. We can see that slotted ALOHA has indeed an advantage with respect to pure ALOHA, as it yields more throughput, for a given offered traffic. In particular, the maximum throughput achievable with slotted ALOHA is $S_{\max} = 0.368$, achieved with $G = 1$. However, slotted ALOHA is more complicated than pure ALOHA, as it requires time synchronization.

Even though slotted ALOHA provides better efficiency than pure ALOHA, its efficiency is still quite low. To improve upon ALOHA, alternative techniques have been proposed, like carrier sense multiple access (CSMA), in the effort of reducing the number of collisions. The study of CSMA methods goes beyond the scope of this book and the interested reader is encouraged to check [2] for an in-depth analysis. It is, nevertheless, useful to recall here that mechanisms like CSMA, which are widely employed and work quite well in wired networks (e.g., Ethernet), are not as efficient in wireless channels, because of the radio channel variability.

Alternative ways to have efficient multiple access schemes, more suitable for wireless links, and with great efficiency, have been proposed in [5], based on a completely different philosophy. In [5], the transmission is slotted, but collisions are not immediately discarded. On the contrary, if K users collide on a slot, they are asked to retransmit their packet over the immediately subsequent $K-1$ slots. In this way, it is sure that there will be K collisions over K consecutive slots. In each retransmission, each user changes some property of the transmitted signal (typically its phase). Different users change their initial phase, on each slot, in a manner different from each other. If, in the meanwhile, no other users attempt to access the channel, the receiver collects K observations that are linear combinations of K unknowns (the data on the K packets). If this linear system admits a solution, the receiver is capable of resolving the K packets, received in K slots, thus implying no loss of efficiency. This is just an example where the combination of link layer strategies properly coupled to physical layer strategies, may give rise to a considerable gain in the overall network efficiency.

If the receiver has n_R antennas, it can handle up to n_R colliding packets, again by solving a system of n_R equations in n_R unknowns. This provides a great increase in throughput even without asking for retransmissions. We will now consider with more detail the physical layer of CDMA systems.

4.3 CODE DIVISION MULTIPLE ACCESS

The basic idea underlying CDMA systems can be traced back to the so-called *spread-spectrum* (SS) systems. To better clarify the basic principles of CDMA, we start our analysis by illustrating the basic principles of spread-spectrum systems. We review only the structure of spread-spectrum systems known as direct sequence

(DS) spread-spectrum, as this constitutes the basic building block of CDMA systems.

4.3.1 Spread-Spectrum Systems

Let us consider a radio terminal that wishes to send symbols at rate $R_s = 1/T_s$ symbols/s, where T_s is the symbol period. This system requires a baseband bandwidth of (at least) $B_s = 1/(2T_s)$ Hz. In SS systems, each symbol to be transmitted is multiplied by a code (vector) composed of N elements, denoted as *chips*. The parameter N is known as the code *spreading factor*. To keep the symbol rate constant, each chip must have a duration $T_c = T_s/N$. The sequences $c(l)$ used in DS-SS systems are typically *pseudonoise* real sequences and have unitary modulus [i.e., $c^2(l) = 1$]. If we denote with $s(m)$ the symbols, with $c(l)$ the code sequence, with $l = 0, \ldots, N-1$, and with $g(t)$ the impulse response of the shaping filter, the transmitted baseband signal is

$$x(t) = \sum_{m=-\infty}^{\infty} s(m) \sum_{l=0}^{N-1} c(l) g(t - mNT_c - lT_c). \tag{4.16}$$

This signal requires a bandwidth of (at least) $W = 1/(2T_c) = NB_s$ Hz. This feature justifies the name of the method, as the systems spreads the transmit power over a spectrum that is larger than the spectrum strictly required to send the original sequence of symbols, by a factor N. At first sight, increasing the transmission bandwidth well beyond the value required for data transmission does not seem to be a useful idea. On the contrary, we will show below that a DS-SS system has several interesting features that justify its application. The baseband transmission scheme

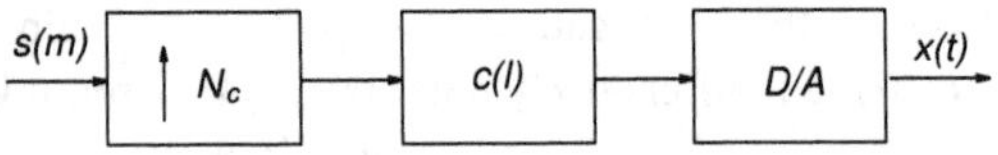

Figure 4.8 Baseband transmission scheme of a DS-SS system.

is sketched in Figure 4.8, and it is composed of the following blocks:

1. The first block performs up-sampling by a factor N (i.e., it inserts $N - 1$ zeros after each symbol); the data rate changes then from the symbol rate R_s, at the input of the up-sampler, to the chip rate $R_c = NR_s$, at the output of the up-sampler;

2. The second block is a discrete-time filter having, as impulse response, the code sequence $c(l)$, with $l = 0, \ldots, N-1$;

3. The third block performs a digital to analog (D/A) conversion, producing the continuous-time baseband signal $x(t)$, given by (4.16).

The baseband receiver is sketched in Figure 4.9, where we can recognize the following blocks:

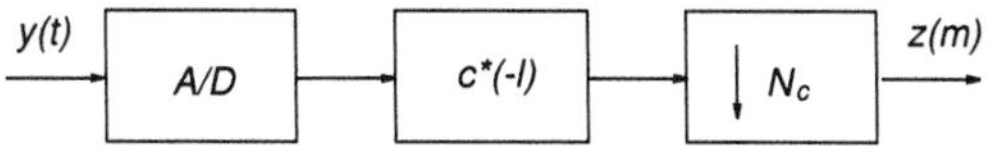

Figure 4.9 Baseband reception scheme of a DS-SS system.

1. The first block performs an analog to digital (A/D) conversion, producing the sequence $y(m)$;
2. The second block is a discrete-time filter having as impulse response the code sequence $c^*(-l)$, i.e. the filter matched to $c(l)$;
3. The third block performs a down-sampling by a factor N, i.e. it discards $N - 1$ samples every N samples; the data rate passes then from R_c, at the input of the down-sampler, to $R_s = R_c/N$, at the output of the down-sampler.

In practice, because of the down-sampling operation, it is not necessary to compute all the values at the output of the receive filter, but only the values with indexes given by integer multiples of N. Furthermore, often the code sequence $c(l)$ is real, so that we may neglect the conjugate appearing in Figure 4.9. In such a case, the matched filter, followed by the down-sampler, at the receiver, can be implemented very simply as follows. The received sequence is parsed into blocks of N consecutive samples. Each block is then multiplied by the sequence $c(l)$, $l = 0, \ldots, N-1$. The N products are then summed up.

Typically, the transmission scheme of a DS-SS system is sketched as in Figure 4.10. However, one has to pay attention to the fact that the data rate before and after the

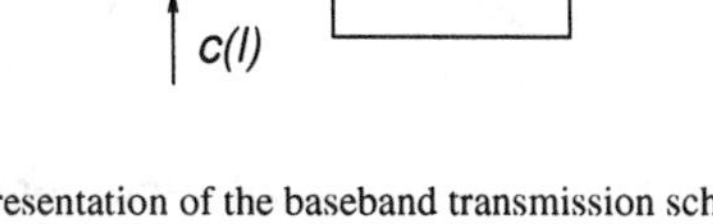

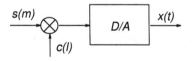

Figure 4.10 Alternative representation of the baseband transmission scheme of a DS-SS system.

multiplication by the code $c(l)$ is different, whereas in the scheme of Figure 4.8, the change of rate is made evident by the presence of the up-sampler. However, the representation of Figure 4.10 is useful to capture the basic operation of the transmitter (i.e., the multiplication of each symbol by the code sequence). With the warning in mind about the rate change, the overall DS-SS system can be represented as in Figure 4.11, where $i(t)$ represents the interference, whereas $n(t)$ is noise. The

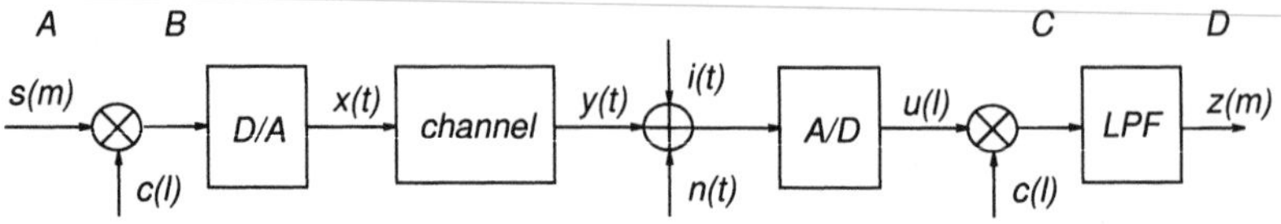

Figure 4.11 Alternative representation of the baseband transmission scheme of a DS-SS system.

final block in the receiving chain is a lowpass filter (LPF), with bandwidth $1/2T_s$.

To have a physical insight into the properties of a DS-SS system, it is useful to analyze how the spectra of useful signals and interference change, as they pass through the system. To this end, in Figure 4.11, we have labeled four points as A, B, C, and D. The power spectral densities in points A and B, denoted simply as $S_A(f)$ and $S_B(f)$, are reported in Figure 4.12 Comparing $S_A(f)$ with $S_B(f)$, we observe

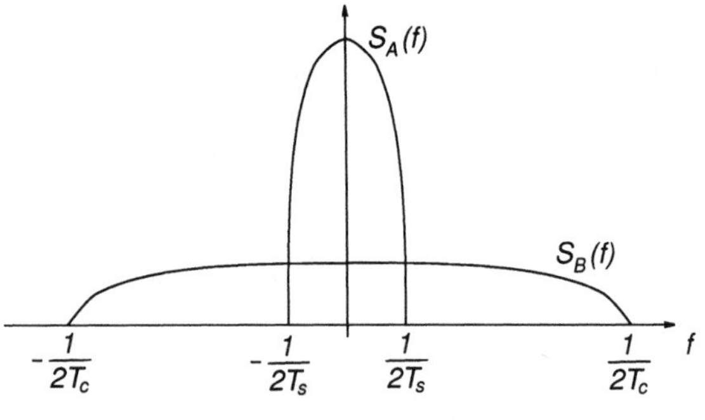

Figure 4.12 Power spectral densities in different points of a DS-SS system.

how the multiplication by the code spreads the spectrum. This is a consequence of using pseudonoise code sequences. On the other hand, since the spreading sequence $c(l)$ has unitary modulus, multiplying the symbol sequence by the spreading code does not alter the signal power. For this reason, the areas subtended by the two power spectral densities in Figure 4.12 are the same. Using large spreading factors

N, the expanded spectrum may have a very low-power spectral density[2]. At the receiver, after the multiplication by the spreading code $c(l)$, exploiting the property that $c^2(l) = 1$, the power spectral density of the useful signal shrinks back into the bandwidth between $-1/2T_S$ and $1/2T_S$. Conversely, all other signals (i.e., interference or noise) have a spectral density whose shape is like $S_B(f)$. At the input of the final lowpass filter, the useful signal has then a spectrum like $S_A(f)$ in Figure 4.12, whereas all other signals (i.e., either interference or noise) have a wideband spectrum, like $S_B(f)$ in Figure 4.12, because of the multiplication by the code sequence, at the receiver. Hence, the useful signal passes unaltered through the final lowpass filter, whereas the interference is attenuated by a power factor N. The overall system yields then a gain N in the signal-to-noise-plus-interference ratio (SINR), as proved with the following derivations.

If we consider the mth transmitted symbol $s(m)$, the corresponding sequence of chips observed at the receiver is:

$$y_m(n) = s(m)c(n) + i(n) + v(n), \quad n = 0, \ldots, N-1, \tag{4.17}$$

where $i(n)$ and $v(n)$ are the interference and noise sequences. Let us denote with P_s^{in}, P_i^{in}, and P_n^{in} the power of signal, interference and noise at the input of the DS-SS receiver, respectively. We compute now the power of these three contributions at the output of the receiver. Multiplying $y_m(n)$ by $c(n)$, and summing over n, we get

$$z(m) = \sum_{n=0}^{N-1} c(n)y_m(n) = Ns(m) + \sum_{n=0}^{N-1} c(n)i(n) + \sum_{n=0}^{N-1} c(n)v(n). \tag{4.18}$$

In (4.18), we have used the property $c^2(n) = 1$. To compute the power of the interference and noise contribution, we exploit now the property that the spreading sequence is a pseudonoise sequence. In particular, considering the elements of $c(n)$ as i.i.d. random variables with zero mean and unit variance, the power of the interference and noise, at the output of the DS-SS receiver, are $P_i^{\text{out}} = NP_i^{\text{in}}$ and $P_n^{\text{out}} = NP_n^{\text{in}}$, respectively. As a consequence, the SINR at the output of the DS-SS receiver is

$$\text{SINR}^{\text{out}} = \frac{N^2 P_s^{\text{in}}}{NP_i^{\text{in}} + NP_n^{\text{in}}} = N \cdot \text{SINR}^{\text{in}}. \tag{4.19}$$

2 This property was indeed the motivation underlying one of the first uses of spread-spectrum systems for low-probability of intercept (LPI) radars. In fact, the pseudonoise nature of the spreading sequence, coupled with the very low-power spectral density may make the transmitted signal very similar to the receive noise, so that this kind of signal has low probability of being intercepted by hostile receivers.

This relationship shows that DS-SS provides a gain factor N in terms of SINR and thus it explains the robustness of DS-SS against interferences.

After this preliminary review of spread spectrum systems, we can now concentrate on CDMA systems. We consider separately the downlink and then the uplink channel.

4.3.2 CDMA Receivers: Downlink Channel

Let us consider a downlink channel where the access point (or base station) sends information to M active users. We denote with $\boldsymbol{f}_k$ the N-size code vector, assigned to the kth user. The elements of each code vector are conventionally called *chips*. The nth block transmitted from the BS is then

$$\boldsymbol{x}(n) = \sum_{m=1}^{M} s_m(n)\boldsymbol{f}_m := \boldsymbol{F}\boldsymbol{s}(n), \tag{4.20}$$

where $\boldsymbol{F} := (\boldsymbol{f}_1, \ldots, \boldsymbol{f}_M)$ is the $N \times M$ matrix whose columns are the codes of all the active users, whereas $\boldsymbol{s}(n) := [s_1(n), \ldots, s_M(n)]$ is the vector containing the symbols pertaining to all users, in the nth block. The user codes are orthogonal to each other. The structure of the signal in (4.20) can be interpreted as the multiplexing of M DS-SS signals. This perspective is useful because it suggests how to choose the user codes. In fact, we know from previous section that, to increase the robustness against interference, it is useful to employ pseudo noise codes. This choice leads to the so-called direct sequence (DS) CDMA systems.

Typically, the codes have constant modulus (i.e., all chips have the same modulus). This is especially useful for transmissions from mobile handsets that have a limited transmit power, as they can make their high-power amplifier operate close to the saturation region. Furthermore, the user codes are orthogonal to each other. Hence, if the amplitude of each chip is one, the codes respect the following relationship

$$\boldsymbol{f}_m^H \boldsymbol{f}_k = N\delta_{mk}. \tag{4.21}$$

Because of the spectrum expansion resulting from the use of spread-spectrum techniques, a channel that could be flat-fading with respect to the symbol rate, might become frequency-selective with respect to the chip rate. In formulas, this means that the expanded bandwidth W becomes greater than the channel coherence bandwidth B_c, even though the symbol rate R_s is less than the channel coherence bandwidth B_c. Indeed, the most typical situation in DS-SS applications is that

$$R_s < B_c < W. \tag{4.22}$$

This means that, typically, spread spectrum systems convert a flat-fading channel into a frequency-selective channel. Even though frequency selectivity could be seen as an annoying factor in wireless communications, in fact, if properly exploited, the channel dispersiveness can be turned into a useful source of diversity, useful to combat the channel fading. Hence, the band expansion can be seen also as a useful way to induce diversity gain to combat fading[3].

We derive now a formal expression for the received signal in the user terminals of the downlink channel. In our derivations, we assume that the code length N is greater than the channel order L. This is, in fact, the most common situation, in practice. We will denote with $h(l)$ the impulse response of the channel between the BS and the user of interest. The nth received block can be written, in matrix form, as

$$\boldsymbol{y}(n) = \overline{\boldsymbol{H}}\boldsymbol{x}(n-1) + \boldsymbol{H}\boldsymbol{x}(n) + \underline{\boldsymbol{H}}\boldsymbol{x}(n+1) + \boldsymbol{v}(n), \tag{4.23}$$

where $\boldsymbol{v}(n)$ is the noise vector, whereas the matrices $\boldsymbol{H}$, $\underline{\boldsymbol{H}}$ and $\overline{\boldsymbol{H}}$ are $(N+L) \times N$ matrices having the following structure

$$\boldsymbol{H} := \begin{pmatrix} h(0) & 0 & 0 & \cdots & 0 \\ \vdots & h(0) & 0 & \ddots & 0 \\ h(L) & \ddots & \ddots & \ddots & \vdots \\ 0 & \ddots & \ddots & \ddots & 0 \\ \vdots & \ddots & h(L) & \ddots & h(0) \\ \vdots & \ddots & \ddots & \ddots & \vdots \\ 0 & \cdots & \cdots & \cdots & h(L) \end{pmatrix}, \tag{4.24}$$

$$\underline{\boldsymbol{H}} := \begin{pmatrix} 0 & \cdots & \cdots & \cdots & 0 \\ \vdots & \ddots & \ddots & \ddots & 0 \\ 0 & \ddots & \ddots & \ddots & 0 \\ h(0) & \ddots & \ddots & \ddots & \vdots \\ \vdots & \ddots & \ddots & \ddots & \vdots \\ h(L-1) & \cdots & h(0) & \cdots & 0 \end{pmatrix}, \tag{4.25}$$

3 The diversity gain will be studied in detail in Chapter 5.

$$\overline{\boldsymbol{H}} := \begin{pmatrix} 0 & \cdots & 0 & h(L) & \cdots & h(1) \\ \vdots & \ddots & \ddots & \ddots & \ddots & \vdots \\ \vdots & \ddots & \ddots & \ddots & \ddots & h(L) \\ \vdots & \ddots & \ddots & \ddots & \ddots & 0 \\ \vdots & \ddots & \ddots & \cdots & \ddots & \vdots \\ 0 & \cdots & \cdots & \cdots & \cdots & 0 \end{pmatrix}. \tag{4.26}$$

The terms $\overline{\boldsymbol{H}}\boldsymbol{x}(n-1)$ and $\underline{\boldsymbol{H}}\boldsymbol{x}(n+1)$ represent the interblock interference (IBI) of the $(n-1)$-th and $(n+1)$-th blocks, respectively, on the nth block.

The channel matrix $\boldsymbol{H}$ can be written, equivalently, as

$$\boldsymbol{H} = \sum_{l=0}^{L} h(l)\boldsymbol{J}_l, \tag{4.27}$$

where $\boldsymbol{J}_l$ is the $(N+L) \times N$ shift matrix having all ones in its lth subdiagonal and zeros elsewhere. Combining (4.23) with (4.27), we can write

$$\boldsymbol{y}(n) = \sum_{m=1}^{M} s_m(n) \sum_{l=0}^{L} h(l)\boldsymbol{J}_l \boldsymbol{f}_m + \boldsymbol{w}(n) := \sum_{m=1}^{M} s_m(n)\boldsymbol{c}_m + \boldsymbol{w}(n), \tag{4.28}$$

where we have incorporated in $\boldsymbol{w}(n)$ all the disturbance terms, namely noise and inter-block interference (IBI). We see, from (4.28), that each symbol $s_m(n)$, after passing through the channel, is carried, equivalently, by a vector $\boldsymbol{c}_m$, of length $N+L$, related to the original user code $\boldsymbol{f}_m$, of size N, by the relationship

$$\boldsymbol{c}_m := \sum_{l=0}^{L} h(l)\boldsymbol{J}_l \boldsymbol{f}_m. \tag{4.29}$$

The main problems with CDMA come exactly from this code transformation: Whereas the original vectors $\boldsymbol{f}_m$ are orthogonal to each other, the vectors $\boldsymbol{c}_m$ are not orthogonal, in general. This loss of orthogonality complicates the user separation, at the receiver. The received codes $\boldsymbol{c}_m$ are orthogonal if the channel is flat-fading (i.e., $L = 0$) but this is typically not the case.

We analyze now the most common receivers for mobile handsets in DS-CDMA systems. The receiver design depends, primarily, on the knowledge, available at the receiver, of the codes assigned to the active users. Typically, each mobile user terminal knows only its own code, but it does not know the codes of the other active users. As a consequence, the receiver in a mobile handset cannot perform a joint multiuser detection. Furthermore, it is particularly important to simplify as much as possible the design of the handset receiver, to simplify the hardware and, most important, to save the battery life. Under such conditions, the most common and simple receiver structure is the so-called *RAKE* receiver.

RAKE receiver

The philosophy of this kind of receiver is, as its name says, to "rake" all the energy pertaining to the symbol of the user of interest. At the same time, MUI or IBI are treated as an additional white noise. This is clearly a suboptimum approach, since in fact MUI has clearly a structure. Nevertheless, the RAKE receiver is very simple to implement, and it can give appreciable results when there are no many interferers superimposed. The RAKE receiver is essentially the matched filter with respect to the vector $\boldsymbol{c}_m$. Assimilating the interference to white noise, the matched filter for the mth user is given by the row vector $\boldsymbol{g}_m^H = \boldsymbol{c}_m^H$, with $\boldsymbol{c}_m$ given by (4.29). The output of the matched filter is then

$$\begin{aligned} z_m(n) &:= \boldsymbol{g}_m^H \boldsymbol{y}(n) \sum_{k=1}^{M} s_k(n) \sum_{i=0}^{L} \sum_{l=0}^{L} h^*(i) h(l) \boldsymbol{f}_m^H \boldsymbol{J}_i^T \boldsymbol{J}_l \boldsymbol{f}_k + w(n) \\ &= s_m(n) \sum_{i=0}^{L} \sum_{l=0}^{L} h^*(i) h(l) \boldsymbol{f}_m^H \boldsymbol{J}_i^T \boldsymbol{J}_l \boldsymbol{f}_m \\ &+ \sum_{\substack{k=1 \\ k \neq m}}^{M} s_k(n) \sum_{i=0}^{L} \sum_{l=0}^{L} h^*(i) h(l) \boldsymbol{f}_m^H \boldsymbol{J}_i^T \boldsymbol{J}_l \boldsymbol{f}_k + w(n), \end{aligned} \tag{4.30}$$

where $w(n) := \boldsymbol{g}_m^H \boldsymbol{w}(n)$. We can single out three contributions from (4.30): The first term is the useful term, the second is MUI, the third is noise. Given the structure of $\boldsymbol{J}_i$, it is straightforward to show that the terms $\boldsymbol{f}_m^H \boldsymbol{J}_i^T \boldsymbol{J}_l \boldsymbol{f}_k$ are the values of the correlation between codes $\boldsymbol{f}_m$ and $\boldsymbol{f}_k$, evaluated at lag $l - i$. Indicating with $R_{mk}(l - i)$ the correlation between codes m and k, at lag $l - i$, we can rewrite

(4.30) as

$$
\begin{aligned}
z_m(n) \quad := \quad & N s_m(n) \sum_{l=0}^{L} |h(l)|^2 + s_m(n) \sum_{\substack{i,l=0 \\ i \neq l}}^{L} h^*(i)h(l)R_{mm}(l-i) \\
+ \quad & \sum_{\substack{k=1 \\ k \neq m}}^{M} s_k(n) \sum_{\substack{i,l=0 \\ i \neq l}}^{L} h^*(i)h(l)R_{mk}(l-i) + w(n),
\end{aligned}
\tag{4.31}
$$

where we have used the property that $R_{mm}(0) = N$. In the last term of (4.31), we dropped the terms with $i = l$, because $R_{mk}(0) = 0$, for $m \neq k$, thanks to the codes' orthogonality (4.21).

The structure of the vector $\boldsymbol{g}_m^H$ that multiplies the received vector $\boldsymbol{y}(n)$ explains the name of the receiver. In fact, multiplying $\boldsymbol{y}(n)$ by $\boldsymbol{g}_m^H$, from the left side, means that the single components of the received signal carrying the symbol $s_m(n)$ are first re-aligned in time, thanks to the multiplication by $\boldsymbol{J}_i^T$, and then passed through a filter matched to both the code structure and the channel. The overall scheme can thus be seen as a rake that gathers all the contributions pertaining to the signal of interest and piles them up.

From (4.31), we can make the following remarks.

- *Diversity gain:* The useful term has an SNR gain, with respect to white noise, equal to $G_c = N \sum_{l=0}^{L} |h(l)|^2$. This gain contains, as factors, the code gain N and the channel gain $\sum_{l=0}^{L} |h(l)|^2$. This last term gives rise to the so-called *diversity gain*, which will be analyzed in detail in Chapter 5. Nevertheless, it is useful to anticipate here some important properties. It will be proved in Chapter 5, in fact, that if the channel coefficients $h(l)$ are i.i.d. complex Gaussian random variables, with zero mean and the same variance (Rayleigh channel model), the bit error rate averaged over the channel statistics behaves, asymptotically at high SNR, as

$$
\overline{\text{BER}} \simeq \frac{G_k}{(G_c \text{SNR})^{G_d}},
\tag{4.32}
$$

  where the exponent G_d is the so-called diversity gain. For the RAKE receiver, the diversity gain is $G_d = L + 1$. Equation (4.32) is quite important as it shows that multipath, if properly exploited, becomes the source of an important gain, as it affects the exponent of the SNR. Without multipath, we would have in fact neither ISI nor MUI, but the exponent of the SNR in (4.32) would be equal to

1. Conversely, in the presence of independent paths, we do have ISI and MUI, but the average BER decreases as $1/\text{SNR}^{L+1}$, instead of $1/\text{SNR}$ (at least as far as the MUI is not too high to put a lower limit to the BER).

- *Code cross-correlation:* The MUI term in (4.31) depends on the *cross-correlation* properties of the codes. In principle, if we were able to design codes with null cross-correlation, the MUI could be nulled. However, this is a problem that it does not permit an exact solution, as the number of degrees of freedom (the number of chips composing the codes) is less than the number of terms that we wish to null. Nevertheless, from (4.31), it is clear that choosing user codes with low cross-correlation properties is evidently beneficial for the RAKE receiver. This motivates the use of codes with low cross-correlation.

- *Code orthogonality:* The orthogonality among the codes allowed us to null, in the last term of (4.31), the contribution $\sum_{i=0}^{L} |h(i)|^2 R_{mk}(0)$, which would have been, otherwise, the most important contribution to the interference.

- *Code auto-correlation:* The second contribution in (4.31) depends on the *auto-correlation* properties of the mth code pertaining to the user of interest, expressed by the terms $R_{mm}(l-i)$. To reduce self-interference, the codes should be pseudonoise codes, with a code autocorrelation as similar to a unit pulse as possible. This motivates the use of pseudonoise codes.

In summary, all previous comments show how important it is, in the design of a DS-CDMA system, to use codes that are, at the same time: orthogonal, pseudonoise, and mutually uncorrelated. Not surprisingly, then, all these properties have all been incorporated, for example, in the UMTS standard, which will be illustrated in Section 4.4.

Some performance curves are useful to illustrate the behavior of a RAKE receiver. In Figure 4.13, it is shown the average symbol error rate (SER) versus the SNR, obtained using a RAKE receiver. The main parameters are: code length $N = 64$; number of active users $M = 16$; and channel order $L_h = 6$. The constellation is QPSK. Each curve refers to a specific user. The user codes used in Figure 4.13 are built as Walsh-Hadamard codes of length 64. The floor in the BER, at high SNR, is due to the MUI. Figure 4.13 shows a great dispersion of the SER relative to different users. This happens because different codes may be more or less affected by the transit through the channel and more or less prone to MUI. More specifically, the major reason for this sensitivity is that some of the Walsh-Hadamard codes may have a very low spectral content. For example, the first two codes of a

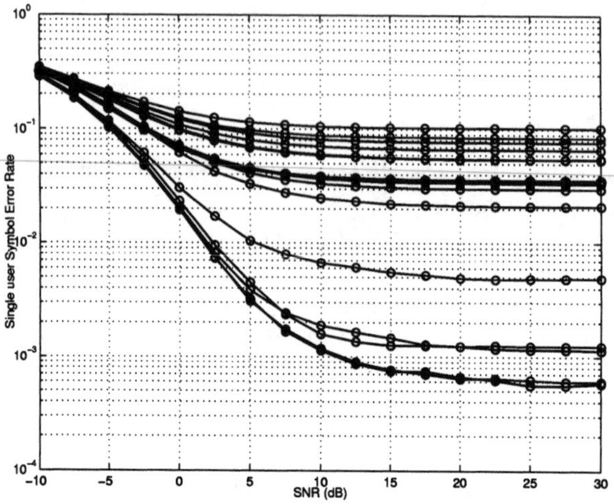

Figure 4.13 Average symbol error rate versus SNR (decibels) for a RAKE receiver with 16 active users with code length $N = 64$.

Walsh-Hadamard (as generated by Matlab, for example), contain only one spectral component. Hence, they are very sensitive to the channel zero location.

From the analysis of the RAKE receiver carried out before, better performance can be obtained by using pseudonoise (orthogonal) codes. As an example, in Figure 4.14, we show the average SER obtained in the same conditions as Figure 4.13, except that the user codes are now obtained as Walsh-Hadamard codes, multiplied by a binary real pseudorandom code. Comparing Figures 4.13 and 4.14, we can see that the SER dispersion is considerably reduced using (orthogonal) pseudonoise codes. This justifies why, in practice, DS-CDMA systems always employ pseudonoise orthogonal user codes. This issue will be further explored in the study of the UMTS system, carried out in Section 4.4.

To improve the receiver performance, one could follow various strategies, such as: 1) incorporating an estimation of the interference covariance matrix, to treat the MUI as colored (as opposed to white) noise and 2) implementing some kind of blind receiver that minimizes the energy of the interference at the output of the receiver, without assuming knowledge of all user codes. The analysis of these techniques goes beyond the scope of this book, and the interested reader may check, for example, [6].

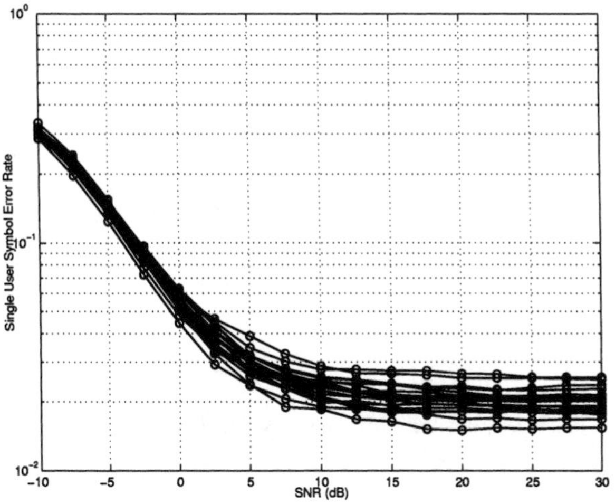

Figure 4.14 Average symbol error rate versus SNR (decibels) for a RAKE receiver with 16 active users with code length $N = 64$.

4.3.3 CDMA Receivers: Uplink Channel

We turn now our attention to the uplink channel, where the receiver has more computational capabilities than the mobile handset. Furthermore, the receiver can exploit the knowledge of all user codes. The situation is more complicated than in the downlink case, for two main reasons: 1) the signals transmitted by each user arrive at the access point through different channels, and 2) because of the different distances between the user terminals and the access point, the signals from different users arrive at the access point asynchronously. However, in practice, the relative delay is usually much smaller than the code duration. If we incorporate the relative delays within the channel impulse response from each user, the received vector can be written as

$$\boldsymbol{y}(n) = \sum_{k=1}^{M} \left[\overline{\boldsymbol{H}}_k \boldsymbol{f}_k s_k(n-1) + \boldsymbol{H}_k \boldsymbol{f}_k s_k(n) + \underline{\boldsymbol{H}}_k \boldsymbol{f}_k s_k(n+1)\right] + \boldsymbol{v}(n), \tag{4.33}$$

where the matrices $\overline{\boldsymbol{H}}_k$, $\boldsymbol{H}_k$, and $\underline{\boldsymbol{H}}_k$ have the same structure as in (4.23), referring to the channel between the kth user and the access point.

Denoting by $\boldsymbol{w}(n)$ the sum of IBI and noise, the received block can be rewritten as

$$\boldsymbol{y}(n) = \sum_{k=1}^{M} \boldsymbol{H}_k \boldsymbol{f}_k s_k(n) + \boldsymbol{w}(n). \tag{4.34}$$

Introducing the matrix

$$\boldsymbol{C} := [\boldsymbol{c}_1, \ldots, \boldsymbol{c}_M] := [\boldsymbol{H}_1 \boldsymbol{f}_1, \ldots, \boldsymbol{H}_M \boldsymbol{f}_M], \tag{4.35}$$

we can rewrite (4.34) as

$$\boldsymbol{y}(n) = \boldsymbol{C}\boldsymbol{s}(n) + \boldsymbol{w}(n). \tag{4.36}$$

Starting from this equation, we can derive several alternative ways to recover the symbols.

Maximum likelihood decoder

If the disturbance vector $\boldsymbol{w}(n)$ may be well modeled as Gaussian noise, with covariance matrix $\boldsymbol{R}_{ww}$, the probability density function of observing a vector $\boldsymbol{y}(n)$, when $\boldsymbol{s}(n)$ has been transmitted, and the channels $\boldsymbol{H}_k$ are known, is

$$p_{Y/S}(\boldsymbol{y}/\boldsymbol{s}) = \frac{1}{|\pi \boldsymbol{R}_{ww}|} e^{-[\boldsymbol{y}(n) - \boldsymbol{C}\boldsymbol{s}(n)]^H \boldsymbol{R}_{ww}^{-1} [\boldsymbol{y}(n) - \boldsymbol{C}\boldsymbol{s}(n)]}. \tag{4.37}$$

The symbols $s_k(n)$ are related to the information bits through a mapping rule specified by the adopted constellation. Let us indicate with $\boldsymbol{b}(n)$ the vector containing the bits transmitted in the nth slot. If we use a constellation of order 2^{n_b} to encode n_b bits on each transmitted symbol, the vector $\boldsymbol{b}(n)$ is composed of Mn_b bits. Let us denote with $\boldsymbol{s}[\boldsymbol{b}(n)]$ the rule that maps the Mn_b bits onto the symbol vector $\boldsymbol{s}(n)$. The maximum likelihood (ML) decoder is the algorithm that searches for the vector of bits $\boldsymbol{b}(n)$ that maximizes $p_{Y/S}\boldsymbol{y}/\boldsymbol{s}[\boldsymbol{b}(n)]$. Given the structure of $p_{Y/S}(\boldsymbol{y}/\boldsymbol{s})$, its maximization is equivalent to the minimization of the quadratic form appearing in the exponent of (4.37) over $\boldsymbol{b}(n)$. Hence, the ML estimate of the transmitted bits is

$$\hat{\boldsymbol{b}}_{ML}(n) = \arg\min_{\boldsymbol{b}(n)} [\boldsymbol{y}(n) - \boldsymbol{C}\boldsymbol{s}(\boldsymbol{b}(n))]^H \boldsymbol{R}_{ww}^{-1} [\boldsymbol{y}(n) - \boldsymbol{C}\boldsymbol{s}(\boldsymbol{b}(n))]. \tag{4.38}$$

Minimizing (4.38) requires, in general, searching over all possible sequences of bits of length Mn_b. The number of such sequences is clearly 2^{Mn_b}. Hence, the straightforward application of the ML decoder is in general too difficult to implement, from the computational cost point of view.

In an effort to simplify the decoder, many suboptimal approaches first derive an estimate of the symbols $s_k(n)$, without exploiting the finite cardinality of the constellation. Typically, the estimate is carried out over the complex field. Then, a decision is taken on the basis of the estimated symbols and, eventually, the information bits are retrieved by using the inverse bit-symbol mapping rule. This strategy is clearly suboptimal, but much simpler to implement than ML decoding. There are different ways to estimate the symbols and the choice among them is the result of a trade-off between complexity and performance. We will now review some of the basic estimation strategies.

ML estimate

The ML estimate is obtained by finding the vector $\hat{\boldsymbol{s}}(n)$ that minimizes the exponent of (4.37) over the complex field. Equating the gradient of the exponent to zero, we obtain[4]

$$\hat{\boldsymbol{s}}_{ML}(n) = (\boldsymbol{C}^H \boldsymbol{R}_{ww}^{-1} \boldsymbol{C})^{-1} \boldsymbol{C}^H \boldsymbol{R}_{ww}^{-1} \boldsymbol{y}(n). \tag{4.39}$$

Zero-forcing receiver

The zero-forcing receiver is the solution of (4.36), neglecting the noise

$$\hat{\boldsymbol{s}}_{ZF}(n) = \boldsymbol{C}^{\dagger} \boldsymbol{y}(n), \tag{4.40}$$

where $\boldsymbol{C}^{\dagger}$ denotes the pseudo-inverse of $\boldsymbol{C}$; that is[5],

$$\boldsymbol{C}^{\dagger} = (\boldsymbol{C}^H \boldsymbol{C})^{-1} \boldsymbol{C}^H. \tag{4.41}$$

The most computationally demanding operation in deriving the ZF detector is the matrix inversion in (4.41). In this regard, it is useful to notice that the matrix to be inverted has dimension $M \times M$, given by the number of active users. Hence, the ZF method is more appealing in case of a low number of active users. The ZF method derives its name from its capability of nulling the interference. In fact, the ZF concentrates its action only on the intersymbol interference. However, the ZF receiver does not incorporate any control on the noise level, so that, depending on the channel conditions, the noise could be enhanced by the multiplication by $\boldsymbol{C}^{\dagger}$.

MMSE receiver

A more robust receiver, incorporating both the effects of intersymbol interference and noise, is the minimum mean square error (MMSE) receiver. Proceeding as in

4 We assume that the inversion operations in (4.39) are admissible.

5 This expression is true if $\boldsymbol{C}$ is full column rank.

Chapter 3, assuming uncorrelated symbols (i.e., $\boldsymbol{R}_{ss} = \sigma_s^2 \boldsymbol{I}$) the MMSE estimate is

$$\hat{\boldsymbol{s}}_{\text{MMSE}}(n) = \sigma_s^2 \boldsymbol{C}^H (\sigma_s^2 \boldsymbol{C}\boldsymbol{C}^H + \boldsymbol{R}_{ww})^{-1} \boldsymbol{y}(n) \tag{4.42}$$

where $\boldsymbol{R}_{ww}$ is the covariance matrix of $\boldsymbol{w}(n)$. Factorizing[6] $\boldsymbol{R}_{ww}$ as $\boldsymbol{R}_{ww} = \boldsymbol{R}_{ww}^{1/2} \boldsymbol{R}_{ww}^{H/2}$, we can rewrite the matrix $(\sigma_s^2 \boldsymbol{C}\boldsymbol{C}^H + \boldsymbol{R}_{ww})^{-1}$ as

$$(\sigma_s^2 \boldsymbol{C}\boldsymbol{C}^H + \boldsymbol{R}_{ww})^{-1} = \boldsymbol{R}_{ww}^{-H/2} \left[\sigma_s^2 (\boldsymbol{R}_{ww}^{-1/2} \boldsymbol{C})(\boldsymbol{R}_{ww}^{-1/2} \boldsymbol{C})^H + \boldsymbol{I} \right]^{-1} \boldsymbol{R}_{ww}^{-1/2}. \tag{4.43}$$

Then, using the matrix identity ($\alpha > 0$)

$$\boldsymbol{A}(\alpha \boldsymbol{I} + \boldsymbol{A}^H \boldsymbol{A})^{-1} = (\boldsymbol{A}\boldsymbol{A}^H + \alpha \boldsymbol{I})^{-1} \boldsymbol{A}, \tag{4.44}$$

(4.42) can be rewritten as

$$\hat{\boldsymbol{s}}_{\text{MMSE}}(n) = \left(\frac{1}{\sigma_s^2} \boldsymbol{I} + \boldsymbol{C}^H \boldsymbol{R}_{ww}^{-1} \boldsymbol{C} \right)^{-1} \boldsymbol{C}^H \boldsymbol{R}_{ww}^{-1} \boldsymbol{y}(n). \tag{4.45}$$

This form is more appealing than (4.42) because it requires the inversion of a matrix of order $M \times M$, instead of $(N + L) \times (N + L)$. In practice, especially when the channel order L is much smaller than the code length N, to simplify the MMSE design, we can model the intersymbol interference as white noise so that the covariance matrix $\boldsymbol{R}_{ww}$ can be expressed as an identity matrix (i.e., $\boldsymbol{R}_{ww} = \sigma_{n+i}^2 \boldsymbol{I}$) where σ_{n+i}^2 is the power of noise plus interference. In such a case, the MMSE estimate becomes

$$\hat{\boldsymbol{s}}_{\text{MMSE}}(n) = \left(\frac{\sigma_{n+i}^2}{\sigma_s^2} \boldsymbol{I} + \boldsymbol{C}^H \boldsymbol{C} \right)^{-1} \boldsymbol{C}^H \boldsymbol{y}(n). \tag{4.46}$$

Interestingly, we can see that the MMSE estimator tends to become equal to the ZF estimator when the noise is negligible (i.e., when the ratio $\sigma_{n+i}^2 / \sigma_s^2$ tends to zero).

Successive decoding

Both ZF and MMSE receivers require a nonnegligible computational cost, because of the computation of a matrix inverse, operation that has to be repeated periodically, with an update rate depending on the channel coherence time. A different approach with respect to all previous schemes consists of using an iterative procedure that combines symbol detection and interference cancelation. Since $N + L$ is certainly

6 This factorization always exists since $\boldsymbol{R}_{ww}$, thanks to the presence of additive white noise, is a positive definite matrix.

greater than M, $\boldsymbol{C}$ can always be decomposed, using the QR factorization, as

$$\boldsymbol{C} = \boldsymbol{U}_R \boldsymbol{R}, \tag{4.47}$$

where $\boldsymbol{U}_R$ is an $(N + L) \times M$ pseudo-unitary matrix (i.e., such that $\boldsymbol{U}_R^H \boldsymbol{U}_R = \boldsymbol{I}$) whereas $\boldsymbol{R}$ is an $M \times M$ upper triangular matrix. Multiplying the received vector $\boldsymbol{y}(n)$ by $\boldsymbol{U}_R^H$ from the left side, we obtain the M-size vector

$$\boldsymbol{z}(n) := \boldsymbol{U}_R^H \boldsymbol{y}(n) = \boldsymbol{R}\boldsymbol{s}(n) + \boldsymbol{v}'(n), \tag{4.48}$$

where $\boldsymbol{v}'(n) := \boldsymbol{U}_R^H \boldsymbol{v}(n)$. Because of the pseudo-unitary structure of $\boldsymbol{U}_R$, if $\boldsymbol{v}(n)$ is white and Gaussian, $\boldsymbol{v}'(n)$ is also white and Gaussian, with the same statistical properties as $\boldsymbol{v}(n)$. The system (4.48) is easier to invert than (4.36), because the coefficient matrix $\boldsymbol{R}$ is triangular. In particular, exploiting the upper triangularity of $\boldsymbol{R}$, we can write (we drop the time index n for simplicity of notation)

$$\begin{aligned} z_M &= R_{M,M} s_M + v'_M, \\ z_{M-1} &= R_{M-1,M-1} s_{M-1} + R_{M-1,M} x_M + v'_{M-1}, \\ &\cdots \\ z_1 &= R_{1,1} s_1 + \sum_{k=2}^{M} R_{1,k} x_k + v'_1. \end{aligned} \tag{4.49}$$

From (4.49), we can set up an iterative decoding and interference cancelation structure that works as follows. We decode s_M first by taking

$$\hat{s}_M = D\left[\frac{z_M}{R_{M,M}}\right], \tag{4.50}$$

where $D[a]$ denotes the decision rule that yields the symbol in the constellation closest to a. The contribution of the decoded symbol s_M can then be canceled from z_{M-1}, so that the symbol s_{M-1} can be decoded as

$$\hat{s}_{M-1} = D\left[\frac{z_{M-1} - R_{M-1,M}\hat{s}_M}{R_{M-1,M-1}}\right]. \tag{4.51}$$

The iterations proceed through successive decoding and cancelations, until s_1 is estimated. In the ith step, we have

$$\hat{s}_i = D\left[\frac{z_i - \sum_{k=i+1}^{M} R_{i,k}\hat{s}_k}{R_{i,i}}\right]. \tag{4.52}$$

If the intermediate decisions are correct, the successive cancelation method can provide advantages with respect to the previous methods. Clearly, this approach is more appealing when the number of users M is small. The successive decoding and cancelation scheme is sensitive to the order with which the data are decoded and canceled. From (4.52), it is evident that the best way to apply the procedure requires one to first order the data, so that the symbol with the highest $R_{i,i}$ is decoded first, and so on.

4.3.4 Power Control

From the performance analysis of the RAKE receiver, it is clear that in a system with nonperfect cancelation of MUI, the user signals arriving at the base station with less power than others suffer from a performance degradation due to high interference. Since the received power depends on the link length, the users that are more distant from the access point achieve worse performance than users close to the access point. This is the so called *near-far* problem. It is then important to readjust the power of each transmitter, so that each user could get satisfactory performance. This is why third generation cellular systems, like UMTS, implement a power control strategy. We will now review some of the principles of power control, recalling the basic work of Yates [7].

A good performance measure, in the presence of interference, is the signal-to-noise plus interference ratio (SINR). Recalling the analysis carried out in Section 1.3.1, the use of pseudonoise orthogonal codes yields a gain N in the SINR. Thus, denoting with α_j the fraction of the power transmitted by the jth user that is received at the base station, the SINR for user j is

$$\text{SINR}_j = N \frac{\alpha_j p_j}{\sum\limits_{\substack{i=1\\ i\neq j}}^{M} \alpha_i p_i + \sigma_n^2}, \tag{4.53}$$

where σ_n^2 is the received noise power. If the number of interferers is sufficiently high, we can invoke the central limit theorem and model the interference as a Gaussian random variable. In such a case, the BER of each user is a function of its own SINR. Hence, requiring that the BER of each user be not greater than a maximum value P_e is equivalent to require that the SINR of each user be greater than a given threshold, let us say γ_0. In formulas, this requires that the following system of inequalities be satisfied:

$$\frac{\alpha_j p_j}{\sum\limits_{\substack{i=1\\ i\neq j}}^{M} \alpha_j p_i + \sigma_n^2} \geq \frac{\gamma_0}{N}. \tag{4.54}$$

This is equivalent to require that the power of each user p_j be greater than a threshold

$$p_j \geq f_j(\boldsymbol{p}) := \frac{\gamma_0}{N\mu_j(\boldsymbol{p})}, \tag{4.55}$$

where $\boldsymbol{p} := (p_1, \ldots, p_M)$ is the vector containing the powers of all users and

$$\mu_j(\boldsymbol{p}) := \frac{\alpha_j}{\sum\limits_{\substack{i=1 \\ i \neq j}}^{M} \alpha_i p_i + \sigma_n^2}. \tag{4.56}$$

Introducing the vector

$$\boldsymbol{f}(\boldsymbol{p}) = \left[\frac{\gamma_0}{N\mu_1(\boldsymbol{p})}, \ldots, \frac{\gamma_0}{N\mu_M(\boldsymbol{p})} \right],$$

the equality in (4.55) is satisfied when

$$\boldsymbol{f}(\boldsymbol{p}) = \boldsymbol{p}. \tag{4.57}$$

Recalling that a *fixed point* of a function $f(x)$ is a point x that is solution of the equation

$$f(x) = x, \tag{4.58}$$

we may say, from (4.57), that the power vector $\boldsymbol{p}$ is a (vector) fixed point of the (vector) function $\boldsymbol{f}(\boldsymbol{p})$. In practice, there might be several fixed points, depending on the function $\boldsymbol{f}(\boldsymbol{p})$ and on the constraints on the power vector $\boldsymbol{p}$. It is then important to prove under which conditions there exists a unique solution $\boldsymbol{p}$ satisfying (4.57) and how to compute the power vector $\boldsymbol{p}$. To this end, it is useful to introduce the concept of *standard* function [7].

A vector function $\boldsymbol{f}(\boldsymbol{p})$ is a *standard* function if, for all $\boldsymbol{p} \geq \boldsymbol{0}$, the following properties are satisfied:

- *Positivity:* $\boldsymbol{f}(\boldsymbol{p}) > \boldsymbol{0}$;
- *Monotonicity*[7]*:* If $\boldsymbol{p} \geq \boldsymbol{p}'$, then $\boldsymbol{f}(\boldsymbol{p}) \geq \boldsymbol{f}(\boldsymbol{p}')$;
- *Scalability:* For all $\alpha > 1$, $\alpha \boldsymbol{f}(\boldsymbol{p}) > \boldsymbol{f}(\alpha \boldsymbol{p})$.

It is straightforward to prove that the function $\boldsymbol{f}(\boldsymbol{p})$, defined in (4.55) is a standard function. The property of being a standard function is important in the power control problem because, if there is a fixed point for $\boldsymbol{f}(\boldsymbol{p})$ and $\boldsymbol{f}(\boldsymbol{p})$ is a standard function,

7 We adopt the convention that the vector inequality $\boldsymbol{p} > \boldsymbol{p}'$ is an inequality in all components.

then *the fixed point is unique* [7].

Furthermore, it is easy to set up a decentralized iterative algorithm that converges to the fixed point. The possibility for a decentralized solution is particularly important as it allows every radio node to adapt its own power without requiring neither the exchange of signaling with the other radio nodes nor the presence of a centralized control unit.

If we build the sequence of vectors

$$\boldsymbol{p}(n+1) = \boldsymbol{f}[\boldsymbol{p}(n)], \tag{4.59}$$

starting from any feasible vector $\boldsymbol{p}(0)$ (feasibility here means that the power vector respects the power constraint), then, iterating (4.59), if the sequence converges, it converges to the unique fixed point.

What is important to remark about the iterative algorithm (4.59), besides its simplicity, is that it can be implemented in a decentralized fashion. In fact, to update its own transmit power, each user requires only the knowledge of its own SINR.

A numerical example is useful to test the convergence speed of the adaptive power control strategy described before. In Figure 4.15, we report the evolution of the SINR for each user, as a function of the iteration index n, for a system composed of $M = 8$ users, each one employing a user code of length $N = 64$. Figure 4.15 shows that after a very few iterations, the powers change in order to guarantee the same SINR for all users. Finally, it can be checked that the solution for the powers that yields the same SINR for all users corresponds to forcing the same received power for all the users. Setting $p_j^{(r)} = p^{(r)}$, we obtain

$$\frac{p^{(r)}}{(M-1)p^{(r)} + \sigma_n^2} \geq \frac{\gamma_0}{N}. \tag{4.60}$$

This requires that the number of users M be less than the following threshold

$$M \leq \frac{N}{\gamma_0} + 1 - \frac{\sigma_n^2}{p^{(r)}} \approx \frac{N}{\gamma_0} + 1, \tag{4.61}$$

where in the last approximation, we have neglected the additive noise, with respect to the interference. Equation (4.61) shows that there is an upper bound on the maximum number of users that can get access. The bound depends on the code length N, as well as on the required minimum SINR.

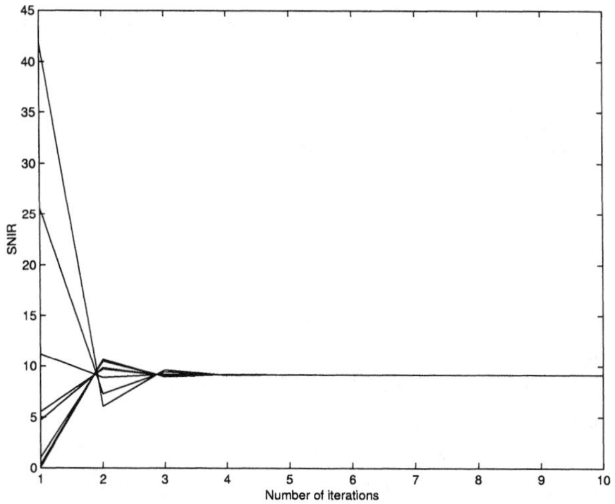

Figure 4.15 Signal-to-interference-plus-noise ratio versus iteration index.

4.4 CASE STUDY: UMTS

In this section, we study an important example of CDMA system, given by the universal mobile telecommunication system (UMTS) planned for the third generation (3G) cellular systems [8].

The main system parameters are listed in Table 4.1. The transmission in the down-

Table 4.1

Key Technical Characteristics of UMTS Physical Layer

Multiple-access scheme	DS-CDMA
Duplexing scheme	TDD / FDD
Chip rate	4.096 Mcps
Bit rate	variable, up to 2 Mbps
Modulation	QPSK
Shaping filter	Root raised cosine (roll-off factor: $\beta = 0.22$)
Slot duration	0.625 ms
Number of slots per frame	16
Spreading factor	Variable, from 4 to 256
Channel coding	Convolutional coding (rate $1/2$ or $1/3$) Optional outer Reed-Solomon coding (rate $4/5$)

link channel is organized in frames composed of 16 time slots. Each time slot has a structure, as sketched in Figure 4.16. The slot is subdivided in two main parts:

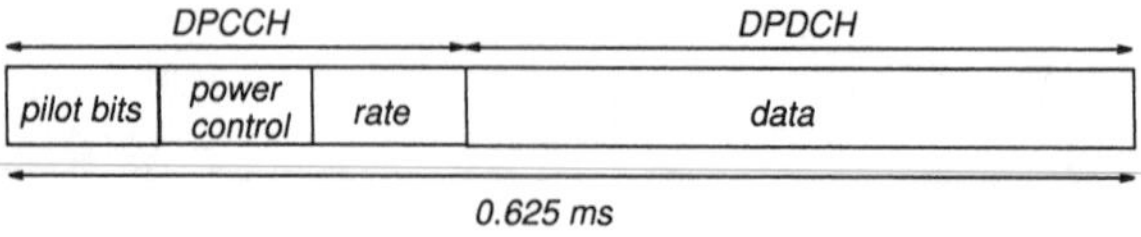

Figure 4.16 Structure of the time slot used in the downlink channel of UMTS.

a dedicated physical control channel (DPCCH), used to transmit control data, and a dedicated physical data channel (DPDCH), used to convey the information bits. The control data contains: 1) pilot bits, to be used at the receiver to estimate the channel; 2) power control bits, with which the base station communicates to the mobile terminal to adjust its power level; and 3) rate information data, used from the transmitter to inform the receiver about the data rate which is being used.

Among the many requirements of UMTS, the system must provide a *multirate* capability. This means that the system has to be able to support the simultaneous access of users transmitting at different rates. This is an important feature of UMTS, which is designed to carry multimedia signals (i.e., voice, images, videos, and data). The multirate requirement is achieved, in UMTS, assigning codes with different lengths to users asking for different rates and combining this property with different correction coding rates. The underlying chip rate is the same for all users.

The set of user codes are built in order to guarantee their orthogonality, even though they have different lengths[8]. The codes are constructed using the iterative procedure illustrated in Figure 4.17. Each branch in the tree contains a code. Each node is the root of two departing branches: The upper branch contains the code present in the root, repeated, whereas the lower branch contains the root code, repeated and with the opposite sign. The potential user codes to be assigned to the users are all the codes with length from 4 to 256. The codes with length 4 are assigned to the users asking for higher rates, whereas the codes with length 256 are given to the users demanding for the slowest rate. Each level in the tree refers to a given code length and then to the same rate. It can be easily checked that codes with the same length

8 Talking about orthogonality between codes with different lengths requires some caution. With references to the codes reported in the branches of the tree diagram sketched in Figure 4.17, the scalar products between a code $\mathbf{c}_1$ of length m and a code $\mathbf{c}_2$, of length $2^k\, m$ must be evaluated by computing the following scalar products:

$$\sum_{n=1}^{m} c_1(n)c_2(2^i + n), \quad i = 0, \ldots, k-1. \tag{4.62}$$

If all these products are null, we say that the codes $\mathbf{c}_1$ and $\mathbf{c}_2$ are orthogonal.

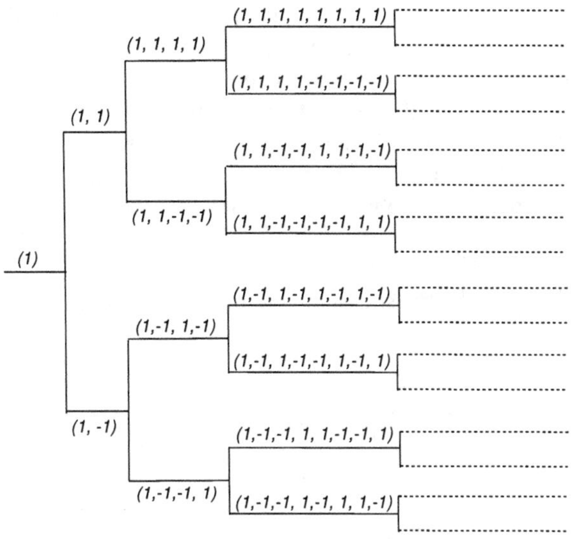

Figure 4.17 Iterative procedure to build orthogonal variable spreading factor user codes.

are orthogonal to each other.

However, a more interesting aspect of the code design sketched in Figure 4.17 is that there is orthogonality also between codes having different lengths, provided that the code assignment satisfies the following rule: *A code on a given node is assigned to a user if and only if no other codes on the path from that node to the root of the tree or in the subtrees below that given code have already been assigned to other users.* This implies that once a code has been assigned, all the codes in the branches departing from that node cannot be assigned to any other user. This means also that the number of codes that can be assigned within a cell is not fixed, but it depends on the rates required by the users: High rate users deprive access to many potential slow rate users (all the users that could get the codes belonging to the branches departing from the given high rate user).

Using only the code orthogonality, the system could accommodate only users asking for rates whose ratio is a power of two. Actually, UMTS accommodates many more rates by combining the variable spreading factor idea described above with different levels of error protection. In UMTS, there is in fact an inner convolutional encoder. For any choice of the spreading factor, the rate of the encoder can be adjusted in order to make available many more rates. Then, given a pool of available rates, if the user requires a rate that is not present in the pool, the system assigns to the user its closest available rate. To match the rate, there are two possibilities, depending on whether the rate asked by the user is lower or greater than one of the available rates. If it is lower, the transmitter inserts dummy bits, which are simply discarded at the receiver, to increase the rate. Otherwise, the transmitter uses *puncturing* to reduce the user rate. Puncturing implies that some of the coded bits are periodically discarded. Convolutional encoders permit correct decoding, even in the presence of puncturing, provided that the puncturing operation is done on a specific pattern of bits.

The block diagram of the transmission scheme is reported in Figure 4.18. In Figure

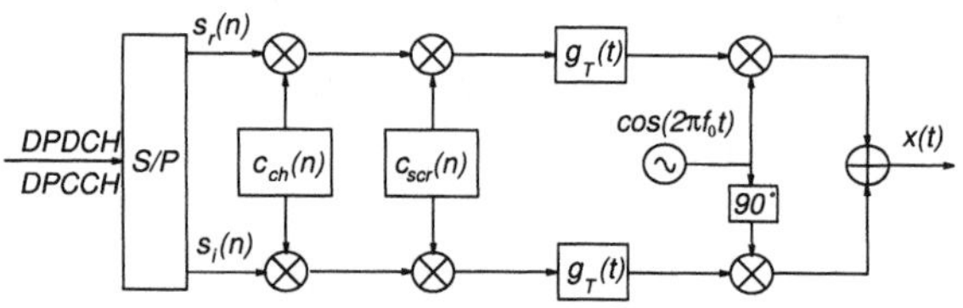

Figure 4.18 Block diagram of the downlink transmission scheme.

4.18, the first block maps the input data into the in-phase and quadrature components of the QPSK modulation used for transmission. Each sequence is then first multiplied by the user (or channelization) code $c_{ch}(n)$ and then multiplied by the scrambling code $c_{scr}(n)$, specific to the cell to which the user has been associated. Specifically, the scrambling code is a $40,960$ chip (10 ms) segment of a length $2^{18} - 1$ Gold code, repeated in each frame. The two in-phase and quadrature sequences pass then through the shaping filter, with impulse response $g_T(t)$. The shaping filters have a transfer function given by the root of a raised cosine, with roll-off factor $\beta = 0.22$. The two in-phase and quadrature components are then modulated by the carrier, at frequency f_0, to produce the output QPSK signal $x(t)$. The multiplication of the orthogonal user codes, built according to the tree diagram of Figure 4.17, by a pseudonoise scrambling code has a twofold implication: 1) it differentiates adjacent cells, by assigning to each cell a different scrambling code and 2) it implements the properties of having codes that are mutually orthogonal, pseudonoise, and allow for multi-rate transmission.

The structure of the time slot used for uplink is sketched in Figure 4.19. Differently from the downlink channel, the control and data streams (DPCCH and DPDCH) travel in parallel. This feature is implemented by sending the two data sequences to the in-phase and quadrature signals.

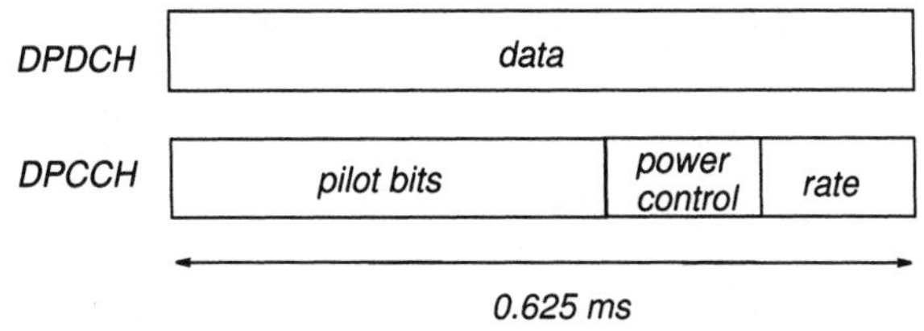

Figure 4.19 Structure of the time slot used in the uplink channel of UMTS.

4.5 SPACE-DIVISION MULTIPLE ACCESS

So far, we have only considered multiple access schemes based on single antenna transceivers. We analyze now what happens when the base station has a multi-antenna transceiver. This yields the possibility for space-division multiple access (SDMA). We will basically recall the main results of [9].

Let us consider the receiving scheme sketched in Figure 4.20, where there are three users sending their signals $s_1(t)$, $s_2(t)$, and $s_3(t)$ to a base station equipped with n_R receive antennas, uniformly spaced, at distance d. We make the following

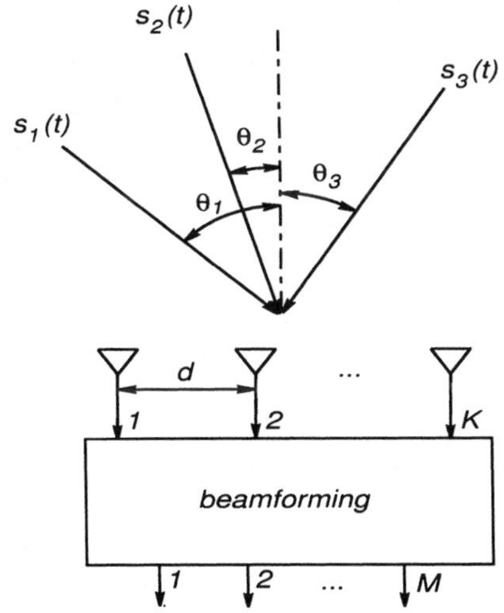

Figure 4.20 Space division multiple access scheme.

assumptions:

1. The sources are sufficiently far apart from the base station, so that all wavefronts impinging on the array are planar;

2. The incoming signal is narrowband, with respect to the array; this means that the amplitude of the baseband signal arriving at the array does not vary over the length of the array, even though the instantaneous phase varies;

3. The signals from different users are independent;

4. Each user signal may arrive at the array through a multipath propagation; however, the relative delays are not distinguishable from each other, at the receiver; this implies that the channels from each user to the base station is flat-fading;

5. The channels are known at the receive side, but they are unknown at the transmit side;

6. The noise components observed at all the receiving antennas are spatially and temporally white complex Gaussian random variables.

Under the previous assumptions, the signals collected by the n_R antennas can be arranged into a vector. In the presence of a single user, let us say user i, and a one-path link, the base station collects n_R signals that form the vector

$$\boldsymbol{x}(t) := h_i[1, e^{j2\pi d \sin\theta_i/\lambda}, \cdots, e^{j2\pi d(n_R-1)\sin\theta_i/\lambda}]^T s_i(t) := h_i \boldsymbol{a}^T(\theta_i) s_i(t), \tag{4.63}$$

where θ_i is the incidence angle of the plane wave coming from user i and λ is the transmit wavelength. The vector $\boldsymbol{a}(\theta)$ is the *steering vector*. In general, the signal $s_i(t)$ may arrive at the receiver through multiple paths. In case of L paths, each arriving with an incidence angle $\theta_{i,l}$, $l = 1, \ldots, L$, and channel coefficient $h_{i,l}$, the array response vector, relative to user i, is

$$\boldsymbol{a}_i = \sum_{l=1}^{L} h_{i,l} \boldsymbol{a}_i(\theta_{i,l}). \tag{4.64}$$

In the presence of M simultaneous users, each one transmitting a signal $s_m(t)$, $m = 1, \ldots, M$, the overall received vector is

$$\boldsymbol{x}(t) = \sum_{m=1}^{M} \boldsymbol{a}_m s_m(t) + \boldsymbol{n}(t) = \boldsymbol{A}\boldsymbol{s}(t) + \boldsymbol{n}(t), \tag{4.65}$$

where $\boldsymbol{A} = (\boldsymbol{a}_1, \cdots, \boldsymbol{a}_M)$ is a $K \times M$ matrix, $\boldsymbol{s}(t) = [s_1(t), \cdots, s_M(t)]$ is the vector of transmitted signals, at time t, and $\boldsymbol{n}(t)$ is the received noise vector. The matrix $\boldsymbol{A}$, whose columns are the array response vectors corresponding to each user, is the so-called *array manifold* .

Normalizing, for simplicity, the noise power to 1, we have the following covariance matrices (assuming zero-mean vectors)

$$\boldsymbol{R}_{ss} = E\{\boldsymbol{s}(t)\boldsymbol{s}^H(t)\} = \text{diag}(\sigma_1^2, \cdots, \sigma_K^2) \tag{4.66}$$

$$\boldsymbol{R}_{nn} = E\{\boldsymbol{n}(t)\boldsymbol{n}^H(t)\} = \boldsymbol{I}. \tag{4.67}$$

Consequently, we also have

$$\boldsymbol{R}_{xx} = E\{\boldsymbol{x}(t)\boldsymbol{x}^H(t)\} = \boldsymbol{A}\boldsymbol{R}_{ss}\boldsymbol{A}^H + \boldsymbol{I}. \tag{4.68}$$

Sampling the vector $\boldsymbol{x}(t)$, we can write the discrete-time input/output relationship

$$\boldsymbol{x}(n) = \sum_{m=1}^{M} \boldsymbol{a}_m s_m(n) + \boldsymbol{n}(n) = \boldsymbol{A}\boldsymbol{s}(n) + \boldsymbol{n}(n). \tag{4.69}$$

This equation is perfectly equivalent to (4.36), even though the entries of $\boldsymbol{x}(n)$ in (4.69) are collected in the space domain, whereas the entries of of $\boldsymbol{y}(n)$ in (4.36) were collected in the time domain. By virtue of this mathematical equivalence, we can apply all the decoding schemes, derived for (4.36), to (4.69) as well. The results will give rise to different ways to combine the signal received by the array in a given time instant. This combination can be seen, in the spatial domain as a *beamforming* technique, where the spatial beams are formed in order to optimize some performance parameter, such as MMSE for example.

As an example, let us consider a base station with a uniformly spaced array with an inter-antenna distance $d = \lambda/2$. The number of users is $M = 3$. The signals impinge the array with incidence angles of -15, 0 and 30 degrees. The SNR is 20 dB, for each user. Using the MMSE approach, each beamformer is built with coefficients given by one row of the matrix

$$\boldsymbol{G}^H = \left(\boldsymbol{A}^H\boldsymbol{A} + \frac{1}{\sigma_s^2}\boldsymbol{I}\right)^{-1} \boldsymbol{A}^H, \tag{4.70}$$

where $\boldsymbol{A}$ is given in (4.68) and $\sigma_s^2 = 100$. In Figure 4.21, we show the modulus of the receive radiation pattern of each of the three beams formed by taking the rows of $\boldsymbol{G}^H$, using a receive array composed of $n_R = 4$ or $n_R = 8$ elements. More specifically, each radiation pattern

$$G(\theta) = F(\theta)C(\theta) \tag{4.71}$$

is the product between the array factor

$$F(\theta) = \sum_{k=0}^{n_R-1} g(k)e^{j2\pi k \sin(\theta)d/\lambda}, \tag{4.72}$$

where $g(k)$ is the (complex) amplitude coefficient of the kth element, and the radiation pattern $C(\theta)$ of each element of the array. We assume that $C(\theta)$ follows a cardioid pattern, that is

$$C(\theta) = \frac{1 + \cos(\theta)}{2}. \tag{4.73}$$

From Figure 4.21, we see that each beam tends to put a null in the direction of the users different from the user of interest. It is also evident that, by increasing the number of array elements, but keeping the interelement distance unchanged, the array has more angular discrimination capabilities. In fact, comparing Figure 4.21(a) and (b), we see that, in case of $n_R = 8$, each beamformer is able to put the maximum in the direction of the user of interest, whereas this does not happen in the case $n_R = 4$.

In this section, the receive array was assumed to combine only samples taken at one time instant, thus working only in the space domain. In such a domain, we may discriminate users coming from different directions. However, the spatial processing alone is not capable of discriminating users whose signals arrive from the same direction, even though the users are not co-located. In such a case, we need to resort to time processing. Indeed, in a more general setup, the receive array can collect sequences of temporal samples from all the receive antennas. In this way, it is possible to design a *joint space-time* processing strategy that takes advantage of both spatial and time structures of the incoming data. The interested reader is invited to check, for example, the excellent tutorial on space-time processing, given in [10]. More details on joint space-time processing will be given in Chapter 6.

4.5.1 Capacity Region of Receive Array Systems

Let us analyze now the maximum information rates achievable when we use a receive array, in the multiple access channel. The most important figure of merit to assess the achievable rates, in case of multiple access, is the so-called *capacity region*, defined as follows. Let us start with a two-user system, for simplicity. A rate pair (R_1, R_2) is said to be *achievable* if there exists a coding strategy that insures arbitrarily low error probabilities for both users, as the length of the codewords employed by the two users goes to infinity. In practice, all users are subject to some physical constraints. Typically, the average transmit power is constrained not to exceed a maximum value. The capacity region is the set of all achievable rates,

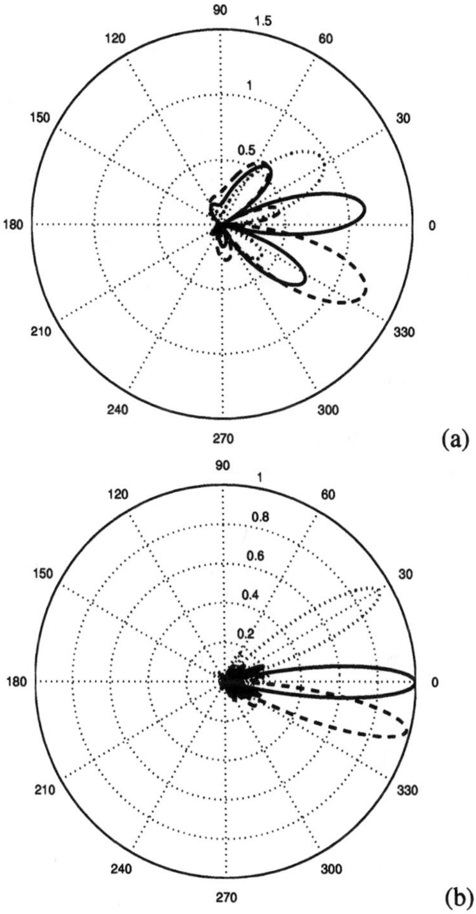

Figure 4.21 MMSE beamforming in case of $M = 3$ users and (a) $K = 4$ or (b) $K = 8$ receive antennas.

satisfying the physical constraints.

If the probability densities of the input sequences are fixed, it is relatively easy to derive the capacity region. In such a case, in fact, given a discrete memoryless multiple access channel, characterized by the channel transition probability $p_{Y/X_1,X_2}(y/x_1,x_2)$ and by the input probability densities $p_{X_1}(x_1)$ and $p_{X_2}(x_2)$, the region of the achievable rates is the pentagon defined by the following three inequalities:

$$R_1 \leq I(X_1;Y/X_2); \quad (4.74)$$

$$R_2 \leq I(X_2;Y/X_1); \quad (4.75)$$

$$R_1 + R_2 \leq I(X_1,X_2;Y), \quad (4.76)$$

where the mutual information values are computed with respect to the assigned probability densities.

Often, in practice, the input pdfs are not fixed. However, there is a constraint, for example, on the average transmit power of each user. In such a case, the capacity region is the *convex hull* of the union of all the pentagons corresponding to the input pdf, for each user, whose variance is less than the assigned user power.

In general, since signals from different users are typically independent, the input distribution must take the form $p_{X_1,X_2}(x_1,x_2) = p_{X_1}(x_1)p_{X_2}(x_2)$. This constraint makes the search for the capacity region of a multiple access channel a difficult task, in general. However, there is an important exception where the capacity region can be found easily. If the input constraints are power constraints, which means that the transmit power of user 1 (2) must be upper bounded by σ_1^2 (σ_2^2), and the channel transition probability $p_{Y/X_1,X_2}(y/x_1,x_2)$ is Gaussian (this happens in the case of additive Gaussian noise) the optimization problem has a simple solution. In such a case, in fact, the optimal input pdfs are Gaussian, with zero mean and variance σ_1^2 and σ_2^2, respectively, and the convex hull operation is not necessary.

The capacity region depends also on the decoding strategy followed at the receiver. The major difference concerns the use of joint decoding, as opposed to independent decoding. We analyze now these two cases, separately.

4.5.2 Joint Decoding

We derive now the capacity region for the case of joint decoding of the signals arriving at the base station, in the general case of M simultaneous users. The additive noise is white Gaussian noise, with zero mean and unit variance (i.e.,

$\sigma_n^2 = 1$). Each user has a constraint on its average transmit power. We denote by σ_k^2 the average power of the kth user. Let us denote with $\mathcal{S}$ a subset of user indexes (i.e., a subset of the set $\{1, 2, \cdots, M\}$) and with $\mathcal{S}_c$ the complement of $\mathcal{S}$, with respect to the set $\{1, 2, \cdots, M\}$. Let us also introduce the matrices $\boldsymbol{A}_{\mathcal{S}}$ and $\boldsymbol{A}_{\mathcal{S}_c}$, composed of the columns of $\boldsymbol{A}$, whose indexes belong to $\mathcal{S}$ or to $\mathcal{S}_c$, respectively. Since $\boldsymbol{R}_{ss}$ is diagonal, using the previous notation, the covariance matrix $\boldsymbol{R}_{xx}$ in (4.68), can be rewritten as

$$\boldsymbol{R}_{xx} = \boldsymbol{A}_{\mathcal{S}}\boldsymbol{R}_{\mathcal{S}}\boldsymbol{A}_{\mathcal{S}}^H + \boldsymbol{A}_{\mathcal{S}_c}\boldsymbol{R}_{\mathcal{S}_c}\boldsymbol{A}_{\mathcal{S}_c}^H + \boldsymbol{I}, \tag{4.77}$$

where $\boldsymbol{R}_{\mathcal{S}}$ (alternatively, $\boldsymbol{R}_{\mathcal{S}_c}$) is the diagonal matrix of the sources whose indexes belong to $\mathcal{S}$ ($\mathcal{S}_c$).

The boundary of the capacity region may be reached through successive decoding and cancelation of the decoded users. Thus, indicating with $\mathcal{S}_c$ the set of indexes of the users that have already been successfully decoded, the maximum mutual information between the remaining users and the received signal is upper-bounded by

$$\begin{aligned} I(\boldsymbol{X}_{\mathcal{S}}; \boldsymbol{Y}/\boldsymbol{X}_{\mathcal{S}_c}) &= H(\boldsymbol{Y}/\boldsymbol{X}_{\mathcal{S}_c}) - H(\boldsymbol{Y}/\boldsymbol{X}_{\mathcal{S}}, \boldsymbol{X}_{\mathcal{S}_c}) \\ &\le \frac{1}{2}\log_2 \left|\boldsymbol{I} + \boldsymbol{A}_{\mathcal{S}}\boldsymbol{R}_{\mathcal{S}}\boldsymbol{A}_{\mathcal{S}}^H\right| \text{ bps.} \end{aligned} \tag{4.78}$$

The equality is achieved when the input constellations are complex random Gaussian variables, with zero mean and variance σ_k^2. Hence, any set of achievable rates must satisfy the following inequalities [9]:

$$\sum_{k\in\mathcal{S}} R_k \le \frac{1}{2}\log_2 \left|\boldsymbol{I} + \boldsymbol{A}_{\mathcal{S}}\boldsymbol{R}_{\mathcal{S}}\boldsymbol{A}_{\mathcal{S}}^H\right| \text{ bps,} \tag{4.79}$$

for any subset $\mathcal{S}$ of $\{1, 2, \ldots, M\}$. As an example, in the two-user case (i.e., $M = 2$) the capacity region is determined by the following set of inequalities[9]

$$\begin{aligned} R_1 &\le \frac{1}{2}\log_2 \left|\boldsymbol{I} + \sigma_1^2\boldsymbol{a}_1\boldsymbol{a}_1^H\right| = \frac{1}{2}\log_2\left(1 + \sigma_1^2\|\boldsymbol{a}_1\|^2\right) \\ R_2 &\le \frac{1}{2}\log_2 \left|\boldsymbol{I} + \sigma_2^2\boldsymbol{a}_2\boldsymbol{a}_2^H\right| = \frac{1}{2}\log_2\left(1 + \sigma_2^2\|\boldsymbol{a}_2\|^2\right) \\ R_1 + R_2 &\le \frac{1}{2}\log_2 \left|\boldsymbol{I} + \sigma_1^2\boldsymbol{a}_1\boldsymbol{a}_1^H + \sigma_2^2\boldsymbol{a}_2\boldsymbol{a}_2^H\right|. \end{aligned} \tag{4.80}$$

In this case, the capacity region resulting from joint decoding is given by the points inside the pentagon sketched in Figure 4.22. The specific shape of the boundary can

9 In the rates expressions, we make often use of the identity $|\boldsymbol{I} + \boldsymbol{AB}| = |\boldsymbol{I} + \boldsymbol{BA}|$.

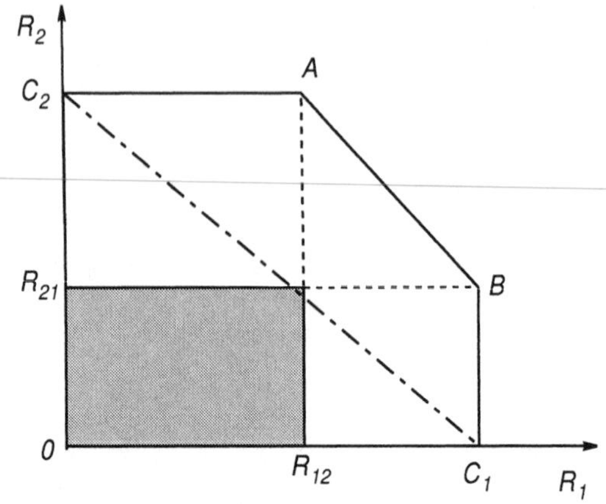

Figure 4.22 Capacity region of a two-user multiple access system.

be justified as follows. The maximum value of R_1 is the capacity C_1 of the link between user 1 and the access point, in the absence of user 2. The same argument applies to user 2, when user 1 does not transmit. The vertical segment from C_1 to B is achieved decoding user 2 first and then canceling the contribution of user 2 from the received signal. As far as decoding of user 2 is done without errors, the cancelation is perfect and then user 1 acts as if user 2 were not present. At the encoding stage, user 1 builds Gaussian codewords assuming no interference present. At the same time, user 2 builds codewords treating user 1 as additive Gaussian noise. When the rate of user 2 exceeds a certain value R_{21}, the receiver is not able to decode user 2 without errors anymore, and thus the rate of user 1 starts to decrease as user 2 acts as an additive interference. Similar arguments apply to explain the segment from C_2 to A. The dashed and dotted line, in Figure 4.22, represents the sum of the rates that would be achieved by time sharing. From Figure 4.22, it is clear that time sharing is suboptimum, as it is always included in the capacity region.

Setting, for simplicity $\boldsymbol{a}_1^H \boldsymbol{a}_1 = \boldsymbol{a}_2^H \boldsymbol{a}_2 = \alpha$, the capacity region can be rewritten as

$$
\begin{aligned}
R_1 &\leq \frac{1}{2}\log_2\left(1+\alpha\sigma_1^2\right) \\
R_2 &\leq \frac{1}{2}\log_2\left(1+\alpha\sigma_2^2\right) \\
R_1 + R_2 &\leq \frac{1}{2}\log_2\left[1+\alpha(\sigma_1^2+\sigma_2^2)+\alpha^2\sigma_1^2\sigma_2^2\sin^2\phi_{12}\right],
\end{aligned}
\qquad (4.81)
$$

where ϕ_{12} is the angle between the vectors $\boldsymbol{a}_1$ and $\boldsymbol{a}_2$; that is,

$$\sin^2 \phi_{12} = 1 - \frac{|\boldsymbol{a}_1^H \boldsymbol{a}_2|^2}{\|\boldsymbol{a}_1\|^2 \|\boldsymbol{a}_2\|^2}. \tag{4.82}$$

From (4.81), we see that when the user positions are such that the array response vectors pertaining to the two users are orthogonal to each other (i.e., $\sin^2 \phi_{12} = 1$) the rate region becomes

$$\begin{aligned} R_1 &\leq \frac{1}{2} \log_2 \left(1 + \alpha\sigma_1^2\right) \\ R_2 &\leq \frac{1}{2} \log_2 \left(1 + \alpha\sigma_2^2\right) \\ R_1 + R_2 &\leq \frac{1}{2} \log_2 \left(1 + \alpha\sigma_1^2\right) + \frac{1}{2} \log_2 \left(1 + \alpha\sigma_2^2\right) \end{aligned} \tag{4.83}$$

This means that the capacity region, in case of orthogonal channels, is a rectangle.

In the three-user scenario, the admissible rates are the values contained in a volume whose typical shape is sketched in Figure 4.23. Each corner on the boundary

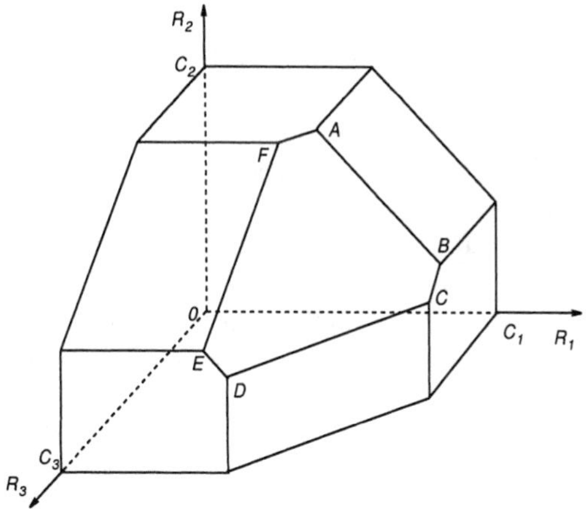

Figure 4.23 Capacity region of a three-user multiple access system.

corresponds to a specific decoding order. Since there are M users, the number of all possible decoding orders is equal to the number of permutation of the M users indexes (i.e., $M!$) .

4.5.3 Independent Decoding

In the case of independent decoding, when decoding the mth user, all other users are treated as additive noise. The capacity region for independent decoding is then

$$R_k \leq \frac{1}{2}\log_2\left(\frac{|\boldsymbol{R}_{xx}|}{|\boldsymbol{R}_{xx} - \sigma_k^2 \boldsymbol{a}_k \boldsymbol{a}_k^H|}\right). \tag{4.84}$$

In the two-user case, the capacity region for independent decoding is a rectangle bounded by

$$R_1 \leq \frac{1}{2}\log_2\left(1 + \frac{\alpha\sigma_1^2 + \alpha^2\sigma_1^2\sigma_2^2\sin^2\phi_{12}}{1+\alpha\sigma_2^2}\right), \tag{4.85}$$

$$R_2 \leq \frac{1}{2}\log_2\left(1 + \frac{\alpha\sigma_2^2 + \alpha^2\sigma_1^2\sigma_2^2\sin^2\phi_{12}}{1+\alpha\sigma_1^2}\right). \tag{4.86}$$

Interestingly, if the array response vectors of the two users are orthogonal to each other, we get the same bounds on R_1 and R_2, as given by the first two lines of (4.83). This means that, when the spatial channels are orthogonal to each other, the capacity region achieved with independent decoding is the same as the one achieved with joint decoding.

The capacity region, in case of independent decoding, is the shaded rectangle sketched in Figure 4.22. The limits R_{12} and R_{21} in Figure 4.22 are the maximum values of the rates R_1 and R_2 given in 4.85 and 4.86, respectively. In general, independent decoding implies an efficiency loss, but it is simpler to implement. The loss depends on the users' position. In particular, if the two channels are orthogonal to each other (i.e., when $\sin^2\phi_{12} = 1$) there are no losses. Conversely, the maximum losses occur when the directions of arrival tend to coincide (i.e., when $\sin^2\phi_{12} = 0$). This is not surprising because, when the directions of arrival coincide, the receiver has no capabilities of separating the two signals working only in the space domain.

4.6 SUMMARY

In this chapter we have reviewed some of the basic issues related to multiple access strategies. The problem has been faced primarily from the physical layer point of view. The reader interested in studying more details of the medium access control strategies as well as the network aspects is invited to read [2]. We have considered fixed and random access strategies. We have shown how the common access schemes, namely TDMA, FDMA, and CDMA, can all be seen as alternative ways to handle the problem working in the time domain. The availability of multiantenna transceivers introduces a great flexibility as it allows us to manage the access problem working in the *joint* space-time domain. In Chapter 6, the access problem will be further analyzed showing that the optimal access strategy, instead of being a priori fixed, is the result of a dynamic optimization based on game theory principles.

References

[1] Pahlavan, K., Levesque, A. H., *Wireless Information Networks*, New York: Wiley Series in Telecommunications, 1995.

[2] Bertsekas, D., Gallager, R., *Data Networks*, 2nd Ed., Upper Saddle River, NJ: Prentice Hall, 1992.

[3] Abramson, N., "The ALOHA System," in *Computer Communication Networks*, Abramson, N., and Kuo, F., (eds.), Englewood Cliffs, NJ: Prentice-Hall, 1973.

[4] Abramson, N., "The throughput of packet broadcasting channels," *IEEE Trans. on Commun.*, Jan. 1977, pp. 117–128.

[5] Tsatsanis, M. K., Zhang, R., "Network-assisted diversity for random access wireless networks," *IEEE Trans. on Signal Processing*, March 2000, pp. 702–711.

[6] Wang, X., Poor, V. H., *Wireless Communication Systems*, Upper Saddle River, NJ: Prentice Hall, 2004.

[7] Yates, R. D., "A framework for uplink power control in cellular radio systems", *IEEE Journal on Selected Areas in Commun.*, Vol. 7, Sept. 1995, pp. 1341–1348.

[8] 3GPP TS 25.302 v4.3.0, *3rd Generation Partnership Project; Technical Specification Group Radio Access Network; Services provided by the Physical Layer*, December 2001.

[9] Suard, B., Xu, G., Liu, H., Kailath, T., "Uplink channel capacity of space-division-multiple access schemes", *IEEE Trans. on Inform. Theory*, Vol. 44, July 1998, pp. 1468–1476.

[10] Paulraj, A. J., Papadias, C. B., "Space-time processing for wireless communications", *IEEE Signal Processing Magazine*, Nov. 1997, pp. 49–83.

Chapter 5

Diversity and Multiplexing Gain in Open-Loop MIMO Systems

5.1 INTRODUCTION

In this chapter we describe the fundamental performance limits of MIMO wireless systems. We assume that there is no feedback from the receiver to the transmitter so that the transmitter does not have any knowledge about the channel. We denote this kind of system as an *open-loop* system to distinguish it from *closed-loop* systems, where there is some kind of feedback, from the receive to the transmit side, that allows the transmitter to acquire knowledge about the channel. This knowledge can be properly exploited to optimize the transmission strategy. Closed-loop systems will be the subject of chapter 6.

Open loops are commonly found in wireless communication systems, where the channel fluctuations make complicated the knowledge (prediction) of the channel at the transmit side. Furthermore, in broadcast systems, where the transmitter sends data to several receivers simultaneously, it does not even make sense to optimize the transmission strategy with respect to a specific channel. We start by reviewing the statistical properties of random matrices, as they have a straightforward impact on the performance of random channels. Then, we will introduce the fundamental performance parameters, in a very general setup. After reviewing the basic properties of SISO systems, we will study in detail the diversity gain and the rate gain of MIMO systems. For most of the chapter, we will assume that the transmitter has no knowledge about the channel but that the receiver knows the channel perfectly. In the last section, we will relax this assumption, reviewing the major losses resulting from imperfect channel knowledge at the receiver.

5.2 STATISTICAL PROPERTIES OF THE EIGENVALUES OF RANDOM MATRICES

This section reviews the statistics of the eigenvalues of random matrices, as they play a fundamental role in the performance of MIMO systems. Let us consider a (channel) matrix $\boldsymbol{H}$ with i.i.d. circularly symmetric, complex Gaussian random variables, with zero mean and unit variance. The dimension of $\boldsymbol{H}$ is $n_R \times n_T$. Let us now introduce the matrix

$$\mathcal{W} := \begin{cases} \boldsymbol{H}\boldsymbol{H}^H & n_R < n_T \\ \boldsymbol{H}^H\boldsymbol{H} & n_R \geq n_T \end{cases} \tag{5.1}$$

and the variables $n = \max(n_R, n_T)$ and $m = \min(n_R, n_T)$. The $m \times m$ matrix $\mathcal{W}$ is a *Wishart* matrix; $\mathcal{W}$ is Hermitian and nonnegative definite and has real non-negative eigenvalues.

Since $\boldsymbol{H}$ is random, the eigenvalues of $\mathcal{W}$ are themselves random. Under the previous assumption, the joint pdf of the *ordered* eigenvalues of $\mathcal{W}$ is [1]

$$p_{\boldsymbol{\lambda}}^{\text{ord}}(\lambda_1, \ldots, \lambda_m) = c\, e^{-\sum_{i=1}^{m} \lambda_i} \prod_{i=1}^{m} \lambda_i^{n-m} \prod_{j=i+1}^{m} (\lambda_i - \lambda_j)^2, \tag{5.2}$$

with $\lambda_1 \geq \lambda_2 \geq \ldots \geq \lambda_m \geq 0$, and c is a normalization factor. From (5.2), one can also derive the joint pdf of the *unordered* eigenvalues

$$p_{\boldsymbol{\lambda}}(\lambda_1, \ldots, \lambda_m) = \frac{c}{m!}\, e^{-\sum_{i=1}^{m} \lambda_i} \prod_{i=1}^{m} \lambda_i^{n-m} \prod_{j=i+1}^{m} (\lambda_i - \lambda_j)^2. \tag{5.3}$$

As an example, in a 2×2 MIMO system, (5.3) becomes

$$p_{\lambda_1,\lambda_2}(\lambda_1, \lambda_2) = c_2 e^{-(\lambda_1+\lambda_2)} (\lambda_2 - \lambda_1)^2, \;\; \lambda_1 \geq 0, \lambda_2 \geq 0. \tag{5.4}$$

In this case, the joint pdf of the unordered eigenvalues is reported in Figure 5.1.

It is in general very difficult to derive the marginal pdf of each eigenvalue. Nevertheless, it is possible to obtain the pdf of the smallest or largest eigenvalue of $\boldsymbol{H}$ in case of zero-mean uncorrelated coefficients [2]. More recently, Kang and Alouini have also derived the expression for the largest eigenvalues in the case of correlated channels with a nonzero average [3]. These pdfs play an important role in the performance of MIMO systems, as we will show next.

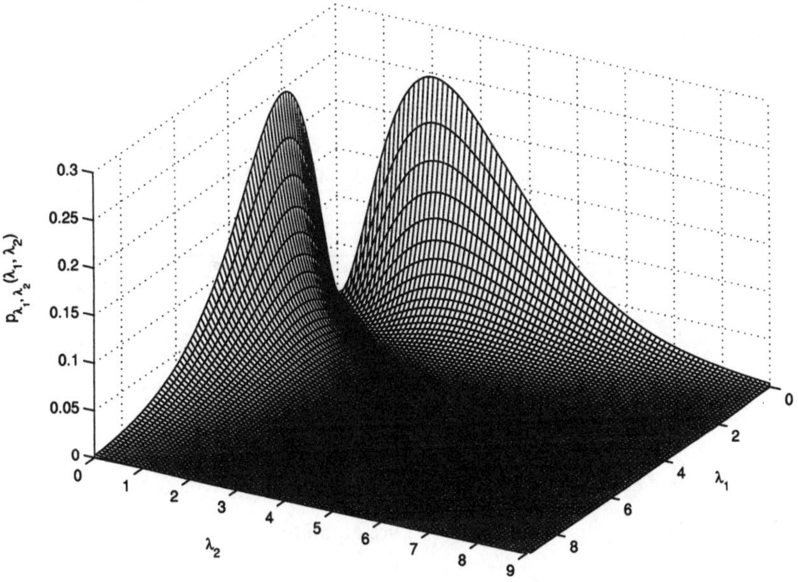

Figure 5.1 Joint pdf $p_{\lambda_1,\lambda_2}(\lambda_1,\lambda_2)$, obtained with $n = m = 2$.

From [2], the cumulative distribution function (CDF) of the largest eigenvalue $\lambda_{\max}$ is

$$\Pr\{\lambda_{\max} \leq x\} = c_1|\mathbf{\Gamma}(x)|, \tag{5.5}$$

where $\mathbf{\Gamma}(x)$ is the matrix whose (i, j) entry is

$$\Gamma_{i,j}(x) := \int_0^x u^{m+i+j-2}\, e^{-u}\, du, \tag{5.6}$$

and c_1 is a coefficient necessary to impose a unit area of the pdf.

Similarly, the CDF of the smallest eigenvalue is

$$\Pr\{\lambda_{\max} \leq x\} = 1 - c_2|\overline{\mathbf{\Gamma}}(x)|, \tag{5.7}$$

where $\overline{\mathbf{\Gamma}}$ is the matrix whose (i, j) entry is

$$\overline{\Gamma}_{i,j}(x) := \int_x^\infty u^{m+i+j-2}\, e^{-u}\, du, \tag{5.8}$$

and c_2 is a coefficient necessary to impose a unit area of the pdf.

To grasp the effect of the channel matrix size on these pdf, we report a few examples in Table 5.1.

Table 5.1

Probability Density Function of the Smallest or Largest Eigenvalues of Rayleigh-Fading MIMO Channels for Different MIMO Configurations.

n_R	n_T	$p_{\lambda_{min}}(x)$	$p_{\lambda_{max}}(x)$
2	1	$x\,e^{-x}$	$x\,e^{-x}$
2	2	$2\,e^{-2x}$	$(2-2x+x^2)\,e^{-x}-2\,e^{-2x}$
3	1	$\frac{1}{2}\,x^2\,e^{-x}$	$\frac{1}{2}\,x^2\,e^{-x}$
3	2	$x\,(x+3)\,e^{-2x}$	$\frac{1}{2}\,x\,(6-4x+x^2)\,e^{-x}-x\,(3+x)\,e^{-2x}$
3	3	$3\,e^{-3x}$	$p_{3,max}(x)$
4	4	$4\,e^{-4x}$	$p_{4,max}(x)$

The two pdf's $p_{3,\max}(x)$ and $p_{4,\max}(x)$ are equal to

$$\begin{aligned} p_{3,\max}(x) &= \frac{1}{4}\,(12-24x+24x^2-8x^3+x^4)\,e^{-x} \\ &- \frac{1}{2}\,(12-12x+6x^2+2x^3+x^4)\,e^{-2x}+3\,e^{-3x} \end{aligned} \tag{5.9}$$

and

$$\begin{aligned} p_{4,\max}(x) &= -4e^{-4x} \\ &+ \frac{1}{12}\,e^{-3x}(144-144x+72x^2+56x^3+46x^4+10x^5+x^6) \\ &- \frac{1}{72}\,e^{-2x}(864-1728x+1728x^2-192x^3+96x^4-96x^5 \\ &\qquad +32x^6-4x^7+x^8) \\ &+ \frac{1}{36}\,e^{-x}(144-432x+648x^2-408x^3+126x^4-18x^5+x^6). \end{aligned} \tag{5.10}$$

To make a fair comparison of MIMO systems with a different number of transmit antennas, it is necessary to normalize the power transmitted from each antenna in order to enforce the same overall transmitted power. This can be achieved by assigning a variance $\sigma_h^2 = 1/n_T$ to each entry of the channel matrix $\boldsymbol{H}$. The pdf of the normalized channel eigenvalues can be then obtained from the previous expressions through the normalization

$$\overline{p}_k(x) = \frac{1}{\sigma_h^2} p_k \left(\frac{x}{\sigma_h^2} \right). \tag{5.11}$$

As an example, in Figure 5.2 we report the marginal pdfs of the four eigenvalues of a 4×4 MIMO channel. Solid lines refer to the theoretical pdf of the smallest and

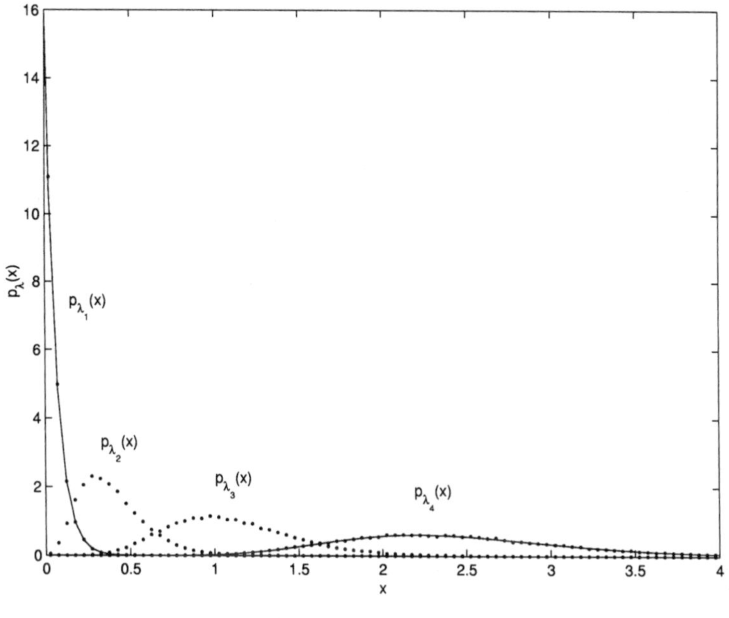

Figure 5.2 Marginal pdf for a 4×4 Rayleigh-fading MIMO channel: theoretical value (solid line) and histograms (dots).

largest eigenvalues, whereas the dots represent the histograms of the eigenvalues. We can see from Figure 5.2 that even though each channel is Rayleigh-fading and then it has an exponential pdf, the eigenvalues of the resulting MIMO channel have pdfs with smaller and smaller probabilities of assuming low values, as their order increases. This is the first fundamental property of MIMO systems:

A MIMO channel creates propagation modes that are more robust against channel fades than each single SISO component of the MIMO channel.

This property will be properly quantified through the introduction of the diversity gain, later on in this chapter. It is also interesting to see what happens when the

MIMO system has more receive antennas than transmit antennas. As an example, Figure 5.3 shows the marginal pdf of the two (nonull) eigenvalues of a 4×2 MIMO channel. Again, solid lines refer to the theoretical pdf of the smallest and largest eigenvalues, whereas the dots refer to the histograms. Comparing Figures 5.2 and

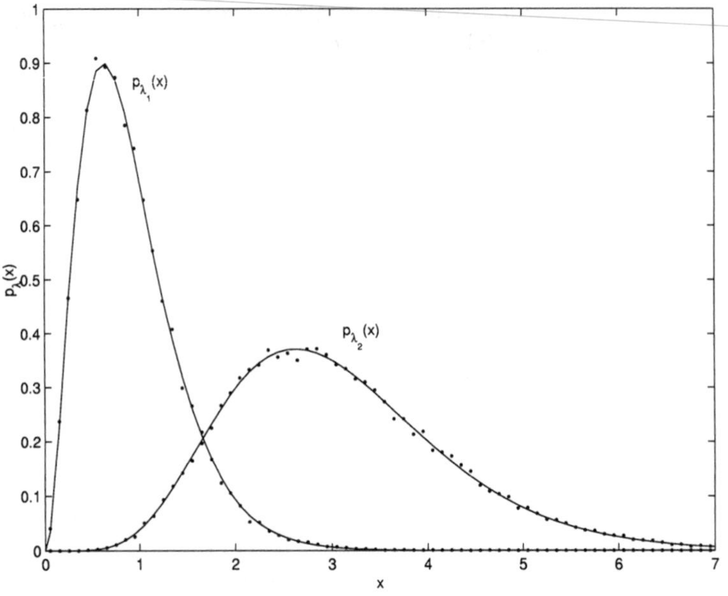

Figure 5.3 Marginal pdf for a 4×2 Rayleigh fading MIMO channel: theoretical value (solid line) and histograms (dots).

5.3, it is interesting to observe that the pdf of the smallest eigenvalue in the 4×2 case has lower probability of assuming values around the origin, than the corresponding eigenvalue of the 4×4 system. This suggests that a 4×2 MIMO system is more robust than a 4×4 system, as its worst mode is more robust than the corresponding mode in the 4×4 configuration. However, a 4×4 system has four modes, whereas a 4×2 system has only two modes. This means that the rate of a 4×4 may be expected to be higher than the rate of a 4×2 system, for a given total power budget.

These considerations introduce a second basic feature of MIMO systems:

A MIMO channel makes possible an increase of transmission rate with respect to a SISO channel, thanks to multimode propagation; however, there exists a fundamental trade-off between the rate gain and the robustness against fading.

Digging further into the behavior of the statistics of the smallest and largest eigenvalues, around the origin, we can draw the following general remarks:

- The pdf of the smallest eigenvalue has a Taylor series expansion around the origin $x = 0$, equal to[1]

$$p_{\lambda_{\min}}(x) = c_{n_T,n_R}\, x^{n_R-n_T} + o(x^{n_R-n_T}), \tag{5.12}$$

  where c_{n_T,n_R} is a constant depending only on n_T and n_R;

- The pdf of the largest eigenvalue has a Taylor series expansion around the origin equal to

$$p_{\lambda_{max}}(x) = k_{n_T,n_R}\, x^{n_R \cdot n_T-1} + o(x^{n_R \cdot n_T-1}). \tag{5.13}$$

These properties have a direct impact on the performance of MIMO systems. One of the first works showing how to make full benefit of the eigenvalue statistics in MIMO systems was carried out by Andersen in [4]. In this chapter, the channel statistics will be used to derive the fundamental performance of MIMO systems. Before starting the analysis, it is useful to anticipate some basic concepts.

A SISO flat-fading channel, with channel coefficient h, gives rise to an error probability that is a function of the channel realization. In case of a SISO channel, the channel power gain (attenuation) coincides with the channel single eigenvalue. Using the same notation as before, we can thus denote with λ the channel power gain. The error probability of a SISO channel is then $P_e = P_e(\lambda)$.

To characterize the channel with a single parameter, we can compute, for example, the average error probability by taking the expected value of $P_e(\lambda)$ with respect to λ. If we denote with $p_\lambda(\lambda)$ the pdf of λ, it has been proved in [5] that the average error probability behaves asymptotically (i.e., as the SNR goes to infinity) as

$$\overline{P}_e := E_\lambda\{P_e(\lambda)\} \propto \frac{c}{\mathrm{SNR}^{G_d}}, \tag{5.14}$$

where c is a constant and the exponent G_d is the *diversity gain*. The value of G_d depends on the behavior of $p_\lambda(\lambda)$ in the neighborhood of the origin. In particular, let us introduce the integer q as the smallest order of the derivative of $p_\lambda(x)$, evaluated in $x = 0$, that is different from zero. In formulas, q is defined through the following relationships

$$\left.\frac{d^k p_\lambda(x)}{dx^k}\right|_{x=0} = 0,\ 0 \le k < q,\ \text{and}\ \left.\frac{d^q p_\lambda(x)}{dx^q}\right|_{x=0} \neq 0. \tag{5.15}$$

1 We assume here, for simplicity of notation, that $n_R \geq n_T$.

The diversity gain is simply [5]

$$G_d = q + 1.$$

The diversity gain is a very important parameter as it describes the decaying rate of the bit error rate with the SNR.

We are now able to quantify the robustness of MIMO systems against fading. If we consider the expression of the pdf of the smallest and largest eigenvalues of Rayleigh-fading MIMO channels, shown in Table 5.1, we may state that the *different channel modes have different diversity gains*. More precisely, the diversity gain, for flat-fading stationary MIMO, varies within the following interval

$$n_R - n_T + 1 \leq G_d \leq n_T n_R, \tag{5.16}$$

where the lower bound is given by the mode associated to the smallest eigenvalue and the upper bound is given by the mode associated to the largest eigenvalue.

These considerations suggest that to increase the system robustness against fading, it would be better to have n_R as larger than n_T as possible, so that the worst channel (the one associated to the smallest eigenvalue) had some diversity gain.

However, it is important to remark that the previous statement holds true for a MIMO system characterized by the I/O relationship

$$\boldsymbol{y} = \boldsymbol{H}\boldsymbol{x}, \tag{5.17}$$

with $\boldsymbol{H}$ having the statistical properties described before. If instead of mapping the information bits into a vector $\boldsymbol{x}$ of size $n_T \times 1$, we map them into *matrix codewords* of size $n_T \times n_B$, composed of n_B consecutive blocks in time, and if the channel is time-invariant, at least within n_B consecutive blocks, we can write the following I/O relationship

$$\boldsymbol{Y} = \boldsymbol{H}\boldsymbol{X}, \tag{5.18}$$

where $\boldsymbol{X}$ is the transmitted codeword and $\boldsymbol{Y}$ is the received codeword. Interestingly, whereas the system (5.17) has a diversity gain $n_R - n_T + 1$, which implies no gain at all when $n_R = n_T$, the system (5.18) may have diversity gain $G_d = n_R n_T$, if $n_B \geq n_T$. This is true provided that the rule for mapping the information bits into the matrix codeword $\boldsymbol{X}$ is chosen appropriately. These considerations anticipate one of the most important concept underlying the use of multiantenna systems: Encoding the information bits into codewords that span over *both space and time domains*, we can reach the maximum possible diversity gain. The strategies for designing such an encoding schemes will be the subject of Chapter 7, devoted to space-time coding techniques.

Before concluding this section, it is worth adding that, knowing the channel at the transmit side, one could discard the worst channels and then transmit only over the modes with the highest diversity gain, or, alternatively, one could load power on each mode adaptively to change the pdf of the resulting mode and get, possibly, the maximum diversity gain over each mode. However, this would require an accurate channel knowledge at the transmitter, and this is not always available. The optimal design of MIMO systems when the transmitter has some kind of channel knowledge will be studied in Chapter 6.

5.3 FUNDAMENTAL PERFORMANCE PARAMETERS

In a digital communication system, the fundamental performance parameters are the bit error rate and the number of bits that we are able to send through the channel, per units of time and frequency (number of bits per second per Hertz or bit/s/Hz), guaranteeing an arbitrarily low error probability. An excellent review of the performance over fading channels is, for example, [6]. In general, every linear system employing block transmission, whether SISO or MIMO, can be described by an I/O relationship assuming the form

$$\boldsymbol{Y} = \boldsymbol{H}\boldsymbol{X} + \boldsymbol{W}, \tag{5.19}$$

where $\boldsymbol{H}$ is the channel matrix, $\boldsymbol{X}$ is the matrix of transmitted symbols, and $\boldsymbol{W}$ is additive noise. The structure of $\boldsymbol{H}$ depends on the number of transmit/receive antennas, on the channel dispersiveness in time/frequency, as well as on the transmission strategy.

The only way to assess the performance of a wireless system globally (i.e., not conditioned to a specific channel realization) is probabilistic, based on random channel modeling, as described in Chapter 2. The set of parameters depends, on one side, on the quality of service (QoS) requirements and, on the other side, on the channel randomness. The two most typical performance parameters are bit (or packet) error rate and number of bits (packets) transmitted correctly, per unit of time and frequency (bit/s/Hz). Depending on the channel time variability with respect to the transmission rate, it is useful to distinguish between deterministic, ergodic, and nonergodic channels, as detailed in the following sections.

Deterministic channel

In this case, the channel matrix $\boldsymbol{H}$ is fixed. This situation exemplifies the case where the channel is very slowly fading, and we are concerned with a specific link. In such a case, the symbol error probability and the capacity are functions of $\boldsymbol{H}$. We will denote them as $P_e(\boldsymbol{H})$ and $C(\boldsymbol{H})$, respectively. In a wireless context, these numbers can be interpreted as the error probability and the capacity, *conditioned* to a specific outcome of the random channel, let us say $\boldsymbol{H}$. The error probability depends on the constellation used at the transmitter as well as on the detection scheme adopted at the receiver. The capacity can be computed as follows.

Let us denote with $H(\boldsymbol{X})$ and $H(\boldsymbol{Y})$ the entropy of the channel input and output, respectively. If $\boldsymbol{X}$ is a matrix composed of discrete random variables, it assumes a finite number of states. Let us call each state as $\boldsymbol{X}_i$ and let us denote with N_X the number of states. We can do the same with the channel output, introducing the states $\boldsymbol{Y}_j$, and their number N_Y. The entropy of $\boldsymbol{X}$ is [7][2]

$$H(\boldsymbol{X}) = \sum_{i=1}^{N_X} P_X(\boldsymbol{X}_i) \log_2 \left[\frac{1}{P_X(\boldsymbol{X}_i)} \right], \tag{5.20}$$

where $P_X(\boldsymbol{X}_i)$ is the probability $\Pr\{X = X_i\}$. The entropy of $\boldsymbol{Y}$ can be computed similarly.

If $\boldsymbol{X}$ is composed of continuous random variables, described by the joint pdf $p_X(\boldsymbol{x})$, we can compute the differential entropy of X as

$$H(\boldsymbol{X}) = \int_{\mathcal{D}_X} p_X(\boldsymbol{x}) \log_2 \left[\frac{1}{p_X(\boldsymbol{x})} \right] d\boldsymbol{x}, \tag{5.21}$$

where the integral is carried out over the multidimensional domain $\mathcal{D}_X$, denoting the domain where the random vector $\boldsymbol{x}$ is defined. The entropy $H(\boldsymbol{X})$ measures the a priori uncertainty about the random matrix $\boldsymbol{X}$.

According to the first Shannon's theorem on source coding [7], if $\boldsymbol{X}$ is discrete, the entropy measures the minimum average number of bits necessary to encode $\boldsymbol{X}$ without any information loss.

We recall now the definition of conditional entropy, as it will be fundamental in the derivation of the mutual information and then the channel capacity. If $\boldsymbol{X}$ and $\boldsymbol{Y}$ are

2 Using 2 as the base of the logarithm, the measurement unit is the bit.

discrete, the entropy of $\boldsymbol{X}$ conditioned to $\boldsymbol{Y}$ is

$$H(\boldsymbol{X}/\boldsymbol{Y}) = \sum_{i=1}^{N_X}\sum_{j=1}^{N_Y} p_{X,Y}(\boldsymbol{X}_i, \boldsymbol{Y}_j) \log_2 \left[\frac{1}{p_{X/Y}(\boldsymbol{X}_i/\boldsymbol{Y}_j)}\right]. \tag{5.22}$$

$H(\boldsymbol{X}/\boldsymbol{Y})$ measures the average uncertainty about $\boldsymbol{X}$, after having observed $\boldsymbol{Y}$. If $\boldsymbol{X}$ and $\boldsymbol{Y}$ are continuous, the differential entropy of $\boldsymbol{X}$, conditioned to $\boldsymbol{Y}$, is

$$H(\boldsymbol{X}/\boldsymbol{Y}) = \int_{\mathcal{D}_X}\int_{\mathcal{D}_Y} p_{X,Y}(\boldsymbol{x}, \boldsymbol{y}) \log_2 \left[\frac{1}{p_{X/Y}(\boldsymbol{x}/\boldsymbol{y})}\right] d\boldsymbol{x}d\boldsymbol{y}. \tag{5.23}$$

The *mutual information* exchanged between $\boldsymbol{X}$ and $\boldsymbol{Y}$ is equal to the average reduction of the a priori uncertainty about a set of variables, let us say $\boldsymbol{X}$ for example, after we have observed the other set, for example $\boldsymbol{Y}$. In formulas, the mutual information between $\boldsymbol{X}$ and $\boldsymbol{Y}$ is

$$I(\boldsymbol{X};\boldsymbol{Y}) = H(\boldsymbol{X}) - H(\boldsymbol{X}/\boldsymbol{Y}) = H(\boldsymbol{Y}) - H(\boldsymbol{Y}/\boldsymbol{X}). \tag{5.24}$$

It is useful to remark that when $\boldsymbol{X}$ and $\boldsymbol{Y}$ are continuous random variables, their differential entropies, as defined in (5.21) and (5.23), do not have a precise physical meaning. Nevertheless, the mutual information exchanged between them, as defined in (5.24), is a well defined quantity, that can be measured in number of bits per symbol.

If $\boldsymbol{X}$ and $\boldsymbol{Y}$ are the input and output of a channel, the *channel capacity* is defined as the maximum value of the mutual information over all possible statistical distributions of $\boldsymbol{X}$. The mutual information would depend, in this case, on the channel matrix $\boldsymbol{H}$ and thus we indicate it as $I(\boldsymbol{X};\boldsymbol{Y}/\boldsymbol{H})$. Therefore, the capacity of a given channel $\boldsymbol{H}$ is

$$C(\boldsymbol{H}) = \max_{p_X(\boldsymbol{x})} I(\boldsymbol{X};\boldsymbol{Y}/\boldsymbol{H}) \text{ bps}. \tag{5.25}$$

The capacity is measured, as the mutual information, in units of bits per symbol. In general, the joint pdf $p_X(\boldsymbol{x})$ of the input must satisfy some constraint (e.g., a constraint on the average power). In his seminal papers [8], Shannon proved that the channel capacity is the maximum number of bits per symbol that can be transmitted through the channel, insuring an arbitrarily low error probability, provided that a sufficiently long code is applied to the transmitted sequence.

Ergodic channel

In this case, we assume a block transmission, where each block experiences a different fading. This model is known as block fading. Denoting with $\boldsymbol{H}_k$ the channel matrix pertaining to the kth block, channel ergodicity implies that any given function $f(\boldsymbol{H}_k)$ of the channel matrix satisfies the following condition

$$E_{\boldsymbol{H}}\{f(\boldsymbol{H}_k)\} = \lim_{N\to\infty} \frac{1}{2N+1} \sum_{k=-N}^{N} f(\boldsymbol{H}_k), \tag{5.26}$$

with probability one. In (5.26), $E_{\boldsymbol{H}}\{f(\boldsymbol{H})\}$ means expected value of $f(\boldsymbol{H})$, with respect to the joint pdf of $\boldsymbol{H}$. Intuitively speaking, a channel is ergodic if, when we transmit over a sufficiently high number of blocks, we experience all the relevant (from a statistical point of view) channel states. This requires the transmission time to be sufficiently greater than the channel coherence time. The ergodicity assumption allows us, for example, to compute the average (in time) error probability over all the received blocks as the expected value of the error probability conditioned to a given channel realization; that is

$$\overline{P}_e = \lim_{N\to\infty} \frac{1}{2N+1} \sum_{k=-N}^{N} P_e(\boldsymbol{H}_k) = E_{\boldsymbol{H}}\{P_e(\boldsymbol{H})\}. \tag{5.27}$$

We derive now the capacity of an ergodic channel. We consider the case where the channel is not known at the transmitter side, but it is perfectly known at the receiver side. We assume also that the transmitted sequence $\boldsymbol{X}$ is statistically independent of $\boldsymbol{H}$. The independence between $\boldsymbol{X}$ and $\boldsymbol{H}$ is certainly verified, as the transmitter does not have any knowledge about $\boldsymbol{H}$.
Since the channel is perfectly known to the receiver, every time we transmit a matrix $\boldsymbol{X}$ we observe the pair of matrices $\boldsymbol{Y}$ and $\boldsymbol{H}$. Hence, the mutual information exchanged between the channel input and the (overall) output is

$$I[\boldsymbol{X};(\boldsymbol{Y},\boldsymbol{H})] = I(\boldsymbol{X};\boldsymbol{H}) + I(\boldsymbol{X};\boldsymbol{Y}/\boldsymbol{H}). \tag{5.28}$$

Since $\boldsymbol{X}$ and $\boldsymbol{H}$ are statistically independent, we do not get any information about $\boldsymbol{X}$, after having observed $\boldsymbol{H}$, i.e., $I(\boldsymbol{X};\boldsymbol{H}) = 0$. Thus, we can write (5.28) as

$$I[\boldsymbol{X};(\boldsymbol{Y},\boldsymbol{H})] = I(\boldsymbol{X};\boldsymbol{Y}/\boldsymbol{H}). \tag{5.29}$$

Using the definition of entropy, we get

$$I[(\boldsymbol{X};\boldsymbol{Y}/\boldsymbol{H})] = E_{\boldsymbol{H}}\left\{I(\boldsymbol{X};\boldsymbol{Y}/\boldsymbol{H})\right\}, \tag{5.30}$$

where the mutual information inside the expected value operator has to be intended as the mutual information between $\boldsymbol{X}$ and $\boldsymbol{Y}$, conditioned to the channel realization $\boldsymbol{H}$. Exploiting the ergodicity property, if we denote by $\boldsymbol{H}_k$ the channel realization over the kth block, the average mutual information is

$$I(\boldsymbol{X};\boldsymbol{Y}/\boldsymbol{H}) = \lim_{N\to\infty} \frac{1}{2N+1} \sum_{k=-N}^{N} I(\boldsymbol{X};\boldsymbol{Y}/\boldsymbol{H}_k) = E_{\boldsymbol{H}}\{I(\boldsymbol{X};\boldsymbol{Y}/\boldsymbol{H})\}. \tag{5.31}$$

The capacity is then the maximum of (5.31), over all the possible input pdfs, that is,

$$C = \max_{p_{\boldsymbol{X}}(\boldsymbol{x})} E_{\boldsymbol{H}}\{I(\boldsymbol{X};\boldsymbol{Y}/\boldsymbol{H})\}. \tag{5.32}$$

This quantity is known as the *ergodic capacity*.
In case of discrete random variables, the input and output alphabets are typically assigned and then the maximization has to be performed over all possible probability distributions of $\boldsymbol{X}$, over the assigned alphabets. Conversely, in the continuous case, typically there is an average power constraint on the transmitted sequence, which translates into a constraint on the second order moments of $\boldsymbol{X}$.

Nonergodic channel

If the ergodicity assumption is not valid, we cannot compute the ergodic capacity or the average error probability, as described before. This happens, for example, when the duration of the transmission is smaller than the channel coherence time, so that the transmission time is not sufficient to explore all relevant information about the channel. In such a case, the error probability $P_e(\boldsymbol{H})$ and the capacity $C(\boldsymbol{H})$ are functions of a set of random variables, and thus they are themselves random variables. Therefore, the only way to describe the channel is through the analysis of the statistics of these random variables, e.g., their moments or their cumulative distribution function (CDF). The CDFs relative to the error probability and to the capacity are defined as follows

$$D_{P_e}(p) := \Pr\{P_e(\boldsymbol{H}) \leq p\}, \tag{5.33}$$

$$D_C(c) := \Pr\{C(\boldsymbol{H}) \leq c\}, \tag{5.34}$$

respectively. Besides being useful statistical tools, these functions have a physical interpretation. In fact, each service delivered through a communication system is characterized by its own QoS parameters (e.g., error probability and delay). So, for example, in third generation cellular systems, it is necessary to deliver voice with an error probability of 10^{-3}, whereas data should be delivered with 10^{-6} error probability. If the system provides an error probability smaller than the target value, let us call it P_{target}, the system is working properly, but, if the actual error rate is greater than P_{target}, we say that the system experiences an *out-of-service* event. It is common practice, in expressing the requirements for a system, to specify the probability of the out-of-service event. This is equal to

$$P_{\text{out}} := \Pr\{P_e(\boldsymbol{H}) > P_{\text{target}}\} = 1 - D_{P_e}(P_{\text{target}}), \tag{5.35}$$

where, in the last equality, we have used (5.33).

Similarly, when we transmit with a certain rate r, measured in terms of bits per symbol, through a random channel, there is a certain probability that the chosen rate is greater than the actual channel capacity $C(\boldsymbol{H})$. In such a case, by the converse of Shannon's channel coding theorem, we know that there does not exist any coding strategy that can make the error probability arbitrarily low. Hence, it is useful to measure the probability of such an event. This probability is known as the *outage probability* and is defined as

$$P_{\text{outage}} := \Pr\{C(\boldsymbol{H}) < r\} = D_C(r). \tag{5.36}$$

5.4 SISO CHANNEL

In this section, we will compute the main system performance parameters for a SISO channel. In the next sections, we will extend the analysis to MIMO channels. We consider a flat-fading channel described by the I/O relationship:

$$y_k = h_k s_k + n_k, \tag{5.37}$$

where s_k is the transmitted symbol, belonging to a constellation of finite order, whose elements belong to an alphabet $\mathcal{A}$ of cardinality $|\mathcal{A}|$; h_k is the channel coefficient at the time corresponding to the transmission of the kth symbol; n_k is additive noise. We assume that the noise sequence, denoted by $\{n_k\}$, is a sequence of independent identically distributed (iid) circularly symmetric complex Gaussian random variables, with zero mean and variance $E\{|n_k|^2\} = \sigma_n^2$. The performance depends on the channel properties. Proceeding according to the general formulation given in the previous section, we have the following cases.

Deterministic channel

The error probability depends on the constellation used to send the information bits, as well as on the decoding strategy. For example, using an M-ary pulse amplitude modulation (PAM), whose constellation is composed of symbols assuming the values $A_m = 2m - 1 - M$, $m = 1, \ldots, M$, the symbol error probability using coherent decoding is [9][3]:

$$p_{(M)}^{\text{PAM}}(h) = \frac{M-1}{M} \operatorname{erfc}\left(\sqrt{\frac{3|h|^2\gamma_{\text{av}}}{(M^2-1)}}\right), \tag{5.38}$$

where $\gamma_{\text{av}} := \sigma_s^2/\sigma_n^2$ is the average signal-to-noise ratio (SNR) per symbol. Since an M-ary PAM symbol conveys $\log_2 M$ bits per symbol, it is useful to introduce the average SNR per bit, defined as

$$\gamma_b = \frac{\gamma_{\text{av}}}{\log_2 M}. \tag{5.39}$$

Combining (5.38) and (5.39), the symbol error probability can be expressed in terms of the bit SNR as

$$p_{(M)}^{\text{PAM}}(h) = \frac{M-1}{M} \operatorname{erfc}\left(\sqrt{\frac{3|h|^2 \log_2 M\gamma_b}{(M^2-1)}}\right). \tag{5.40}$$

If we use a square M-ary QAM, the symbol error probability is equal to [9]

$$\begin{aligned} p_{(M)}^{\text{QAM}}(h) &= 1-(1-p_{\sqrt{M}}^{\text{PAM}})^2 = 2p_{\sqrt{M}}^{\text{PAM}} - p_{\sqrt{M}}^{\text{PAM}^2} \approx 2p_{\sqrt{M}}^{\text{PAM}} \\ &= 2\frac{\sqrt{M}-1}{\sqrt{M}} \operatorname{erfc}\left(\sqrt{\frac{3|h|^2}{M-1}\frac{\gamma_{av}}{2}}\right), \end{aligned} \tag{5.41}$$

where the last approximation is valid for low values of $2p_{(M)}^{\text{PAM}}$.

Given the symbol error probability, it is not trivial, in general, to find out the bit error probability. Clearly, the relationship between symbol and bit error probabilities depends on the rule used for mapping bits into symbols. The most typical mapping is the so-called Gray bit mapping, where any pair of adjacent symbols in the QAM constellation map sequences of bits that differ by no more than one bit. Gray mapping implies that symbols at the minimum Euclidean distance are also at the minimum Hamming distance[4].

The reason for using Gray encoding is to minimize the BER, for a given symbol error rate. Using a Gray encoder, the bit error probability for QAM constellations,

3 We drop the dependence of h_k on the index k for convenience of notation.

4 The Euclidean distance between two symbols $\boldsymbol{s}_i$ and $\boldsymbol{s}_j$ is $|\boldsymbol{s}_i - \boldsymbol{s}_j|$; their Hamming distance is the number of distinct bits between the binary strings encoding $\boldsymbol{s}_i$ and $\boldsymbol{s}_j$.

conditioned to h, can be well approximated as [10]

$$\mathrm{BER}(h) \simeq \frac{\sqrt{M}-1}{\sqrt{M}\log_2\sqrt{M}}\,\mathrm{erfc}\left(\sqrt{\frac{\beta|h|^2}{\sigma_n^2}}\right) + \frac{\sqrt{M}-2}{\sqrt{M}\log_2\sqrt{M}}\,\mathrm{erfc}\left(3\sqrt{\frac{\beta|h|^2}{\sigma_n^2}}\right), \tag{5.42}$$

where $\beta = \frac{3\,\sigma_s^2}{2\,(M-1)}$.

Let us compute the capacity now, assuming the general case where all variables involved are complex. If the noise is circularly symmetric Gaussian and the input sequence is a sequence of continuous random variables with a constraint on the average transmit power, independent of the noise, the input pdf that maximizes the mutual information is circularly symmetric Gaussian. In such a case, the entropies appearing in (5.24) are [7]

$$H(X) = \log_2(\pi e \sigma_s^2), \quad H(Y) = \log_2[\pi e(|h|^2\sigma_s^2 + \sigma_n^2)] \tag{5.43}$$

and

$$H(Y/X) = \log_2(\pi e \sigma_n^2). \tag{5.44}$$

Therefore, the maximum value of the mutual information is

$$C = \log\left(1 + |h|^2\frac{\sigma_s^2}{\sigma_n^2}\right) \text{ bps.} \tag{5.45}$$

Ergodic channel

Since the error probability, for a flat-fading channel, depends on the channel only through the variable $|h|^2$, it is useful to introduce the random variable $\alpha = |h|^2$. Denoting with $p_{\mathcal{A}}(\alpha)$ the pdf of α, the average value of the error probability is

$$P_{\mathrm{av}} = \int_0^\infty p_{\mathcal{A}}(\alpha)P_{\mathrm{e}}(\alpha)d\alpha, \tag{5.46}$$

where $P_{\mathrm{e}}(\alpha)$ denotes the error probability, conditioned to α. It could coincide with (5.38), for PAM transmission, or with (5.41), for QAM systems. Using the rather general pdf

$$p_{\mathcal{A}}(\alpha) = \frac{1}{(L-1)!\,\bar{\alpha}^L}\,\alpha^{L-1}\,e^{-\alpha/\bar{\alpha}}, \tag{5.47}$$

we can make use of the following integration formula

$$\int_0^\infty \frac{1}{2}\,\mathrm{erfc}\left(\sqrt{x}\right)\frac{1}{(L-1)!m^L}\,x^{L-1}e^{-x/m}dx$$
$$= \left(\frac{1-\mu}{2}\right)^L \sum_{k=0}^{L-1}\binom{L-1+k}{k}\left(\frac{1+\mu}{2}\right)^k, \tag{5.48}$$

where

$$\mu := \sqrt{\frac{m}{1+m}}. \tag{5.49}$$

The Rayleigh channel, with $\alpha = |h|^2$, is a particular case of (5.47), corresponding to $L = 1$. Hence, for a Rayleigh-fading ergodic channel, using (5.48), the error probability for an M-ary PAM transmission system is

$$P_e = \left(1 - \frac{1}{M}\right)(1-\mu), \tag{5.50}$$

where μ is given by (5.49) and

$$m = \frac{3\gamma_{\text{av}}\sigma_h^2}{M^2-1}. \tag{5.51}$$

At high SNR, i.e., for $3\gamma_{\text{av}}\sigma_h^2/(M^2-1) \gg 1$, we have $1-\mu \approx 1/(2m)$ and then

$$P_e \approx \frac{M-1}{M}\frac{M^2-1}{6\gamma_{\text{av}}\sigma_h^2}. \tag{5.52}$$

Comparing (5.52) with (5.38), we notice that the average error probability of a flat-fading Rayleigh channel decreases much more slowly than in the case of deterministic channels. In fact, whereas in a deterministic channel the error probability decreases exponentially with the SNR, in case of a flat-fading Rayleigh channel, the average error rate decreases only as 1/SNR. This is indeed one of the major drawbacks of SISO wireless channels.

A numerical example is shown in Figure 5.4, where we compare the symbol error rate, in case of QPSK constellation, of a deterministic channel with the average symbol error rate of a Rayleigh-fading channel: The dotted line refers to (5.38), with $|h| = 1$ for the deterministic channel, the solid line (curve labeled with $n_R = 1$) refers to the average error probability (5.50) for Rayleigh-fading channels, having $\sigma_h = 1$. We can clearly observe that even at moderate symbol error rates, like 10^{-3}, the loss is about 18 dB and it increases as the error rate decreases.This is indeed a huge loss that justifies the use of multiantenna transceivers, as a way to reduce the loss without increasing neither the bandwidth nor the transmit power, as will be shown in the ensuing sections.

As far as the ergodic capacity is concerned, from the general setup established in (5.32), we have

$$C = E_\alpha\left\{\frac{1}{2}\log_2\left(1+\alpha\frac{\sigma_s^2}{\sigma_n^2}\right)\right\}. \tag{5.53}$$

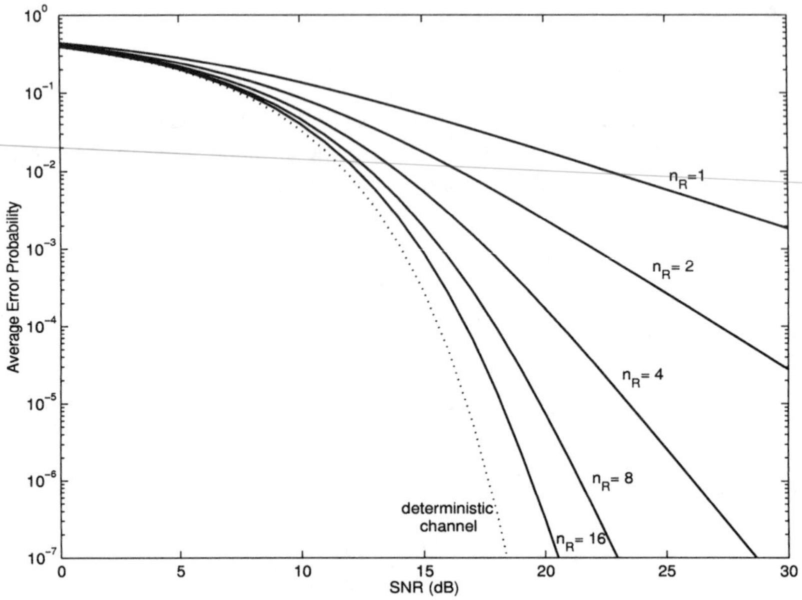

Figure 5.4 Symbol error probability for deterministic channel (dotted line) and fading channel (solid line) for different values of n_R.

In case of Rayleigh fading, with $p_{\mathcal{A}}(\alpha) = \frac{1}{\sigma_h^2}\, e^{-\alpha/\sigma_h^2}$, this integral is equal to

$$C = \frac{1}{2\log(2)} e^{\sigma_n^2/\sigma_s^2\sigma_h^2} \Gamma\left(0, \frac{\sigma_n^2}{\sigma_s^2\sigma_h^2}\right), \tag{5.54}$$

where $\Gamma(a, z)$ is the incomplete Gamma function, defined as

$$\Gamma(a, z) := \int_z^\infty t^{a-1} e^{-t} dt. \tag{5.55}$$

A comparison between the capacity of a deterministic channel and the ergodic capacity is reported in Figure 5.5. To make the comparison, we assumed that the deterministic channel has $|h| = 1$ and the Rayleigh-fading channel has $\sigma_h = 1$.

Nonergodic channel

In case of a nonergodic Rayleigh-fading channel, the error probability is a random variable itself. From (5.38), the out-of-service probability, with reference to a target

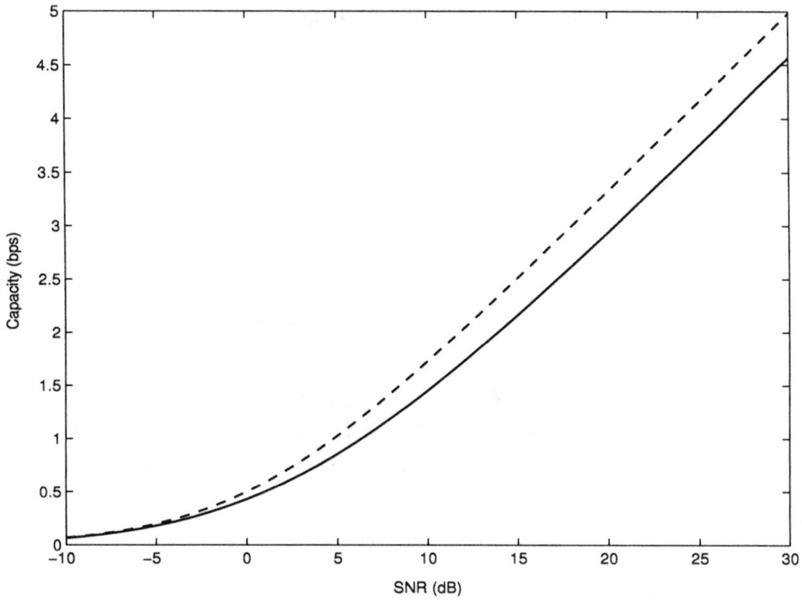

Figure 5.5 Ergodic capacity (solid line) and capacity of a deterministic channel (dotted line) versus SNR (decibels).

probability P_{target}, using a PAM constellation of order M, is

$$\begin{aligned} P_{\text{out}} &:= \Pr\left\{\left(1-\frac{1}{M}\right)\operatorname{erfc}\left(\sqrt{\frac{3\alpha}{M^2-1}\gamma_{\text{av}}}\right) > P_{\text{target}}\right\} \\ &= \Pr\left\{\alpha < \frac{M^2-1}{3\gamma_{\text{av}}}\left[\operatorname{erfc}^{-1}\left(\frac{MP_{\text{target}}}{M-1}\right)\right]^2\right\}. \end{aligned} \quad (5.56)$$

In case of a Rayleigh-fading channel, where $D_{\mathcal{A}}(\alpha) = 1 - \exp(\alpha/\sigma_h^2)$, we have

$$P_{\text{out}} = 1 - e^{-\frac{M^2-1}{3\gamma_{\text{av}}\sigma_h^2}\left[\operatorname{erfc}^{-1}\left(\frac{Mp_T}{M-1}\right)\right]^2}. \quad (5.57)$$

Proceeding similarly, we can derive the out-of-service probability for the QAM case.

As far as the outage probability of a SISO Rayleigh-fading system is concerned, we have

$$\begin{aligned} P_{\text{outage}} &= \Pr\left\{\frac{1}{2}\log_2\left(1+\alpha\frac{\sigma_s^2}{\sigma_n^2}\right) < r\right\} \\ &= \Pr\left\{\alpha < \left(2^{2r}-1\right)\frac{\sigma_n^2}{\sigma_s^2}\right\} \\ &= D_{\mathcal{A}}\left((2^{2r}-1)\frac{\sigma_n^2}{\sigma_s^2}\right) \\ &= 1 - e^{-(2^{2r}-1)\sigma_n^2/\sigma_s^2\sigma_h^2}. \end{aligned} \tag{5.58}$$

Note that if we wish to make $P_{\text{outage}} = 0$, we need to take $r = 0$. This means that no reliable transmission can take place over a SISO Rayleigh fading channel, with zero outage probability! On such a kind of channel, if we wish to transmit with a rate $r > 0$, we must necessarily tolerate an outage probability strictly greater than zero.

In summary, in this section we have seen that SISO wireless systems suffer from two major impairments concerning the average error rate and the maximum transmission rate that still allows for an arbitrarily low error probability. In the next sections, we will show how multiple antenna systems can provide a considerable performance improvement with respect to the SISO case.

5.5 DIVERSITY GAIN

In this section, we start deriving a general expression for the average error probability. We distinguish between slowly varying channels where many successive transmitted blocks experience the same channel values, and fast varying channels, where successive blocks see different channel realizations. The general expressions will be used to introduce the important concepts of *diversity gain* and *coding gain*, which play a fundamental role in multiple antenna systems.

A useful formula that we will exploit throughout this section is the following. Given a complex, circularly symmetric, Gaussian random vector $\boldsymbol{z}$ with expected value $\boldsymbol{\mu}$ and covariance matrix $\boldsymbol{\Sigma}$, that is $\boldsymbol{z} \sim \mathcal{N}(\boldsymbol{\mu}, \boldsymbol{\Sigma})$, and a positive semidefinite Hermitian matrix $\boldsymbol{A}$, we have

$$E_{\boldsymbol{z}}\left\{e^{-\boldsymbol{z}^H\boldsymbol{A}\boldsymbol{z}}\right\} = \frac{e^{-\boldsymbol{\mu}^H\boldsymbol{A}(\boldsymbol{I}+\boldsymbol{\Sigma}\boldsymbol{A})^{-1}\boldsymbol{\mu}}}{|\boldsymbol{I}+\boldsymbol{\Sigma}\boldsymbol{A}|}. \tag{5.59}$$

5.5.1 Slowly Fading Channels

Let us consider the slowly fading model

$$\boldsymbol{y}_j = \boldsymbol{H}\boldsymbol{x}_j + \boldsymbol{w}_j,\ j = 1, \ldots, p, \tag{5.60}$$

where the channel matrix $\boldsymbol{H}$ is assumed to be constant over, at least, p consecutive blocks. We assume that the input column vectors $\boldsymbol{x}_j$ have size n, whereas the output column vectors $\boldsymbol{y}_j$ have size q; $\boldsymbol{w}_j$ is additive noise, assumed to be white Gaussian with zero mean and variance σ_w^2. The channel matrix $\boldsymbol{H}$ is $q \times n$. For the moment, we do not relate the dimensions p and q to any physical quantity, to leave the analytical derivations as general as possible. At the end of this general formulation, we will relate the performance to physically meaningful quantities.

Stacking p column vectors one after the other, we collect the $q \times p$ matrix

$$\boldsymbol{Y} = \boldsymbol{H}\,\boldsymbol{X} + \boldsymbol{W}, \tag{5.61}$$

where $\boldsymbol{X}$ is $n \times p$ and $\boldsymbol{W}$ is $q \times p$.

We assume that the sequence of information bits to be transmitted is parsed into blocks and that each block is mapped into the matrix (codeword) $\boldsymbol{X}$. This means that the information is spread over the *joint space-time* domain. We use this general formulation that includes, as special cases, the situations of *space-only* coding, where $\boldsymbol{X}$ is a column vector (i.e., $p = 1$), *time-only* coding, where $\boldsymbol{X}$ is a row vector (i.e., $n = 1$), and *joint space-time* coding, where both n and p are greater than one.

We compute now the pairwise error probability of transmitting the codeword $\boldsymbol{X}$ and deciding for the codeword $\hat{\boldsymbol{X}}$, conditioned to the channel $\boldsymbol{H}$, assuming maximum likelihood detection.

In the presence of AWGN, the error of deciding for $\hat{\boldsymbol{X}}$, when $\boldsymbol{X}$ is transmitted, can be formulated as the following event

$$\|\mathbf{Y} - \boldsymbol{H}\,\boldsymbol{X}\|^2 \geq \left\|\mathbf{Y} - \boldsymbol{H}\,\hat{\boldsymbol{X}}\right\|^2. \tag{5.62}$$

Plugging (5.61) in both members of (5.62), we can rewrite (5.62) as

$$\begin{aligned}\|\boldsymbol{W}\|^2 &\geq \left\|\boldsymbol{H}\,\boldsymbol{X} + \boldsymbol{W} - \boldsymbol{H}\,\hat{\boldsymbol{X}}\right\|^2 = \left\|\boldsymbol{W} - \boldsymbol{H}\,(\hat{\mathbf{X}} - \boldsymbol{X})\right\|^2 \\ &= \|\boldsymbol{W}\|^2 + \left\|\boldsymbol{H}\,(\hat{\boldsymbol{X}} - \boldsymbol{X})\right\|^2 - 2\mathrm{Re}\left\{\mathrm{tr}\left[\boldsymbol{W}^H\boldsymbol{H}\,(\hat{\boldsymbol{X}} - \boldsymbol{X})\right]\right\}.\end{aligned} \tag{5.63}$$

Simplifying, the error event is

$$2\mathrm{Re}\left\{\mathrm{tr}\left[\boldsymbol{W}^H\boldsymbol{H}(\hat{\boldsymbol{X}}-\boldsymbol{X})\right]\right\} \geq \left\|\boldsymbol{H}(\hat{\boldsymbol{X}}-\boldsymbol{X})\right\|^2 := d_H^2\left(\boldsymbol{X},\hat{\boldsymbol{X}}\right), \tag{5.64}$$

where we have introduced the distance $d_H\left(\boldsymbol{X},\hat{\boldsymbol{X}}\right)$ between the two codewords $\boldsymbol{X}$ and $\hat{\boldsymbol{X}}$, conditioned to $\boldsymbol{H}$.

The term on the left-hand side of (5.64) is a Gaussian real random variable, with zero mean and variance equal to $2\,\sigma_w^2\,d_H^2(\boldsymbol{X},\hat{\boldsymbol{X}})$. Therefore, the pairwise error probability (PEP), conditioned to the channel $\boldsymbol{H}$, is

$$\mathrm{PEP}(\boldsymbol{X},\hat{\boldsymbol{X}}/\boldsymbol{H}) = Q\left(\sqrt{\frac{1}{2\sigma_w^2}\,d_H^2(\boldsymbol{X},\hat{\boldsymbol{X}})}\right), \tag{5.65}$$

having introduced the Q-function

$$Q(x) = \frac{1}{\sqrt{2\pi}}\int_x^\infty e^{-\frac{u^2}{2}}\,du. \tag{5.66}$$

Using the inequality

$$Q(x) \leq \frac{1}{2}\,e^{-\frac{x^2}{2}},\quad x \geq 0, \tag{5.67}$$

the PEP, conditioned to the channel, can be upper-bounded as follows

$$\mathrm{PEP}(\boldsymbol{X},\hat{\boldsymbol{X}}/\boldsymbol{H}) \leq \frac{1}{2}\,e^{-d_H^2(\boldsymbol{X},\hat{\boldsymbol{X}})/4\sigma_w^2}. \tag{5.68}$$

Introducing the matrix $\boldsymbol{\Xi} = (\hat{\boldsymbol{X}} - \boldsymbol{X})$, we can rewrite the squared distance $d_H^2\left(\boldsymbol{X},\hat{\boldsymbol{X}}\right)$ as a quadratic form of the channel coefficients, as[5]

$$\begin{aligned} d_H^2\left(\boldsymbol{X},\hat{\boldsymbol{X}}\right) &= \mathrm{tr}\left(\boldsymbol{\Xi}^H\boldsymbol{H}^H\boldsymbol{H}\boldsymbol{\Xi}\right) = \mathrm{tr}\left(\boldsymbol{H}\boldsymbol{\Xi}\boldsymbol{\Xi}^H\boldsymbol{H}^H\right) \\ &= \mathrm{vec}(\boldsymbol{H}^H)^H\left(\mathbf{I}_q\otimes\boldsymbol{\Xi}\boldsymbol{\Xi}^H\right)\mathrm{vec}(\boldsymbol{H}^H). \end{aligned}$$

We can now use (5.59) to compute the average value of the upper bound for the PEP, setting $z := \mathrm{vec}(\boldsymbol{H}^H)$. From (5.59) we realize that if the channel has a line

5 We used the property of the Kronecker products that, given three matrices $\mathbf{A}(n\times p)$, $\boldsymbol{X}(p\times q)$, and $\mathbf{B}(q\times n)$, the following identity holds true

$$\mathrm{tr}\left(\mathbf{A}\boldsymbol{X}\mathbf{B}\right) = \mathrm{vec}(\mathbf{A}^T)^T(\boldsymbol{I}_n\otimes\boldsymbol{X})\,\mathrm{vec}(\mathbf{B}) = \mathrm{vec}(\mathbf{A}^H)^H(\boldsymbol{I}_n\otimes\boldsymbol{X})\,\mathrm{vec}(\mathbf{B}).$$

of sight, its mean value μ is different from zero and then, according to (5.59), the average error rate has an exponential decrease. However, in the absence of any line of sight propagation, or more generally any nonnull mean value, the average error has a polynomial decrease, as we will detail next. In particular, assuming that the vector $\text{vec}(\boldsymbol{H}^H)$ has zero mean and covariance matrix $\boldsymbol{\Sigma}_{\boldsymbol{H}}$, and using (5.59), the upper bound of the pairwise error is

$$\overline{\text{PEP}}(\boldsymbol{X}, \hat{\boldsymbol{X}}) \leq \frac{1}{\left| \boldsymbol{I}_{qn} + \frac{1}{4\sigma_w^2} \boldsymbol{\Sigma}_{\boldsymbol{H}} \left(\boldsymbol{I}_q \otimes \boldsymbol{\Xi}\boldsymbol{\Xi}^H \right) \right|}. \tag{5.69}$$

In case of channels with uncorrelated coefficients, this expression simplifies further. In particular, if the channel coefficients have the same variance σ_h^2 (i.e., $\boldsymbol{\Sigma}_{\boldsymbol{H}} = \sigma_h^2 \boldsymbol{I}_{qn}$), the bound becomes

$$\begin{aligned} \overline{\text{PEP}}(\boldsymbol{X}, \hat{\boldsymbol{X}}) &\leq \frac{1}{\left| \boldsymbol{I}_{qn} + \frac{\sigma_h^2}{4\sigma_w^2} \left(\boldsymbol{I}_q \otimes \boldsymbol{\Xi}\boldsymbol{\Xi}^H \right) \right|} = \frac{1}{\left| \boldsymbol{I}_n + \frac{\sigma_h^2}{4\sigma_w^2} \boldsymbol{\Xi}\boldsymbol{\Xi}^H \right|^q} \\ &= \frac{1}{\prod_{i=1}^{r} \left(1 + \frac{\sigma_h^2 \sigma_s^2}{4\sigma_w^2} \Delta_i \right)^q} \end{aligned} \tag{5.70}$$

where r is the rank of $\boldsymbol{\Xi}\boldsymbol{\Xi}^H$ and $\Delta_i, i = 1, \ldots, r$, denote the nonnull eigenvalues of $\boldsymbol{\Xi}\boldsymbol{\Xi}^H / \sigma_s^2$, where σ_s^2 is the symbol variance[6].

Equation (5.70) is important as it shows how the bound is related to the choice of the codewords $\boldsymbol{X}$, through the properties of the error matrix $\boldsymbol{\Xi}$. For example, for a given SNR, one could consider all pairwise error probabilities, for all possible pairs $\boldsymbol{X} \neq \hat{\boldsymbol{X}}$, and minimize the maximum bound over all the pairs. Unfortunately, this optimization problem would be very difficult to solve. One of the major difficulties comes from the fact that the codewords are defined over a discrete set. In such a case, the problem is not convex and thus it does not admit, in general, a simple method to reach the global optimum and an exhaustive search would be too complex to implement[7]. Furthermore, the best coding strategy for a given SNR is not necessarily optimal for a different SNR.

It is useful to simplify the bound, seeing for example what happens at the extremes of the SNR (i.e., for low or high SNR). Expanding the product in (5.70), and setting

6 The normalization by the symbol variance σ_s^2 is used to single out the coding gain, as it will be shown in (5.74)

7 The basic principles of convex optimization will be reviewed in Chapter 6.

$\gamma = \sigma_h^2 \sigma_s^2 / (4\sigma_w^2)$, the denominator of (5.70) can be rewritten as

$$\begin{aligned} \prod_{i=1}^{r} (1+\gamma\Delta_i)^q &= \left(1+\sum_{i=1}^{r} \Delta_i \gamma + \cdots + \prod_{i=1}^{r} \Delta_i \gamma^r\right)^q \\ &= \left[1+\operatorname{tr}\left(\Xi\Xi^H\right)\gamma + \cdots + \left|\Xi\Xi^H\right|\gamma^r\right]^q . \end{aligned} \quad (5.71)$$

This expansion shows that, at low SNR (i.e., when $\gamma \ll 1$), the most important parameter is the trace of $\Xi\Xi^H$, whereas, at high SNR, the most important parameter is the determinant of $\Xi\Xi^H$. In particular, at high SNR (i.e., when $\gamma\Delta_i \gg 1$, $\forall i$) the bound on the average PEP decreases, asymptotically, as

$$\overline{\mathrm{PEP}}(\boldsymbol{X}, \hat{\boldsymbol{X}}) \leq \frac{1}{\left[\left(\prod_{i=1}^{r} \Delta_i\right)^{1/r} \gamma\right]^{rq}}. \quad (5.72)$$

It is useful to introduce the quantities (the dependence of r and Δ_i on the pair $\boldsymbol{X}$ and $\hat{\boldsymbol{X}}$ is now indicated explicitly)

$$g_d(\boldsymbol{X}, \hat{\boldsymbol{X}}) := r(\boldsymbol{X}, \hat{\boldsymbol{X}})\, q \quad (5.73)$$

and

$$g_c(\boldsymbol{X}, \hat{\boldsymbol{X}}) := \left[\prod_{i=1}^{r} \Delta_i(\boldsymbol{X}, \hat{\boldsymbol{X}})\right]^{1/r} . \quad (5.74)$$

Hence, we can rewrite (5.72) as

$$\overline{\mathrm{PEP}}(\boldsymbol{X}, \hat{\boldsymbol{X}}) \leq \frac{1}{\left[g_c(\boldsymbol{X}, \hat{\boldsymbol{X}})\,\gamma\right]^{g_d(\boldsymbol{X}, \hat{\boldsymbol{X}})}}. \quad (5.75)$$

Given the average pairwise error probabilities $\overline{\mathrm{PEP}}(\boldsymbol{X}, \hat{\boldsymbol{X}})$, for all distinct pairs $(\boldsymbol{X}, \hat{\boldsymbol{X}})$, and assuming that all matrices $\boldsymbol{X}$ are equiprobable, we can upper-bound the average error probability as follows. Let us denote with $\mathcal{X}$ the alphabet containing all possible matrices $\boldsymbol{X}$. Then, the upper bound on the average error probability is

$$P_{\text{av}} \leq \frac{1}{|\mathcal{X}|} \sum_{\boldsymbol{X}\in\mathcal{X}} \sum_{\hat{\boldsymbol{X}}\in\mathcal{X}\setminus\boldsymbol{X}} \overline{\mathrm{PEP}}(\boldsymbol{X}, \hat{\boldsymbol{X}}), \quad (5.76)$$

where $|\mathcal{X}|$ is the cardinality of $\mathcal{X}$ and the symbol $\mathcal{X}\setminus\boldsymbol{X}$ denotes the set $\mathcal{X}$, deprived of the element $\boldsymbol{X}$.

Asymptotically, as the SNR goes to infinity, the dominant terms in the summation (5.76) are the ones that decrease more slowly, i.e., the terms with the smallest value of $g_d(\boldsymbol{X}, \hat{\boldsymbol{X}})$. Hence, if we introduce the variables

$$G_d = \min_{\boldsymbol{X} \neq \hat{\boldsymbol{X}}} g_d(\boldsymbol{X}, \hat{\boldsymbol{X}}) \tag{5.77}$$

$$G_c = \min_{\boldsymbol{X} \neq \hat{\boldsymbol{X}}} g_c(\boldsymbol{X}, \hat{\boldsymbol{X}}), \quad \text{when } g_d(\boldsymbol{X}, \hat{\boldsymbol{X}}) = G_d \tag{5.78}$$

and we denote with G_n the number of pairs such that $g_c = G_c$ and $g_d = G_d$, we can finally write

$$P_{\text{av}} \leq \frac{G_n}{|\mathcal{X}|} \frac{1}{\left[G_c \frac{\sigma_h^2 \sigma_s^2}{4\sigma_w^2}\right]^{G_d}}. \tag{5.79}$$

The quantity G_d is known as the *diversity gain*, and G_c is the *coding gain*.

From (5.79), it is evident that, at least asymptotically, the diversity gain G_d is the most important parameter, because G_d is the exponent of the SNR and then it determines the slope of the average error probability as a function of the SNR, on a log – log scale. Therefore, increasing the diversity gain, we increase the slope with which the average error probability decreases. Conversely, the coding gain G_c acts as a multiplying factor of the SNR. Therefore, increasing G_c we have a gain on the SNR, which manifests itself as a shift of the average error probability curve, again on a log – log scale, toward lower SNR values.

The previous analysis is not only useful to evaluate the performance of a MIMO system. It is also especially valuable to find out the "best" coding strategies, to build the matrices $\boldsymbol{X}$ in order to optimize some performance criterion, for example to minimize the average error probability. Given the role of the diversity and coding gain, the most common criterion (see, e.g., [11]) consists of choosing the coding matrices that maximize the diversity gain and, for the given diversity gain, maximize the coding gain. In asymptotic terms (i.e., when the SNR tends to infinity) this is indeed the best strategy. However, for practical BER values, especially when the maximum diversity gain is potentially high, we can sacrifice part of the diversity gain to increase the coding gain. In this way, we can achieve a target error probability with lower SNR using a method that yields a suboptimum diversity gain, but a higher coding gain. An example illustrating such an approach was shown, for example, in [12].

It is useful to consider now the application of the previous derivations to some cases of interest.

5.5.1.1 *SIMO Flat-Fading Channels*

Let us consider a SIMO system with one transmit and n_R receive antennas. The channels are zero-mean and flat-fading. This case is a particular case of (5.61) with $q = n_R$ and $n = 1$. Let us consider a single transmitted block (i.e., $p = 1$). In such a case, $\Xi\Xi^H$ is a scalar and thus $r = 1$. From (5.73) and (5.77), we deduce that the maximum diversity gain is

$$G_d = n_R. \tag{5.80}$$

This result is achieved if the n_R channels are independent. Therefore, we recovered one of the first key properties of multiple receive antenna systems, known since a long time: The diversity gain of a multiple receive antenna system is equal to the number of (independent) channels. Indeed, for SIMO systems with independent flat fading it is possible to derive the average symbol error rate in closed form, without passing through the bound on the PEP. In fact, given a SIMO flat-fading system with n_R antennas, where each link is characterized by the coefficient h_i, if we transmit the symbol s, we receive, on each antenna, the value $y_i = h_i s + w_i$, with $i = 1, \ldots, n_R$. If we combine these n_R samples linearly, multiplying the ith sample by h_i^*, we obtain the random variable

$$z = \sum_{i=1}^{n_R} |h_i|^2 s + \sum_{i=1}^{n_R} h_i^* w_i. \tag{5.81}$$

This kind of receiver is known as *maximal ratio combining* (MRC) receiver because it maximizes the signal-to-noise ratio [9]. Equation (5.81) shows that, using an MRC, we have, equivalently a SISO flat-fading channel characterized by a flat-fading coefficient $\alpha := \sum_{i=1}^{n_R} |h_i|^2$ plus additive noise with zero mean and variance $\sigma_v^2 = \sum_{i=1}^{n_R} |h_i|^2 \sigma_n^2$, if σ_n^2 is the variance of the input noise samples. If the channels are Rayleigh-fading and independent, the variable α has the following pdf

$$p_{\mathcal{A}}(\alpha) = \frac{1}{(n_R - 1)! \sigma_h^{2n_R}} \alpha^{n_R - 1} e^{-\alpha/\sigma_h^2}, \quad \alpha > 0. \tag{5.82}$$

Hence, using (5.46), (5.82), and (5.48), we obtain

$$P_{av} = 2\left(1 - \frac{1}{M}\right)\left(\frac{1-\mu}{2}\right)^{n_R} \sum_{k=0}^{n_R - 1} \binom{n_R - 1 + k}{k} \left(\frac{1+\mu}{2}\right)^k, \tag{5.83}$$

where

$$\mu := \sqrt{\frac{\bar{\gamma}}{1+\bar{\gamma}}}, \quad \text{with } \bar{\gamma} := \frac{3\sigma_s^2 \sigma_h^2}{(M^2 - 1)\sigma_n^2}. \tag{5.84}$$

At high SNR, $1 - \mu \approx 1/(2\bar{\gamma})$, whereas $1 + \mu \approx 2$ and the asymptotic behavior of (5.83) is

$$P_e \approx 2\left(1 - \frac{1}{M}\right)\left(\frac{1}{4\bar{\gamma}}\right)^{n_R}\binom{2n_R - 1}{n_R}. \tag{5.85}$$

Hence, at high SNR, the average error probability shows a diversity gain equal to n_R, as predicted by the bound (5.80). A numerical example of average error rate, obtained transmitting QPSK symbols, using a parametric number of antennas is reported in Figure 5.4. For a fair comparison, we set $\sum_{i=1}^{n_R} E\{|h_i|^2\} = 1$, for every value of n_R. The dotted line shows the error rate of a deterministic SISO channel with coefficient $|h| = 1$. We can see that, as n_R increases, the SIMO system tends to recover most of the losses with respect to the deterministic channel.

5.5.1.2 *MISO Flat-Fading Channels*

Let us consider now a MISO system, where we have n_T transmit antennas and one receive antenna. This case is particularly important in the downlink of cellular systems. In such a case, in fact, it is not a real problem to install multiple transmit antennas at the base station. Conversely, it is certainly more complicated to have multiple antennas on the portable handset, especially because we should put the antennas separated enough to have sufficient decorrelation between the channels. The MISO case is also a particular case of (5.61), pertaining to the choice $q = 1$ and $n = n_T$. Let us consider, as before, the transmission of only one block (i.e., $p = 1$). In such a case, $\Xi\Xi^H$ is again rank one, hence $r = 1$, whichever is the value of n_T, so that

$$G_d = 1.$$

This is indeed quite an annoying result: *A MISO system transmitting codewords spanning over the space domain, but not over the time domain, has no diversity gain!*

However, the performance changes completely if, for the same MISO configuration, we span the information bits into codewords $\boldsymbol{X}$ lasting Q consecutive blocks. In such a case, we must use (5.61), with $q = 1$, $n = n_T$, and $p = Q$. If $Q \geq n_T$, the maximum value that the rank of $\Xi\Xi^H$ may assume is $r = n_T$. Therefore, using again (5.73) and (5.77), the maximum diversity gain becomes

$$G_d = n_T.$$

This shows that *a MISO system may have full diversity $G_d = n_T$, as a SIMO system with the same number of antennas, provided that the codewords span the joint space-time domain, for at least n_T blocks.* Mapping information bits onto codewords spanning the space-time domain is known as *space-time coding*. Therefore, *to achieve maximum diversity gain with MISO systems, it is necessary to use space-time coding.*

This is indeed a very good result, because it shows that we can have the maximum diversity gain also with a small cellular handset, having only one antenna, in both uplink and downlink, provided that the base station has multiple antennas. Space-time coding will be the subject of Chapters 7, 8, and 9.

5.5.1.3 *MIMO Flat-Fading Channels*

If we have a real MIMO channel with n_T transmit and n_R receive antennas, and we concatenate a number of blocks $Q \geq n_T$, we can use (5.61) with $q = n_R$, $n = n_T$, and $p = Q$. In such a case, the maximum rank of $\Xi\Xi^H$ is $r = n_T$ and thus the maximum diversity gain is

$$G_d = n_R n_T.$$

The maximum diversity gain of a flat-fading system is then equal to the number of independently fading channels, which in this case is equal to the product between the number of transmit and receive antennas. It is worth remarking that, again as in the MISO case, to achieve full diversity it is necessary to distribute the codewords over a number of consecutive blocks equal, at least, to n_T. Interestingly, this result shows that an $n_R \times n_T$ MIMO system can get diversity gain $G_d = n_R n_T$, even if the channel modes (especially the modes associated to the smallest eigenvalues) have diversity smaller than $n_R n_T$ (see, e.g., Section 5.2). This feature is made available by space-time coding.

5.5.1.4 *MIMO Frequency-Selective Fading Channels*

Let us consider now a time-dispersive (or frequency-selective) MIMO system where each channel has an impulse response of order L (length $L+1$), with i.i.d. complex Gaussian coefficients. In such a case, to simplify the decoding strategy, we assume the use of block transmissions incorporating a guard interval among successive blocks. In particular, using blocks of length N and cyclic prefixes of length L, with $L < N$, as described in Chapter 3, each channel is described by a circulant Toeplitz matrix. Let us denote by $\boldsymbol{H}_{ij}$ the channel matrix characterizing the (temporal) channel between the ith transmit and the jth receive antenna. The structure of each matrix $\boldsymbol{H}_{ij}$ is as in (3.17), so that $\boldsymbol{H}_{ij}$ is $N \times N$, circulant and Toeplitz. If we stack the N-size received blocks received by the n_R receive antennas, one on top of each other, we can write the input-output relationship as

$$\boldsymbol{y} := \begin{pmatrix} y_1 \\ y_2 \\ \vdots \\ y_{n_R} \end{pmatrix} = \begin{pmatrix} \boldsymbol{H}_{11} & \boldsymbol{H}_{11} & \cdots & \boldsymbol{H}_{1n_T} \\ \boldsymbol{H}_{21} & \boldsymbol{H}_{21} & \cdots & \boldsymbol{H}_{2n_T} \\ \vdots & \ddots & \ddots & \vdots \\ \boldsymbol{H}_{n_R1} & \boldsymbol{H}_{n_R1} & \cdots & \boldsymbol{H}_{n_Rn_T} \end{pmatrix} \begin{pmatrix} \boldsymbol{x}_1 \\ \boldsymbol{x}_2 \\ \vdots \\ \boldsymbol{x}_{n_T} \end{pmatrix} + \boldsymbol{w}. \tag{5.86}$$

Stacking now Q consecutive blocks, we end up with an I/O relationship like (5.61), with $q = n_R N$, $n = n_T N$, and $p = Q$. It is now important to study the rank of $\boldsymbol{\Sigma_H}$. Let us consider the covariance matrix pertaining to each matrix $\boldsymbol{H}_{ij}$, i.e. $\boldsymbol{C}_{ij} := E\{\text{vec}(\boldsymbol{H}_{ij}^H)\text{vec}^H(\boldsymbol{H}_{ij}^H)\}$. It turns out that the maximum rank of $\boldsymbol{C}_{ij}$ is $L+1$. Therefore, even assuming that the matrix $\boldsymbol{\Xi\Xi}^H$ had the maximum rank, now the number of nonnull entries of $\boldsymbol{\Sigma_H \Delta}$ cannot be greater than $(L+1)n_R n_T$. Therefore, the maximum diversity gain is now

$$G_d = n_T n_R (L+1). \tag{5.87}$$

Hence, a frequency-selective MIMO system is capable of collecting a diversity gain equal to the product of the spatial diversity $n_T n_R$ times the *multipath* diversity $L+1$. This is perfectly justifiable from a physical point of view, as it states that the maximum diversity gain is equal to the number of independently fading paths, which is now equal to the number of MIMO channels times the number of paths for each channel. It is important to emphasize that, thanks to space-time coding, instead of being an annoying factor, multipath becomes a beneficial source of diversity, if properly exploited.

5.5.2 Diversity Gain of Fast Varying Channels

Let us consider now the fast varying channel model

$$\boldsymbol{y}_j = \boldsymbol{H}_j \boldsymbol{x}_j + \boldsymbol{w}_j, \; j = 1, \ldots, p, \tag{5.88}$$

where the channel is assumed to vary from block to block. This situation could be obtained through block interleaving. Stacking together p consecutive blocks, we can write

$$\boldsymbol{Y} = [\boldsymbol{H}_1\boldsymbol{x}_1 \;\; \boldsymbol{H}_2\boldsymbol{x}_2 \;\; \cdots \;\; \boldsymbol{H}_p\boldsymbol{x}_p] + \boldsymbol{W}, \tag{5.89}$$

where $\boldsymbol{Y}$ is $q \times p$, $\boldsymbol{W}$ is $q \times p$, $\boldsymbol{H}_j$ is $q \times n$, and $\boldsymbol{x}_j$ is $n \times 1$.

The pairwise error event between the transmitted codeword $\boldsymbol{X} = [\boldsymbol{x}_1 \; \boldsymbol{x}_2 \; \cdots \; \boldsymbol{x}_p]$ and the decoded codeword $\hat{\boldsymbol{X}} = [\hat{\boldsymbol{x}}_1 \; \hat{\boldsymbol{x}}_2 \; \cdots \; \hat{\boldsymbol{x}}_p]$ can now be written as

$$\|\boldsymbol{Y} - [\boldsymbol{H}_1\boldsymbol{x}_1 \;\; \boldsymbol{H}_2\boldsymbol{x}_2 \;\; \cdots \;\; \boldsymbol{H}_p\boldsymbol{x}_p]\|^2 \geq \|\boldsymbol{Y} - [\boldsymbol{H}_1\hat{\boldsymbol{x}}_1 \;\; \boldsymbol{H}_2\hat{\boldsymbol{x}}_2 \;\; \cdots \;\; \boldsymbol{H}_p\hat{\boldsymbol{x}}_p]\|^2.$$

Proceeding as in the previous section, introducing the vectors $\boldsymbol{\xi}_j = (\hat{\boldsymbol{x}}_j - \boldsymbol{x}_j)$, $j = 1, \ldots, p$, this condition can be rewritten as

$$2\mathrm{Re}\left\{\mathrm{tr}\left(\boldsymbol{W}^{\mathrm{H}}\left[\boldsymbol{H}_1\boldsymbol{\xi}_1 \quad \cdots \quad \boldsymbol{H}_p\boldsymbol{\xi}_p\right]\right)\right\} \geq \left\|\left[\boldsymbol{H}_1\boldsymbol{\xi}_1 \quad \cdots \quad \boldsymbol{H}_p\boldsymbol{\xi}_p\right]\right\|^2. \quad (5.90)$$

The right-hand side term of (5.90) can be written as

$$\begin{aligned} d^2(\boldsymbol{X}, \hat{\boldsymbol{X}}) &= \left\|\left[\boldsymbol{H}_1\boldsymbol{\xi}_1 \; \boldsymbol{H}_2\boldsymbol{\xi}_2 \quad \cdots \quad \boldsymbol{H}_p\boldsymbol{\xi}_p\right]\right\|^2 \\ &= \mathrm{tr}\left(\begin{bmatrix} \boldsymbol{\xi}_1^H\boldsymbol{H}_1^H \\ \boldsymbol{\xi}_2^H\boldsymbol{H}_2^H \\ \vdots \\ \boldsymbol{\xi}_p^H\boldsymbol{H}_p^H \end{bmatrix} \left[\boldsymbol{H}_1\boldsymbol{\xi}_1 \; \boldsymbol{H}_2\boldsymbol{\xi}_2 \quad \cdots \quad \boldsymbol{H}_p\boldsymbol{\xi}_p\right]\right) \\ &= \sum_{k=1}^{p} \mathrm{tr}\left(\boldsymbol{H}_k\boldsymbol{\xi}_k\boldsymbol{\xi}_k^H\boldsymbol{H}_k^H\right) \\ &= \sum_{k=1}^{p} \mathrm{vec}(\boldsymbol{H}_k^H)^H\left(\boldsymbol{I}_q \otimes \boldsymbol{\xi}_k\boldsymbol{\xi}_k^H\right)\mathrm{vec}(\boldsymbol{H}_k^H). \end{aligned}$$

Denoting by $\boldsymbol{\Sigma}_{\boldsymbol{H}}^{(k)}$ the covariance matrix of $\mathrm{vec}(\boldsymbol{H}_k^H)$, and using again (5.59), we can upper-bound the average pairwise error probability, in case of zero mean channels, as

$$\overline{\mathrm{PEP}}(\boldsymbol{X}, \hat{\boldsymbol{X}}) \leq \frac{1}{\prod_{k=1}^{p}\left|\boldsymbol{I}_{qn} + \frac{1}{4\sigma_w^2}\boldsymbol{\Sigma}_{\boldsymbol{H}}^{(k)}\left(\boldsymbol{I}_q \otimes \boldsymbol{\xi}_k\boldsymbol{\xi}_k^H\right)\right|}.$$

If the channel coefficients are uncorrelated, with equal variance σ_h^2 (i.e., $\boldsymbol{\Sigma}_{\boldsymbol{H}}^{(k)} = \sigma_h^2\boldsymbol{I}_{qn}$, for $k = 1, \ldots, p$), we obtain

$$\begin{aligned} \overline{\mathrm{PEP}}(\boldsymbol{X}, \hat{\boldsymbol{X}}) &\leq \frac{1}{\prod_{k=1}^{p}\left|\boldsymbol{I}_{qn} + \frac{\sigma_h^2}{4\sigma_w^2}\left(\boldsymbol{I}_q \otimes \boldsymbol{\xi}_k\boldsymbol{\xi}_k^H\right)\right|} = \frac{1}{\prod_{k=1}^{p}\left|\boldsymbol{I}_n + \frac{\sigma_h^2}{4\sigma_w^2}\boldsymbol{\xi}_k\boldsymbol{\xi}_k^H\right|^q} \\ &= \frac{1}{\prod_{k\in\mathcal{A}}\left(1 + \frac{\sigma_h^2}{4\sigma_w^2}\left\|\boldsymbol{\xi}_k\right\|^2\right)^q} \approx \frac{1}{\left[g_c\left(\frac{\sigma_h^2\sigma_s^2}{4\sigma_w^2}\right)\right]^{q\,\delta_H}}, \end{aligned} \quad (5.91)$$

where, for any given pair $(\boldsymbol{X}, \hat{\boldsymbol{X}})$, $\mathcal{A}$ is the set of indices k such that the terms $\|\boldsymbol{\xi}_k\|^2$ are different from zero and $\delta_H = \delta_H(\boldsymbol{X}, \hat{\boldsymbol{X}})$ denotes the cardinality of $\mathcal{A}$.

The last approximation in (5.91) holds at high SNR. The parameter

$$g_c = g_c(\boldsymbol{X}, \hat{\boldsymbol{X}}) := \left(\frac{1}{\sigma_s^2} \prod_{k \in \mathcal{A}} \|\boldsymbol{\xi}_k\|^2 \right)^{\frac{1}{\delta_H}} \tag{5.92}$$

is the coding gain, for the given pair $(\boldsymbol{X}, \hat{\boldsymbol{X}})$. Proceeding as in Section 5.5.1, we obtain the following diversity and coding gains, respectively:

$$G_d = \min_{\mathbf{X} \neq \hat{\mathbf{X}}} \delta_H(\boldsymbol{X}, \hat{\boldsymbol{X}}) \cdot q, \tag{5.93}$$

$$G_c = \min_{\mathbf{X} \neq \hat{\mathbf{X}}} g_c(\boldsymbol{X}, \hat{\boldsymbol{X}}), \quad \text{when } \delta_H(\boldsymbol{X}, \hat{\boldsymbol{X}}) \cdot q = G_d. \tag{5.94}$$

MIMO flat-fading channels

In the case of a MIMO system with n_T transmit and n_R receive antennas, where we transmit Q blocks over independently fading slots, the maximum possible value of δ_H is Q. Hence, from (5.94), the maximum diversity gain is

$$G_d = Q\, n_R. \tag{5.95}$$

Remark 1: Equation (5.95) shows that, in case of fast flat-fading channel, a MIMO system has the potential for a diversity gain equal to the product between the receive diversity n_R and the time diversity Q, equal to the number of independently fading blocks. This gain is achievable if the coding strategy guarantees that the minimum distance δ_H is equal to Q.

Remark 2: In (5.95), there is no gain factor proportional to the transmit diversity n_T. This is a considerable loss and it is due to the channel's fast variability.

5.5.3 Diversity Gain of Block Fading Channels

Let us see now if there is any way to get all forms of diversity, namely the transmit, receive, and time diversity.

The cases studied before of slow and fast fading channels can be seen as special cases of the following channel model

$$\boldsymbol{y}_j = \boldsymbol{H}_{f(j)} \boldsymbol{x}_j + \boldsymbol{w}_j, \quad j = 1, \ldots, p, \tag{5.96}$$

where the function $f(j)$ characterizes the channel variability. Setting $f(j) = \lceil j/b \rceil$[8], where b is an integer factor of p, both slow and fast fading cases can be

8 The symbol $\lceil x \rceil$ denotes the nearest integer greater than or equal to x.

seen as special cases of (5.96), corresponding to $b = p$ or $b = 1$, respectively.

Taking any integer value of b such that $1 < b < p$, the channel is constant in every set of b consecutive blocks and then it varies from a set of b blocks to the next one. For example, let us assume that $p = b \cdot g$, where b is the number of blocks within which the channel is assumed to be constant, whereas g is the number of independent channel fading states, within a set of p blocks. Stacking p consecutive blocks, over the g channel states, the overall received matrix can be expressed as

$$\boldsymbol{Y} = [\boldsymbol{H}_1\boldsymbol{X}_1 \ \ \boldsymbol{H}_2\boldsymbol{X}_2 \ \ \cdots \ \ \boldsymbol{H}_g\boldsymbol{X}_g] + \boldsymbol{W}, \tag{5.97}$$

where $\boldsymbol{Y}$ is $q \times p$, $\boldsymbol{W}$ is $q \times p$, $\boldsymbol{H}_k$ is $q \times n$, and $\boldsymbol{X}_k$ is $n \times b$, with $k = 1, \ldots, g$. Introducing the error matrices $\Xi_k := \hat{\boldsymbol{X}}_k - \boldsymbol{X}_k$, with $k = 1, \ldots, g$, the error event is now

$$2\mathrm{Re}\left\{\mathrm{tr}\left(\boldsymbol{W}^H [\boldsymbol{H}_1\Xi_1 \ \ \cdots \ \ \boldsymbol{H}_g\Xi_g]\right)\right\} \geq \| [\boldsymbol{H}_1\Xi_1 \ \ \cdots \ \ \boldsymbol{H}_g\Xi_g] \|^2 . \tag{5.98}$$

Proceeding as before, the squared distance assumes now the form

$$\begin{aligned}
d^2(\boldsymbol{X}, \hat{\boldsymbol{X}}) &= \| [\boldsymbol{H}_1\Xi_1 \ \ \boldsymbol{H}_2\Xi_2 \ \ \cdots \ \ \boldsymbol{H}_g\Xi_g] \|^2 \\
&= \mathrm{tr}\left(\begin{bmatrix} \Xi_1^H \boldsymbol{H}_1^H \\ \Xi_2^H \boldsymbol{H}_2^H \\ \vdots \\ \Xi_g^H \boldsymbol{H}_g^H \end{bmatrix} [\boldsymbol{H}_1\Xi_1 \ \ \boldsymbol{H}_2\Xi_2 \ \ \cdots \ \ \boldsymbol{H}_g\Xi_g]\right) \\
&= \sum_{k=1}^{g} \mathrm{tr}\left(\Xi_k^H \boldsymbol{H}_k^H \boldsymbol{H}_k \Xi_k\right) = \sum_{k=1}^{g} \mathrm{tr}\left(\boldsymbol{H}_k \Xi_k \Xi_k^H \boldsymbol{H}_k^{\mathrm{H}}\right) \\
&= \sum_{k=1}^{g} \mathrm{vec}(\boldsymbol{H}^H)^H \left(\boldsymbol{I}_q \otimes \Xi_k \Xi_k^H\right) \mathrm{vec}(\boldsymbol{H}^H).
\end{aligned}$$

Denoting by $\boldsymbol{\Sigma}_{\boldsymbol{H}}^{(k)}$ the covariance matrix of the vector $\mathrm{vec}(\boldsymbol{H}_k^H)$, the bound on the pairwise error probability is

$$\overline{\mathrm{PEP}}(\mathbf{X}, \hat{\mathbf{X}}) \leq \frac{1}{\prod_{k=1}^{g} \left| \boldsymbol{I}_{qn} + \frac{1}{4\sigma_w^2} \boldsymbol{\Sigma}_{\boldsymbol{H}}^{(k)} \left(\boldsymbol{I}_q \otimes \Xi_k \Xi_k^H\right) \right|}. \tag{5.99}$$

In the case of channels with uncorrelated coefficients, with equal variance σ_h^2 (i.e., $\mathbf{\Sigma}_{\boldsymbol{H}}^{(k)} = \sigma_h^2 \boldsymbol{I}_{qn}$, for $k = 1, \ldots, g$), we obtain

$$\begin{aligned} \overline{\mathrm{PEP}}(\mathbf{X}, \hat{\mathbf{X}}) &\leq \frac{1}{\prod_{k=1}^{g} \left| \boldsymbol{I}_{qn} + \frac{\sigma_h^2}{4\sigma_w^2} \left(\boldsymbol{I}_q \otimes \Xi_k \Xi_k^H \right) \right|} \\ &= \frac{1}{\prod_{k=1}^{g} \prod_{i=1}^{r^{(k)}} \left(1 + \frac{\sigma_s^2 \sigma_h^2}{4\sigma_w^2} \Delta_i^{(k)}\right)^q}, \end{aligned}$$

where $r^{(k)}$ is the rank of $\Xi_k \Xi_k^H$ and $\Delta_i^{(k)}$, with $i = 1, \ldots, r^{(k)}$, are the nonnull eigenvalues of $\Xi_k \Xi_k^H / \sigma_s^2$. Given the size of the matrices Ξ_k, the maximum value for $r^{(k)}$ is $\min(n, b)$.

MIMO flat-fading channel

Let us consider a MIMO system with n_T transmit and n_R receive antennas. The codewords span $Q\, n_T$ blocks. Successive blocks are interleaved with an interleaver memory greater than the channel coherence time. Thanks to the interleaver, the channel is seen as constant over n_T blocks, but it varies from each set of n_T blocks to the next. This situation can then be modeled as in (5.96), setting $q = n_R$, $n = n_T$, $b = n_T$, $g = Q$, and $p = n_T Q$. In this case, the maximum value that the rank of $\Xi_k \Xi_k^H$ can achieve is n_T. As a consequence, the diversity gain is

$$G_d = n_R\, n_T\, Q. \tag{5.100}$$

Remark 3: In this case, the diversity gain is the product between the receive diversity n_R, the transmit diversity n_T, and the time diversity Q. To achieve this gain, it is necessary: 1) to transmit codewords composed of $n_T\, Q$ blocks; 2) to use a block interleaver such that every set of consecutive n_T sub-blocks composing one codeword sees a constant channel, but successive sub-blocks are distant enough to observe independent channel realizations; 3) to use codewords such that each difference matrix Ξ_k is full rank.

The interleaver plays a fundamental role here and it must work as follows. Given N code matrices $\boldsymbol{X}_1, \boldsymbol{X}_2, \ldots, \boldsymbol{X}_N$, each one of size $q \times p$, we can subdivide each matrix into p/b blocks of size $q \times b$, interleave these blocks and send them in succession, as[9] $\boldsymbol{X}_1(:,1), \ldots, \boldsymbol{X}_1(:,b)$, $\boldsymbol{X}_2(:,1), \ldots, \boldsymbol{X}_2(:,b)$, $\boldsymbol{X}_N(:,1), \ldots, \boldsymbol{X}_N(:,b)$, $\boldsymbol{X}_1(:,b+1), \ldots, \boldsymbol{X}_1(:,2b), \boldsymbol{X}_2(:,b+1)$, and so on. In this way, given a transmission rate $R = 1/T$, if bT is less than the channel coherence time T_c, but NT

9 We use Matlab notation here, so that the symbol $\boldsymbol{X}(:,l)$ indicates the lth column of $\boldsymbol{X}$.

is greater than T_c, the first b blocks of each code matrix see a constant channel. However, any two successive sets of b blocks of the *same* code experience an uncorrelated fading. Hence this interleaving strategy meets the assumptions underlying (5.100).

In summary, in this section, we have seen which are the bounds on the diversity gain and under which conditions it is possible to achieve full time and spatial diversity. In particular, *spreading the codewords across both space and time domain* has been proved to be the key strategy to achieve full diversity. Interestingly, we have seen that full diversity can be reached even without having any channel knowledge at the transmit side, provided that the codewords are built appropriately. We have not seen explicitly how to construct the codewords yet. But this will be the subject of Chapters 7 and 8, entirely devoted to space-time coding.

5.6 MULTIPLEXING GAIN

The diversity gain studied in the previous section is certainly one of the key aspects of multiantenna systems. Nevertheless, the most striking advantage coming from the availability of multiple antennas is the considerable potential increase of the transmission rate, for the same total transmitted power and the same bandwidth, with respect to a single antenna system. This gain is known as *multiplexing gain.* Given the scarcity of the available spectrum, together with the need of keeping the transmission power as low possible, it is not surprising that this unique capability of MIMO systems has sparkled so much interest in both industries and academies. Multiplexing gain is the subject of this section. We start again with deterministic channels and then we will analyze ergodic and nonergodic channels.

5.6.1 Deterministic Channel

Let us start again with the I/O relationship

$$\boldsymbol{y} = \boldsymbol{H}\boldsymbol{x} + \boldsymbol{w}, \tag{5.101}$$

where $\boldsymbol{H}$, of size $n_R \times n_T$, is fixed, the noise vector $\boldsymbol{w}$ is white, circularly symmetric complex Gaussian with zero mean and covariance $\boldsymbol{C}_w = \sigma_w^2 \boldsymbol{I}$, and the input vector $\boldsymbol{x}$ has zero mean and power constraint $E\{\boldsymbol{x}^H \boldsymbol{x}\} = \text{tr}\{\boldsymbol{x}\boldsymbol{x}^H\} = p_T$. It is well known [7] that a random vector $\boldsymbol{x}$ having covariance matrix $\boldsymbol{Q}$ has maximum entropy if it is circularly symmetric complex Gaussian and, in such a case, the maximum entropy is

$$H(\boldsymbol{x}) = \log_2 |\pi e \boldsymbol{Q}|. \tag{5.102}$$

The mutual information for a deterministic channel is

$$I(\boldsymbol{x};\boldsymbol{y}) = H(\boldsymbol{y}) - H(\boldsymbol{y}/\boldsymbol{x}) = H(\boldsymbol{y}) - H(\boldsymbol{w}). \tag{5.103}$$

The channel capacity is the maximum of the mutual information over all the input probability distributions that satisfy the power constraint. Under the above assumptions, the mutual information is maximum when $\boldsymbol{y}$ is also Gaussian. Since the covariance matrix of $\boldsymbol{y}$ is $\boldsymbol{C}_y = \boldsymbol{HQH}^H + \sigma_w^2\boldsymbol{I}$, the maximum entropy of $\boldsymbol{y}$ is $H(\boldsymbol{y}) = \log_2|\pi e\boldsymbol{C}_y| = \log_2|\pi e(\sigma_w^2\boldsymbol{I} + \boldsymbol{HQH}^H)|$. Hence, the channel capacity is

$$C(\boldsymbol{H}) = \max_{\boldsymbol{Q}} \log_2\left|\boldsymbol{I}_{n_R} + \frac{1}{\sigma_w^2}\boldsymbol{HQH}^H\right| \text{ bps}, \tag{5.104}$$

where the maximum is now taken over all the nonnegative definite Hermitian matrices $\boldsymbol{Q}$ that satisfy the power constraint.

To get more insight into the capacity expression, it is useful to introduce the eigendecomposition $\boldsymbol{H}^H\boldsymbol{H} = \boldsymbol{U}^H\boldsymbol{\Lambda U}$, where $\boldsymbol{U}$ is unitary and $\boldsymbol{\Lambda}$ is diagonal, with diagonal elements sorted in decreasing order. Using this eigendecomposition and the identity $|\boldsymbol{I} + \boldsymbol{AB}| = |\boldsymbol{I} + \boldsymbol{BA}|$, valid for any pair of matrices $\boldsymbol{A}$ and $\boldsymbol{B}$ whose sizes are congruent with the previous multiplications, we can rewrite (5.104) as

$$\begin{aligned} C(\boldsymbol{H}) &= \max_{\boldsymbol{Q}} \log_2\left|\boldsymbol{I}_{n_T} + \frac{1}{\sigma_w^2}\boldsymbol{H}^H\boldsymbol{HQ}\right| \\ &= \max_{\boldsymbol{Q}} \log_2\left|\boldsymbol{I}_{n_T} + \frac{1}{\sigma_w^2}\boldsymbol{U}^H\boldsymbol{\Lambda UQ}\right| \\ &= \max_{\boldsymbol{Q}} \log_2\left|\boldsymbol{I}_{n_T} + \frac{1}{\sigma_w^2}\boldsymbol{\Lambda}^{1/2}(\boldsymbol{UQU}^H)\boldsymbol{\Lambda}^{1/2}\right|. \end{aligned} \tag{5.105}$$

The matrix $\tilde{\boldsymbol{Q}} := \boldsymbol{UQU}^H$ is nonnegative definite when and only when $\boldsymbol{Q}$ is nonnegative definite. Moreover, since $\boldsymbol{U}$ is unitary, $\text{tr}(\tilde{\boldsymbol{Q}}) = \text{tr}(\boldsymbol{Q})$. Hence, the matrix $\tilde{\boldsymbol{Q}}$ satisfies the same constraints as $\boldsymbol{Q}$. As a consequence, there is no loss of optimality in maximizing the capacity with respect to $\tilde{\boldsymbol{Q}}$ instead of $\boldsymbol{Q}$. In formulas, we can solve, equivalently, the following maximization problem

$$C(\boldsymbol{H}) = \max_{\tilde{\boldsymbol{Q}}} \log_2\left|\boldsymbol{I}_{n_T} + \frac{1}{\sigma_w^2}\boldsymbol{\Lambda}\tilde{\boldsymbol{Q}}\right|. \tag{5.106}$$

This is interesting because it shows that we can consider, instead of the (mixing) channel $\boldsymbol{H}$, the corresponding (nonmixing) diagonal matrix $\boldsymbol{\Lambda}^{1/2}$. Furthermore, since $\boldsymbol{I}_{n_T} + \boldsymbol{\Lambda}\tilde{\boldsymbol{Q}}/\sigma_w^2$ is certainly positive definite, we can use Hadamard's inequality [7]. Hadamard's inequality states that, for any positive definite matrix $\boldsymbol{A}$,

$|\boldsymbol{A}| \leq \prod_i A_{ii}$, with equality if and only if $\boldsymbol{A}$ is diagonal.

As a consequence of Hadamard's inequality, there is no loss of generality in assuming that $\tilde{\boldsymbol{Q}}$ is diagonal. Thus, denoting with r the rank of $\boldsymbol{H}$ and introducing the vector $\boldsymbol{q} := (q_1, \ldots, q_{n_T})$ containing the diagonal entries of $\tilde{\boldsymbol{Q}}$, the optimal values of $q_i, i = 1, \ldots, n_T$, are obtained by solving the following constrained optimization problem

$$\max_{\boldsymbol{q}} \left\{ \sum_{i=1}^{n_T} \log_2 \left(1 + \frac{\lambda_i q_i}{\sigma_w^2} \right) \right\} \quad \text{subject to}$$
$$\sum_{i=1}^{n_T} q_i = p_T, \; q_i \geq 0. \tag{5.107}$$

Knowing the channel at the transmit side, one can find the optimal coefficients q_i as a function of the channel. Lacking such information at the transmitter side, the only meaningful power distribution is uniform, so that we can take $q_k = p_T/n_T$ or, in matrix form, we may set $\boldsymbol{Q} = p_T \boldsymbol{I}/n_T$. As a consequence, from (5.105), the mutual information is

$$\begin{aligned} I(\boldsymbol{H}) &= \log_2 \left| \boldsymbol{I}_{n_T} + \frac{p_T}{n_T \sigma_w^2} \boldsymbol{H}^H \boldsymbol{H} \right| \\ &= \log_2 \left| \boldsymbol{I}_{n_T} + \frac{p_T}{n_T \sigma_w^2} \boldsymbol{\Lambda} \right| = \sum_{i=1}^{r} \log_2 \left(1 + \frac{\lambda_i p_T}{n_T \sigma_w^2} \right), \end{aligned} \tag{5.108}$$

where r is the rank of $\boldsymbol{H}$ and λ_i, $i = 1, \ldots, r$, are the non null eigenvalues[10] of $\boldsymbol{H}^H \boldsymbol{H}$. This quantity is not the capacity, as we have not maximized the mutual information with respect to the input covariance matrix. Nevertheless, even though $I(\boldsymbol{H})$ is certainly less than the capacity, (5.108) shows that, at high SNR (i.e., when $\text{SNR} := p_T/\sigma_w^2 \gg n_T/\lambda_i$) the mutual information behaves as

$$I(\boldsymbol{H}) \approx r \log_2(\text{SNR}) + \sum_{i=1}^{r} \log_2 \left(\frac{\lambda_i}{n_T} \right). \tag{5.109}$$

Compared to the analogous formula, valid for the SISO case, where $C(\text{SNR})$ goes, asymptotically, like $\log_2(\text{SNR})$, (5.109) shows that, at high SNR, a MIMO channel behaves as a set of r independent channels, where r is the rank of $\boldsymbol{H}$. Denoting by $I_{n_T, n_R}(\text{SNR})$ the mutual information pertaining to a MIMO channel with n_T transmit and n_R receive antennas, we can define the *multiplexing gain* for a

10 The non null eigenvalues of $\boldsymbol{H}^H \boldsymbol{H}$ are equal to the non null singular values of $\boldsymbol{H}$.

deterministic channel with $\boldsymbol{H}$ known at the receive side but unknown at the transmit side, as

$$G_m = \lim_{\mathrm{SNR}\to\infty} \frac{I_{n_T,n_R}(\mathrm{SNR})}{I_{1,1}(\mathrm{SNR})}. \tag{5.110}$$

Based on (5.108), the multiplexing gain of a deterministic MIMO channel is

$$G_m = \mathrm{rank}(\boldsymbol{H}) \leq \min(n_T, n_R). \tag{5.111}$$

This is one of the key properties of MIMO systems: The capacity of a MIMO system may increase by a factor r equal to the channel rank, without increasing neither the bandwidth nor the overall transmit power. However, the previous result has been obtained over a deterministic channel. Let us consider now what happens for random channels.

5.6.2 Ergodic Channels

Let us consider now Rayleigh block flat-fading channel model, where $\boldsymbol{H}$ is $n_R \times n_T$. We review now the basic properties about the capacity of such channels, as derived by Telatar [13]. In general, the ergodic capacity is given by (5.32). We compute (5.32) in the case of Rayleigh-fading.

We know, from the previous section, that if $\boldsymbol{x}$ is constrained to have covariance matrix $\boldsymbol{Q}$ and the additive noise is white and Gaussian, the maximum mutual information, conditioned to the channel realization $\boldsymbol{H}$, is equal to the function $\Psi(\boldsymbol{Q}, \boldsymbol{H}) := \log_2 |\boldsymbol{I} + \boldsymbol{H}\boldsymbol{Q}\boldsymbol{H}^H/\sigma_w^2|$. Hence, using (5.32), to find the capacity, we need to maximize the function

$$\Psi(\boldsymbol{Q}) := E_{\boldsymbol{H}} \left\{ \log_2 \left| \boldsymbol{I} + \frac{1}{\sigma_w^2} \boldsymbol{H}\boldsymbol{Q}\boldsymbol{H}^H \right| \right\} \tag{5.112}$$

over all Hermitian, nonnegative definite matrices $\boldsymbol{Q}$ satisfying the power constraint $\mathrm{tr}(\boldsymbol{Q}) = p_T$. Since $\boldsymbol{Q}$ is Hermitian nonnegative definite, it can be written as $\boldsymbol{Q} = \boldsymbol{V}\boldsymbol{\Gamma}\boldsymbol{V}^H$, where $\boldsymbol{V}$ is unitary and $\boldsymbol{\Gamma}$ is diagonal, with nonnegative real entries. Thus, (5.112) can be rewritten as

$$\Psi(\boldsymbol{\Gamma}) := E_{\boldsymbol{H}} \left\{ \log_2 \left| \boldsymbol{I} + \frac{1}{\sigma_w^2} (\boldsymbol{H}\boldsymbol{V})\boldsymbol{\Gamma}(\boldsymbol{H}\boldsymbol{V})^H \right| \right\}. \tag{5.113}$$

Since $\boldsymbol{V}$ is unitary and $\boldsymbol{H}$, in the Rayleigh model, is composed of circularly symmetric complex Gaussian random variables, the probability distribution of $\tilde{\boldsymbol{H}} := \boldsymbol{H}\boldsymbol{V}$ is the same as $\boldsymbol{H}$. Thus, we can rewrite (5.113) as

$$\Psi(\boldsymbol{\Gamma}) := E_{\tilde{\boldsymbol{H}}} \left\{ \log_2 \left| \boldsymbol{I} + \frac{1}{\sigma_w^2} \tilde{\boldsymbol{H}}\boldsymbol{\Gamma}\tilde{\boldsymbol{H}}^H \right| \right\}. \tag{5.114}$$

This allows us to restrict our attention, without any loss of optimality, to diagonal symbol covariance matrices, setting $\boldsymbol{Q} = \boldsymbol{\Gamma}$.

We prove now that, among all diagonal matrices $\boldsymbol{\Gamma}$ satisfying the power constraint, the optimal $\boldsymbol{\Gamma}$ is simply proportional to the identity matrix. Specifically, let us introduce the permutation matrix $\boldsymbol{\Pi}_i$ of dimension $n_T \times n_T$. There are $n_T!$ distinct permutation matrices of size $n_T \times n_T$. Any permutation[11] of the (diagonal) matrix $\boldsymbol{\Gamma}$ can be written as $\boldsymbol{\Gamma}_i = \boldsymbol{\Pi}_i \boldsymbol{\Gamma} \boldsymbol{\Pi}_i^H$. Obviously, the matrix $\boldsymbol{\Gamma}_i$ satisfies the same power constraint as $\boldsymbol{\Gamma}$. We can also check that $\Psi(\boldsymbol{\Gamma}_i) = \Psi(\boldsymbol{\Gamma})$, for every permutation index i. In fact,

$$\Psi(\boldsymbol{\Gamma}_i) = E_{\tilde{\boldsymbol{H}}} \left\{ \log_2 \left| \boldsymbol{I} + \frac{1}{\sigma_w^2} \tilde{\boldsymbol{H}} \boldsymbol{\Pi}_i \boldsymbol{\Gamma} \boldsymbol{\Pi}_i^H \tilde{\boldsymbol{H}}^H \right| \right\}. \tag{5.115}$$

For a Rayleigh-fading channel, the matrix $\tilde{\boldsymbol{H}} \boldsymbol{\Pi}_i$ has exactly the same distribution as $\tilde{\boldsymbol{H}}$. Hence, the expected value in (5.115) coincides with (5.113).

Now, we exploit the property that the function $f(\boldsymbol{A}) := \log_2 |\boldsymbol{A}|$ is concave over the set of positive definite matrices $\boldsymbol{A}$. Since the argument of the determinant in (5.115) is positive definite by construction, the concavity property allows us to write

$$\frac{1}{n_T!} \sum_{i=1}^{n_T!} \log_2 \left| \boldsymbol{I} + \frac{1}{\sigma_w^2} \tilde{\boldsymbol{H}} \boldsymbol{\Gamma}_i \tilde{\boldsymbol{H}}^H \right| \leq \log_2 \left| \boldsymbol{I} + \frac{1}{\sigma_w^2} \tilde{\boldsymbol{H}} \left(\frac{1}{n_T!} \sum_{i=1}^{n_T!} \boldsymbol{\Gamma}_i \right) \tilde{\boldsymbol{H}}^H \right|. \tag{5.116}$$

If we take the expected value of the two sides of this inequality, on the left side we still have $\Psi(\Gamma)$, since each term of the summation, by virtue of (5.115), coincides with $\Psi(\Gamma)$. On the right-hand side, the summation $\frac{1}{n_T!} \sum_{i=1}^{n_T!} \boldsymbol{\Gamma}_i$ is proportional to the identity matrix. More precisely,

$$\frac{1}{n_T!} \sum_{i=1}^{n_T!} \boldsymbol{\Gamma}_i = \frac{p_T}{n_T} \boldsymbol{I}. \tag{5.117}$$

Hence, we have

$$\Psi(\boldsymbol{\Gamma}) \leq \Psi \left(\frac{p_T}{n_T} \boldsymbol{I} \right). \tag{5.118}$$

where the equality holds if

$$\boldsymbol{\Gamma} = \frac{p_T}{n_T} \boldsymbol{I}. \tag{5.119}$$

Therefore, the capacity is achieved using the matrix $\boldsymbol{\Gamma}$ given by (5.119), i.e., uniform power loading.

11 We consider only permutations affecting the order of the diagonal entries.

Using the optimal $\mathbf{\Gamma}$, the capacity of a flat-fading Rayleigh channel is

$$C = E_{\boldsymbol{H}} \left\{ \log_2 \left| \boldsymbol{I} + \frac{p_T}{n_T\, \sigma_w^2} \boldsymbol{H}\boldsymbol{H}^H \right| \right\}. \tag{5.120}$$

The capacity can be rewritten in terms of the eigenvalues of $\boldsymbol{H}\boldsymbol{H}^H$. Denoting by λ_i the eigenvalues of the matrix $\mathcal{W}$ introduced in (5.1), the ergodic capacity can be written as

$$C = E \left\{ \sum_{i=1}^{m} \log_2 \left(1 + \frac{p_T \lambda_i}{n_T\, \sigma_w^2} \right) \right\}, \tag{5.121}$$

where $m = \min(n_R, n_T)$. Using the unordered eigenvalue distribution (5.3), we can compute (5.121) as

$$C = m\, E \left\{ \log \left(1 + \frac{p_T\, \lambda_1}{n_T \sigma_w^2} \right) \right\}, \tag{5.122}$$

where we have used the property that the first order marginal pdf of the unordered joint pdf is the same for all eigenvalues λ_i. The marginal pdf of λ_1 has been derived in [13] and it is equal to

$$p_{\lambda_1}(\lambda_1) = \frac{1}{m} \sum_{i=1}^{m} \phi_i^2(\lambda_1) \lambda_1^{|n_R - n_T|} e^{-\lambda_1}, \tag{5.123}$$

where

$$\phi_{i+1}(\lambda_1) = \left[\frac{i!}{(i + |n_R - n_T|\,)!} \right]^{1/2} L_i^{|n_R - n_T|}(\lambda_1),$$

and $L_i^{|n_R - n_T|}(x)$ is the Laguerre polynomial of order i and degree $|n_R - n_T|$; that is,

$$L_i^{|n_R - n_T|}(x) := \frac{1}{i!}\, e^x\, x^{-|n_R - n_T|} \frac{d^i}{dx^i} \left(e^{-x} x^{|n_R - n_T| + i} \right).$$

Therefore, the capacity of a flat-fading Rayleigh channel is

$$C = \int_0^{\infty} \log_2 \left(1 + \frac{p_T\, \lambda}{n_T \sigma_w^2} \right) \sum_{k=0}^{m-1} \frac{k!}{(k+d)!} \left[L_k^d(\lambda) \right]^2 \lambda^d e^{-\lambda} d\lambda\,, \tag{5.124}$$

where $d = |n_R - n_T|$. An example of capacity for Rayleigh-fading channel is reported in Fig. 5.6, in the case of $n_T = n_R$.

Figure 5.6 shows that the capacity increases by a factor proportional to n_T (or n_R).

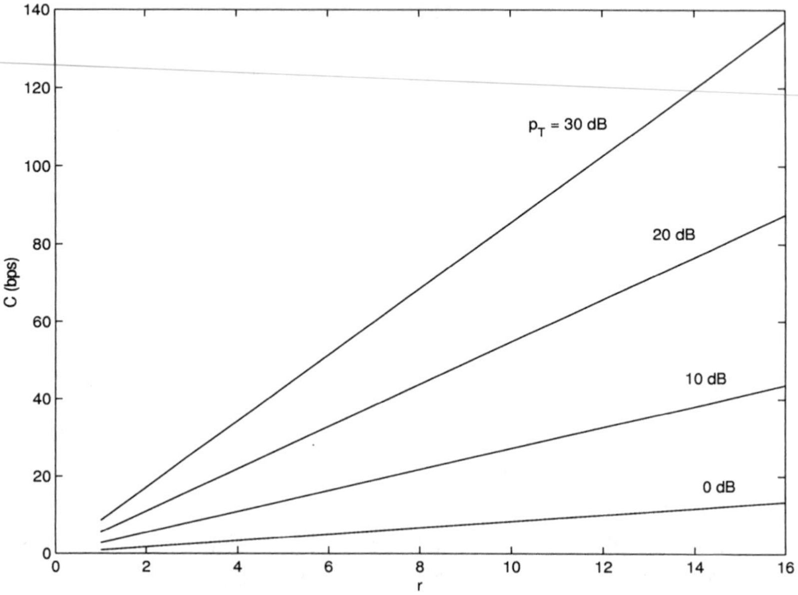

Figure 5.6 Ergodic capacity for Rayleigh flat fading channel versus $r = n_T = n_R$.

Indeed, this is just a particular case of a more general property that *the capacity increases by a factor equal to* $\min(n_T, n_R)$. This is indeed an extraordinary advantage because it states that a MIMO system provides a considerable increase of the transmission rate, without requiring any increase neither or bandwidth nor of transmit power.

It is worth adding a remark that the previous results have been derived under the assumption of independent Rayleigh-fading channels. The capacity increase is much less evident in cases where there is a line of sight and/or the channels are correlated.

In fact, the gain depends on the rank of $\boldsymbol{H}$: The maximum rate gain requires $\boldsymbol{H}$ to be full rank. In practice, if transmit and receive arrays are in line of sight and there is no back scattering from the environment, the channel matrix is rank one with probability one. In such a case, there is no multiplexing gain. However, such a condition is not really critical because a strong LOS MIMO propagation, with no random scattering, is a relatively fortunate circumstance, from the performance point of view. Conversely, the most critical situation occurs when there is no LOS and the channel is strongly affected by multipath propagation. Fortunately, these are the cases where MIMO systems offer most of their advantages.

Frequency-selective channels

Let us consider now a frequency-selective MIMO channel, where we use a block transmission strategy. Let us denote with N the length of each transmitted block and with L the (maximum) channel order. The input-output relationship is still as in (5.61), with

$$\boldsymbol{H} := \begin{pmatrix} \boldsymbol{H}_{11} & \boldsymbol{H}_{11} & \cdots & \boldsymbol{H}_{1n_T} \\ \boldsymbol{H}_{21} & \boldsymbol{H}_{21} & \cdots & \boldsymbol{H}_{2n_T} \\ \vdots & \ddots & \ddots & \vdots \\ \boldsymbol{H}_{n_R 1} & \boldsymbol{H}_{n_R 1} & \cdots & \boldsymbol{H}_{n_R n_T} \end{pmatrix}, \tag{5.125}$$

where each matrix $\boldsymbol{H}_{ij}$ is $N \times N$ and Toeplitz. Incorporating a CP or a ZP of sufficient length, the matrices $\boldsymbol{H}_{ij}$ can be made also circulant. In such a case, each matrix $\boldsymbol{H}_{ij}$ is diagonalized as $\boldsymbol{H}_{ij} = \boldsymbol{W}\boldsymbol{\Lambda}_{ij}\boldsymbol{W}^H$, where $\boldsymbol{W}$ is the (normalized) DFT matrix (i.e., $W_{kl} = e^{j2\pi kl/N}/\sqrt{N}$). The overall matrix $\boldsymbol{H}$ can then be diagonalized as follows

$$\boldsymbol{H} = (\boldsymbol{I} \otimes \boldsymbol{W})\boldsymbol{\Lambda}(\boldsymbol{I} \otimes \boldsymbol{W})^H, \tag{5.126}$$

where $\otimes$ indicates the Kronecker product and

$$\boldsymbol{\Lambda} := \begin{pmatrix} \boldsymbol{\Lambda}_{11} & \boldsymbol{\Lambda}_{12} & \cdots & \boldsymbol{\Lambda}_{1n_T} \\ \boldsymbol{\Lambda}_{21} & \boldsymbol{\Lambda}_{22} & \cdots & \boldsymbol{\Lambda}_{2n_T} \\ \vdots & \ddots & \ddots & \vdots \\ \boldsymbol{\Lambda}_{n_R 1} & \boldsymbol{\Lambda}_{n_R 2} & \cdots & \boldsymbol{\Lambda}_{n_R n_T} \end{pmatrix}. \tag{5.127}$$

It is easy to verify that $(\boldsymbol{I} \otimes \boldsymbol{W})$ is itself a unitary matrix. Following the same arguments as with flat-fading channels, the capacity of an ergodic frequency-selective channel is then

$$C = \frac{1}{N} E_{\boldsymbol{H}} \left\{ \log_2 |\boldsymbol{I} + \alpha \boldsymbol{\Lambda}\boldsymbol{\Lambda}^H| \right\}, \tag{5.128}$$

where $\alpha := p_T / n_T \sigma_w^2$. The normalization by N has been introduced to measure the capacity in terms of number of bits per transmitted symbol. Each matrix $\boldsymbol{\Lambda}_{ij}$ in (5.127) is diagonal, so that $\boldsymbol{\Lambda}$ is a block matrix with diagonal blocks. Using permutations of rows and columns of $\boldsymbol{\Lambda}$ in (5.127), we can reduce $\boldsymbol{\Lambda}$ to a diagonal matrix as

$$\boldsymbol{\Lambda} = \begin{pmatrix} \boldsymbol{\Lambda}(0) & \mathbf{0} & \cdots & \mathbf{0} \\ \mathbf{0} & \boldsymbol{\Lambda}(1) & \ddots & \mathbf{0} \\ \vdots & \ddots & \ddots & \vdots \\ \mathbf{0} & \cdots & \cdots & \boldsymbol{\Lambda}(N-1) \end{pmatrix}, \tag{5.129}$$

with

$$\boldsymbol{\Lambda}(k) := \begin{pmatrix} H_{11}(k) & H_{12}(k) & \cdots & H_{1n_T}(k) \\ H_{21}(k) & H_{22}(k) & \cdots & H_{2n_T}(k) \\ \vdots & \ddots & \ddots & \vdots \\ H_{n_R1}(k) & H_{n_R2}(k) & \cdots & H_{n_Rn_T}(k) \end{pmatrix}, \tag{5.130}$$

where $H_{qp}(k) := \sum_{i=0}^{L} h_{qp}(i)e^{-j2\pi ik/N}$ is the transfer function of the channel between the p-th transmit and the q-th receive antennas, evaluated on the kth subcarrier. Introducing (5.127) in (5.128), we obtain

$$\begin{aligned} C &= \frac{1}{N}E_{\boldsymbol{H}}\left\{\log_2 \prod_{k=0}^{N-1} |\boldsymbol{I} + \alpha\boldsymbol{\Lambda}(k)\boldsymbol{\Lambda}^H(k)|\right\} \\ &= \frac{1}{N}\sum_{k=0}^{N-1} E_{\boldsymbol{H}}\left\{\log_2 |\boldsymbol{I} + \alpha\boldsymbol{\Lambda}(k)\boldsymbol{\Lambda}^H(k)|\right\}, \end{aligned} \tag{5.131}$$

In the Rayleigh-fading model case, each entry $H_{ij}(k)$ is a complex Gaussian random variable. If we normalize, for comparison purposes, all the channel impulse responses $\boldsymbol{h}_{ij} := [h_{ij}(0), \ldots, h_{ij}(L)]$, so that $\|\boldsymbol{h}_{ij}\|^2 = 1$, the entries of each matrix $\boldsymbol{\Lambda}(k)$ have the same statistical distribution as $\boldsymbol{H}$ in the flat-fading case. Therefore, each term in the summation in (5.131) gives rise to the same result as in (5.124). Hence, summing up the N terms and dividing by N, we arrive at an expression having the same form as the ergodic capacity (5.124) of flat-fading channels.

5.6.3 Nonergodic Channels

If the channel is not ergodic, we can only characterize the mutual information in probabilistic terms. Since the channel is not known at the transmitter side, we can only distribute the power uniformly across the channel modes. Hence, proceeding as with (5.108), denoting by r the rank of $\boldsymbol{H}$, the mutual information is

$$\begin{aligned} I(\boldsymbol{H}) &= \log_2 \left| \boldsymbol{I}_{n_R} + \frac{p_T}{n_T \sigma_w^2} \boldsymbol{H}\boldsymbol{H}^H \right| \\ &= \log_2 \left| \boldsymbol{I}_{n_R} + \frac{p_T}{n_T \sigma_w^2} \boldsymbol{\Lambda} \right| = \sum_{i=1}^{r} \log_2 \left(1 + \frac{\lambda_i p_T}{n_T \sigma_w^2} \right). \end{aligned} \quad (5.132)$$

This quantity is not equal to the capacity, as the input covariance matrix is not the optimal one. The difference with respect to (5.108) is that (5.132) is a random variable. In general, it is not easy to derive the statistics of $I(\boldsymbol{H})$. However, in [14, 15], it was shown that, in the asymptotic case, when n tends to infinity, it is possible to find out the limiting distribution of $I(\boldsymbol{H})$. We will now review the basic results of [14, 15]. Assuming that the entries H_{ij} are i.i.d. random variables with zero mean and unit variance, it is known the limiting distribution of the eigenvalues λ_i of $\boldsymbol{H}^H\boldsymbol{H}$, in the asymptotic case when the matrix dimensions tend to infinity.

Let us denote the empirical distribution of the eigenvalues of $\boldsymbol{H}^H\boldsymbol{H}$, for a given size n_T, as

$$F_{n_T}(\lambda) := \frac{1}{n_T} \left| \{ i : \lambda_i \leq \lambda \} \right|, \quad (5.133)$$

where the symbol $|\mathcal{A}|$ indicates the cardinality of the set $\mathcal{A}$. The empirical distribution $F_{n_T}(\lambda)$ gives the percentage of eigenvalues of $\boldsymbol{H}^H\boldsymbol{H}$ less than λ. As n_T tends to infinity, the function $G_{n_T}(\lambda) := F_{n_T}(n_T\lambda)$ converges to the so called Marčenko-Pastur CDF $G(\lambda)$, whose associated pdf is [17]

$$g(\lambda) = (1-c)^+\delta(\lambda) + \frac{\sqrt{(\lambda - a)(b - \lambda)}}{2\pi\lambda} 1_{[a,b]}(\lambda), \quad (5.134)$$

where $c := n_R/n_T$, $a := (\sqrt{c} - 1)^2$, $b := (\sqrt{c} + 1)^2$ and $1_{[a,b]}(\lambda)$ is the indicator functions, defined as

$$1_{[a,b]}(x) := \begin{cases} 1 & x \in [a,b] \\ 0 & \text{elsewhere.} \end{cases} \quad (5.135)$$

In particular, when the channel matrix is square (i.e., $n_T = n_R$) we have $c = 1$, $a = 0$, and $b = 4$, so that the limiting pdf is

$$g(\lambda) = \frac{1}{2\pi\lambda} \sqrt{\lambda(4 - \lambda)}\, 1_{[0,4]}(\lambda). \quad (5.136)$$

Using this asymptotic distribution, it is possible to compute the mutual information as (for simplicity of notation, we assume $\sigma_n^2 = 1$)

$$\frac{I_{n_T}(\boldsymbol{H})}{n_T} \to \int_0^4 \log_2 \left(1 + \frac{p_T}{n_T} \lambda \right) g(\lambda) d\lambda. \quad (5.137)$$

This result, already pointed out in [18], is rather remarkable, as it states that the mutual information, under the assumption of uniform power allocation, is *independent of the pdf of the entries* H_{ij} *of* $\boldsymbol{H}$, provided only that the random variables H_{ij} are i.i.d., with zero mean and unit variance.

The integral in (5.137) was computed in closed form in [20] (even though in a different context) and the result, in case of $c = 1$, is

$$I(p_T) := \lim_{n_T \to \infty} \frac{I_{n_T}(\boldsymbol{H})}{n_T} = 2\log_2(1 + \sqrt{4p_T + 1}) - \frac{\log_2 e}{4p_T}\left(\sqrt{4p_T + 1} - 1\right)^2. \tag{5.138}$$

Correlated fading

All previous expressions assumed that the channel entries are uncorrelated. However, in practice, as we have already seen in Chapter 2, there might be a nonnegligible correlation among the channel entries H_{ij}. The problem has been studied, for example, in [14, 15, 21]. In [15], a very elegant closed form solution was derived, even though in a specific case of correlation matrix. The solution given in [15] was based on *free probability* theory. We will only recall the main results of [15] here. The channel matrix $\boldsymbol{H}$ is assumed to be modeled either as

$$\boldsymbol{H} = \boldsymbol{C}^{1/2}\boldsymbol{U}, \tag{5.139}$$

which models a correlation only between the receive antennas, or as

$$\boldsymbol{H} = \boldsymbol{U}\boldsymbol{C}^{1/2}, \tag{5.140}$$

which models a correlation only between the transmit antennas. In both cases, the matrix $\boldsymbol{U}$ is $q \times n$ and its entries are i.i.d. circularly symmetric complex Gaussian random variables, with zero mean and unit variance. The matrix $\boldsymbol{C}$ is an Hermitian Toeplitz matrix that impresses the desired correlation. The key observation of [15], recalling [22], was that as the dimensions of the matrices $\boldsymbol{H}$ and $\boldsymbol{C}$ increase indefinitely, the matrices $\boldsymbol{C}$ and $\boldsymbol{U}\boldsymbol{U}^H/n_T$ become asymptotically *free*. This makes it possible to derive the asymptotic distribution of the eigenvalues of $\boldsymbol{H}\boldsymbol{H}^H$ from the asymptotic distribution of the eigenvalues of $\boldsymbol{C}$ and $\boldsymbol{U}\boldsymbol{U}^H$. Deriving a closed-form expression is rather difficult. Nevertheless, assuming that the the correlating matrix $\boldsymbol{C}$ had an asymptotic eigenvalue distribution

$$p_C(x) = \frac{2ab}{\pi(\sqrt{b} - \sqrt{a})^2 x^2}\sqrt{\left(\frac{x}{a} - 1\right)\left(1 - \frac{x}{b}\right)}\,1_{[a,b]}, \tag{5.141}$$

where a and b denote the boundaries of the support for $p_C(x)$, Mestre [15] was able to derive the asymptotic distribution of the eigenvalues of $\boldsymbol{H}\boldsymbol{H}^H$

$$g(\lambda) = (1-c)^+\delta(\lambda) + \frac{\sqrt{(\lambda-a)(b-\lambda)}}{2\pi\lambda(1+\mu\lambda)} 1_{[a,b]}(\lambda), \tag{5.142}$$

where

$$\begin{aligned} a &= 1+c+2\mu c - 2\sqrt{c(1+\mu)(1+\mu c)}, \\ b &= 1+c+2\mu c + 2\sqrt{c(1+\mu)(1+\mu c)}, \end{aligned} \tag{5.143}$$

with

$$\mu := \frac{(\sqrt{b}-\sqrt{a})^2}{4ab}. \tag{5.144}$$

The coefficient μ controls the correlation between the antenna elements. In particular, $\mu = 0$ gives rise to a pdf $p_C(x)$ concentrated in 1, meaning that the correlation matrix $\boldsymbol{C}$ is an identity matrix (i.e., the channels are uncorrelated). Conversely, μ tending to infinity describes the fully correlated case. An example of distribution of the eigenvalues of $\boldsymbol{H}\boldsymbol{H}^H$ is reported in Figure 5.7, showing $g(\lambda)$, for different values of μ, in case of square channel matrix (i.e., $c = 1$).

We can observe that, as μ increases (i.e., as the correlation increases) the eigenvalues tend to be assume, with higher probabilities, lower and higher values, with respect to the uncorrelated case.

Using (5.142), Mestre derived the mutual information for the case of correlated fading at the receiver (cf. 5.139):

$$I = \log_2|\beta w(\beta,c,\mu)| + \frac{1}{\mu}\log_2|1-\mu c v(\beta,c,\mu)| - (c-1)\log_2|u(\beta,c,\mu)|, \tag{5.145}$$

with

$$\begin{aligned} u(\beta,c,\mu) &= \begin{cases} \frac{c+(1-c)\beta+2\mu c^2-\sqrt{R(\beta,c,\mu)}}{2c(\beta-\mu c)} & \beta \neq \mu c \\ -\frac{c(1+\beta)}{c+(1+c)\beta} & \beta = \mu c \end{cases} \\ v(\beta,c,\mu) &= \frac{c+(1+c)\beta+2\mu c\beta-\sqrt{R(\beta,c,\mu)}}{2c(1+\mu)(1+\mu c)\beta} \\ w(\beta,c,\mu) &= \frac{(1+(1+c)\mu)[c+\beta(1+c)+\sqrt{R(\beta,c,\mu)}]-2\mu c(\beta-\mu c)}{2c(1+\mu)(1+\mu c)\beta}, \end{aligned}$$

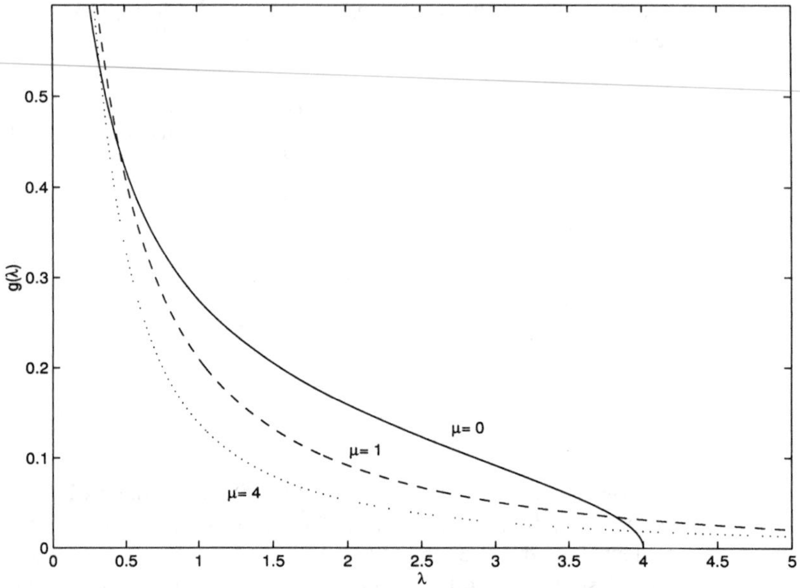

Figure 5.7 Asymptotic pdf of the eigenvalues of $\boldsymbol{H}\boldsymbol{H}^H/n$, for different values of the correlation parameter μ.

with

$$R(\beta, c, \mu) := [c + (1 + c)\beta]^2 - 4c\beta(\beta - \mu c). \tag{5.146}$$

An example of mutual information is reported in Figure 5.8, relative to the square matrix case (i.e., $c = 1$), for different values of μ. From Figure 5.8 we can notice how, as expected, increasing the correlation, the mutual information decreases. In [14] it was shown that, working with physically meaningful data obtained by simulating indoor propagation, the capacity of a correlated flat-fading MIMO channel still increases proportionally to $\min(n_R, n_T)$, but with a loss in growth rate in the order of $10 - 20\%$.

Before concluding this section, it is necessary to remark that the maximum diversity increase is directly related to the rank of the channel matrix. Most of the rate increase is possible only if the channel introduces sufficient angular spread of the radiated signal. This is often the case in urban environments, where it is rare to have a dominant line of sight propagation contribution. It is also true that if there is a dominant line of sight, it is less necessary to have a capacity boost, through the MIMO system, because the channel is inherently better than the no line of sight.

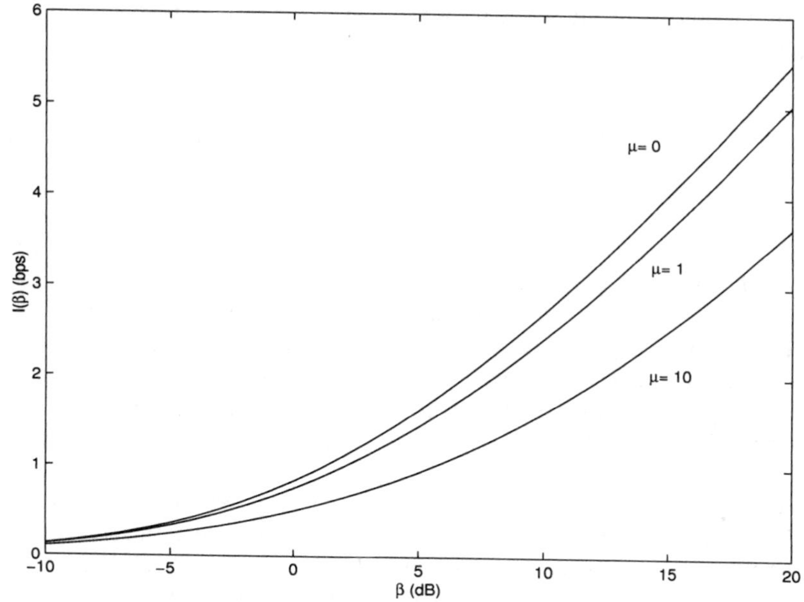

Figure 5.8 Mutual information versus SNR (decibels), for different values of correlation (e.g., μ).

For a better understanding of the limits on channel capacity resulting from specific channel characteristics, it is useful to read [16].

5.7 RATE-DIVERSITY TRADE-OFF

As anticipated at the beginning of this chapter, there is a fundamental trade-off between diversity gain and multiplexing gain, so that it is not possible to reach, at the same time, maximum diversity gain and maximum multiplexing gain. This fundamental trade-off was derived by Zheng and Tse in [23] and we will review their basic results in this section.

Let us start again with the ergodic capacity formula

$$C(\text{SNR}) = E\left\{\log_2\left|\boldsymbol{I} + \frac{\text{SNR}}{n_T}\boldsymbol{H}\boldsymbol{H}^H\right|\right\}. \tag{5.147}$$

At sufficiently high SNR, where the identity matrix in (5.147) can be neglected, we may write, approximately

$$C(\mathrm{SNR}) = m \log_2(\mathrm{SNR}) + \sum_{i=1}^{m} E\{\log_2(\lambda_i)\}, \tag{5.148}$$

where m is the rank of $\boldsymbol{H}\boldsymbol{H}^H$ and $\lambda_i, i = 1, \ldots, m$ are the nonnull eigenvalues of $\boldsymbol{H}\boldsymbol{H}^H$. From (5.148), it is evident that at high SNR, the capacity increases as $m \log_2(\mathrm{SNR})$. Compared to analogous formula for the SISO case, where $C(\mathrm{SNR})$ goes, asymptotically, like $\log_2(\mathrm{SNR})$, (5.148) states that, at high SNR, a MIMO channel behaves as a set of m independent channels. Equivalently, a MIMO channel offers the possibility for a multiplexing gain equal to m, the rank of $\boldsymbol{H}\boldsymbol{H}^H$. In case of uncorrelated flat fading, the gain is $m = \min(n_T, n_R)$.

Since the capacity increases as $m \log_2(\mathrm{SNR})$, to take advantage of this opportunity, it is advisable, in practice, to consider a modulation scheme that increases the constellation order, as the SNR increases. Let us denote with $R(\mathrm{SNR})$ the rate, in bits per symbol, supported at a given SNR. In [23], a scheme was said to achieve multiplexing gain r if

$$\lim_{\mathrm{SNR}\to\infty} \frac{R(\mathrm{SNR})}{\log_2(\mathrm{SNR})} = r. \tag{5.149}$$

Similarly, a scheme achieves a diversity gain d, if

$$\lim_{\mathrm{SNR}\to\infty} \frac{\log_2[P_e(\mathrm{SNR})]}{\log_2(\mathrm{SNR})} = -d, \tag{5.150}$$

where $P_e(\mathrm{SNR})$ is the SER at a given SNR. In [23] it was shown that, in the case of Rayleigh flat-fading channels, indicating with k the multiplexing gain, with $0 \leq k \leq m$, the maximum diversity gain achievable, for each value of k, is given by the piecewise-linear function connecting the points $(k, d(k))$, where

$$d(k) = (n_R - k)(n_T - k). \tag{5.151}$$

An example is reported in Figure 5.9, relative to the case of $n_T = n_R = 4$. We can see that, as it could be guessed intuitively, as we increase the multiplexing gain, the diversity gain decreases and viceversa. How we can approach these fundamental trade-offs will be the subject of Chapters 7 and 8 on space-time codes.

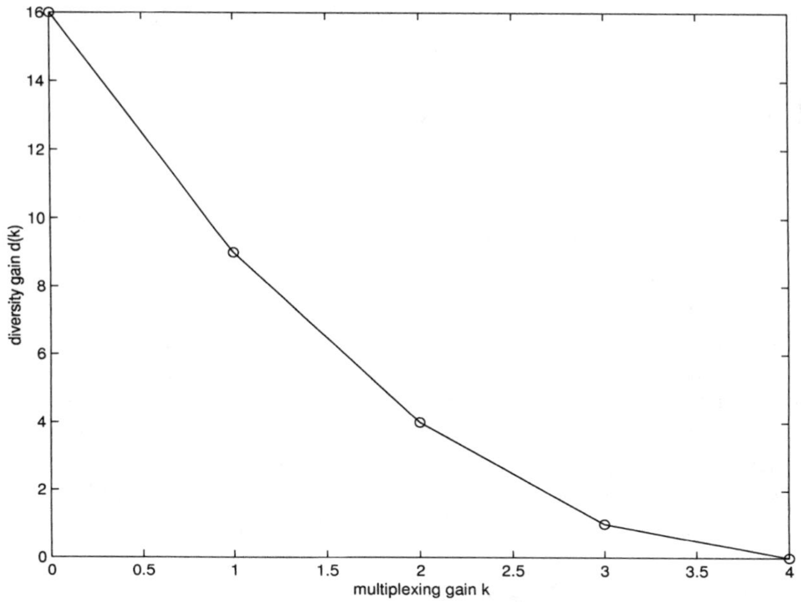

Figure 5.9 Multiplexing/diversity trade-off.

5.8 CAPACITY LIMITS IN CASE OF PARTIAL CHANNEL KNOWLEDGE AT THE RECEIVER

The property that the capacity of a MIMO channel increases by a factor proportional to $\min(n_T, n_R)$ has been derived under the assumption that the receiver has perfect knowledge of the channel. However, in a wireless context, because of the channel variability, the channel is only estimated. Therefore, its knowledge is affected by estimation errors. In this section, we review the fundamental bounds on the channel capacity resulting from imperfect channel knowledge. We assume, as throughout this chapter, that the transmitter has no channel knowledge, or that it has only knowledge about the statistical distribution about the channel, but not of the channel realization. An excellent tutorial on the capacity limits under different degrees of channel knowledge is [24].

The degree of knowledge can vary, depending on the channel variability as well as on the receiver estimation capabilities. Channel estimation requires also some kind of training, which implies that the transmitter has to send training sequences that also reduce the system capacity. The degree of channel knowledge can vary from knowledge of the channel distribution information (CDI), where only the statistical distribution of the channel is known, but not the specific realization, to perfect channel state information (CSI), where the channel realization is supposed to be perfectly known. In the middle, we may have different degrees of knowledge, like knowledge of the channel mean or covariance.

One of the first papers that addressed the problem of evaluating the capacity of MIMO systems having only CDI at the receiver (CDIR) was by Marzetta and Hochwald [25]. In [25], the channel was modeled as i.i.d. complex Gaussian rv's assumed to be constant for T symbols, after which they change to another independent realization. In such a case, it was shown that, for a fixed number of antennas, as the length of the coherence interval T increases, the capacity approaches the limit obtainable when the receiver knows the channel exactly. However, there is a major distinction with respect to the perfect CSI case: The capacity dos not increase at all as the number of transmit antennas increases beyond T. This same problem was further explored by Zheng and Tse in [26], where it was proved that, at high SNR, capacity is achieved using no more than $M^* = \min[n_T, n_R, \lfloor T/2 \rfloor]$ transmit antennas, where $T \geq 2$. Zheng and Tse prove, in particular, that at high SNR, the capacity increases only as

$$C(\text{SNR}) \simeq M^*(1 - M^*/T) \log_2(\text{SNR}). \tag{5.152}$$

This is in stark contrast with the capacity increase in case of perfect CSI at the receiver, where, asymptotically, we have

$$C(\text{SNR}) \simeq \min(n_T, n_R) \log_2(\text{SNR}). \tag{5.153}$$

Hence, these considerations show that the block fading assumption is indeed crucial. If the channel fades very abruptly from block to block, and it is unknown at the receiver, we cannot expect the capacity rate of increase promised by (5.153). In case of independent fading from block to block (i.e., when $T = 1$), Lapidoth and Moser proved in [27] that the capacity increases only double-logarithmically with the SNR and there is no gain factor due to the multiple antennas!

Fortunately, in most communication systems, the channel does not vary so fast to assume independent values over consecutive blocks. However, the results of [26] and [27] are fundamental as they relate the capacity of MIMO channels with the channel fluctuation rate. If the channel variability is so high that the receiver is not able to estimate the channel, increasing the number of antennas does not induce the benefits that one would expect form MIMO systems with perfect CSI.

5.9 SUMMARY

In this chapter we have shown how a MIMO system can improve on the performance of a SISO system. The two major performance factors are the *diversity gain* and the *multiplexing gain*. In summary, we have proved the following properties about the diversity gain of a flat-fading MIMO channel with n_T transmit and n_R receive antennas. It is useful to distinguish between the cases where the channels are stationary (i.e., they assume the same value over the length of the transmitted code) or not. If the channels are stationary and flat-fading, the diversity gain is: 1) $G_d = n_R$, if the codewords spans only the space domain or 2) $G_d = n_R n_T$, if the codewords span *both* space and time, occupying, at least, n_T symbols. If the channels are frequency-selective (i.e.. they introduce a delay spread of $L + 1$ symbols) the maximum diversity gain is $G_d = n_R n_T (L + 1)$.

If the channel is nonstationary, the diversity gain depends on the channel variability. In particular, if the channel is stationary over Q blocks, but it varies from each set of Q blocks to the next, the maximum diversity gain is $G_d = n_T n_R Q$. These gains are achieving only through an appropriate code selection. In this section, we have shown the properties that a code design must respect to achieve the maximum diversity gain. We have not seen explicitly, how to design the codes. This will be the subject of Chapter 7.

The other major gain of MIMO systems is the rate increase factor, which depends on the number of transmit/receive antennas. If the transmitter has no knowledge about the channel, but the receiver knows the channel perfectly, the capacity increases logarithmically with the SNR (at high SNR), with a gain factor equal to the rank of the channel matrix. This gain factor, in case of independent fading channels, can reach the value $\min(n_T, n_R)$. This is probably the most striking advantage of MIMO systems. If the receiver has no knowledge about the channel realization, the advantage decreases considerably. If the channel is stationary, the loss is rather limited, but the loss may become considerable if the channel is fast fading.

The interested reader is encouraged to read the seminal works of Foschini et al. [18] and Telatar [13] on MIMO channel capacity. Both Foschini and Telatar considered the uncorrelated fading case. The more challenging correlated case was considered in [14, 15, 21]. A very good tutorial on diversity was given by Ma and Giannakis in [28]. To study the diversity gain, the performance has been given in terms of a bound, valid asymptotically, as the SNR goes to infinity. However, a more precise performance analysis was derived in [29, 30] (see also the references therein). In this chapter, we have basically considered only the Rayleigh-fading case. For the analysis of the non-Gaussian case, one interesting and tutorial paper is [28].

References

[1] James, A. T., "Distribution of matrix variates and latent roots derived from normal samples," *Annals of Mathematical Statistics*, vol. 35, 1964, pp. 475–501.

[2] Khatri, C. G., "Distribution of the largest or the smallest characteristic root under null hypothesis concerning multivariate normal populations," *Annals of Mathematical Statistics*, Vol. 35, Dec. 1964, pp. 1807–1810.

[3] Kang, M., Alouini, M.-S., "Largest eigenvalue of complex Wishart matrices and performance analysis of MIMO MRC systems," *IEEE Journal on Selected Areas in Communications*, Vol. 21, April 2003, pp. 418–426.

[4] Andersen, J. B., "Array gain and capacity for known random channels with multiple element arrays at both ends," *IEEE Journal on Selected Areas in Communications*, Vol. 11, Nov. 2000, pp. 2172–2178.

[5] Wang, Z., Giannakis, G.B., "A simple and general parameterization quantifying performance in fading channels," *IEEE Transactions on Communications*, Aug. 2003, pp. 1389–1398.

[6] Biglieri, E., Proakis, J., Shamai (Shitz), S., "Fading channels: Information theoretic and communications aspects," *IEEE Trans. on Information Theory*, Vol. 44, Oct. 1998, pp. 2619–2692.

[7] Cover, T. M., Thomas, J. A., *Elements of Information Theory*, New York: John Wiley & Sons, 1991.

[8] Shannon, C., "A mathematical theory of communications," *Bell Labs Tech. J.*, July and October 1948, pp. 379–423, 623–656.

[9] Proakis, J. , *Digital Communications*, (4th edition), New York: McGraw-Hill, 2000.

[10] Cho, K., Yoon, D., "On the general BER expression of one- and two-dimensional amplitude modulations," *IEEE Trans. on Commun.*, Vol. 50, July 2002, pp. 1074 – 1080.

[11] Tarokh, V., Seshadri, N., Calderbank, A.R., "Space-time codes for high data rate wireless communications: Perfromance criterion and code construction," *IEEE Trans. on Information Theory*, Vol. 44, Mar. 1998, pp. 744–765.

[12] Scutari, G., Paccapeli, G., Barbarossa, S., "Concatenated space-time coding with optimal trade-off between diversity and coding gain," *Proc. of ICASSP 2003*, Hong Kong (China), Vol. 4, April 2003, pp. 333–336.

[13] Telatar, I., E., "Capacity of multi-antenna Gaussian channels," *AT&T Technical Memo*, 1995; see also *European Trans. on Telecommun.*, Vol. 10, no. 6, 1999, pp. 586–595.

[14] Chuah, C.-N., et al. "Capacity scaling in MIMO wireless systems under correlated fading," *IEEE Trans. on Information Theory*, Vol. 48, March 2002, pp. 637–650.

[15] Mestre, X., Fonollosa, J.R., Pages-Zamora, A., "Capacity of MIMO channels: asymptotic evaluation under correlated fading," *IEEE Journal on Selected Areas in Commun.*, Vol. 21, June 2003, pp. 829–838.

[16] Chizhik, D., et al. "Keyholes, correlations and capacities of multielement transmit and receive antennas," *IEEE Trans. on Wireless Commun.*, Vol. 1, Apr. 2002, pp. 361–368.

[17] Silverstein, J. W., Bai, Z., D., "On the empirical distribution of eigenvalues of a class of large dimensional random matrices," *J. Multivariate Anal.*, Vol. 54, no. 2, 1995, pp. 175–192.

[18] Foschini, G., J., Gans, M., J., "On limits of wireless communication in a fading environment when using multiple antennas," *Wireless Personal Communications*, Vol. 6, March 1998, pp. 311–335.

[19] Foschini, G., J., Gans, M., J., "Capacity when using diversity at transmit and receive sitess and the Rayleigh-faded matrix channel is unknown at the transmitter," *WINLAB Workshop on Wireless Information Networks*, March 1996.

[20] Verdú, S., Shamai (Shitz), S., "Spectral efficiency of CDMA with random spreading," *IEEE Trans. on Information Theory*, Vol. 45, Mar. 1999, pp. 622–640.

[21] Rapajic, P. B., Popescu, D., "Information capacity of a random signature multiple-input multiple-output channel," *IEEE Trans. on Commun.*, Vol. 48, Aug. 2000, pp. 1245–1248.

[22] Hiai, F., Petz, D., *The semicircle law-Free random varables and entropy*, Providence, RI: American Mathematical Society, 2000.

[23] Zheng, L., Tse, D.N.C., "Diversity and multiplexing: A fundamental tradeoff in multiple antenna channels," *IEEE Trans. on Information Theory*, Vol. 49, May 2003, pp. 1073–1096.

[24] Goldsmith, et al. " *IEEE Journal on Selected Areas in Communications*, Vol. 21, June 2003, pp. 684–702.

[25] Marzetta, T., Hochwald, B., "Capacity of a mobile multiple-antenna communication link in Rayleugh flat fading," *IEEE Trans. on Information Theory*, Vol. 45, Jan. 1999, pp. 139–157.

[26] Zheng, L., Tse, D.N.C., "Packing spheres in the Grassmann manifold: A geometric approach to the noncoherent multi-antenna channel," *IEEE Trans. on Information Theory*, Vol. 48, Feb. 2002, pp. 359–383.

[27] Lapidoth, A., Moser, S. M., "Capacity bounds via duality with applications to multiple-antenna systems on flat fading channels," *IEEE Trans. on Information Theory*, Vol. 49, Oct. 2003, pp. 2426–2467.

[28] Ma, X., Giannakis, G.B., "Complex field coded MIMO systems: performance, rate, and trade-offs," *Wireless Communications and Computing*, Vol. 2, Feb. 2002, pp. 693–717.

[29] Biglieri, E., Taricco, G., Tulino, A., "Performance of space-time codes for a large number of antennas," *IEEE Trans. on Information Theory*, Vol. 48, July 2002, pp. 1794–1803.

[30] Taricco, G., Biglieri, E., "Exact pairwise error probability of space-time codes," *IEEE Trans. on Information Theory*, Vol. 48, Feb. 2002, pp. 510–513.

Chapter 6

Optimal Design of Closed-Loop MIMO Systems

6.1 INTRODUCTION

We turn now our attention to *closed-loop* systems, where the transmitter has some knowledge about the channel. This knowledge, whether partial or complete, can be advantageously exploited to design the transmission strategy in order to optimize the system performance. The idea of using (partial or full) channel knowledge in multiantenna systems to synthesize the radiation pattern accordingly is not novel. That information can be used in fact to steer the antenna beam electronically toward the direction of interest or by nulling the radiation pattern toward some specific directions. This approach is known as *beamforming*. However, beamforming is only a specific example, as it works only in the space domain and it assumes that the system is either MISO (transmit beamforming) or SIMO (receive beamforming). Improvements with respect to classical beamforming can be achieved by incorporating time as a further independent variable, that is by working in the *joint* space-time domain, and generalizing the beamforming idea to MIMO systems.

In this chapter, we provide a very general framework for designing the optimal transmission strategy, according to alternative criteria and constraints. In section 6.2 we provide the basic mathematical tools, namely convex optimization and majorization theory. Then, we formulate the optimization problem in a very general framework, in Section 6.3, where we provide alternative solutions. In Section 6.4, 6.5, and 6.6, we specialize the solution to SISO time-varying channels, MISO frequency-selective, and MIMO systems, respectively. We consider the impact of imperfect channel knowledge in Section 6.7. In particular, since in practice the transmitter knows the channel only within some inevitable estimation error, we show how the transmitter can incorporate partial channel knowledge in the optimization strategy.

Finally, in Section 6.8 and 6.9 we consider the multiuser context. We start with the optimal access methods for the multiple access and the broadcast channel. Then, we consider the more challenging case of networks with no pre-existing infrastructure, as in ad-hoc or sensor networks. In such a context, we address the fundamental question about multiple access systems: What is the best multiplexing (modulation) strategy? This is known as the cocktail party problem. We formulate this problem as a multi-objective optimization problem. We show that its solution requires some kind of centralized control and it is then quite complex to implement. Then, we formulate the cocktail party problem as a strategic game where players (radio nodes) compete with each other for the use of the available resources. This second approach is more appealing than the multiobjective strategy as it leads to decentralized solutions.

6.2 MATHEMATICAL TOOLS

We start by reviewing some basic tools that will be helpful in deriving the optimal coding strategies. A very elegant solution to the joint linear transmit/receiver optimization problem was given by Palomar et al. in [1, 2], where the optimization problems were greatly simplified by using two basic mathematical tools, namely convex optimization and majorization theory. Two basic references for convex optimization and majorization theory are [3] and [4], respectively. Here, we will only review the basic properties that are useful for our optimization problems.

6.2.1 Schur-Convexity and Schur-Concavity

An nth dimensional set $\mathcal{A} \subseteq \mathbb{R}^n$ is *convex* if, for any pair of vectors $\boldsymbol{x}, \boldsymbol{y} \in \mathcal{A}$ and any real parameter $\theta \in [0, 1]$, the vector $\boldsymbol{z} = \theta\boldsymbol{x} + (1-\theta)\boldsymbol{y} \in \mathcal{A}$. An example of convex and nonconvex sets, defined on a two-dimensional space, is sketched in Figure 6.1.

6.2.1.1 Convex functions

A function $f(\boldsymbol{x})$ of an n-size vector $\boldsymbol{x}$, defined over a set $\mathcal{A}$, is *convex* over $\mathcal{A}$ if, for all $\boldsymbol{x}, \boldsymbol{y} \in \mathcal{A}$ and any real parameter $\theta \in [0, 1]$, it satisfies the following inequality

$$f(\theta\boldsymbol{x} + (1-\theta)\boldsymbol{y}) \leq \theta f(\boldsymbol{x}) + (1-\theta) f(\boldsymbol{y}). \tag{6.1}$$

We generalize now the concept of convexity.

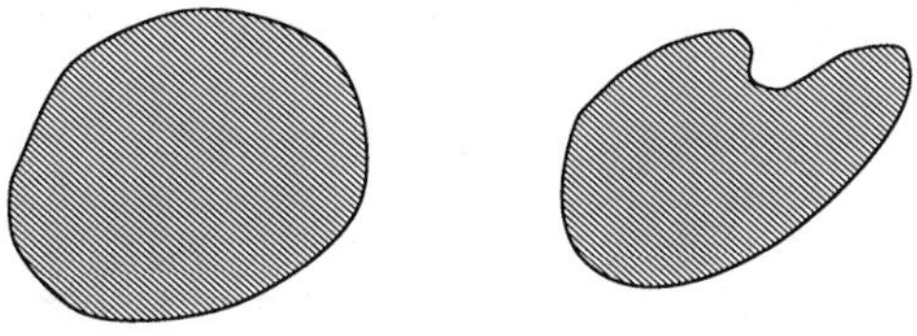

Figure 6.1 Examples of convexity: (a) convex set and (b) nonconvex set.

Given a real vector $\boldsymbol{x}$ of size n, we denote with

$$x_{[1]} \geq \cdots \geq x_{[n]} \tag{6.2}$$

the components of $\boldsymbol{x}$ in decreasing order and with

$$x_{(1)} \leq \cdots \leq x_{(n)} \tag{6.3}$$

the components of $\boldsymbol{x}$ in increasing order.

Definition: Given two real vectors $\boldsymbol{x}$ and $\boldsymbol{y}$, of size n, $\boldsymbol{x}$ is *majorized* by $\boldsymbol{y}$ or, equivalently, $\boldsymbol{y}$ *majorizes* $\boldsymbol{x}$, if

$$\begin{aligned}\sum_{i=1}^{k} x_{[i]} &\leq \sum_{i=1}^{k} y_{[i]}, \quad 1 \leq k \leq n-1, \quad \text{and} \\ \sum_{i=1}^{n} x_{[i]} &= \sum_{i=1}^{n} y_{[i]}.\end{aligned} \tag{6.4}$$

This property is also identified through the symbol $\boldsymbol{x} \prec \boldsymbol{y}$. Alternatively, the previous conditions can be expressed as follows

$$\begin{aligned}\sum_{i=1}^{k} x_{(i)} &\geq \sum_{i=1}^{k} y_{(i)}, \quad 1 \leq k \leq n-1, \\ \sum_{i=1}^{n} x_{(i)} &= \sum_{i=1}^{n} y_{(i)}.\end{aligned} \tag{6.5}$$

Definition: A real-valued function $f(\boldsymbol{x})$, defined over a set $\mathcal{A}$, is *Schur-convex* on $\mathcal{A}$ if

$$\boldsymbol{x} \prec \boldsymbol{y} \ \text{ on } \mathcal{A} \ \Rightarrow f(\boldsymbol{x}) \leq f(\boldsymbol{y}). \tag{6.6}$$

Similarly, a real-valued function $f(\boldsymbol{x})$, defined over a convex set $\mathcal{A}$, is *Schur-concave* on $\mathcal{A}$ if

$$\boldsymbol{x} \prec \boldsymbol{y} \text{ on } \mathcal{A} \Rightarrow f(\boldsymbol{x}) \geq f(\boldsymbol{y}). \tag{6.7}$$

As a consequence of these definitions, if a function $f(\boldsymbol{x})$ is Schur-convex over $\mathcal{A}$, the opposite function $-f(\boldsymbol{x})$ is Schur-concave. Some basic properties about Schur-convexity or Schur-concavity are listed here below.

1. Given any real vector $\boldsymbol{x}$ and the corresponding average vector $\boldsymbol{m}_x$, whose entries are the average value of $\boldsymbol{x}$, i.e., $m_{x_i} = \sum_{k=1}^{n} x_k/n, \forall i$, then

$$\boldsymbol{m}_x \prec \boldsymbol{x}. \tag{6.8}$$

2. Given two real vectors $\boldsymbol{x}$ and $\boldsymbol{y}$ of size n, with $\boldsymbol{x} \prec \boldsymbol{y}$, there always exists a sequence of linear transformations, represented by the matrices $\boldsymbol{T}_1, \ldots, \boldsymbol{T}_K$, such that $\boldsymbol{x} = \boldsymbol{T}_K \cdots \boldsymbol{T}_1 \boldsymbol{y}$, with $K < n$.

3. Given an $n \times n$ Hermitian matrix $\boldsymbol{A}$ with diagonal entries denoted by the vector $\boldsymbol{d}(\boldsymbol{A})$ and (real) eigenvalues $\boldsymbol{\lambda}(\boldsymbol{A})$, then

$$\boldsymbol{\lambda}(\boldsymbol{A}) \succ \boldsymbol{d}(\boldsymbol{A}). \tag{6.9}$$

4. Given a vector $\boldsymbol{x} = (x_1, \ldots, x_n)$ and a convex function $f(x)$, the function $g(\boldsymbol{x}) = \sum_i f(x_i)$ is Schur-convex.

6.2.2 Convex Problems

Let us consider a constrained optimization problem, with m inequality constraints and p equality constraints. The problem can be expressed mathematically as

$$\begin{aligned} \min_{\boldsymbol{x}} \quad & f_0(\boldsymbol{x}) \\ \text{subject to} \quad & \boldsymbol{g}_i(\boldsymbol{x}) \leq 0, 1 \leq i \leq m, \\ & \boldsymbol{h}_i(\boldsymbol{x}) = 0, 1 \leq i \leq p, \end{aligned} \tag{6.10}$$

where $f_0(\boldsymbol{x})$ is the objective (or cost) function, the m functions $\boldsymbol{g}_i(\boldsymbol{x}), i = 1, \ldots, m$ express the inequality constraints and the p functions $\boldsymbol{h}_i(\boldsymbol{x}), i = 1, \ldots, p$ express the equality constraints. If the functions $f_0(\boldsymbol{x})$ and $\boldsymbol{g}_i(\boldsymbol{x}), i = 1, \ldots, m$, are all convex and the functions $\boldsymbol{h}_i(\boldsymbol{x}), i = 1, \ldots, m$, are affine, then the *problem is convex.* The set $\mathcal{D}$ given by the intersection of the domains of all the functions involved in the formulation (6.10) is called the *domain* of the problem. A point $\boldsymbol{x} \in \mathcal{D}$ is *feasible* if it satisfies all the constraints.

In practice, not all problems are expressible in convex form. However, sometimes it is possible to reformulate the problems in order to put them in a convex form (this operation is sometimes called "convexification"). Why is it important to have a problem expressed in convex form? Because in such a case there are very efficient methods to solve the problem and, most important, the solution is unique. An optimal reference for studying convex optimization problems, with a special emphasis on convexification methods, is [3].

Before starting our analysis, it is useful to recall a few matrix properties:

5. Given the matrices $\boldsymbol{A}$, $\boldsymbol{B}$, $\boldsymbol{C}$, and $\boldsymbol{D}$, whose dimensions are consistent with the ensuing multiplications, with $\boldsymbol{A}$ and $\boldsymbol{C}$ invertible, the following identity holds true (matrix inversion lemma):

$$(\boldsymbol{A}+\boldsymbol{BCD})^{-1} = \boldsymbol{A}^{-1} - \boldsymbol{A}^{-1}\boldsymbol{B}(\boldsymbol{DA}^{-1}\boldsymbol{B}+\boldsymbol{C}^{-1})^{-1}\boldsymbol{DA}^{-1}; \tag{6.11}$$

6. Given two matrices $\boldsymbol{X}$ and $\boldsymbol{Y}$, the symbol $\boldsymbol{X} \geq \boldsymbol{Y}$ means that $\boldsymbol{X} - \boldsymbol{Y}$ is a positive semidefinite matrix. The following property holds true:

$$\boldsymbol{X} \geq \boldsymbol{Y} \Rightarrow |\boldsymbol{X}| \geq |\boldsymbol{Y}|; \tag{6.12}$$

7. Given any two matrices $\boldsymbol{X}$ and $\boldsymbol{Y}$ of sizes congruent with the following operations, the following property always holds true:

$$|\boldsymbol{I}+\boldsymbol{XY}| = |\boldsymbol{I}+\boldsymbol{YX}|; \tag{6.13}$$

8. Given two matrices $\boldsymbol{A}$, of size $n \times m$, and $\boldsymbol{B}$, of size $m \times n$, and a square $m \times m$ invertible, positive definite matrix $\boldsymbol{R}$, the following identity holds true:

$$\boldsymbol{A}(\boldsymbol{BA}+\boldsymbol{R})^{-1} = (\boldsymbol{AR}^{-1}\boldsymbol{B}+\boldsymbol{I})^{-1}\boldsymbol{AR}^{-1}, \tag{6.14}$$

provided that the inversion operations are admissible.

6.3 OPTIMIZATION STRATEGIES

Let us consider the linear system sketched in Figure 6.2. Assuming perfect knowledge of the channel matrix $\boldsymbol{H}$ at *both* transmit and receive sides, in this chapter we will show how to derive the optimal coding matrices $\boldsymbol{F}$ and $\boldsymbol{G}$, under different optimality criteria and operative constraints. To leave the formulation and solution as general as possible, we do not impose any structure on $\boldsymbol{H}$, so that the solution can be equally well applied to SISO or MIMO systems. In the ensuing sections, we will specialize the solutions to different transmit/receive structures and propagation characteristics in order to gain physical insight into the optimal strategies.

Denoting by $\hat{s}$ the estimated vector, the I/O relationship is

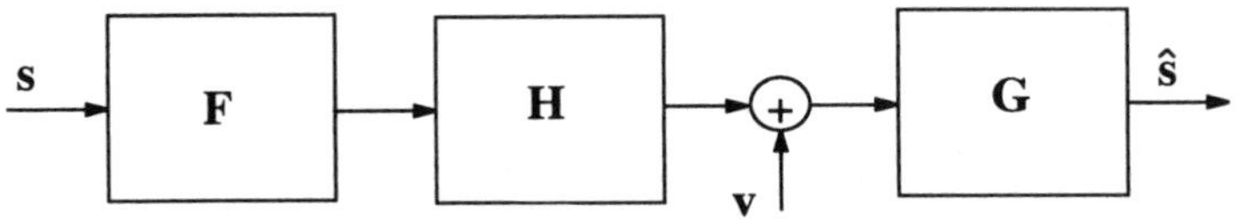

Figure 6.2 Block linear system.

$$\hat{s} = GHFs + Gv, \tag{6.15}$$

where s and $\hat{s}$ are column vectors of size M, F is $N \times M$, H is $P \times N$, G is $M \times P$ and, finally, the noise vector v has size $P \times 1$. To make our formulation as general as possible, to be able to use it for both SISO and MIMO systems, we do not impose any restriction on the sizes of the matrices involved in (6.15). It is important to introduce the mean square error (MSE) matrix, defined as

$$\text{MSE}(F, G) := E\{(s - \hat{s})(s - \hat{s})^H\}. \tag{6.16}$$

Using (6.15), and denoting with R_{ss} and R_{vv} the covariance matrices of the vectors s and v, respectively, (6.16) becomes[1]

$$\text{MSE}(F, G) = (GHF - I)R_{ss}(GHF - I)^H + GR_{vv}G^H. \tag{6.17}$$

In our derivations, we assume that the symbols are uncorrelated with unit variance, so that $R_{ss} = I$. There is no loss of generality in making this assumption, as we could always incorporate a whitening matrix as a right factor of F. Setting $R_{ss} = I$ in (6.17), we obtain

$$\text{MSE}(F, G) = G(HFF^H H^H + R_{vv})G^H + I - GHF - F^H H^H G^H. \tag{6.18}$$

From the MSE matrix MSE(F, G), we can compute the mean square error in the symbol estimate as the trace of MSE(F, G), that is

$$\text{MSE} = E\{||\hat{s} - s||^2\} = \text{tr}[\text{MSE}(F, G)]. \tag{6.19}$$

1 The vectors s and v are assumed to be zero mean.

Among all matrices $\boldsymbol{G}$, there is an optimum one, namely $\boldsymbol{G}_{\text{mmse}}$, that minimizes the mean square error matrix $\text{MSE}(\boldsymbol{F}, \boldsymbol{G})$, in the sense that the difference matrix $\text{MSE}(\boldsymbol{F}, \boldsymbol{G}) - \text{MSE}(\boldsymbol{F}, \boldsymbol{G}_{\text{mmse}})$ is positive semidefinite, for any pair of matrices $\boldsymbol{F}$ and $\boldsymbol{G}$. This optimal matrix $\boldsymbol{G}_{\text{mmse}}$ is the matrix that minimizes the variance of each entry of the error vector $\boldsymbol{s} - \hat{\boldsymbol{s}}$.

From (6.18), each error $s_i - \hat{s}_i$ has a variance equal to the (i, i) entry of $\text{MSE}(\boldsymbol{F}, \boldsymbol{G})$

$$E\{|s_i - \hat{s}_i|^2\} = \boldsymbol{g}_i(\boldsymbol{H}\boldsymbol{F}\boldsymbol{F}^H\boldsymbol{H}^H + \boldsymbol{R}_{vv})\boldsymbol{g}_i^H + 1 - \boldsymbol{g}_i\boldsymbol{H}\boldsymbol{f}_i - \boldsymbol{f}_i^H\boldsymbol{H}^H\boldsymbol{g}_i^H, \quad (6.20)$$

where we have indicated with $\boldsymbol{g}_i$ the ith row of $\boldsymbol{G}$ and with $\boldsymbol{f}_i$ the ith column of $\boldsymbol{F}$. This function is a convex function of $\boldsymbol{g}_i$ and thus its minimum value can be obtained by equating the gradient of $\text{MSE}(\boldsymbol{F}, \boldsymbol{G})$ with respect to $\boldsymbol{g}_i^H$ to $\boldsymbol{0}$. The result is

$$\boldsymbol{g}_i^{\text{mmse}} = \boldsymbol{f}_i^H\boldsymbol{H}^H(\boldsymbol{H}\boldsymbol{F}\boldsymbol{F}^H\boldsymbol{H}^H + \boldsymbol{R}_{vv})^{-1}, \; i = 1, \ldots, N. \quad (6.21)$$

Stacking all these N row vectors one above the other, we obtain the MMSE solution

$$\begin{aligned} \boldsymbol{G}_{\text{mmse}} &= \boldsymbol{F}^H\boldsymbol{H}^H(\boldsymbol{H}\boldsymbol{F}\boldsymbol{F}^H\boldsymbol{H}^H + \boldsymbol{R}_{vv})^{-1} & (6.22)\\ &= (\boldsymbol{I} + \boldsymbol{F}^H\boldsymbol{H}^H\boldsymbol{R}_{vv}^{-1}\boldsymbol{H}\boldsymbol{F})^{-1}\boldsymbol{F}^H\boldsymbol{H}^H\boldsymbol{R}_{vv}^{-1}, & (6.23) \end{aligned}$$

where in the second equality, we have exploited the property (6.14). We prove now that any other choice of $\boldsymbol{G}$ leads to an MSE matrix $\text{MSE}(\boldsymbol{F}, \boldsymbol{G})$ such that the difference matrix $\text{MSE}(\boldsymbol{F}, \boldsymbol{G}) - \text{MSE}(\boldsymbol{F}, \boldsymbol{G}_{\text{mmse}})$ is positive semidefinite. In fact, setting $\boldsymbol{A} = (\boldsymbol{H}\boldsymbol{F}\boldsymbol{F}^H\boldsymbol{H}^H + \boldsymbol{R}_{vv})$ for simplicity of notation, so that $\boldsymbol{G}_{\text{mmse}} = \boldsymbol{F}^H\boldsymbol{H}^H\boldsymbol{A}^{-1}$, we may verify that

$$\text{MSE}(\boldsymbol{F}, \boldsymbol{G}) - \text{MSE}(\boldsymbol{F}, \boldsymbol{G}_{\text{mmse}}) = (\boldsymbol{G} - \boldsymbol{F}^H\boldsymbol{H}^H\boldsymbol{A}^{-1})\boldsymbol{A}(\boldsymbol{G}^H - \boldsymbol{A}^{-1}\boldsymbol{H}\boldsymbol{F}). \quad (6.24)$$

The matrix on the right-hand side of (6.24) is certainly a positive semidefinite matrix, as the matrix $\boldsymbol{A}$ is positive semidefinite. It is easy to check that the minimum value of (6.24) is reached when $\boldsymbol{G} = \boldsymbol{G}_{\text{mmse}} = \boldsymbol{F}^H\boldsymbol{H}^H\boldsymbol{A}^{-1}$.

Using (6.23), we may write the MSE matrix as a function of $\boldsymbol{F}$ only, that is,

$$\begin{aligned} \text{MSE}(\boldsymbol{F}) &:= \text{MSE}(\boldsymbol{F}, \boldsymbol{G}_{\text{mmse}})\\ &= \boldsymbol{I} - \boldsymbol{F}^H\boldsymbol{H}^H(\boldsymbol{H}\boldsymbol{F}\boldsymbol{F}^H\boldsymbol{H}^H + \boldsymbol{R}_{vv})^{-1}\boldsymbol{H}\boldsymbol{F}. & (6.25) \end{aligned}$$

Using the matrix inversion lemma (6.11), we can also rewrite this expression as

$$\text{MSE}(\boldsymbol{F}) = (\boldsymbol{I} + \boldsymbol{F}^H\boldsymbol{H}^H\boldsymbol{R}_{vv}^{-1}\boldsymbol{H}\boldsymbol{F})^{-1} = (\boldsymbol{I} + \boldsymbol{F}^H\boldsymbol{R}_H\boldsymbol{F})^{-1}, \quad (6.26)$$

having set, in the second equality $\boldsymbol{R}_H := \boldsymbol{H}^H \boldsymbol{R}_{vv}^{-1} \boldsymbol{H}$.

Many criteria for deriving the optimal coding strategy assume a cost function which depends only on the entries on the main diagonal of MSE$(\boldsymbol{F})$, containing the mean square errors between the symbols and their estimates. In general, the optimization involves the minimization (or maximization) of a cost function, let us call it $f_0(\boldsymbol{x})$, which may represent for example average BER, MSE, or SNR. At the same time, the transmitter must satisfy some constraints. One of the most common is on the maximum average transmit power, stating that $\boldsymbol{F}$ must obey the following inequality

$$\text{tr}(\boldsymbol{F}\boldsymbol{F}^H) \leq p_T. \tag{6.27}$$

Denoting by $\boldsymbol{d}(\boldsymbol{A})$ the main diagonal of matrix $\boldsymbol{A}$, many optimization criteria can be formulated as follows

$$\begin{array}{rll} \boldsymbol{F} = & \text{argmin} & f_0\{\boldsymbol{d}[\text{MSE}(\boldsymbol{F})]\}, \\ & \text{subject to} & tr(\boldsymbol{F}\boldsymbol{F}^H) \leq p_T. \end{array} \tag{6.28}$$

A very elegant and general solution of this optimization problem was given by Palomar et al. in [1, 2], where it was shown how, using majorization theory, the problem can be scalarized and then simplified considerably. Now we recall the main result of [1, 2].

We consider separately the cases where the cost function is Schur-concave or Schur-convex. Here we still keep the formulation as general as possible so as to give the reader the basic notions about a powerful tool. Then, in the ensuing sections, we will show examples of applications of Schur-convexity and concavity to problems of interest for wireless communications.

Schur-concave cost functions

If the objective function $f_0(\boldsymbol{x})$ is Schur-concave, combining (6.7) and (6.9), we may say that

$$f_0\{\boldsymbol{\lambda}[\text{MSE}(\boldsymbol{F})]\} \leq f_0\{\boldsymbol{d}[\text{MSE}(\boldsymbol{F})]\}, \tag{6.29}$$

where $\boldsymbol{\lambda}(\text{MSE}(\boldsymbol{F}))$ is the vector containing the eigenvalues of MSE$(\boldsymbol{F})$, in decreasing order, whereas $\boldsymbol{d}(\text{MSE}(\boldsymbol{F}))$ is the main diagonal of MSE$(\boldsymbol{F})$. We assume, without any loss of generality, that the elements of $\boldsymbol{d}(\text{MSE}(\boldsymbol{F}))$ are in decreasing order. Then, from (6.29), it turns out that the minimum of $f_0\{\boldsymbol{d}[\text{MSE}(\boldsymbol{F})]\}$ is reached when MSE$(\boldsymbol{F})$ is diagonal with diagonal elements in decreasing order, so that $\boldsymbol{\lambda}[\text{MSE}(\boldsymbol{F})] = \boldsymbol{d}[\text{MSE}(\boldsymbol{F})]$. This means that the optimal coding matrix $\boldsymbol{F}$ must be such that MSE$(\boldsymbol{F}) = (\boldsymbol{I} + \boldsymbol{F}^H \boldsymbol{R}_H \boldsymbol{F})^{-1}$ be a diagonal matrix, with diagonal entries in decreasing order. Equivalently, $\boldsymbol{F}^H \boldsymbol{R}_H \boldsymbol{F}$ must be a diagonal matrix, with diagonal entries in increasing order.

To solve this problem, it is useful to introduce the diagonalization of $\boldsymbol{R}_H := \boldsymbol{H}^H \boldsymbol{R}_{vv}^{-1} \boldsymbol{H}$, that is,

$$\boldsymbol{R}_H = \widetilde{\boldsymbol{V}} \widetilde{\boldsymbol{\Lambda}} \widetilde{\boldsymbol{V}}^H, \tag{6.30}$$

where $\widetilde{\boldsymbol{V}}$ is unitary and $\widetilde{\boldsymbol{\Lambda}}$ is diagonal with diagonal elements in increasing order. The matrix $\boldsymbol{R}_H$ is $N \times N$ and its rank, denoted by $r = \text{rank}(\boldsymbol{R}_H)$, is at most equal to N. At this stage, we do not impose any constraint on the number M of transmitted symbols in each block, and for the ensuing derivations, we introduce the parameter $n = \min(M, r)$. The most general structure of the eigendecomposition (6.30) is then

$$\boldsymbol{R}_H = (\boldsymbol{V}_0 \;\; \boldsymbol{V}_1 \;\; \boldsymbol{V}) \begin{pmatrix} \boldsymbol{0} & \boldsymbol{0} & \boldsymbol{0} \\ \boldsymbol{0} & \boldsymbol{\Lambda}_1 & \boldsymbol{0} \\ \boldsymbol{0} & \boldsymbol{0} & \boldsymbol{\Lambda} \end{pmatrix} \begin{pmatrix} \boldsymbol{V}_0^H \\ \boldsymbol{V}_1^H \\ \boldsymbol{V}^H \end{pmatrix}, \tag{6.31}$$

where $\boldsymbol{V}$ is $N \times n$, $\boldsymbol{V}_1$ is $N \times (r-n)$, $\boldsymbol{\Lambda}$ is diagonal $n \times n$ and its diagonal elements are in increasing order, and the same holds for $\boldsymbol{\Lambda}_1$, which is $(r-n) \times (r-n)$. Note that $\boldsymbol{V}^H \boldsymbol{V} = \boldsymbol{I}$, and $\boldsymbol{V}_k^H \boldsymbol{V} = \boldsymbol{0}$, for $k = 0, 1$.

Using (6.31), MSE($\boldsymbol{F}$) is diagonalized by using an $\boldsymbol{F}$ with the following structure [2]

$$\boldsymbol{F} = \boldsymbol{V} \bar{\boldsymbol{\Phi}}, \tag{6.32}$$

where $\boldsymbol{V}$ is given in (6.31), whereas $\bar{\boldsymbol{\Phi}}$ is an $n \times M$ matrix whose structure is $\bar{\boldsymbol{\Phi}} = [\boldsymbol{0}, \boldsymbol{\Phi}]$, where $\boldsymbol{\Phi}$ is an $n \times n$ diagonal matrix. The presence of the null block in $\bar{\boldsymbol{\Phi}}$ occurs only when we try to transmit a number of symbols M greater than the number of modes n of the channel i.e., when $M > \text{rank}(\boldsymbol{R}_H)$. When this occurs $(M - r)$ symbols are lost due to the multiplication by zero in $\bar{\boldsymbol{\Phi}}$. Since this is a condition that should be avoided, in the following we will assume $M \leq \text{rank}(\boldsymbol{R}_H)$, that is $n = \min(M, \text{rank}(\boldsymbol{R}_H)) = M$.

We may now verify that the matrix MSE($\boldsymbol{F}$) is diagonal. In fact, combining (6.30) and (6.32), MSE($\boldsymbol{F}$) becomes

$$\text{MSE}(\boldsymbol{F}) = (\boldsymbol{I} + \boldsymbol{F}^H \boldsymbol{R}_H \boldsymbol{F})^{-1} = (\boldsymbol{I} + |\boldsymbol{\Phi}|^2 \boldsymbol{\Lambda})^{-1}, \tag{6.33}$$

where $|\boldsymbol{\Phi}|^2 = \boldsymbol{\Phi}\boldsymbol{\Phi}^*$.

Remark R.1: With $\boldsymbol{F}$ having the structure in (6.32), the channel composed by the cascade of $\boldsymbol{F}$, $\boldsymbol{H}$, and $\boldsymbol{G}$ is equivalent to the *parallel of n flat-fading* channels. In particular, using (6.23), we have[2]

$$\begin{aligned} G_{\text{mmse}} H F &= (I + F^H R_H F)^{-1} F^H R_H F \\ &= |\Phi|^2 \Lambda (I + |\Phi|^2 \Lambda)^{-1}. \end{aligned} \tag{6.34}$$

This implies that the I/O relationship (6.15) reduces to a set of scalar equations of the form[3]

$$\hat{s}_i = \frac{\Phi_i^2 \lambda_i}{1 + \Phi_i^2 \lambda_i} s_i + w_i, \quad i = 1, \ldots, n. \tag{6.35}$$

Remark R.2: It is important to point out that the entries w_i of the final output noise $w := G_{\text{mmse}} v$ are uncorrelated. In fact, the output noise vector has a covariance matrix

$$\begin{aligned} R_{ww} &= (I + F^H H^H R_{vv}^{-1} H F)^{-1} F^H H^H R_{vv}^{-1} H F (I + F^H H^H R_{vv}^{-1} H F)^{-1} \\ &= (I + |\Phi|^2 \Lambda)^{-2} |\Phi|^2 \Lambda, \end{aligned} \tag{6.36}$$

which is diagonal. This means that, if the noise is Gaussian, the components w_i in (6.35) are statistically independent.

In summary, if the cost function is Schur-concave, the overall channel of Figure 6.2 becomes equivalent to the set of parallel *independent* channels, where each subchannel has a flat fading coefficient

$$\alpha_i = \frac{\Phi_i^2 \lambda_i}{1 + \Phi_i^2 \lambda_i} \tag{6.37}$$

that is strictly less than one, and the noise on the ith subchannel has variance

$$\sigma_{w_i}^2 = \frac{\Phi_i^2 \lambda_i}{(1 + \Phi_i^2 \lambda_i)^2}. \tag{6.38}$$

The equivalent scheme is reported in Figure6.3.

The SNR on the ith subchannel is

$$\text{SNR}_i = \Phi_i^2 \lambda_i. \tag{6.39}$$

So far, we have not specified the value assumed by the diagonal entries of Φ yet. These values depend on the optimization criterion. We will see some examples in the next section, after having seen what happens when the optimization problem is Schur-convex.

2 The ordering in the matrix product in (6.34) is irrelevant, since all matrices are diagonal.

3 We may assume without any loss of generality that the coefficients Φ_i are real and nonnegative, since any phase shift can be incorporated in the columns of V.

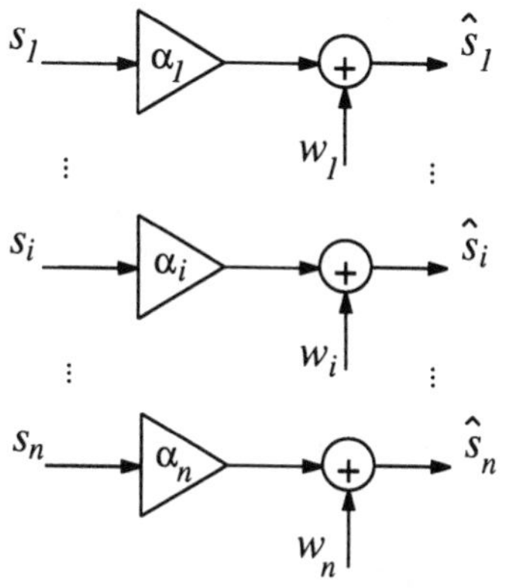

Figure 6.3 Optimal equivalent channel for Schur-concave cost functions.

Schur-convex cost functions

Alternatively, if the cost function $f_0(\boldsymbol{x})$ is Schur-convex, using (6.6) and (6.8), we have

$$f_0(\boldsymbol{m}_d) \leq f_0\{\boldsymbol{d}[\mathrm{MSE}(\boldsymbol{F})]\}, \tag{6.40}$$

where $\boldsymbol{m}_d$ has as elements the average value of $\boldsymbol{d}[\mathrm{MSE}(\boldsymbol{F})]$. Hence, the minimum is reached if we design $\boldsymbol{F}$ so that the matrix $\mathrm{MSE}(\boldsymbol{F})$ has a constant main diagonal, with elements equal to the mean value of $\boldsymbol{d}[\mathrm{MSE}(\boldsymbol{F})]$. This condition can be reached, for example, by choosing $\boldsymbol{F}$ as [2]

$$\boldsymbol{F} = \boldsymbol{V}\bar{\boldsymbol{\Phi}}\boldsymbol{\Theta}^H, \tag{6.41}$$

where $\boldsymbol{V}$ and $\bar{\boldsymbol{\Phi}}$ are as in (6.32), whereas $\boldsymbol{\Theta}$ is an $M \times M$ unitary (rotation) matrix such that $(\boldsymbol{I} + \boldsymbol{F}^H\boldsymbol{R}_H\boldsymbol{F})^{-1}$ has identical diagonal elements. To achieve this condition, it is sufficient to use any unitary matrix that satisfies $|\Theta(i,k)| = |\Theta(i,l)|, \forall i,k,l$. For example, we can use $\boldsymbol{\Theta} = \boldsymbol{W}$, where $\boldsymbol{W}$ is the normalized DFT matrix, with coefficients

$$W(k,l) = \frac{1}{\sqrt{M}}\, e^{j2\pi kl/M}, \tag{6.42}$$

or, if M is an integer multiple of 4, we can take $\boldsymbol{\Theta}$ as the Hadamard matrix, normalized by $\sqrt{M}$.

Using the structure (6.41) in (6.23), the overall channel becomes, equivalently

$$\boldsymbol{G}_{\mathrm{mmse}}\boldsymbol{H}\boldsymbol{F} = \boldsymbol{\Theta}(\boldsymbol{I} + |\boldsymbol{\Phi}|^2\boldsymbol{\Lambda})^{-1}|\boldsymbol{\Phi}|^2\boldsymbol{\Lambda}\boldsymbol{\Theta}^H. \tag{6.43}$$

The mean square error matrix becomes

$$\text{MSE}(\boldsymbol{F}) = (\boldsymbol{I} + \boldsymbol{\Theta}|\boldsymbol{\Phi}|^2\boldsymbol{\Lambda}\boldsymbol{\Theta}^H)^{-1}. \tag{6.44}$$

The effect of the rotation matrix $\boldsymbol{\Theta}$ is to make the diagonal of $\boldsymbol{\Theta}|\boldsymbol{\Phi}|^2\boldsymbol{\Lambda}\boldsymbol{\Theta}^H$ (and thus of $\text{MSE}(\boldsymbol{F})$) constant. However, differently from the Schur-concave case, $\text{MSE}(\boldsymbol{F})$ is not diagonal, in general.

Remark R.3: In case of Schur-convex functions, the optimal solution gives rise to the equivalent structure sketched in Figure 6.4. We can recognize an inner structure

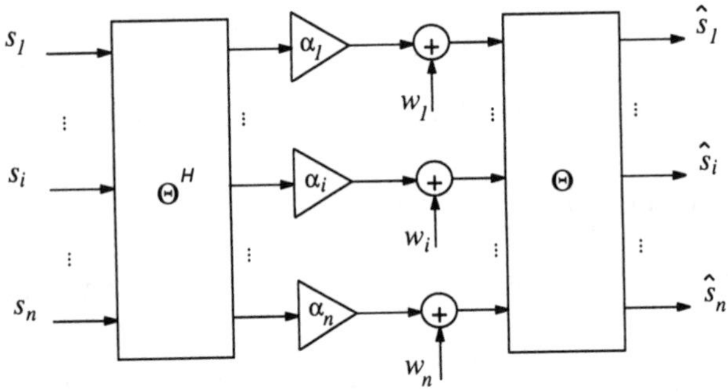

Figure 6.4 Optimal equivalent channel for Schur-convex cost functions.

composed of parallel channels, which is perfectly equivalent to the solution found for Schur-concave problems, placed between a pre-rotation matrix $\boldsymbol{\Theta}^H$ and a post-rotation matrix $\boldsymbol{\Theta}$.

Remark R.4: With Schur-convex functions, differently from Schur-concave functions, the overall channel is not diagonal, in general. This implies that, whereas the Schur-concave case system is ISI-free, with Schur-convex functions, the system is not immune of ISI.

In the following, we consider different optimality criteria and derive the solutions exploiting the general results shown so far. In all cases, we assume an average power constraint at the transmitter.

6.3.1 Minimum Average MSE

The minimum average MSE solution was derived, for example, in [5, 6] and we review it here under the more general perspective offered by the Schur-concavity property. The average MSE is, by definition,

$$\text{MSE}_{ave} := \frac{1}{n}\sum_{i=1}^{n} \text{MSE}(\boldsymbol{F})_{ii}. \tag{6.45}$$

This function is both Schur-convex and Schur-concave. Using, for example, the Schur-concave formulation, the average MSE is[4]

$$\text{MSE}_{ave} := \frac{1}{n}\sum_{i=1}^{n} \frac{1}{1+\Phi_i^2\lambda_i}. \tag{6.46}$$

Imposing the power constraint $\sum_{i=1}^{n}\Phi_i^2 = p_T$, we can solve this optimization problem using the Lagrange multipliers' method. The function to be minimized is

$$\mathcal{J} := \frac{1}{n}\sum_{i=1}^{n} \frac{1}{1+\Phi_i^2\lambda_i} + \mu\left(\sum_{i=1}^{n}\Phi_i^2 - p_T\right), \tag{6.47}$$

where μ is the Lagrange's multiplier. Equating the derivatives of $\mathcal{J}$ with respect to Φ_k to zero, we obtain either $\Phi_k = 0$ or

$$\Phi_k^2 = \left(\frac{1}{\sqrt{\mu n}}\frac{1}{\sqrt{\lambda_k}} - \frac{1}{\lambda_k}\right). \tag{6.48}$$

Since Φ_k^2 cannot be negative, Φ_k^2 can be rewritten in compact form as

$$\Phi_k^2 = \left(\frac{1}{\sqrt{\mu n}}\frac{1}{\sqrt{\lambda_k}} - \frac{1}{\lambda_k}\right)^+, \tag{6.49}$$

where $(x)^+ = \max(0, x)$. The coefficient μ is chosen in order to satisfy the power constraint $\sum_{i=1}^{n}\Phi_i^2 = p_T$. Some channels may therefore be discarded if their coefficient Φ_k is equal to zero. The procedure to decide upon which subchannels are to be discarded is the following. At the beginning, we sort the eigenvalues λ_k in decreasing order (i.e., $\lambda_j \geq \lambda_k$ if $j < k$); we set the number m of effectively used subchannels equal to n and compute μ in order to impose the power constraint, as

$$\mu = \frac{1}{n}\left(\frac{\sum_{k=1}^{m}\frac{1}{\sqrt{\lambda_k}}}{p_T + \sum_{k=1}^{m}\frac{1}{\lambda_k}}\right)^2. \tag{6.50}$$

Using this value of μ we compute all the arguments of $(\cdot)^+$ in (6.49). If the results are all nonnegative, the algorithm stops. If the some values are negative, we set $m = m - 1$ and we iterate the same procedure, until we find that all arguments are nonnegative.

4 We assume Φ_i real (see note 3).

6.3.2 Minimum Average MSE with ZF Constraint

If, besides the power constraint, we impose also that the overall channel respects the ZF constraint i.e., $\boldsymbol{GHF} = \boldsymbol{I}$, the MMSE solution differs from (6.49). The optimal solution was pointed out in [5]. In this case, the matrix $\boldsymbol{G}$ is related to $\boldsymbol{F}$ by the ZF condition, so that

$$\boldsymbol{G} = (\boldsymbol{HF})^{\dagger} \tag{6.51}$$

and the optimal $\boldsymbol{F}$ is

$$\boldsymbol{F} = \boldsymbol{V\Phi U}, \tag{6.52}$$

where $\boldsymbol{V}$ is as in (6.32), $\boldsymbol{U}$ is an arbitrary unitary matrix, and

$$\Phi_i^2 = \sqrt{\frac{p_T}{\mathrm{tr}(\boldsymbol{\Lambda}^{-1})}\frac{1}{\sqrt{\lambda_i}}}. \tag{6.53}$$

Although immune of ISI, this solution is very critical, as it assigns most of the power to the most faded channels, and thus it is not practically appealing.

6.3.3 Maximum Information Rate

The mutual information of the channel reported in Figure 6.2 is [7]

$$I(\hat{\boldsymbol{s}}\,;\boldsymbol{s}) = H(\hat{\boldsymbol{s}}) - H(\hat{\boldsymbol{s}}/\boldsymbol{s}) \tag{6.54}$$

If the additive noise is Gaussian and we impose an average power constraint on the input distribution, the mutual information is maximum when the input vector $\boldsymbol{s}$ is circularly symmetric complex Gaussian. In such a case, the entropies are [7][5]

$$H(\hat{\boldsymbol{s}}) = \log_2 |\,\pi e\,(\boldsymbol{GHFF}^H\boldsymbol{H}^H\boldsymbol{G}^H + \boldsymbol{GR}_{vv}\boldsymbol{G}^H)\,| \tag{6.55}$$

$$H(\hat{\boldsymbol{s}}/\boldsymbol{s}) = \log_2 |\,\pi e\,\boldsymbol{GR}_{vv}\boldsymbol{G}^H\,|. \tag{6.56}$$

Inserting (6.55) and (6.56) in (6.54), we obtain

$$I(\hat{\boldsymbol{s}}\,;\boldsymbol{s}) = \log_2 |\boldsymbol{I} + (\boldsymbol{GR}_{vv}\boldsymbol{G}^H)^{-1}\boldsymbol{GHFF}^H\boldsymbol{H}^H\boldsymbol{G}^H||. \tag{6.57}$$

Using the identity $|\boldsymbol{I}+\boldsymbol{AB}| = |\boldsymbol{I}+\boldsymbol{BA}|$, we can rewrite (6.57) as [6]

$$\begin{aligned} I(\hat{\boldsymbol{s}}\,;\boldsymbol{s}) &= \log_2 |\boldsymbol{I} + \boldsymbol{F}^H\boldsymbol{H}^H\boldsymbol{G}^H(\boldsymbol{GR}_{vv}\boldsymbol{G}^H)^{-1}\boldsymbol{GHF}\,| \\ &= \log_2 |\boldsymbol{I} + \boldsymbol{F}^H\boldsymbol{H}^H\boldsymbol{R}_{vv}^{-1/2}\boldsymbol{R}_{vv}^{1/2}\boldsymbol{G}^H(\boldsymbol{GR}_{vv}\boldsymbol{G}^H)^{-1}\boldsymbol{GR}_{vv}^{1/2}\boldsymbol{R}_{vv}^{-1/2}\boldsymbol{HF}\,| \\ &= \log_2 |\boldsymbol{I} + \boldsymbol{F}^H\boldsymbol{H}^H\boldsymbol{R}_{vv}^{-1/2}\boldsymbol{\Pi}\boldsymbol{R}_{vv}^{-1/2}\boldsymbol{HF}|, \end{aligned} \tag{6.58}$$

5 In the ensuing derivations, the determinant $|\boldsymbol{A}|$ of a matrix $\boldsymbol{A}$ is to be intended as the *effective* determinant, i.e., the product of the nonnull eigenvalues of $\boldsymbol{A}$.

where we have set $\boldsymbol{\Pi} := \boldsymbol{R}_{vv}^{1/2}\boldsymbol{G}^H(\boldsymbol{G}\boldsymbol{R}_{vv}\boldsymbol{G}^H)^{-1}\boldsymbol{G}\boldsymbol{R}_{vv}^{1/2}$. The matrix $\boldsymbol{\Pi}$ is the orthogonal projector onto the range space of $\boldsymbol{R}_{vv}^{1/2}\boldsymbol{G}^H$. Since $\boldsymbol{\Pi}$ is an orthogonal projector, we have $\boldsymbol{\Pi} \leq \boldsymbol{I}$, in the sense that the difference matrix $\boldsymbol{\Pi} - \boldsymbol{I}$ is negative semidefinite. Hence, using (6.12), we have

$$I(\hat{\boldsymbol{s}}\,;\boldsymbol{s}) \leq \log_2|\boldsymbol{I} + \boldsymbol{F}^H\boldsymbol{H}^H\boldsymbol{R}_{vv}^{-1}\boldsymbol{H}\boldsymbol{F},| \qquad (6.59)$$

and the upper bound is reached if and only if $\boldsymbol{R}_{vv}^{1/2}\boldsymbol{G}^H = \boldsymbol{R}_{vv}^{-1/2}\boldsymbol{H}\boldsymbol{F}\boldsymbol{\Gamma}$, where $\boldsymbol{\Gamma}$ is any invertible square matrix. In fact, substituting this choice of $\boldsymbol{G}$ in (6.58), we obtain

$$I(\hat{\boldsymbol{s}}\,;\boldsymbol{s}) = log_2|\boldsymbol{I} + \boldsymbol{F}^H\boldsymbol{H}^H\boldsymbol{R}_{vv}^{-1}\boldsymbol{H}\boldsymbol{F}|. \qquad (6.60)$$

Furthermore, setting $\boldsymbol{\Gamma} = (\boldsymbol{I} + \boldsymbol{F}^H\boldsymbol{H}^H\boldsymbol{R}_{vv}^{-1}\boldsymbol{H}\boldsymbol{F})^{-1}$, we obtain $\boldsymbol{G} = \boldsymbol{G}_{\text{mmse}}$. This proves that there is no loss of optimality in maximizing the mutual information if we take $\boldsymbol{G} = \boldsymbol{G}_{\text{mmse}}$. In other words, the MMSE estimate is a *sufficient statistics* and then it preserves the mutual information.

Comparing the argument of (6.59) with (6.33), the maximum mutual information can be written as

$$I(\hat{\boldsymbol{s}}\,;\boldsymbol{s}) = -\log_2|\,\text{MSE}(\boldsymbol{F})\,|. \qquad (6.61)$$

Therefore, the maximization of the mutual information is equivalent to the minimization of the determinant of the mean square error matrix $\text{MSE}(\boldsymbol{F})$. Using either (6.33) or (6.44) (as the pre- and postmultiplication by a unitary matrix and its adjoint does not affect the determinant), $I(\hat{\boldsymbol{s}}\,;\boldsymbol{s})$ can be written as

$$I(\hat{\boldsymbol{s}}\,;\boldsymbol{s}) = \sum_{i=1}^{n} \log_2(1 + \Phi_i^2\lambda_i). \qquad (6.62)$$

Imposing the power constraint $\sum_{i=1}^{n}\Phi_i^2 = p_T$, we can use the Lagrange multipliers' approach and maximize the function

$$\mathcal{J} = \sum_{i=1}^{n} \log_2(1 + \Phi_i^2\lambda_i) - \mu\left(\sum_{i=1}^{n}\Phi_i^2 - p_T\right). \qquad (6.63)$$

Equating the partial derivatives of $\mathcal{J}$ with respect to Φ_k to zero, we obtain, similarly to (6.49),

$$\Phi_k^2 = \left(\frac{1}{\mu} - \frac{1}{\lambda_k}\right)^+, \qquad (6.64)$$

where μ is computed in order to satisfy the transmit power constraint.

To evaluate which channels are effectively used, i.e., which are the indices k over which $\Phi_k > 0$, we must use the following iterative algorithm.

The eigenvalues are sorted in decreasing order. Then, the algorithm proceeds as follows:

1. Set $m = n$*;*

2. Compute

$$\frac{1}{\mu(m)} = \frac{p_T}{m} + \frac{1}{m}\sum_{i=1}^{m}\frac{1}{\lambda_i}; \tag{6.65}$$

3. Compute

$$p_k = \frac{1}{\mu(m)} - \frac{1}{\lambda_k}, \ \text{for} k = 1, \ldots, m \tag{6.66}$$

4. If $p_m > 0$*, set* $\Phi_k^2 = p_k$, $k = 1, \ldots, m$ *and stop, otherwise set* $\Phi_m = 0$, $m = m - 1$*, and go to step 2.*

It is possible to prove that this algorithm always converges.

The solution (6.64) is the well known *water-filling* solution [7]. The name of the algorithm can be understood by thinking of the way water (power) distributes over a set of communicating vessels, where each vessel lies at a height $1/\lambda_k$ with respect to a common ground. The final level of the water, with respect to the ground, is $1/\mu$. If the amount of water (power) is sufficiently large, we will have, eventually, water in every vessel (we will transmit symbols over all subchannels); otherwise, we may have some empty vessels (where we do not transmit any power), corresponding to the largest $1/\lambda_k$ (most attenuated subchannels) or some partially filled vessels, corresponding to moderate $1/\lambda_k$ (where we transmit an amount of power equal to the corresponding water level).

Using the water-filling solution (6.64), and denoting by $\bar{N}$ the number of channels effectively used, the maximum mutual information is

$$I(\hat{\boldsymbol{s}}, \boldsymbol{s}) = \sum_{i=1}^{\bar{N}} \log_2(1 + \Phi_i^2\lambda_i) = \sum_{i=1}^{\bar{N}} \log_2(\lambda_i/\mu). \tag{6.67}$$

Each term in the summation represents the (maximum) number of bits that can be carried by the relative subchannel. The value has to be intended only as a bound, as it has been derived under the assumption of Gaussian-distributed symbols. In practice, we transmit symbols belonging to finite alphabets. Nevertheless, the above formulas provide a criterion for deciding the number of bits to be sent over each subchannel. From [8], the error probability for an M-QAM constellation is approximately

$$P_e \approx 2\,e^{-\frac{3}{2(M-1)}\mathrm{SNR}}. \tag{6.68}$$

Inverting such expression, we can find M as a function of the desired P_e:

$$M \approx 1 + \frac{3\,\mathrm{SNR}}{2\,\log(2/P_e)}. \tag{6.69}$$

The number of bits carried by a QAM symbol, guaranteeing (approximately) a desired P_e, for a given SNR is then[6]

$$n_b := \lfloor \log_2(M) \rfloor \approx \left\lfloor \log_2\left(1 + \frac{\mathrm{SNR}}{\Gamma}\right)\right\rfloor, \tag{6.70}$$

where $\Gamma := 2\,\log(2/P_e)/3$ is known as the SNR *gap* or *margin*. There are two kinds of approximations here: the gap approximation and the rounding operation in (6.69).

Comparing (6.70) with (6.67), we realize that the number of bits to be sent over each subchannel is given by

$$n_b(k) = \lfloor \log_2(M_k) \rfloor \approx \log_2\left(1 + \frac{\Phi_k^2 \lambda_k}{\Gamma}\right), \tag{6.71}$$

where the coefficient $\Gamma := 2\,\log(2/P_e)/3$ is introduced to guarantee the desired error probability P_e. It is important to remark that the gap is the same over all subchannels (at least within the approximations of (6.69) and (6.70)).

6.3.4 Minimum BER with ZF Constraint

In the previous section, we have seen how to maximize the bit rate, for a given power and a target bit error rate. In this section and in the next one, we will see how to minimize the BER, for a given power and a given transmission rate. In this section, we enforce a zero-forcing constraint, to eliminate ISI. In the next section, we will minimize the mean square error.

The problem of minimum BER precoding, with the ZF constraint, was solved by Ding et al., in [9]. We will now briefly recall the basic results of [9]. The symbol

6 $\lfloor x \rfloor$ denotes integer part of x.

constellation is assumed to be BPSK, but the same results hold true for QPSK. Denoting by P_{e_k} the bit error rate (BER) on the kth subchannel, the average BER is

$$P_e = \frac{1}{\bar{N}} \sum_{i=1}^{\bar{N}} P_{e_k} \tag{6.72}$$

where $\bar{N}$ denotes the number of subchannels where we allocate a non-null power. The BER on each subchannel assumes the form

$$P_{e_k} = \frac{1}{2} \operatorname{erfc}\left(\sqrt{\frac{\text{SNIR}_k}{2}}\right), \tag{6.73}$$

where SNIR_k is the signal-to-interference plus noise ratio over the kth subchannel. Enforcing the ZF constraint, there is no ISI, and thus SNIR_k coincides with the SNR on the kth subchannel, i.e.,

$$\text{SNR}_k = \frac{1}{\sigma_n^2 [\boldsymbol{G}\boldsymbol{G}^H]_{kk}}. \tag{6.74}$$

Hence, the function that we wish to minimize is

$$P_e = \frac{1}{2\bar{N}} \sum_{i=1}^{\bar{N}} \operatorname{erfc}\left(\frac{1}{\sqrt{2\sigma_n^2 [\boldsymbol{G}\boldsymbol{G}^H]_{kk}}}\right). \tag{6.75}$$

The key observation of [9] was that the function $f(x) := \operatorname{erfc}\left(1/\sqrt{2\sigma_n^2 x}\right)$ is convex when x is sufficiently small. More precisely, taking the second order derivative of $f(x) := \operatorname{erfc}\left(1/\sqrt{2\sigma_n^2 x}\right)$, we get

$$f''(x) = \frac{1}{\sqrt{2\pi\sigma_n^2}} x^{-5/2} e^{-\frac{1}{2\sigma_n^2 x}} \left(\frac{1}{2\sigma_n^2 x} - \frac{3}{2}\right). \tag{6.76}$$

From (6.76), the function $f(x)$ is convex if $x < 1/(3\sigma_n^2)$. If this condition holds true for all subchannels, we can apply Jensen's inequality to write

$$P_e = \frac{1}{2\bar{N}} \sum_{i=1}^{\bar{N}} \operatorname{erfc}\left(\frac{1}{\sqrt{2\sigma_n^2 [\boldsymbol{G}\boldsymbol{G}^H]_{ii}}}\right) \geq \frac{1}{2} \operatorname{erfc}\left(\frac{1}{\sqrt{2\sigma_n^2 tr[\boldsymbol{G}\boldsymbol{G}^H]/\bar{N}}}\right), \tag{6.77}$$

where the equality sign holds when all terms are equal to each other.

Imposing the ZF constraint $\boldsymbol{GHF} = \boldsymbol{I}$, $\boldsymbol{G}$ and $\boldsymbol{F}$ become related by the following expression

$$\boldsymbol{G} = (\boldsymbol{HF})^{\dagger}, \tag{6.78}$$

so that[7]

$$\boldsymbol{GG}^H = (\boldsymbol{F}^H\boldsymbol{H}^H\boldsymbol{HF})^{\dagger} = \boldsymbol{F}^{-1}(\boldsymbol{H}^H\boldsymbol{H})^{-1}\boldsymbol{F}^{-H}. \tag{6.79}$$

The optimization problem to solve is then the following

$$\begin{aligned} \boldsymbol{F} &= \text{argmin } \operatorname{tr}(\boldsymbol{F}^{-1}(\boldsymbol{H}^H\boldsymbol{H})^{-1}\boldsymbol{F}^{-H}) \\ &\quad \text{subject to } \operatorname{tr}(\boldsymbol{FF}^H) \le p_T \text{ and} \\ &\quad [\boldsymbol{F}^{-1}(\boldsymbol{H}^H\boldsymbol{H})^{-1}\boldsymbol{F}^{-H}]_{ii} \le \frac{1}{3\sigma_n^2}, \quad i = 1, \ldots, \bar{N}. \end{aligned} \tag{6.80}$$

This is not an easy problem to solve, because of the constraint (6.80). However, if (6.80) could be removed, the solution could be found in closed form and it would coincide with (6.52), where $\boldsymbol{V}$ is as in (6.32) and $\boldsymbol{\Phi}$ is given by (6.53). The unitary matrix $\boldsymbol{U}$ must be such to impose the same value of the SNR_k over all subchannels. This property may be achieved using a matrix $\boldsymbol{U}$ having constant modulus entries. Two possible choices for $\boldsymbol{U}$ are, for example, the DFT matrix or the Hadamard matrix (valid choice only for the case where M is an integer multiple of 4). To check optimality of (6.52) in the current case, we have to check if (6.80) holds true for all subchannels: If (6.80) holds true, the solution is minimum BER, otherwise we cannot guarantee the minimality of the BER.

In [9], the following two-step approach was proposed. The solution is initially found using (6.52) and (6.78). Then, the coefficients $[\boldsymbol{GG}^H]_{ii}$ are ordered in decreasing sense. If the minimum value of $[\boldsymbol{GG}^H]_{ii}$ satisfies the inequality

$$\min_i [\boldsymbol{GG}^H]_{ii} \le \frac{1}{3\sigma_n^2} \tag{6.81}$$

the solution is accepted, otherwise the subchannel with the minimum value of $[\boldsymbol{GG}^H]_{ii}$ is discarded and the associated power is redistributed over the remaining subchannels. The iterative process is repeated until condition (6.81) is satisfied.

6.3.5 Minimum BER with MMSE Receiver

The ZF constraint allows us to write the BER in closed form, as it removes ISI exactly. However, in practice, the ZF constraint could unnecessarily enhance the

7 In the second equality of (6.79), we assume that $\boldsymbol{F}$ and $\boldsymbol{H}^H\boldsymbol{H}$ are invertible.

noise. In [1], [2], Palomar et al. showed how to minimize the average BER, without imposing the ZF constraint. Removing the ZF constraint, it is more difficult to write down a simple closed form expression for the symbol error rate (SER), because of the ISI. Nevertheless, if the number of interfering symbols is sufficiently high, we can invoke the central limit theorem to obtain a simplified, although approximate, closed form for the SER. To be more specific, the I/O relationship (6.15) can be rewritten in order to evidence the intersymbol interference (ISI). In fact, if we introduce the diagonal matrix $\boldsymbol{D}$, whose main diagonal is equal to the main diagonal of the matrix $\boldsymbol{GHF}$, and the matrix $\boldsymbol{D}_{off}$, defined as $\boldsymbol{D}_{off} := \boldsymbol{GHF} - \boldsymbol{D}$, we can rewrite (6.15) as

$$\hat{\boldsymbol{s}} = \boldsymbol{D}\boldsymbol{s} + \boldsymbol{D}_{off}\boldsymbol{s} + \boldsymbol{G}\boldsymbol{v}. \tag{6.82}$$

On the right-hand side of this equation, the first term is the useful term, the second term is ISI and the third term is output noise. If the ISI is composed of the combination of a large number of symbols, invoking the central limit theorem we may approximate it as a complex Gaussian random variable. Under the limit of validity of such an approximation, the ISI can be assimilated to noise and the BER can be derived in closed form. Considering that the ith term on the diagonal of $\boldsymbol{D}$ is $d_{ii} = \boldsymbol{g}_i^T \boldsymbol{H} \boldsymbol{f}_i$, where $\boldsymbol{g}_i^T$ is the ith row of $\boldsymbol{G}$ and $\boldsymbol{f}_i$ is the ith column of $\boldsymbol{F}$, we can express the signal-plus-noise-and-interference ratio (SNIR$_i$) on the ith subchannel as

$$\text{SNIR}_i = \frac{E\{|d_{ii}s_i|^2\}}{E\{|(\boldsymbol{g}_i^T \boldsymbol{HF} - d_{ii}\boldsymbol{e}_i^T)\boldsymbol{s} + \boldsymbol{g}_i^T \boldsymbol{v}|^2\}}$$

having introduced the vector $\boldsymbol{e}_i$ having all entries equal to 0, except the ith one, equal to 1.

Introducing now the Hermitian positive definite matrix $\boldsymbol{R}_i := \boldsymbol{HFF}^H\boldsymbol{H}^H - \boldsymbol{H}\boldsymbol{f}_i\boldsymbol{f}_i^H\boldsymbol{H}^H + \boldsymbol{R}_{vv}$ we can write SNIR$_i$ as

$$\text{SNIR}_i = \frac{|\boldsymbol{g}_i^T \boldsymbol{H} \boldsymbol{f}_i|^2}{\boldsymbol{g}_i^T(\boldsymbol{HFF}^H\boldsymbol{H}^H - \boldsymbol{H}\boldsymbol{f}_i\boldsymbol{f}_i^H\boldsymbol{H}^H + \boldsymbol{R}_{vv})\boldsymbol{g}_i^*} = \frac{|\boldsymbol{g}_i^T \boldsymbol{H} \boldsymbol{f}_i|^2}{\boldsymbol{g}_i^T \boldsymbol{R}_i \boldsymbol{g}_i^*}. \tag{6.83}$$

To find the maximum value of SNIR$_i$ it is useful to define the scalar product between two vectors $\boldsymbol{x}$ and $\boldsymbol{y}$ as $(\boldsymbol{x}, \boldsymbol{y}) := \boldsymbol{x}^H \boldsymbol{R}_i \boldsymbol{y}$. With this definition of scalar product, using Schwartz's inequality, the ratio in (6.83) can be upper bounded as

$$\text{SNIR}_i = \frac{|(\boldsymbol{g}^*, \boldsymbol{R}_i^{-1}\boldsymbol{H}\boldsymbol{f}_i)|^2}{(\boldsymbol{g}_i^*, \boldsymbol{g}_i^*)} \leq \boldsymbol{f}_i^H \boldsymbol{H}^H \boldsymbol{R}_i^{-1} \boldsymbol{H} \boldsymbol{f}_i. \tag{6.84}$$

The upper bound is achieved when the vectors $\boldsymbol{g}_i^*$ and $\boldsymbol{R}_i^{-1}\boldsymbol{H}\boldsymbol{f}_i$ are parallel, i.e., $\boldsymbol{g}_i^* = \alpha \boldsymbol{R}_i^{-1}\boldsymbol{H}\boldsymbol{f}_i$, with $\alpha \in \mathbb{C}$. Let us now establish the correspondence between SNIR$_i$ and the mean square error on the ith subchannel, MSE$_i$. From (6.25), we can rewrite the ith row of $\mathrm{MSE}(\boldsymbol{F})$ as

$$\mathrm{MSE}_i(\boldsymbol{F}) = 1 - \boldsymbol{f}_i^H \boldsymbol{H}^H (\boldsymbol{H}\boldsymbol{F}\boldsymbol{F}^H\boldsymbol{H}^H + \boldsymbol{R}_{vv})^{-1} \boldsymbol{H}\boldsymbol{f}_i$$

$$= 1 - \boldsymbol{f}_i^H \boldsymbol{H}^H (\boldsymbol{H}\boldsymbol{F}\boldsymbol{F}^H\boldsymbol{H}^H - \boldsymbol{H}\boldsymbol{f}_i\boldsymbol{f}_i^H\boldsymbol{H}^H + \boldsymbol{R}_{vv} + \boldsymbol{H}\boldsymbol{f}_i\boldsymbol{f}_i^H\boldsymbol{H}^H)^{-1} \boldsymbol{H}\boldsymbol{f}_i \quad (6.85)$$

Using the matrix inversion lemma (6.11), setting $\boldsymbol{A} = 1, \boldsymbol{B} = \boldsymbol{f}_i^H \boldsymbol{H}^H, \boldsymbol{D} = \boldsymbol{H}\boldsymbol{f}_i$, and $\boldsymbol{C} = (\boldsymbol{H}\boldsymbol{F}\boldsymbol{F}^H\boldsymbol{H}^H - \boldsymbol{H}\boldsymbol{f}_i\boldsymbol{f}_i^H\boldsymbol{H}^H + \boldsymbol{R}_{vv})^{-1}$, we can rewrite $\mathrm{MSE}_i(\boldsymbol{F})$ as

$$\mathrm{MSE}_i(\boldsymbol{F}) = (1 + \boldsymbol{f}_i^H \boldsymbol{H}^H \boldsymbol{R}_i^{-1} \boldsymbol{H}\boldsymbol{f}_i)^{-1} = \frac{1}{1 + \mathrm{SNIR}_i}. \quad (6.86)$$

Inverting this equation, we are also able to express the maximum SNIR_i as a function of the mean square error

$$\mathrm{SNIR}_i = \frac{1}{\mathrm{MSE}_i} - 1. \quad (6.87)$$

Hence, from (6.87), it turns out that maximizing the SINR_i is equivalent to minimizing MSE_i.

We are now able to express the BER for the MMSE receiver. Within the limit of validity of the central limit theorem, we may use the same expression as in (6.73), substituting SNR_k with SNIR_k. In doing so, we get

$$P_{e_k} = \frac{1}{2}\operatorname{erfc}\left[\sqrt{\frac{1}{2}\left(\frac{1}{\mathrm{MSE}_k} - 1\right)}\right]. \quad (6.88)$$

The interesting property of (6.88) is that it is a convex function of MSE_k, for any value of MSE_k. This means that, regardless of the value of MSE_k (at least for BPSK or QPSK constellations), the individual error probabilities on each subchannel are convex functions and thus, the overall objective function, i.e., (6.72) with P_{e_k} as in (6.88), is a Schur-convex function of the MSE achieved on each subchannel. The solution yields the precoding matrix given by (6.41), where

$$\Phi_k^2 = \left(\frac{1}{\sqrt{\mu n}}\frac{1}{\sqrt{\lambda_k}} - \frac{1}{\lambda_k}\right)^+, \quad (6.89)$$

and $\boldsymbol{\Theta}$ is, similarly to $\boldsymbol{U}$ in the MinBER/ZF case, a unitary matrix with constant modulus entries. The decoding matrix $\boldsymbol{G}$ is the MMSE matrix $\boldsymbol{G}_{\mathrm{mmse}}$ given by (6.23).

6.4 SISO CHANNELS

The optimal coding and decoding matrices have been obtained using a completely general setting. The optimal solutions will be now particularized to specific operative conditions, in order to associate a physical meaning with the solutions found before.

6.4.1 Time-Invariant Channels

In case of SISO time-invariant channels, using block transmissions, with blocks of size N, and a CP of length at least equal to the channel order, the channel matrix assumes the structure of (3.17). Then, it admits an eigen-decomposition as in (3.18), with eigenvalues equal to the values of the channel transfer function. In case of white noise, we have

$$\lambda_k = |H(k)|^2 \tag{6.90}$$

where $H(k) = \sum_{l=0}^{L} h(l)e^{-j2\pi kl/N}$.

6.4.1.1 *Optimal transmit power spectral densities*

Equation (6.90) allows us to interpret the power allocation $\{\Phi_k^2\}_{k=0,\ldots,N-1}$ resulting from the optimization strategies studied before as a power distribution function of the discrete frequency k/N. In the limit as the block length increases, the discrete power allocation $\{\Phi_k^2\}_{k=0,\ldots,N-1}$ tends to a function $\mathcal{P}(f)$ of the continuous variable f that assumes the meaning of a frequency. Hence, using (6.90), we can write the asymptotic (as the blocklength goes to infinity) power spectral density $\mathcal{P}(f)$ of the transmitted signals, according to different optimization criteria as follows.

1. *MMSE with average power constraint*

   From (6.49), we get

$$\mathcal{P}_{\mathrm{MMSE}}(f) = \left[\frac{1}{\sqrt{\mu'}}\frac{1}{|H(f)|} - \frac{1}{|H(f)|^2}\right]^+, \tag{6.91}$$

   where $[x]^+ := \max(x, 0)$, and the coefficient μ' is such that the transmitted power is equal to p_T, i.e.,

$$\int_{-\infty}^{\infty} \mathcal{P}_{\mathrm{MMSE}}(f)\, df = p_T. \tag{6.92}$$

2. *Maximum information rate* (MIR) *with average power constraint*

   From (6.64), we obtain

$$\mathcal{P}_{\mathrm{MIR}}(f) = \left[\frac{1}{\mu'} - \frac{1}{|H(f)|^2}\right]^+, \tag{6.93}$$

   where the coefficient μ' is such that the transmitted power is equal to p_T, that is,

$$\int_{-\infty}^{\infty} \mathcal{P}_{\mathrm{MIR}}(f)\, df = p_T. \tag{6.94}$$

3. *Minimum BER with MMSE constraint* (MinBER/MMSE)

   From (6.89), we get

$$\mathcal{P}_{\mathrm{MinBER}}(f) = \left[\frac{1}{\sqrt{\mu'}}\frac{1}{|H(f)|} - \frac{1}{|H(f)|^2}\right]^+, \tag{6.95}$$

   where μ' is such that the transmitted power is equal to p_T, that is,

$$\int_{-\infty}^{\infty} \mathcal{P}_{\mathrm{MinBER}}(f)\, df = p_T. \tag{6.96}$$

In summary, the available power is allocated only over the frequencies where $\mathcal{P}(f) > 0$; the power distribution varies as a function of frequency according to the adopted optimization criterion.

6.4.1.2 *Performance comparison*

It is interesting to compare the performance of the alternative coding strategies studied so far. As an example, in Figures 6.5 and 6.6 we report, respectively, the average BER and mutual information of three different coding strategies seen before, namely: 1) minimum BER with MMSE receiver and average power constraint (MinBER/MMSE); 2) minimum BER with ZF and average power constraints(MinBER/ZF); and 3) maximum mutual information with MMSE receiver and average power constraint (MaxINF/MMSE). The BER and mutual information shown in these figures have been averaged over 200 independent channel realizations. The channels are all FIR, of order $L = 5$, with i.i.d. complex Gaussian random variables with zero mean and unit variance. The BER has been computed for a BPSK constellation. The block length is $N = 64$.

We can observe a good agreement between theory and simulation for all methods. There is a slight mismatch (less than one dB, at high SNR) in the MinBER/MMSE case, due to the approximations used to compute the BER. Comparing the different methods, we see that the MinBER/MMSE method yields a lower BER than the MinBER/ZF method. In fact, the ZF constraint is too restrictive and it can enhance the noise excessively. Conversely, since the MinBER/MMSE method takes into account both noise and ISI, it can never enhance excessively any of these contributions. From Figure 6.5 we also notice that the MaxINF method exhibits a considerable loss, in terms of BER, especially at high SNR. In particular, the average BER corresponding to the MaxINF/MMSE transceiver behaves as $1/\text{SNR}$, at high SNR. This means that the MaxINF/MMSE transceiver does not have any diversity gain.

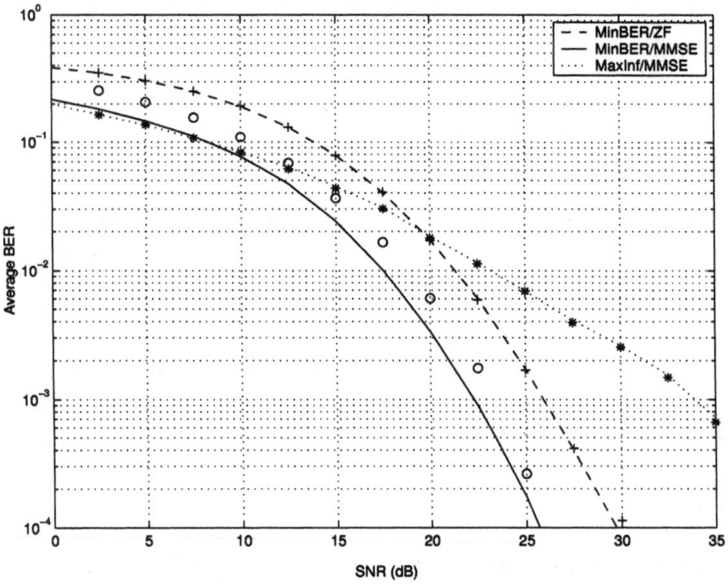

Figure 6.5 Average BER versus σ_s^2/σ_n^2: a) MinBER/MMSE - theory (solid line) and simulations ('o'); b) MinBER/ZF - theory (dashed line) and simulations ('+'); c) MaxINF/MMSE - theory (dotted line) and simulations ('*').

This happens because of the diagonal structure of the equivalent channel induced by MaxINF/MMSE coding. Moreover, at high SNR, water-filling tends to allocate a constant power over all the subchannels. This implies that each symbol is transmitted, equivalently, over a flat fading (Rayleigh) channel. As a consequence, there is no diversity gain. Conversely, the MinBER transceivers, both ZF and MMSE, thanks to the pre- and post-rotation matrices, achieve some diversity gain, as testified by the slopes of the average BER curves shown in Figure 6.5. However, the MaxINF method yields the highest average information rate, as shown in Figure 6.6.

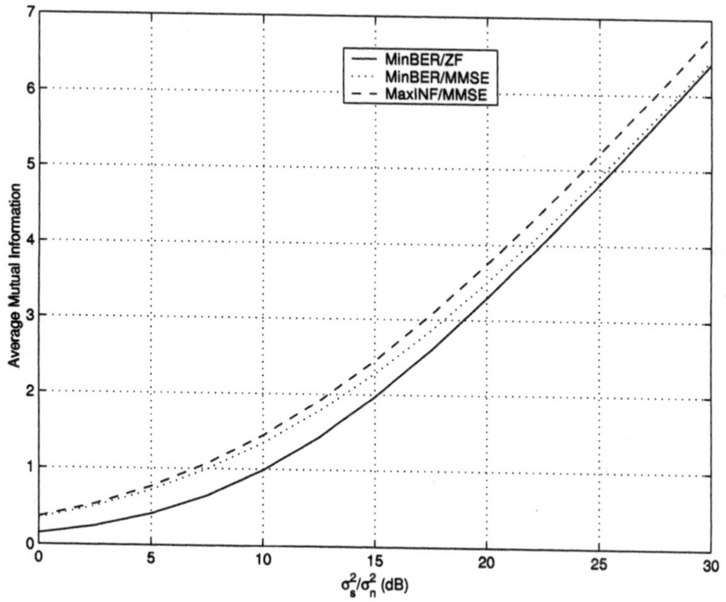

Figure 6.6 Average mutual information versus σ_s^2/σ_n^2 - a) Minimum BER with MMSE receiver and average power constraint (solid line); b) Minimum BER with ZF and average power constraints (dashed line); c) Maximum mutual information with MMSE receiver and average power constraint (dotted line).

6.4.2 Time-Varying Channels

If the channel is time-varying, it is more complicated, in general, to associate a physical meaning with the optimal coding strategy. Nevertheless, since practically all communication channels are underspread, we can exploit the approximate relationship (2.111) between the channel eigenvalues and the channel time-varying transfer function. Using (2.111), we can repeat the same derivations as before writing an (approximate) expression of the time-varying power spectral density of the transmitted signal, as a function of *both* time and frequency. As a result, we have the following behaviors.

1. *MMSE with average power constraint*

Combining (2.109) with (6.48), we obtain[8]

$$\mathcal{P}_{\rm MMSE}(t,f) = \left[\frac{1}{\sqrt{\mu'}}\frac{1}{|H(t,f)|} - \frac{1}{|H(t,f)|^2}\right]^+, \tag{6.97}$$

where μ' is such that the transmitted power is equal to p_T, that is,

$$\int_{-\infty}^{\infty}\int_{-\infty}^{\infty} \mathcal{P}_{\rm MMSE}(t,f)\,dt\,df = p_T. \tag{6.98}$$

2. *Maximum information rate with average power constraint*

Combining (2.109) with (6.64), we get

$$\mathcal{P}_{\rm MIR}(t,f) = \left[\frac{1}{\mu'} - \frac{1}{|H(t,f)|^2}\right]^+, \tag{6.99}$$

where μ' is such that

$$\int_{-\infty}^{\infty}\int_{-\infty}^{\infty} \mathcal{P}_{\rm MIR}(t,f)\,dt\,df = p_T. \tag{6.100}$$

3. *Minimum BER with MMSE constraint*

Combining (2.109) with (6.89), we obtain

$$\mathcal{P}_{\rm MinBER}(t,f) = \left[\frac{1}{\sqrt{\mu'}}\frac{1}{|H(t,f)|} - \frac{1}{|H(t,f)|^2}\right]^+, \tag{6.101}$$

where μ' is such that

$$\int_{-\infty}^{\infty}\int_{-\infty}^{\infty} \mathcal{P}_{\rm MinBER}(t,f)\,dt\,df = p_T. \tag{6.102}$$

It is interesting to compare the different power loading strategies. To this purpose, we report in Figure 6.7 an example of a time-varying transfer function, drawn from a Rayleigh fading channel with delay spread $\Delta_\tau = 8T_s$ and Doppler spread $\Delta_f = 5/NT_s$, where $N = 64$ is the blocklength. Given this transfer function, we report in Figure 6.8 the corresponding power spectral density obtained with the MaxINF (a) and with the MinBER/MMSE (b) criteria. The average power is

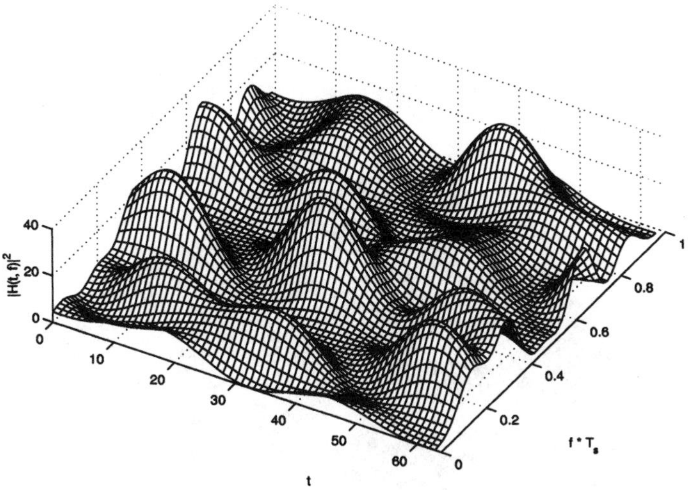

Figure 6.7 Example of time-varying transfer function $|H(t, f)|^2$.

$p_T = 4\sigma_n^2$. It is interesting to observe how the water-filling criterion does not allocate any power over the regions, in the time-frequency domain, where the channel is more attenuated, whereas it allocates an approximately constant power over the remaining domain. Conversely, the MinBER/MMSE criterion allocates more power over the most faded regions.

6.5 MISO CHANNELS

Let us consider now a MISO system with n_T transmit antennas, where the channels are time-invariant FIR filters. Using a block transmission with CP, in this case it is possible to derive the optimal transmission strategy in closed form. More specifically, assuming as optimization criterion the maximization of the mutual information, the optimal coding strategy was derived in [10] and is reported here below. We denote with N the block length (without the CP) and with L the channel order (equal to the CP length).

1. The optimal coding strategy for LTI channels is OFDM, with proper power/bit allocation across the sub-carriers;

8 The relationship used in this chapter is $\lambda_k = |H(t, f)|^2$. This position applies of to all the methods described in this section.

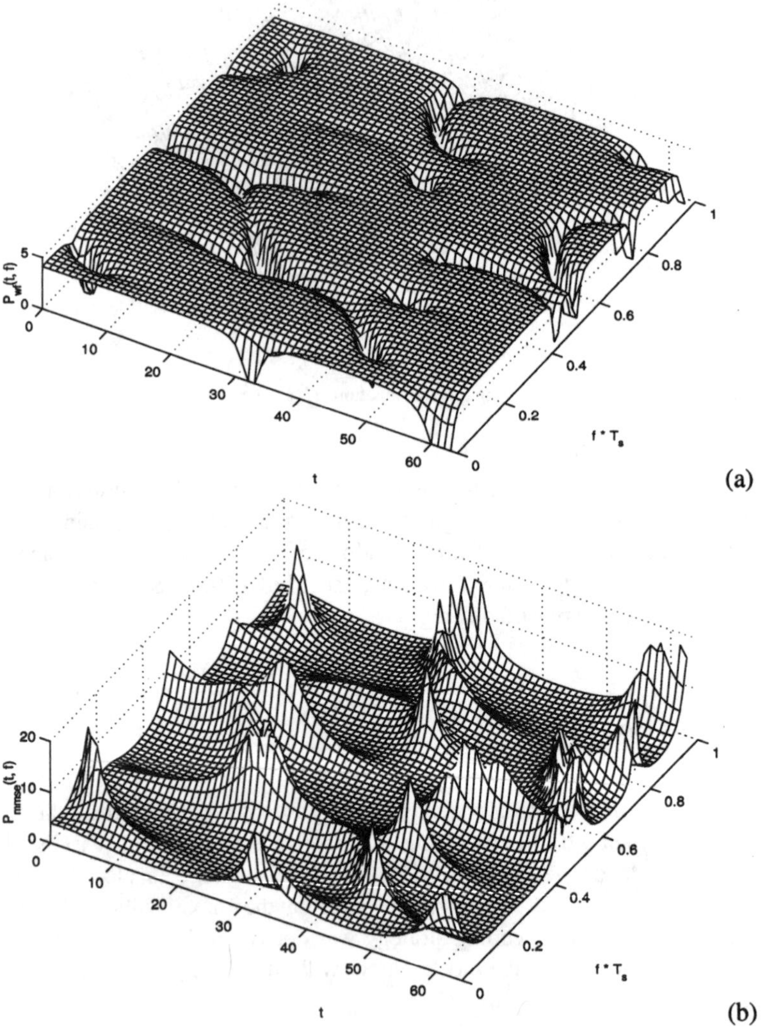

Figure 6.8 Optimal power allocation according to: (a) MaxINF and (b) MinBER/MMSE criteria.

2. Denoting by $H_j(k)$ the value assumed by the jth channel transfer function over the kth frequency bin, I_u the set of sub-carriers where a nonnull power is loaded, and $|\Phi_j(k)|^2$ the power allocated on the kth subcarrier in the jth transmit antenna, the power $|\Phi_j(k)|^2$ is equal to

$$|\Phi_j(k)|^2 = \frac{|H_j(k)|^2}{\sum_{j=1}^{N_T} |H_j(k)|^2} \left(\mathcal{K} - \frac{\sigma_n^2/\sigma_s^2}{\sum_{j=1}^{N_T} |H_j(k)|^2} \right)^+, \qquad \forall k \in I_u, \quad (6.103)$$

where $(x)^+ := \max(x, 0)$. The set of indices I_u and the constant $\mathcal{K}$ are such that the average power constraint is satisfied, i.e., $\sum_{j=1}^{n_T} \sum_{k=0}^{N-1} |\Phi_j(k)|^2 = p_T$;

3. If no channels have zeros on the grid $z_k = e^{j\,2\pi\, k/N}$, with k integer, all antennas transmit over the same portion of the available spectrum (in other words, if one antenna does not allocate any power on the kth subchannel, all other antennas do not send any power on that subchannel);

4. Since (6.103) gives indications only about the square modulus of $\Phi_j(k)$, there is full freedom to choose the phase of $\Phi_j(k)$. This freedom can be exploited to select the phase of $\Phi_j(k)$ in order to maximize the SNR at the receiver. In formulas, we have

$$\Phi_j(k) = \frac{H_j^*(k)}{\sqrt{\sum_{j=1}^{n_T} |H_j(k)|^2}} \sqrt{\left(\mathcal{K} - \frac{\sigma_n^2/\sigma_s^2}{\sum_{j=1}^{n_T} |H_j(k)|^2} \right)^+}, \forall k \in I_u. \quad (6.104)$$

In this way, denoting by $\mathcal{I}_k$ the set of channels where a nonnull power is allocated on the kth sub-carrier, the received kth symbol of the nth block is[9], $\forall k \in I_u$

$$\begin{aligned} y_k(n) &= \sum_{j=1}^{n_T} H_j(k)\Phi_j(k)s_k(n) \\ &= \sqrt{\sum_{j\in\mathcal{I}_k} |H_j(k)|^2 \left(\mathcal{K} - \frac{\sigma_n^2/\sigma_s^2}{\sum_{j\in\mathcal{I}_k} |H_j(k)|^2} \right)} s_k(n) \\ &= \sqrt{\mathcal{K} \sum_{j\in\mathcal{I}_k} |H_j(k)|^2 - \frac{\sigma_n^2}{\sigma_s^2}} s_k(n). \end{aligned} \quad (6.105)$$

9 Since $k \in I_u$, all the arguments of the square root in the (6.104) are strictly positive.

Therefore the signal-to-noise ratio (SNR) at the receiver, on the kth sub-carrier, with $k \in I_u$, is

$$\mathrm{SNR}(k) = \frac{\mathcal{K}\sum_{j\in\mathcal{I}_k}|H_j(k)|^2\sigma_s^2 - \sigma_n^2}{\sigma_n^2} \approx \mathcal{K}\sum_{j\in\mathcal{I}_k}|H_j(k)|^2\frac{\sigma_s^2}{\sigma_n^2}. \qquad (6.106)$$

This shows that, within the limit of validity of the last approximation (that is, when the noise is negligible with respect to the sum of the square moduli), the system maximizing mutual information is equivalent to a maximal ratio combining scheme with order given by the cardinality of $\mathcal{I}_k$, on each sub-carrier.

It is interesting to remark that the solution (6.104) is a generalization of the water-filling solution found for the SISO case, which is a particular case of (6.104) for $N_T = 1$. The power loading (6.103) gives rise to an interesting physical interpretation, as detailed in the following remarks.

- If we exclude the limiting case where the channels have zeros exactly over the set of complex points $\{z_k := \exp(j2\pi k/N); k = 0, 1, \ldots, N-1\}$, the criterion for allocating a nonnull power over the kth sub-carrier depends on the *the sum* of the square moduli of *all* the channel transfer functions. If a subcarrier is discarded from one antenna, it is discarded from all the antennas. In other words, all antennas transmit over the same portion of the spectrum. Furthermore, with respect to the SISO case, it is less likely that a sub-carrier is discarded because, to discard a sub-carrier it is necessary that all (most of) the channels have a high attenuation over that sub-carrier;
- The only exception to the previous remark is represented by the case where some channels have a zero exactly in the set $\{z_k := \exp j2\pi k/N, k = 0, 1, \ldots, N-1\}$. In such a case, the power allocated by the corresponding transmit antenna is null, but all other antennas are allowed to allocate power over that sub-carrier;
- If a nonnull power is allocated over the kth sub-carrier, each antenna transmits a portion of such a power equal to its *relative spatial weight*, quantified by the ratio $|H_j(k)|^2/\sum_{j=1}^{n_T}|H_j(k)|^2$;
- Equation (6.104) shows that the amplitude distribution across the transmit array performs a beamforming, different for each sub-carrier, whose amplitude tapering depends on the SNR and on the global available power. Asymptotically, for n_T going to infinity, assuming that the channels between each transmit antenna and the receiver are realizations of a set of independent ergodic processes, having the same statistical properties, the summations $\sum_{j=1}^{n_T}|H_j(k)|^2$ tend to a value independent of k. Hence, asymptotically, the maximum mutual information distribution tends to a classical beamforming, or MRC at the transmitter, with

$\Phi_j(k) = cH_j^*(k)$, where c is a constant dictated only by the available average power.

6.6 MIMO SYSTEMS

The diversity gain achievable with multiple antenna systems can be obtained also with a MISO or a SIMO system. The unique feature of real MIMO systems is their capability of increasing the information rate for a given bandwidth and total transmit power. In this section we consider the performance of MIMO systems, in terms of average BER and mutual information, for different MIMO configurations. We will then assess the resilience of MIMO systems to interference and channel estimation errors.

6.6.1 MIMO versus SISO Systems

We compare now the performance of MIMO systems in terms of average BER and mutual information, for different coding strategies, namely MinBER/ZF, MinBER/MMSE and MaxINF for different MIMO configurations. As an example, we considered multipath Rayleigh fading channels. Each channel is FIR of order $L = 4$ and the paths are statistically independent complex Gaussian random variables, with zero mean and unit variance. The average BER and information rate (IR), averaged over 100 independent channel realizations, are reported in Figures 6.9 and 6.10, respectively.

It is interesting to observe that the BER improves by increasing the number of antennas, but not as much as in the case analyzed in Chapter 5, where the transmitter had no channel knowledge. This happens because in case of perfect channel knowledge, the system has the capacity of getting multipath diversity gain also in case of SISO systems. This happens to both methods minimizing the BER, thanks to the pre- and post-rotation matrices $\mathbf{\Theta}$ in the scheme of Figure 6.4. This property will be further explored in Chapter 8, where we will show how to construct the rotation matrices in order to get the multipath diversity gain. Increasing the number of antennas induces a spatial diversity gain, but all MIMO configurations have, potentially, multipath diversity gain. This is why the average BER curves corresponding to all MinBER methods have a rapid decrease of the average BER, at high SNR. Conversely, the method maximizing the information rate does not have any multipath diversity gain. For this reason, the curve relative to the SISO case, for the MaxINF method, shows no diversity gain at all. Using MaxINF, an increase of the number of antennas produces a more appreciable gain in terms of average BER, as it provides spatial diversity gain to a system (SISO) with no diversity at all.

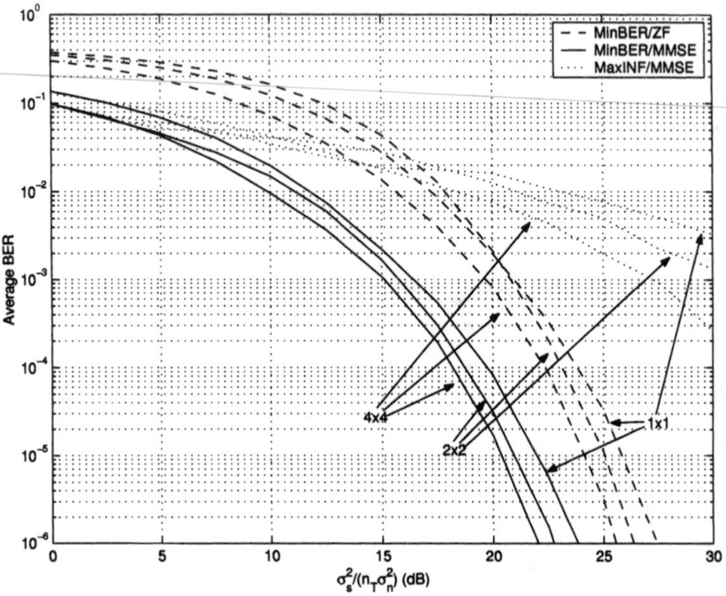

Figure 6.9 Average BER versus SNR (dB) for SISO and MIMO systems: MinBER/MMSE (solid line), MinBER/ZF (dashed line), MaxINF/MMSE (dotted line).

As evidenced in Figure 6.10, the most striking advantage of MIMO systems, in case of perfect channel knowledge, is the increase in information rate (number of bit/s/Hz).

From Figure 6.10, we also notice how the MinBER/MMSE method is quite close to the MaxINF/MMSE method at low SNR. However, at high SNR, the Max-INF/MMSE method clearly outperforms the other methods and the advantage is more and more evident as the number of antennas increases.

6.6.2 Interference Mitigation

It is important to check the performance of the different optimization strategies in the presence of interference. We may have narrowband or wideband interference, meaning, in practice, that the spectrum of the interferer is small or large with respect to the receiver bandwidth. However, in both cases the interference impinges on the receive array from a certain angle (or from a few angles if the interference is subject to multipath). In other words, an interference may be (temporally) narrowband or wideband, but it is typically "narrowband" in the angular (spatial)

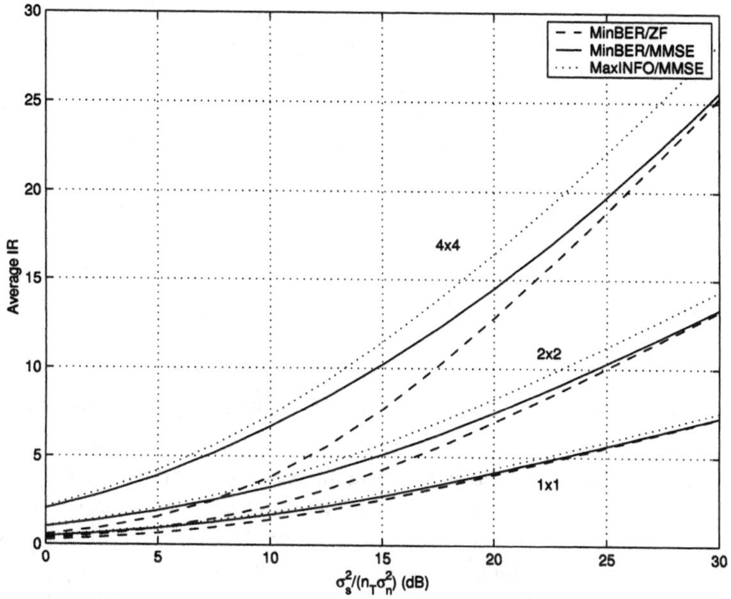

Figure 6.10 Average information rate versus SNR (dB) for SISO and MIMO systems: MinBER/MMSE (solid line), MinBER/ZF (dashed line), MaxINF/MMSE (dotted line).

domain. This observation implies that an interferer can be canceled at the receiver, even if it is wideband, provided that the receiver has a sufficient number of antenna elements, i.e., sufficient spatial filtering capabilities. We explore now the interference mitigation properties of the optimal coding strategies studied so far. To assess their performance, it is necessary to specify the structure of the interference covariance matrix. In a receiver composed of n_R antennas, where each antenna receives a block of N data, the covariance matrix of the interference can be represented through the following block structure

$$\boldsymbol{C}_I = \begin{pmatrix} \boldsymbol{C}_I(1,1) & \boldsymbol{C}_I(1,2) & \cdots & \boldsymbol{C}_I(1,n_R) \\ \boldsymbol{C}_I(2,1) & \boldsymbol{C}_I(2,2) & \cdots & \boldsymbol{C}_I(2,n_R) \\ \vdots & \ddots & \ddots & \vdots \\ \boldsymbol{C}_I(n_R,1) & \boldsymbol{C}_I(n_R,2) & \cdots & \boldsymbol{C}_I(n_R,n_R) \end{pmatrix}, \tag{6.107}$$

where $\boldsymbol{C}_I(k,l) := E\{\boldsymbol{i}_k\boldsymbol{i}_l^H\}$, is the cross-correlation between the interference (time) vectors $\boldsymbol{i}_k$ and $\boldsymbol{i}_l$, received by the kth and lth antennas. Let us suppose to have a linear uniform receive array, as depicted in Figure 6.11. We assume that the source of interference is in the far field of the receiving antenna, so that the

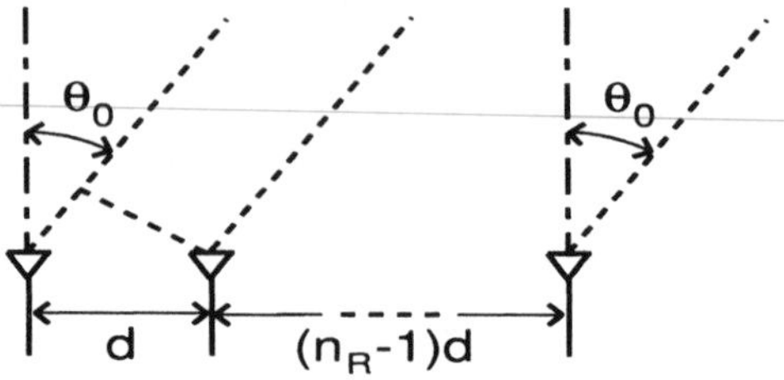

Figure 6.11 Receive array.

impinging wave is well approximated by a plane wave and the array is narrowband, meaning that the analytic signals arriving at different array elements differ only by a phase shift due to the different space traveled. In formulas, if we denote by $x_0(t)$ the analytic signal describing the plane wave arriving at the right element of the array depicted in Figure 6.11, we may always express $x_0(t)$ in terms of its complex (baseband) envelope $\tilde{x}_0(t)$ as $x_0(t) = \tilde{x}_0(t)e^{j\omega_0 t}$, where $f_0 = \omega_0/2\pi$ is the carrier frequency. Denoting byθ_0 the incidence angle of the interference and with d the space between two consecutive elements of the array, the analytic signal $x_k(t)$ arriving at the kth element is $x_k(t) = x_0(t - k\tau_0)$, where $\tau_0 = d\sin(\theta_0)/c$ is the delay between the signals arriving at two consecutive array elements (c is the speed of light). If $x_0(t)$ is a narrowband signal, i.e., if its bandwidth is much smaller than its carrier frequency, we may use the following narrowband approximation

$$x_k(t) = x_0(t - k\tau_0) = \tilde{x}_0(t - k\tau_0)e^{j\omega_0(t-k\tau_0)} \approx [\tilde{x}_0(t)e^{-jk\omega_0\tau_0}]e^{j\omega_0 t}.$$

Denoting by λ the wavelength corresponding to the carrier f_0, i.e., such that $c = f_0\lambda$, the baseband signals arriving at adjacent array elements are subject to a phase shift equal to $\phi_0 = (2\pi\lambda/d)\sin(\theta_0)$.

Narrowband interference (NBI)

The narrowband interference may be modelled as a complex exponential as a function of time, with frequency f_I. Its time covariance matrix is then equal to $\boldsymbol{C}_t = A_I^2 \boldsymbol{c}_I \boldsymbol{c}_I^H$, where $\boldsymbol{c}_I := (1, e^{j2\pi f_I T, \ldots, e^{j2\pi(N-1)f_I T}})$. If θ_0 denotes the angle of incidence of the interference, the overall (space-time) interference covariance matrix is

$$\boldsymbol{C}_I = A_I^2 \boldsymbol{c}_I \boldsymbol{c}_t^H \otimes \boldsymbol{C}_s, \tag{6.108}$$

where $\otimes$ denotes Kronecker product and

$$\boldsymbol{C}_s = \begin{pmatrix} 1 & e^{j\phi_0} & \dots & e^{j(n_R-1)\phi_0} \\ e^{-j\phi_0} & 1 & \dots & e^{j(n_R-2)\phi_0} \\ \vdots & \ddots & \ddots & \vdots \\ e^{-j(n_R-1)\phi_0} & \ddots & \ddots & 1 \end{pmatrix} \tag{6.109}$$

is the space covariance.

Wideband interference

The wideband interference may be modelled as a white noise along the time domain. Denoting by θ_0 the angle of incidence of the interference impinging on the receive antenna, we have

$$\boldsymbol{C}_I = \sigma_I^2 \boldsymbol{I} \otimes \boldsymbol{C}_s, \tag{6.110}$$

with $\boldsymbol{C}_s$ given by (6.109).

As an example of performance, in Figure 6.12 we report the average BER obtained with the MinBER/MMSE method, under the following scenario. The channels are all FIR of order $L = 5$ and there is a wideband interference arriving with the same power as the useful signal with an incidence angle of 15 degrees. The three curves refer to three system configurations: SISO, SIMO (1×2), and MIMO (2×2). The transmit power is the same in all cases. From Figure 6.12 we notice that the SISO system performs very poorly, as it has no capabilities to suppress the wideband interference that acts then exactly like an additive noise. Conversely, as the number of receive antennas increases, the gain is evident, as the receiver acquires the capability of suppressing the interference working in the spatial domain.

It is also interesting to compare the average information rate of SISO and MIMO systems in the presence of interference. As an example, we report in Figure 6.13 the average mutual information obtained with the MinBER/MMSE and the MaxINF methods, for three MIMO configurations: 1×1, 2×2, and 4×4. We can check that a SISO system performs very poorly also interms of information rate. However, as the number of the MIMO dimension increases, we can see a clear performance improvement, also in the presence of a strong interference. To quantify the resilience of optimal MIMO systems against wideband interferences, in Figure 6.14 we compare the average mutual information obtained with the MaxINF method, for three MIMO configurations: 1×1, 2×2, and 4×4. The dotted lines refer to the case of no interference, whereas the solid lines refer to a situation with a wideband interference having the same power as the useful signal. We can see that the loss

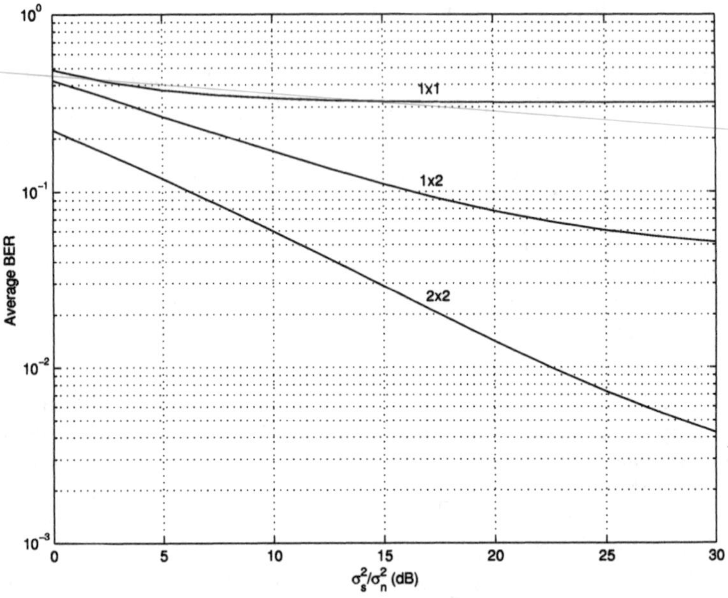

Figure 6.12 Average BER versus SNR (dB) for SISO and MIMO systems obtained with the MinBER/MMSE method.

due to the presence of the interference is less and less evident as we increase the size of the MIMO system.

6.6.3 Resilience to Channel Estimation Errors

The joint optimal design of the encoding/decoding schemes seen so far assumed perfect channel knowledge at *both* the transmit and receive sides. Clearly, this is an idealistic assumption as the channel has to be estimated and the estimate is inevitably affected by errors. To quantify the robustness of the different coding strategies against channel estimation errors, for both SISO and MIMO systems, we have generated the estimated channel impulse response as the true channel response plus a zero mean Gaussian random process, with parametric power. As an example, in Figure 6.15, we report the average BER for a SISO system, obtained using three methods: MinBER/MMSE, MinBER/ZF and MaxINF. In the example, we have generated the true channel response as an FIR filter of order $L = 5$, with i.i.d. taps with zero mean and unit variance. The estimation error has zero mean and variance σ_e^2. From Figure 6.15, we notice that all methods exhibit a BER floor,

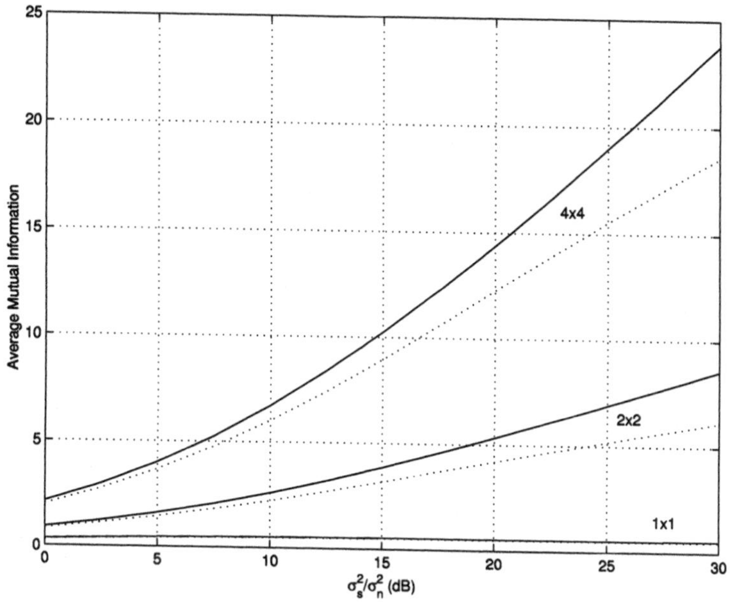

Figure 6.13 Average mutual information versus SNR (dB) for SISO and MIMO systems obtained with the the MaxINF (solid line) and MinBER/MMSE (dotted line) methods.

due to the ISI generated by the channel estimation errors. It is interesting to notice that the MinBER/MMSE method exhibits greater robustness than MinBER/ZF against estimation errors. This happens because the MinBER/ZF method allocates power in a way inversely proportional to the channel transfer function and then it is more prone to errors on the subchannels having more attenuation. In this sense, the MaxINF method is rather robust because it tends to allocate no power over the most faded subchannels. However, it is also true that the MaxINF method exhibits the worst BER values and thus it has less to loose. Conversely, the MinBER/MMSE method, in spite of having the minimum BER values, at least at high SNR, shows a good resilience to the channel estimation errors. It might appear quite surprising that the MaxINF method has a lower average BER than the MinBER/MMSE method, at low SNR. Indeed this happens only because, at low SNR, the MaxINF method discards more subchannels than MinBER/MMSE and thus the two methods operate with different bit rates. Conversely, at high SNR, the rate is (approximately) the same and thus it is fair to compare the three methods only at high SNR.

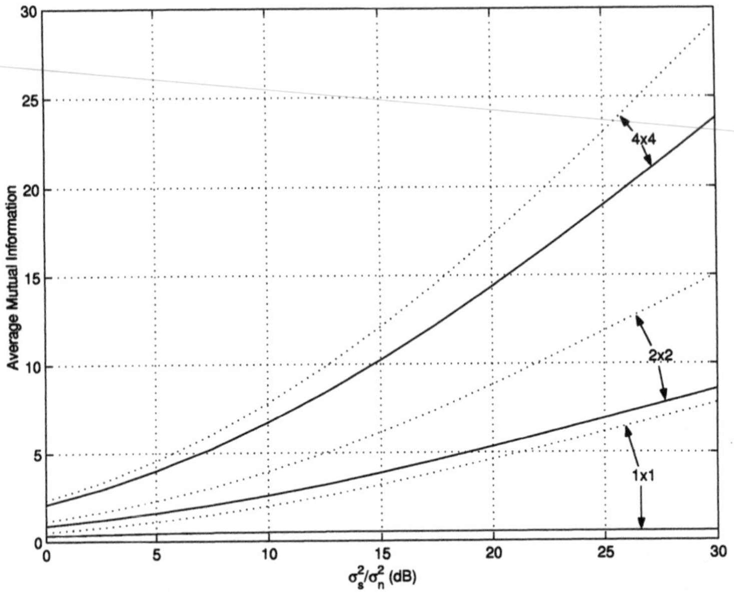

Figure 6.14 Average mutual information versus SNR (dB) for SISO and MIMO systems obtained with the the MaxINF method in case of strong interference case (solid line) and of no interference (dotted line).

6.7 OPTIMAL DESIGN IN CASE OF PARTIAL CHANNEL KNOWLEDGE

In the preceding sections, we have assumed that the channel is perfectly known at the transmitter. In practice, this knowledge is available only within some inevitable estimation errors. In this section, we relax this assumption and we review the methodologies proposed for optimizing the transmission strategy, in the presence of partial channel information.

Information about the channel at the transmitter side can be acquired in two basic manners. In systems employing TDD, transmission and reception occur over the same frequency band. In such a case, the reception channel coincides with the transmission channel. Then, by reciprocity, the same channel estimate carried out at the receiver to retrieve the transmitted information can be used to design the transmission strategy. The situation is more complicated with FDD systems, where transmission and reception occur over non-overlapping frequency bands. In such a case, the only possibility for the transmitter to know the channel is that the receiver

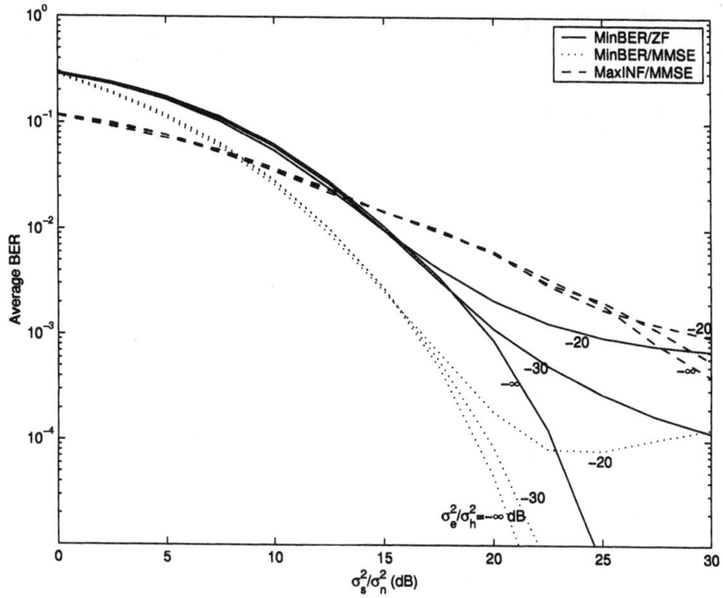

Figure 6.15 Average BER versus SNR (dB) for SISO system, incorporating channel estimation errors with parametric variance, for three methods: MinBER/MMSE (dotted line), MinBER/ZF (solid line), and MaxINF (dashed line).

sends its channel estimate back to the transmitter.

In practice, the channel estimate available at the transmitter does not coincide with the true channel, for the following reasons: 1) any channel estimate, in the presence of noise, is inevitably affected by estimation errors; 2) in case of FDD systems, the feedback channel is itself affected by transmission errors and the channel estimate needs to be properly quantized before transmission; 3) in case of time-varying channel, the channel behavior to be used to optimize the transmission, must be predicted from its past estimates and this operation introduces an inevitable prediction error. This error is more or less important depending on the ratio between the time interval T_0, over which we wish to optimize the transmission, and the channel coherence time T_c. If $T_0 < T_c$, we may assume that the transmit channel is approximately equal to the receive channel (besides estimation errors). But, if $T_0 > T_c$, we need to predict the channel evolution from its past.

In general, we may model the channel information at the transmitter as a set of random variables, whose parameters depend on the true channel. The most common model is the Gaussian model, where the second order statistics contain all the information. Typically, if the channel is slow fading, we may assume to know the channel mean, whereas if the channel is fast fading, we may assume to know only the channel covariance. The two situations are known as *mean-feedback* or *covariance feedback*, respectively.

Mean-feedback model

In the mean-feedback case, the channel estimate available at the transmitter is modeled as follows:

$$\boldsymbol{H} = \sqrt{\frac{K}{K+1}}\boldsymbol{H}_{spec} + \sqrt{\frac{1}{K+1}}\boldsymbol{H}_w, \tag{6.111}$$

where the mean component matrix $\boldsymbol{H}_{spec}$ and the coefficient K are assumed to be known, whereas $\boldsymbol{H}_w$ is modeled as a matrix with i.i.d. zero mean, unit variance, complex circularly symmetric Gaussian random variables.

A typical example is given by the Rice channel model (2.80), where the transmitter is assumed to know the specular component. In such a case, the mean component is the rank-one matrix

$$\boldsymbol{H}_{spec} = \boldsymbol{a}_R\boldsymbol{a}_T^T, \tag{6.112}$$

where $\boldsymbol{a}_T$ and $\boldsymbol{a}_R$ are the specular array response vectors at the transmitter and receiver, respectively. In the MISO case, (6.112) reduces to the n_T-size row vector $\boldsymbol{a}_T^T$, whereas in the SIMO case, (6.112) reduces to the n_R-size column vector $\boldsymbol{a}_R$.

Covariance-feedback model

In the covariance-feedback case, the channel information available at the transmitter is modeled as a random matrix having zero mean value and assuming the following structure

$$\boldsymbol{H} = \boldsymbol{\Sigma}_R^{1/2}\boldsymbol{H}_w\boldsymbol{\Sigma}_T^{1/2}, \tag{6.113}$$

where $\boldsymbol{H}_w$ is a matrix with i.i.d. zero mean, unit variance complex circularly symmetric Gaussian random variables, $\boldsymbol{\Sigma}_T$ is the covariance between the coefficients corresponding to different transmit antennas, and $\boldsymbol{\Sigma}_R$ is the covariance between the coefficients corresponding to different receive antennas. The matrices $\boldsymbol{\Sigma}_R$ and $\boldsymbol{\Sigma}_T$ are assumed to be known at the transmitter side. As a particular case of this model, there is the zero-mean spatially white (ZMSW) model where $\boldsymbol{\Sigma}_R = \boldsymbol{\Sigma}_T = \boldsymbol{I}$.

There have been several works about the optimization of the transmission strategy, in the presence of information about the channel mean or the channel covariance, with respect to information rate or to the bit error rate [11, 12, 13, 14, 15, 16, 17, 18, 19].

For any given input covariance matrix $\boldsymbol{Q}$, the input distribution that achieves the *ergodic* capacity is the vector Gaussian distribution. Mathematically, the problem is to find out the optimum $\boldsymbol{Q}$ to maximize

$$C = \max_{\boldsymbol{Q}:\mathrm{tr}(\boldsymbol{Q})=p_T} C(\boldsymbol{Q}) \tag{6.114}$$

where

$$C(\boldsymbol{Q}) := E_{\boldsymbol{H}}\left\{\log\left|\boldsymbol{I}_{n_R} + \frac{\boldsymbol{HQH}^H}{\sigma_n^2}\right|\right\} \tag{6.115}$$

is the mutual information, assuming that the input vector $\boldsymbol{x}$ has covariance matrix $\boldsymbol{Q}$. The maximum mutual information $(\boldsymbol{Q})$ is achieved by transmitting independent complex circularly symmetric Gaussian symbols along the eigenvectors of $\boldsymbol{Q}$. The problem is then to find the covariance matrix $\boldsymbol{Q}$ that solves (6.114).

Mean feedback

Visotsky and Madhow derided the transmission strategy that maximizes the mutual information for MISO systems, assuming either mean or covariance feedback [12]. In case of mean feedback, modeling the channel information at the transmitter as $\boldsymbol{h} \sim \mathcal{N}(\boldsymbol{a}_T, \sigma_e^2\boldsymbol{I})$, where $\boldsymbol{a}_T$ is the average cannel and σ_e^2 is the variance of the channel estimate (supposed to be the same for all coefficients); both $\boldsymbol{a}_T$ and σ_e^2 are assumed to be known at the transmitter. The covariance matrix that maximizes the information rate is the Hermitian positive semidefinite matrix having eigendecomposition

$$\boldsymbol{Q} = \boldsymbol{U\Lambda U}^H \tag{6.116}$$

where the first column $\boldsymbol{u}_1$ of $\boldsymbol{U}$ is

$$\boldsymbol{u}_1 = \frac{\boldsymbol{a}_T}{\|\boldsymbol{a}_T\|} \tag{6.117}$$

and all other columns of $\boldsymbol{U}$ are arbitrarily chosen, except for the restriction that they are orthonormal between each other and orthogonal to $\boldsymbol{a}_T$; all eigenvalues, except the one associated with $\boldsymbol{u}_1$, are equal to each other and, in particular, equal to

$$\lambda_2 = \ldots \lambda_{n_T} = \frac{p_T - \lambda_1}{n_T - 1}; \tag{6.118}$$

the eigenvalue λ_1 associated to $\boldsymbol{u}_1$ is not known in closed form, but it can be derived numerically without any ambiguity as the unique solution of a convex optimization problem [12].

This solution is intuitively clear. The transmitter performs beamforming in the direction specified by the steering vector $\boldsymbol{a}_T$. The power allocated to beamforming depends on the quality of the channel estimate (i.e., on the ratio $\|\boldsymbol{a}_T\|/\alpha$): if the variance of the estimation error is null, i.e., $\alpha = 0$, all transmit power is allocated to beamforming; if the channel estimate is affected by errors, i.e., $\alpha > 0$, a portion of the available power is devoted to beamforming, while the rest is spread uniformly over the directions specified by the subspace orthogonal to that of beamforming.

This result was generalized to MIMO systems in [14], where the optimal solution is as before, except for using instead of vector $\boldsymbol{a}_T$ in (6.117) the transposed conjugate of the first row of $\boldsymbol{H}_{spec}$, properly normalized to have unit norm.

Covariance feedback

In [12] it was also shown what is the optimal solution for MISO systems, in case of covariance feedback. Assuming that $\boldsymbol{h} \sim \mathcal{N}(\boldsymbol{0}, \boldsymbol{\Sigma})$, with $\boldsymbol{\Sigma}$ known at the transmitter, the optimal covariance matrix is the Hermitian positive semidefinite matrix having eigendecomposition

$$\boldsymbol{Q} = \boldsymbol{U}_\Sigma \boldsymbol{\Lambda}_Q \boldsymbol{U}_\Sigma^H \tag{6.119}$$

where $\boldsymbol{U}_\Sigma$ is the unitary matrix whose columns are the eigenvectors of $\boldsymbol{\Sigma}$ and $\boldsymbol{\Lambda}_Q$ is a diagonal matrix, whose diagonal entries need to be determined through numerical maximization techniques, as a function of the eigenvalue of $\boldsymbol{\Sigma}$. In practice, higher power should be transmitted along the directions specified by the largest eigenvalues of $\boldsymbol{\Sigma}$. This result was then generalized to MIMO systems in [13] and [14].

The previous results have been obtained for MIMO systems, but without taking into account the possibility for space-time coding, i.e., coding across both space and time domains. This problem was investigated in [16], where it was shown that the encoder can be split into a space-time encoder, that depends only on the information to be transmitted, followed by a beamformer, that depends only on the channel information available at the transmitter, without capacity losses. Further works [15, 17, 18], suggested how to combine space-time coding and beamforming in order to minimize a bound on the average error probability.

6.8 OPTIMAL ACCESS METHODS

All systems examined in the previous sections were single user systems. We consider now the optimal transmission strategies for the multiuser case. In this section, we study the multiple access and the broadcast channels. In the next section, we will consider the so called "cocktail party problem", where there are many sources that wish to talk to as many destinations and it is necessary to find out the optimal access strategy. In all cases, we assume, as in the previous sections, that the channels are known at the transmitter side.

6.8.1 Capacity Region of the Multiple Access Channel

In this section, we evaluate the maximum information rates achievable in the multiple access channel. We consider the situation where both access point and user terminals have single antenna transceivers. The extension to the multiantenna case was considered in [20]. Within the single-antenna scenario, we consider the scalar case, where each user transmits one symbol per each channel use, and the vector case, where each user adopts a block transmission strategy.

Scalar case

Let us start with the Gaussian multiple access channel, where each received sample Y is the superposition of M independent symbols X_m, coming from as many users, plus additive white Gaussian noise (we drop the dependence on time, as we consider the stationary case):

$$Y = \sum_{m=1}^{M} h_m X_m + V, \tag{6.120}$$

where h_m is the flat-fading coefficient between the mth user and the access point; V is additive Gaussian noise. Each user is subject to a constraint on its average transmit power. Let us denote with $\boldsymbol{p} := (p_1, \ldots, p_M)$ the vector with the M powers assigned to all the users. From [7], the capacity region is the region of rates $\boldsymbol{R} := (R_1, \ldots, R_M)$ given by the following set:

$$C_{MA}(\boldsymbol{p}; \boldsymbol{R}) = \left\{ \boldsymbol{R} : \sum_{j \in \mathcal{S}} R_j \leq \frac{1}{2} \log(1 + \frac{1}{\sigma_n^2} \sum_{j \in \mathcal{S}} |h_j|^2 p_j), \forall \mathcal{S} \subseteq \{1, \ldots, M\} \right\}. \tag{6.121}$$

The boundary of the capacity region is an M-dimensional polyhedron. The best rates are the values lying on that polyhedron. To reach the corner points of the polyhedron, it is necessary to apply successive decoding. This means that a first user is decoded, its contribution is canceled from the received signal, a second user is then decoded, its contribution is subtracted, and so on. Each corner point on the polyhedron corresponds to a specific decoding order. Since there are $M!$ ways to choose the decoding order of M users (all the permutations of the user index set), the polyhedron has, in general, $M!$ corner points.

Denoting by $(\pi(1), \ldots, \pi(M))$ a specific decoding order, implying that user $\pi(1)$ is decoded first, $\pi(2)$ is decoded second, and so on, the rates of the corresponding corner points are

$$R_{\pi(j)} = \frac{1}{2} \log \left(1 + \frac{\left|h_{\pi(j)}\right|^2 p_{\pi(j)}}{\sigma_n^2 + \sum_{i=j+1}^{M} \left|h_{\pi(i)}\right|^2 p_{\pi(i)}} \right), \; j = 1, \ldots, M. \tag{6.122}$$

Equation (6.122) can be physically justified as follows. When decoding user $\pi(j)$, it is assumed that all users with indexes $\pi(i), i = 1, \ldots, \pi(j-1)$ have been previously successfully decoded and then subtracted. At the moment of decoding user $\pi(j)$, the users with indexes $\pi(i), i = 1, \ldots, \pi(j-1)$ are then absent, whereas the users with indexes $\pi(i), i = j+1, \ldots, \pi(M)$ act as additive white Gaussian noise.

Vector case

Let us consider now the more interesting case where all user terminals and access points still have single antenna transceivers, as in the previous section, but each user adopts a block transmission strategy and the channels are time-dispersive. We report here the basic results derived by Wei Yu [21]. We start by considering the two-user case. Then, we will consider the generalization to any number of users. In the two-user case, at the receiver, we collect the vector

$$\boldsymbol{y} = \boldsymbol{H}_1 \boldsymbol{x}_1 + \boldsymbol{H}_2 \boldsymbol{x}_2 + \boldsymbol{v}, \tag{6.123}$$

where $\boldsymbol{H}_1$ and $\boldsymbol{H}_2$ are the two channel matrices characterizing the channels from users 1 and 2 towards the access point. We suppose that the noise $\boldsymbol{v}$ is Gaussian, with zero mean and covariance matrix $\boldsymbol{R}_{vv}$. We denote with $\boldsymbol{R}_1$ and $\boldsymbol{R}_2$ the covariance matrices of the vectors $\boldsymbol{x}_1$ and $\boldsymbol{x}_2$, respectively. We also assume, as in the scalar case, that there is a constraint on the average transmit power of each user. In particular, indicating with p_1 and p_2 the average powers transmitted by the two users, we have the two constraints:

$$\begin{aligned} \text{tr}(\boldsymbol{R}_1) &\leq p_1, \\ \text{tr}(\boldsymbol{R}_2) &\leq p_2. \end{aligned} \tag{6.124}$$

Following the same approach as in the scalar case, the mutual information expressions must satisfy the inequalities

$$
\begin{aligned}
I(\boldsymbol{X}_1;\boldsymbol{Y}/\boldsymbol{X}_2) &\leq \frac{1}{2}\log_2\left(\frac{|\boldsymbol{H}_1\boldsymbol{R}_1\boldsymbol{H}_1^H+\boldsymbol{R}_{vv}|}{|\boldsymbol{R}_{vv}|}\right),\\
I(\boldsymbol{X}_2;\boldsymbol{Y}/\boldsymbol{X}_1) &\leq \frac{1}{2}\log_2\left(\frac{|\boldsymbol{H}_2\boldsymbol{R}_2\boldsymbol{H}_2^H+\boldsymbol{R}_{vv}|}{|\boldsymbol{R}_{vv}|}\right),\\
I(\boldsymbol{X}_1,\boldsymbol{X}_2;\boldsymbol{Y}) &\leq \frac{1}{2}\log_2\left(\frac{|\boldsymbol{H}_1\boldsymbol{R}_1\boldsymbol{H}_1^H+\boldsymbol{H}_2\boldsymbol{R}_2\boldsymbol{H}_2^H+\boldsymbol{R}_{vv}|}{|\boldsymbol{R}_{vv}|}\right),
\end{aligned}
\tag{6.125}
$$

where the inequalities result from the property that the maximum mutual information is achieved when the input distribution, under a covariance constraint, is Gaussian. The achievable rate region is then:

$$
\mathcal{A}(p_1,p_2)=\bigcup_{\substack{\mathrm{tr}(\boldsymbol{R}_1)\leq p_1,\mathrm{tr}(\boldsymbol{R}_2)\leq p_2\\ \boldsymbol{R}_1,\boldsymbol{R}_2\geq \boldsymbol{0}}}\left\{(R_1,R_2):\begin{array}{c}R_1\leq C_1(\boldsymbol{R}_1)\\ R_2\leq C_2(\boldsymbol{R}_2)\\ R_1+R_2\leq C_{12}(\boldsymbol{R}_1,\boldsymbol{R}_2)\end{array}\right\},
\tag{6.126}
$$

where the quantities $C_1(\boldsymbol{R}_1)$, $C_2(\boldsymbol{R}_2)$, and $C_{12}(\boldsymbol{R}_1,\boldsymbol{R}_2)$ are the capacities given by the right-hand sides of (6.125).

For Gaussian vector channels, with separate constraints on the powers available to the two users, the achievable region (6.126) coincides with the capacity region. Furthermore, the capacity region is convex and the extreme points of the region can be found by maximizing a weighted sum of the rates $\mu R_1 + (1-\mu)R_2$, with $0 \leq \mu \leq 1$. The maximization must be carried out over all possible positive definite covariance matrices $\boldsymbol{R}_1$ and $\boldsymbol{R}_2$ that satisfy the power constraints (6.124).

In practice, it is quite difficult to find out the optimal covariance matrices. The complexity increases also with the number of users and with the block length. Nevertheless, the maximization of the sum-rate can be achieved with a relatively simple algorithm that generalizes the water-filling principle, described in Section 6.3.3 to the multiuser case. The extension, proposed by Yu *et al.* in [20], known as the *iterative water-filling* algorithm, is based on the following basic result:

A set of covariance matrices $\boldsymbol{R}_m$, $m = 1,\ldots,M$, is the solution of the sum-rate maximization problem, under the single-power constraints on each user power, if and only if each matrix $\boldsymbol{R}_m$ is the optimal covariance matrix for the single-user channel characterized by the channel $\boldsymbol{H}_m$ and noise covariance matrix

$$\boldsymbol{Z}_m = \boldsymbol{R}_{vv} + \sum_{\substack{k=1 \\ k \neq m}}^{M} \boldsymbol{H}_k \boldsymbol{R}_k \boldsymbol{H}_k^H. \tag{6.127}$$

Each matrix $\boldsymbol{Z}_m$ is the covariance matrix of the noise plus interference observed at the access point, considering as interference all the users except the mth one. The optimal covariance matrices $\boldsymbol{R}_m$ are obtained through the following algorithm..

Iterative water-filling algorithm:

Initialize $\boldsymbol{R}_i = \boldsymbol{0}$ (or to any nonnegative definite matrix), $i = 1, \ldots, M$;

repeat

for $m = 1$ to M,

$$\boldsymbol{Z}_m = \boldsymbol{R}_{vv} + \sum_{\substack{k=1 \\ k \neq m}}^{M} \boldsymbol{H}_k \boldsymbol{R}_k \boldsymbol{H}_k^H;$$

$$\boldsymbol{R}_m = \arg\max_{\boldsymbol{R}} \tfrac{1}{2} \log |\boldsymbol{H}_m \boldsymbol{R} \boldsymbol{H}_m^H + \boldsymbol{Z}_m|, \text{subject to } \operatorname{tr}(\boldsymbol{R}) \leq p_m$$

end

until the desired accuracy is reached.

The intermediate matrix $\boldsymbol{R}_m$ is found by using the single-user water-filling described in Section 6.3.3. The iterative water-filling algorithm has been proven to converge to the unique sum-rate, from any initial assignments of $\boldsymbol{R}_i$. It is important to remark that, while the maximum sum-rate is unique, the optimal final covariance matrices are not unique.

Example: Let us consider a multiple-access channel with three users. Each transmitter employs block transmission, with block length $N = 64$. The channels are stationary and modeled as FIR filters of order $L = 5$. A cyclic prefix of length 5 is appended at the beginning of each block, to make the channel matrix circulant. The modulus of the three transfer functions are reported in Figure 6.16(a). After running the iterative water-filling algorithm, the optimal power spectral densities are the ones reported in Figure 6.16(b). It is interesting to see that distinct users transmit over nonoverlapping frequency bands. Interestingly, each user "discovers" that the multiplexing strategy that maximizes the sum-rate is OFDMA. It is also interesting to verify, comparing Figure 6.16(a) and (b), that each user decides to transmit over

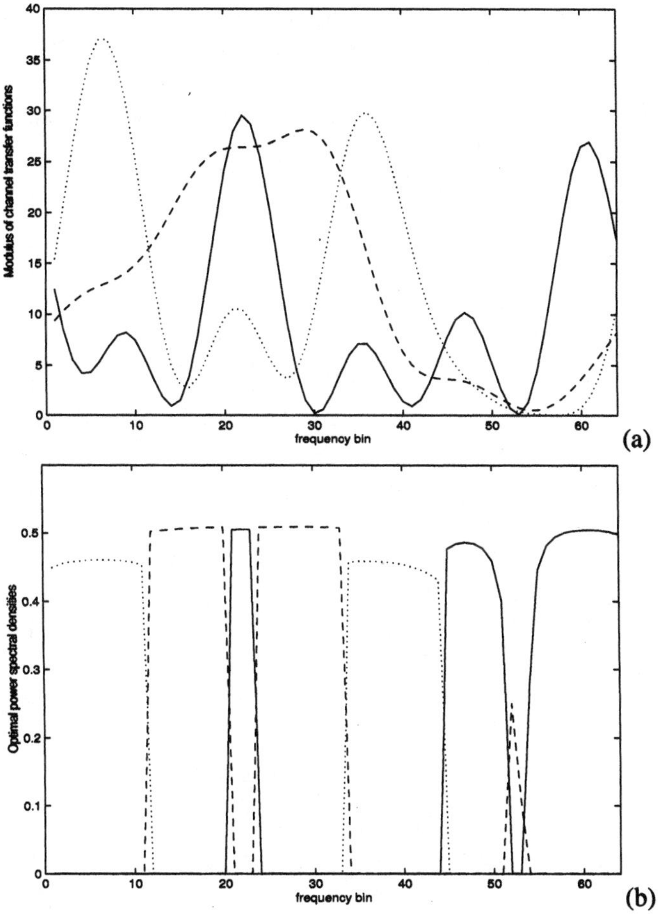

Figure 6.16 Optimal coding for multiple access channel: (a) channel transfer functions and (b) optimal power spectral densities.

the frequency bands where it has the best channel, with respect to the other users[10]. This result had been anticipated by Chen and Verdú in [23].

Using duality arguments, if the channels were all multiplicative channels, the optimal multiplexing strategy would have been TDMA, where each user transmits over the time slot where it has the best channel. Within each time slot, the user who gets the slot distributes its power (bits) as a function of time, according to the water-filling principle.

What is really striking about the iterative water-filling algorithm, is its convergence speed. In the example considered above, the rate of each user, as a function of the iteration number is reported in Figure 6.17. We can see that, after a very few

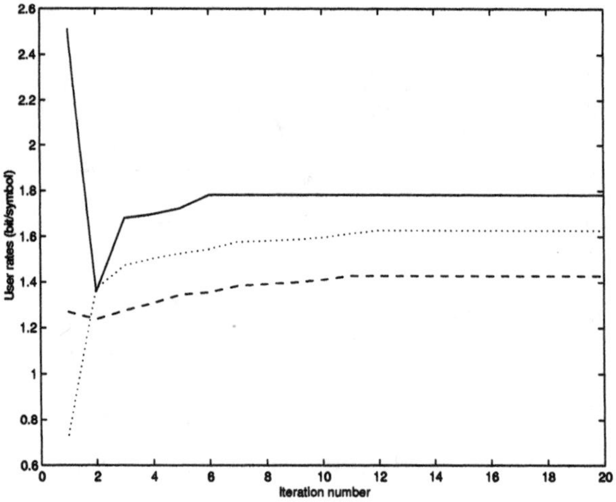

Figure 6.17 Information rates (bps) versus number of iterations in a multiple access system with three users.

iterations, iterative water-filling has already converged to the final optimal rates.

6.8.2 Capacity Region of the Broadcast Channel

In the broadcast channel, the access point transmits information to all active users, simultaneously. The broadcast channel differs from the multiple access channel for the following reasons: 1) there is only one power constraint, given by the power

10 Clearly, the sum-rate maximization does not prevent the situation where a user could get a null rate, because its channel is never better than the other channels. In [22] it was proposed a method that, for a given constraint on the total power, enforces a desired rate profile among the users.

available at the access point, as opposed to the M power constraints referring to the users of the multiple access channel; 2) the decoding operation is performed at each user terminal, *independently* of the other users; 3) whereas, in the multiple access case, the received signals arrive at the receiver through as many different channels as the number of users, in the broadcast case, the entire signal, containing all users symbols, arrives at each receiver through a single channel.

The analysis of the broadcast channel requires the introduction of the concept of *degraded* broadcast channel. Let us consider, for simplicity, the two-user vector broadcast channel:

$$\boldsymbol{y}_1 = \boldsymbol{H}_1\boldsymbol{x} + \boldsymbol{v}_1 \tag{6.128}$$

$$\boldsymbol{y}_2 = \boldsymbol{H}_2\boldsymbol{x} + \boldsymbol{v}_2 \tag{6.129}$$

where $\boldsymbol{x}$ is the vector transmitted by the broadcasting node, $\boldsymbol{y}_1$ and $\boldsymbol{y}_2$ are the vectors received from user 1 and 2, $\boldsymbol{H}_1$ and $\boldsymbol{H}_2$ are the channel matrices between the transmitter and the two receivers, $\boldsymbol{v}_1$ and $\boldsymbol{v}_2$ are the noise vectors at the two receivers. A broadcast channel is degraded if

$$p_{\mathbf{Y}_1,\mathbf{Y}_2/\mathbf{X}}(\boldsymbol{y}_1, \boldsymbol{y}_2/\boldsymbol{x}) = p_{\mathbf{Y}_1/\mathbf{X}}(\boldsymbol{y}_1/\boldsymbol{x})\, p_{\mathbf{Y}_2/\mathbf{Y}_1}(\boldsymbol{y}_2/\boldsymbol{y}_1). \tag{6.130}$$

This is equivalent to say that

$$p_{\mathbf{Y}_2/\mathbf{Y}_1,\mathbf{X}}(\boldsymbol{y}_2/\boldsymbol{y}_1, \boldsymbol{x}) = p_{\mathbf{Y}_2/\mathbf{Y}_1}(\boldsymbol{y}_2/\boldsymbol{y}_1). \tag{6.131}$$

Intuitively speaking, this is equivalent to say that the signal received by one user is a noisy version of the signal received by the other user. When condition (6.130) does not hold true, the channel is said to be a nondegraded broadcast channel.

A Gaussian scalar broadcast channel is a particular case of (6.129) corresponding to scalar inputs and outputs and additive Gaussian noise:

$$y_1 = h_1 x + v_1 \tag{6.132}$$

$$y_2 = h_2 x + v_2. \tag{6.133}$$

The SNR at the two receivers are $\mathrm{SNR}_1 = |h_1|^2 p_T/\sigma_{n_1}^2$ and $\mathrm{SNR}_2 = |h_2|^2 p_T/\sigma_{n_2}^2$, where p_T is the variance of transmitted symbol x and $\sigma_{n_1}^2$ and $\sigma_{n_2}^2$ are the variances of the noise at receiver 1 and 2. If $\mathrm{SNR}_2 < \mathrm{SNR}_1$, y_2 is a degraded (noisier) version of y_1. Hence, a Gaussian *scalar* broadcast channel is *always* a degraded channel. Conversely, a Gaussian *vector* broadcast channel is, in general, *nondegraded*.

Let us consider now the capacity of the Gaussian scalar broadcast channel. A straightforward strategy for broadcasting could seem to be a multiplexing scheme that assigns orthogonal channels (distinct time slots, for example) to distinct users. However, this is, in general, a suboptimal approach. Cover showed that a proper strategy that superimposes the users' information symbols on top of each other performs strictly better than a strategy that send different users' symbols through orthogonal channels. The arguments for deriving this, apparently counter-intuitive statement, are the following. If y_2 is a degraded version of y_1, the information transmitted to user 2 can be decoded by user 1 without error. Hence, user 1 can subtract the interference due to the signal intended for user 2 and retrieve its own information, as if it were in the absence of any interference. Conversely, user 2 "sees" the signal sent to user 1 as an interference. Dividing the available power p_T between $P_1 = \alpha p_T$ for the first user and $P_2 = (1 - \alpha)p_T$ for the second user, with $0 \leq \alpha \leq 1$, the following rate pair is then achievable

$$R_1 = \frac{1}{2}\log_2\left(1 + \frac{\alpha p_T |h_1|^2}{\sigma_{n_1}^2}\right), \tag{6.134}$$

$$R_2 = \frac{1}{2}\log_2\left(1 + \frac{(1-\alpha)p_T |h_2|^2}{\sigma_{n_2}^2 + \alpha p_T |h_2|^2}\right), \tag{6.135}$$

where R_1 is the information rate for user 1, who does not see any interference, and R_2 is the rate for user 2, who sees the interference from user 1. An example of capacity region is shown in Figure 6.18, referring to a two-user scenario, with user 2 having $\text{SNR}_2 = 10$ dB and user 1 having SNR_1 of $10, 13, 20,$ or 30 dB. Solid lines report the rates (6.134) and (6.135), whereas the dashed lines show the rates achievable with time sharing. From Figure 6.18, we notice that, when the two users have the same SNR, the boundary of the capacity region coincides with the rates obtainable with time sharing. However, as the SNR of one user increases with respect to the other, time sharing becomes more and more suboptimal with respect to the superposition strategy. This suggests that when a transmitter is broadcasting towards users located in a relatively wide area, where the users' SNR may be very different from each other, superposition coding can be rather advantageous with respect to time sharing or, more generally, orthogonal coding.

The superposition coding idea leads to the so called *embedded modulation* scheme, where the transmitted symbols are the superposition of, for example, symbols belonging to (potentially different) QAM constellations of different amplitude. An example is shown in Figure 6.19, that shows a constellation composed of two 4-QAM on top of each other. Nearby each symbol, there is the associated Gray coded string of 4 bits. The first two bits are associated to the quadrant, whereas the last two bits are associated to the symbol within the quadrant. A user with an SNR sufficiently high to discriminate symbols at distance d_1 decodes the whole 16-QAM.

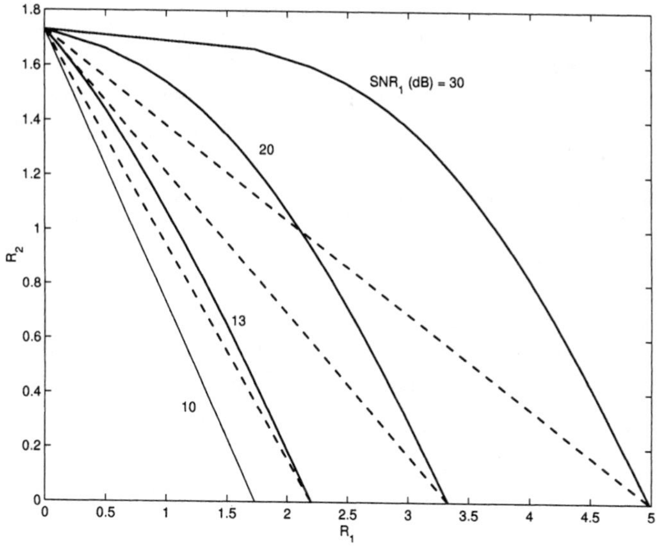

Figure 6.18 Capacity region of two-user scalar broadcast channel; $\mathrm{SNR}_1 = 10$ dB and $\mathrm{SNR}_2 = 10, 13, 20$, or 30 dB.

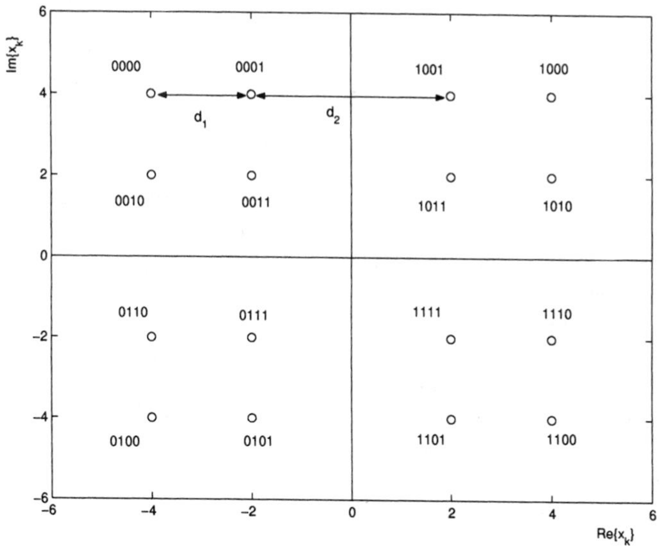

Figure 6.19 Embedded modulation: superimposed QAM.

A user with an SNR unable to distinguish symbols at distance d_1, but still able to discriminate symbols at distance d_2, takes a decision only about which quadrant contains the received sample. In this way, a low SNR receiver retrieves only the first two bits of the transmitted constellation. According to this strategy, the same transmitted signal carries bits at a rate of 4 bit/symbol for high SNR users and at a rate 2 bit/symbol for low SNR users, *at the same time*. This is an example of multirate modulation that is currently employed in digital terrestrial television broadcasting (DTTB), where different SNR users can get the same TV channel, but at different resolution levels [24].

If we consider now the general case of a scalar Gaussian broadcast channel, with M users, Cover derived the capacity region of such a channel channel. Denoting by $\boldsymbol{h} := (h_1, \ldots, h_M)$ the channel gains and by P the transmit power, the capacity is [7][11]

$$C_B(P; \boldsymbol{h}) = \left\{ \boldsymbol{R} : R_j \leq \frac{1}{2} \log \left(1 + \frac{|h_j|^2 p_j^B}{\sigma_n^2 + |h_j|^2 \sum_{m=1}^{M} p_m^B \mathbf{1}[|h_m|^2 > |h_j|^2]} \right) \right\}, \tag{6.136}$$

with $j = 1, \ldots, M$. The set in (6.136) is computed over all the power vectors $(p_1^B, \ldots, p_M^B)$ satisfying the global power constraint, that is $\sum_{m=1}^{M} p_m^B = P$. Any set of rates belonging to the capacity region is achievable through successive decoding and cancelation.

The rate region (6.135) has also a nice interpretation in terms of what is known as *dirty-paper coding* (DPC) [25]. Let us consider the channel

$$y_k = x_k + i_k + v_k \tag{6.137}$$

where x_k is the transmitted signal, i_k is interference, v_k is noise, and y_k is the received signal. Costa proved a rather surprising result [25]: If both i_k and v_k are i.i.d. Gaussian and if i_k is perfectly known at the transmitter prior to transmission, the capacity of the channel (6.137) is the *same* as if i_k were not present. In addition, the optimal transmit signal x_k is *statistically independent* of s_k.

This result suggests an alternative way to derive the capacity region of the Gaussian broadcast channel. Let us denote with $\mathcal{C}_1$ and $\mathcal{C}_2$ the two codebooks, with average power P_1 and P_2, used to send information to user 1 and 2, respectively. Since the transmitter knows both codewords, it can design $\mathcal{C}_1$ so that $\mathcal{C}_2$ is seen as an interference[12]. As a consequence, the rate towards user 1 is (6.134) and the rate

11 $\mathbf{1}[A]$ denotes the indicator function, which is a function of a logical variable A, which assumes the value 1 if A is true, whereas it is equal to 0, if A is false.

12 In other words, $\mathcal{C}_1$ is designed in order to "pre-subtract" the interference given by $\mathcal{C}_2$.

towards user 2 is (6.135).

Let us consider now the *vector* broadcast channel, where the transmitter is a multiantenna system. Denoting with n_T the number of transmit antennas, in the case of n_R single antenna receivers, we have the following n_R channels:

$$\begin{aligned} y_1 &= \boldsymbol{h}_1^T \boldsymbol{x} + v_1 \\ y_2 &= \boldsymbol{h}_2^T \boldsymbol{x} + v_2 \\ &\cdots \\ y_{n_R} &= \boldsymbol{h}_{n_R} \boldsymbol{x} + v_{n_R} \end{aligned} \tag{6.138}$$

where $\boldsymbol{h}_k,\ k = 1, \cdots, n_R$, is $n_T \times 1$, $\boldsymbol{x}$ is $n_T \times 1$, y_k and v_k, with $k = 1, \ldots, n_R$, are scalar variables. The overall channel may then be written in matrix form as

$$\boldsymbol{y} = \boldsymbol{H}\boldsymbol{x} + \boldsymbol{v} \tag{6.139}$$

where $\boldsymbol{H}$ is $n_R \times n_T$, and $\boldsymbol{v}$ is $n_R \times 1$.

The capacity of the nondegraded vector broadcast channel is still an open problem. Nevertheless, there are several recent approaches proposed to "enlarge" the region of the achievable rates as much as possible (even though it is still unknown how close are the achievable regions to the capacity region). An interesting approach is the one proposed by Caire and Shamai [26] (see also [27]) and Ginis and Cioffi [28], who suggested to exploit again, as in the scalar case, the dirty-paper coding idea. We briefly review the approach of [27] here.

Let us introduce the QR decomposition of $\boldsymbol{H}$:

$$\boldsymbol{H} = \boldsymbol{G}\boldsymbol{Q}, \tag{6.140}$$

where $\boldsymbol{G}$ is $n_R \times m$ lower triangular and $\boldsymbol{Q}$ is $m \times n_T$ and it has orthonormal rows; m denotes the rank of $\boldsymbol{H}$. Let $\boldsymbol{x} = \boldsymbol{F}\boldsymbol{u}$ be the transmitted vector, where $\boldsymbol{F}$ is a precoding matrix and $\boldsymbol{u}$ is a vector whose components are generated by successive dirty-paper encoding, as detailed next. By letting $\boldsymbol{F} = \boldsymbol{Q}^H$, we have the equivalent channel

$$\boldsymbol{y} = \boldsymbol{H}\boldsymbol{x} + \boldsymbol{v} = \boldsymbol{G}\boldsymbol{u} + \boldsymbol{v}. \tag{6.141}$$

Since $\boldsymbol{G}$ is lower triangular, (6.141) can be rewritten in scalar form as

$$y_i = G_{ii}u_i + \sum_{j<i} G_{ij}u_j + v_i, \quad i = 1, \ldots, m. \tag{6.142}$$

No information is sent to the users of index $m+1, \ldots, n_R$ (if $n_R > m$).

The entries of $\boldsymbol{u}$ are obtained by successive dirty-paper coding, where, for each i, the (known) interference signal is given by $\sum_{j<i} G_{ij} u_j$. Thus, a codebook $\mathcal{C}_1$ is used to transmit the information for user 1, as if it were without interference, a codebook $\mathcal{C}_2$ is assigned to user 2, pre-subtracting the interference brought by codebook $\mathcal{C}_1$, and so on. This encoding scheme is called zero-forcing dirty-paper (ZFDP) coding. The resulting aggregate rate is [27]

$$r_{ZFDP} = \sum_{i=1}^{m} \left[\log_2\left(\kappa |G_{ii}|^2\right)\right]^+ \tag{6.143}$$

where κ is the solution of the water-filling equation [27]:

$$\sum_{i=1}^{m} \left[\kappa - \frac{1}{|G_{ii}|^2}\right]^+ = p_T, \tag{6.144}$$

and p_T is the transmit power.

6.9 THE COCKTAIL PARTY PROBLEM: A GAME-THEORETIC FORMULATION

In the previous section we have studied the cases of structured networks, composed of users transmitting towards an access point (multiple access channel) or a base station transmitting towards many users (broadcast channel). In this section, we consider the more challenging case of an unstructured network, where there are Q sources communicating with as many destinations, without any intermediate access point. The fundamental question, in such a context, concerns the choice of the multiplexing strategy. Typically, the strategy is chosen a priori. However, this rigidity induces constraints that limit the possibility for the system to optimize the use of the available resources.

In this section, we show how to find out the coding and multiplexing strategy that maximizes the rates of all links, under a constraint on the average transmit power of each terminal. We start formulating the problem as a multiobjective optimization problem. Then, we show how to achieve suboptimal solutions by casting the problem as a game among players (radio nodes) that compete with each other to optimize their own utility functions.

We make a few assumptions, which are useful to simplify the problem. The assumptions are the following:

a1 All channels are modeled as FIR filters of maximum order L_h;

a2 Each terminal transmits blocks of length $N + L$, with a cyclic prefix of length L, where L incorporates the maximum channel order L_h plus the relative delay between the transmission from each source terminal to its own destination terminal; the transmitted block is obtained through linear precoding of the information bits;

a3 The overall system is quasisynchronous;

a4 Coding and decoding operations are performed on each source/destination pair, independently of all other pairs;

a5 The noise is Gaussian and each transmitter is constrained to transmit with a maximum average power;

a6 The channel from each source to its own destination is known to *both* transmit and receive terminals, but not to the other terminals;

a7 Each receiver is able to estimate the covariance matrix of the disturbance coming from other links and to feed it back to its own transmitter, with no errors.

Some of these assumptions, like e.g. **a6** and **a7**, are rather optimistic in practice. It is important to incorporate them to make the problem mathematically tractable. But, as a consequence, the results have to be intended as theoretical bounds.

6.9.1 Multiobjective Formulation

We adopt as optimality criterion the maximization of all the rates in each link, under an average power constraint on each transmitter. This is a *multiobjective* constrained optimization problem. No multiplexing strategy is imposed a priori, to make the system free to select the optimal solution. Hence, in principle, each user interferes with each other. Also, because of **a4**, neither joint detection nor interference cancelation are possible at each receiver.

Before starting, it is necessary to clarify the meaning of the multiobjective optimization. We adopt the Pareto definition: A vector of rates is *Pareto optimal* if *it is not possible to make the rate of some link better off without making at least one of the other links rates worse off.*

Formally speaking, let us consider Q objective functions $f_1(\boldsymbol{x}), \ldots, f_Q(\boldsymbol{x})$. We denote with $\mathcal{X}$ the set of feasible vectors, i.e. the vectors satisfying the problem constraints. Given two vectors $\boldsymbol{x}^{(1)}, \boldsymbol{x}^{(2)} \in \mathcal{X}$, we say that $\boldsymbol{x}^{(1)}$ *dominates* $\boldsymbol{x}^{(2)}$, with respect to the set of functions $\{f_q(\boldsymbol{x})\}_{q=1}^{Q}$, if $f_q(\boldsymbol{x}^{(1)}) \geq f_q(\boldsymbol{x}^{(2)})$, for all q, and there exists at least one index r such that $f_r(\boldsymbol{x}^{(1)}) > f_r(\boldsymbol{x}^{(2)})$. In words,

$x^{(1)}$ dominates $x^{(2)}$ if the corresponding set of functions is better in at least one component function, and it is not worse in all the others.

A vector $x^* \in \mathcal{X}$ is *Pareto optimal*, or optimal in the Pareto sense, if there exists no vector $x \in \mathcal{X}$ that dominates x^*. In general, a multiobjective optimization problem yields several Pareto optimal points, which are all *equally optimal*, in the Pareto sense. The choice among these points has to be carried out according to some extra criterion. For example, talking about transmission rates, one could assign different priorities to different users. The set of all Pareto optimal points is called the *optimal trade-off surface*.

As an example, let us consider the case of two generic functions, where the admissible values that the pair of functions $f_1(x)$ and $f_2(x)$ assume, for $x \in \mathcal{X}$, are given by the shaded area sketched in Figure 6.20. The Pareto optimal trade-off surface is given, in this case, by the two thick curves reported in Figure 6.20.

Let us apply now the Pareto optimality criterion to solve the cocktail party problem.

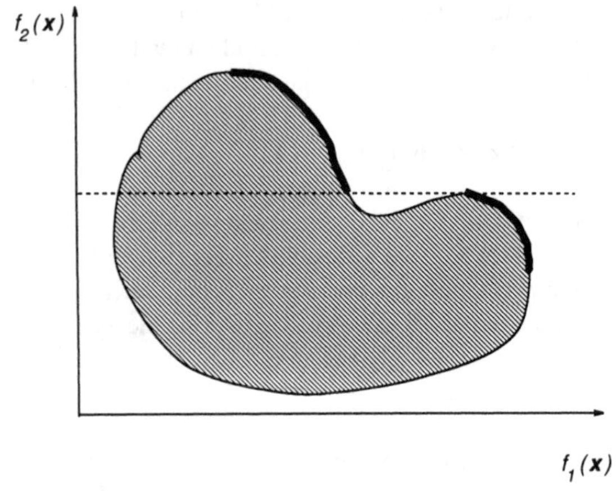

Figure 6.20 Example of Pareto optimality.

The scenario of interest is depicted, for example, in Figure 6.21, where we observe three sources and destinations. Each source transmits to its own destination. Since no multiplexing strategy is assumed a priori, each destination receives interference (dashed lines) from all sources different from its own.

Because of **a2** and **a3**, each source transmits a block $x_q = F_q s_q$, where F_q is the precoding matrix and s_q is the information symbol vector. The block received by

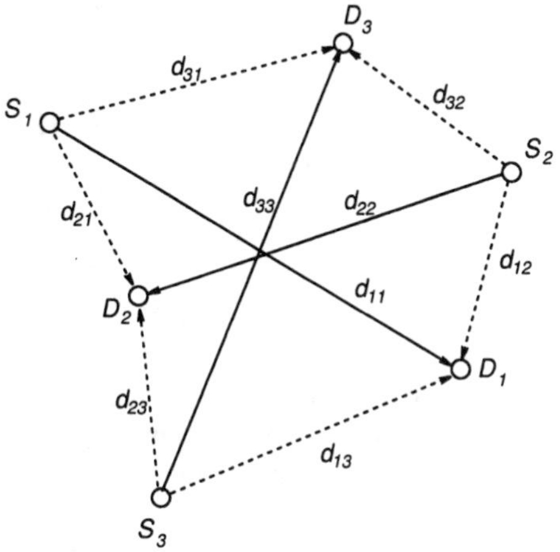

Figure 6.21 Fully-connected network; useful links (solid lines) and interference links (dashed lines).

the qth destination is then (we drop the index block)

$$\boldsymbol{y}_q = \sum_{r=1}^{Q} \boldsymbol{H}_{rq}\boldsymbol{F}_r\boldsymbol{s}_r + \boldsymbol{v}_q, \tag{6.145}$$

where, because of **a2** and **a3**, the channel matrices $\boldsymbol{H}_{rq}$, connecting source r to destination q, are Toeplitz circulant matrices. Under **a5**, the pdf of the symbols transmitted from each source that maximizes the information rate on each link is a Gaussian pdf. Furthermore, because of **a4**, each destination sees the signals received from all other sources as interference signals (as it is not able to decode them). As a consequence, the maximum information rate on the link between source q and its own destination is[13]

$$R_q = \frac{1}{N}\log_2\left(\frac{|\sigma_n^2\boldsymbol{I} + \sum_{r=1}^{Q}\boldsymbol{H}_{rq}\boldsymbol{F}_r\boldsymbol{F}_r^H\boldsymbol{H}_{rq}^H|}{|\sigma_n^2\boldsymbol{I} + \sum_{\substack{r=1\\ r\neq q}}^{Q}\boldsymbol{H}_{rq}\boldsymbol{F}_r\boldsymbol{F}_r^H\boldsymbol{H}_{rq}^H|}\right) \text{ bps.} \tag{6.146}$$

The optimization problem can thus be formulated as the search for the matrices $\boldsymbol{F}_r$, satisfying the power constraints $\text{tr}(\boldsymbol{F}_r\boldsymbol{F}_r^H) = P_r$, that maximize, in the Pareto

13 The normalization by N is used here to evaluate the rate in terms of bit/symbol (bps).

sense, the vector of rates $(R_1, R_2, \ldots, R_Q)$. In formulas,

$$\{\boldsymbol{F}_1, \ldots, \boldsymbol{F}_Q\} = \arg\max_{\boldsymbol{F}_r} \frac{1}{N} \log_2 \left(\frac{|\sigma_n^2 \boldsymbol{I} + \sum_{r=1}^{Q} \boldsymbol{H}_{rq} \boldsymbol{F}_r \boldsymbol{F}_r^H \boldsymbol{H}_{rq}^H|}{|\sigma_n^2 \boldsymbol{I} + \sum_{\substack{r=1 \\ r \neq q}}^{Q} \boldsymbol{H}_{rq} \boldsymbol{F}_r \boldsymbol{F}_r^H \boldsymbol{H}_{rq}^H|} \right),$$
$$\text{subject to } \operatorname{tr}(\boldsymbol{F}_r \boldsymbol{F}_r^H) = P_r, \tag{6.147}$$

where P_r is the power available to the rth transmitter. This is, in general, a formidable problem to solve, as it is not a convex problem. Nevertheless, the assumptions made before can help to simplify the problem. The rates R_q in (6.146) can be rewritten, equivalently, as

$$\begin{aligned} R_q &= \frac{1}{N} \log_2 |(\sigma_n^2 \boldsymbol{I} + \sum_{\substack{r=1 \\ r \neq q}}^{Q} \boldsymbol{H}_{rq} \boldsymbol{F}_r \boldsymbol{F}_r^H \boldsymbol{H}_{rq}^H)^{-1} (\sigma_n^2 \boldsymbol{I} + \sum_{r=1}^{Q} \boldsymbol{H}_{rq} \boldsymbol{F}_r \boldsymbol{F}_r^H \boldsymbol{H}_{rq}^H)| \\ &= \frac{1}{N} \log_2 |\boldsymbol{I} + (\sigma_n^2 \boldsymbol{I} + \sum_{\substack{r=1 \\ r \neq q}}^{Q} \boldsymbol{H}_{rq} \boldsymbol{F}_r \boldsymbol{F}_r^H \boldsymbol{H}_{rq}^H)^{-1} \boldsymbol{H}_{qq} \boldsymbol{F}_q \boldsymbol{F}_q^H \boldsymbol{H}_{qq}^H|. \end{aligned} \tag{6.148}$$

Introducing the so-called *effective* channel matrix

$$\boldsymbol{H}_{qq}^e := \left(\sigma_n^2 \boldsymbol{I} + \sum_{\substack{r=1 \\ r \neq q}}^{Q} \boldsymbol{H}_{rq} \boldsymbol{F}_r \boldsymbol{F}_r^H \boldsymbol{H}_{rq}^H \right)^{-1/2} \boldsymbol{H}_{qq}, \tag{6.149}$$

(6.148) can be simplified into

$$R_q = \frac{1}{N} \log_2 |\boldsymbol{I} + \boldsymbol{H}_{qq}^e \boldsymbol{F}_q \boldsymbol{F}_q^H \boldsymbol{H}_{qq}^{eH}|. \tag{6.150}$$

The meaning of effective channel is evident from the comparison of (6.148) and (6.150): A channel with channel matrix $\boldsymbol{H}_{qq}$ and additive Gaussian colored noise with covariance matrix

$$\boldsymbol{R}_q = \sigma_n^2 \boldsymbol{I} + \sum_{\substack{r=1 \\ r \neq q}}^{Q} \boldsymbol{H}_{rq} \boldsymbol{F}_r \boldsymbol{F}_r^H \boldsymbol{H}_{rq}^H, \tag{6.151}$$

is perfectly equivalent, from the capacity point of view, to a channel with white Gaussian noise, with unit variance, and channel matrix $\boldsymbol{H}^e_{qq}$ given by (6.149).

Under the initial assumptions, the matrix $\boldsymbol{H}^e_{qq}$ can be approximately diagonalized, for large N, as

$$\boldsymbol{H}^e_{qq} \approx \boldsymbol{W}\boldsymbol{\Lambda}^e_{qq}\boldsymbol{W}^H, \tag{6.152}$$

where $\boldsymbol{W}$ is the FFT matrix, with entries $W_{kl} = e^{j2\pi kl/N}/\sqrt{N}$. Within the limits of validity of this approximation, the rates R_q can be rewritten as

$$R_q = \frac{1}{N}\log_2|\boldsymbol{I} + \boldsymbol{\Lambda}^e_{qq}\tilde{\boldsymbol{F}}_q\tilde{\boldsymbol{F}}^H_q\boldsymbol{\Lambda}^{e^H}_{qq}|, \tag{6.153}$$

where $\tilde{\boldsymbol{F}}_q := \boldsymbol{W}^H\boldsymbol{F}_q$. Since $\mathrm{tr}(\tilde{\boldsymbol{F}}_q\tilde{\boldsymbol{F}}^H_q) = \mathrm{tr}(\boldsymbol{F}_q\boldsymbol{F}^H_q)$, $\tilde{\boldsymbol{F}}_q$ satisfies the same constraint as $\boldsymbol{F}_q$. Hence, we can maximize R_q, with respect to $\tilde{\boldsymbol{F}}_q$, for $q = 1, \ldots, Q$, subject to the same power constraint as $\boldsymbol{F}_q$. The maximization of each term R_q in (6.153) can be carried out by noticing that, since the argument of the determinant in (6.153) is a positive definite matrix, by virtue of Hadamard's inequality [7], R_q in (6.153) is maximum when $\tilde{\boldsymbol{F}}_q$ is a diagonal matrix. Specifically, setting $\tilde{\boldsymbol{F}}_q = \boldsymbol{\Phi}_q$, with $\boldsymbol{\Phi}_q$ diagonal, the optimal structure for each $\boldsymbol{F}_q$ is

$$\boldsymbol{F}_q = \boldsymbol{W}\tilde{\boldsymbol{F}}_q = \boldsymbol{W}\boldsymbol{\Phi}_q. \tag{6.154}$$

Hence, recalling the main results of Chapter 3, within the limits of validity of all previous assumptions and approximations, *the optimal multiplexing strategy, in case of stationary channels, is a multicarrier strategy*. However, whether distinct users get non-overlapping bands or not is to be decided yet.

Using the coding structure (6.154), the rate expressions can be considerably simplified. In fact, denoting with $p_q(i) := |\Phi_q(i)|^2$, $i = 0, \ldots, N-1$, the power allocated by the qth source on the ith subcarrier, introducing (6.154) in (6.153), we can rewrite each rate R_q as a function of the vector $\boldsymbol{p} := (\boldsymbol{p}_1, \ldots, \boldsymbol{p}_Q)$, containing the power vectors $\boldsymbol{p}_q = [p_q(0), \ldots, p_q(N-1)]$ pertaining to all the transmitters. Denoting by $H_{rq}(i)$ the transfer function of the channel between the rth source and the qth destination, that is $H_{rq}(i) = \sum_{k=0}^{L} h_{rq}(q)\, e^{-j2\pi ik/N}$, the rates can be rewritten as

$$R_q(\boldsymbol{p}) = \frac{1}{N}\sum_{i=0}^{N-1}\log\left(1 + \frac{1}{\Gamma}\frac{|H_{qq}(i)|^2 p_q(i)}{\sigma_n^2 + \sum\limits_{\substack{r=1\\r\neq q}}^{Q}|H_{rq}(i)|^2 p_r(i)}\right). \tag{6.155}$$

In (6.155) we have also incorporated the SNR gap Γ, chosen in order to enforce a target bit error rate (see, e.g., Section 6.3.3).

In summary, under the assumptions **a1÷a7**, the initial optimization problem is converted into the search for the power vector $\boldsymbol{p}^* := (\boldsymbol{p}_1^*, \ldots, \boldsymbol{p}_Q^*)$ that maximizes the rate vector $(R_1, \ldots, R_Q)$, in the Pareto sense. In formulas,

$$\begin{array}{ll} \{\boldsymbol{p}_1^*, \cdots, \boldsymbol{p}_Q^*\} = \text{argmax}\{R_1(\boldsymbol{p}), R_2(\boldsymbol{p}), \ldots, R_Q(\boldsymbol{p})\}, \\ \text{subject to} \quad \sum_{i=0}^{N-1} p_k(i) \leq P_k, \;\; k = 1, \ldots, Q, \end{array} \tag{6.156}$$

with $R_q(\boldsymbol{p})$ as given in (6.155). This is still a formidable optimization problem to solve, as, in general, it is non convex. Furthermore, besides the specific difficulties related to the specific objective functions to be maximized, there is one more practical aspect to be considered. The solution of (6.156) requires, in general, a centralized unit that, knowing all channels and interferences, computes the optimal coding matrices. This is a rather unrealistic assumption, especially for wireless channel communications.

6.9.2 Basic Game Theory Concepts

To avoid the difficulties of the centralized solution, we formulate now the optimization problem (6.156) as a strategic game. The game theory formulation offers in fact a powerful tool to search for a decentralized solution of the optimal multiplexing problem. In this section, we review first the basic concepts of game theory. The interested reader is invited to check [29, 30, 31, 32] for a more in-depth overview of game theory.

Game theory is, in general, the mathematical representation of the interaction among individuals (players) whose actions affect other individuals. Games can be formulated in different alternatives ways, depending on the model of interaction among players. The most known forms are the *extensive form* and the *strategic form.* A strategic form game is completely specified by three sets of objects: the set of players, the set of strategies for each player, and the set of utility (or payoff) functions for each player. Formally speaking, a strategic game is completely defined by the following structure $\mathscr{G}$

$$\mathscr{G} = \{\Omega, \{\mathcal{X}_q\}_{q\in\Omega}, \{\varphi_q\}_{q\in\Omega}\},$$

where

$\Omega \equiv \{1, 2, \ldots, Q\}$ is the set of *players*;

$\{\mathcal{X}_q\}_{q\in\Omega}$ is the set of feasible *strategies* $\mathbf{x}_q$, for each player;

$\{\varphi_q\}_{q\in\Omega}$ is the set of *payoff* functions that map the strategies onto real numbers that quantify the utility of each player.

The values assumed by the payoff functions depend, in general, on the strategies $\mathbf{x} := [\mathbf{x}_1^T, \ldots, \mathbf{x}_Q^T]^T \in \mathcal{X}$ of *all* players; the cartesian product $\mathcal{X} = \mathcal{X}_1 \times \cdots \times \mathcal{X}_Q$ is the set of all the strategy profiles.

It is important to remark that a strategic game is a *static model* that ignores questions of timing and treats players as if they choose their strategies simultaneously. Conversely, an *extensive form* game is a *dynamic model* that includes a full description of the sequence of moves and events that occur over time in an actual play of the game. In general, strategic games can be seen as a substantial simplification of real games. Nevertheless, von Neumann and Morgenstern suggested a procedure for constructing a game in strategic form, given any extensive form game [33]. In this book, we will consider only strategic games. Furthermore, we assume that there is no cooperation among the radio nodes.

In a strategic *noncooperative* game, each player competes with the others and it chooses its strategy in order to maximize its own payoff, regardless of what happens to the other players. Mathematically speaking, denoting with $\mathbf{x}_{-q} := [\mathbf{x}_1^T, \ldots, \mathbf{x}_{q-1}^T, \mathbf{x}_{q+1}^T, \ldots, \mathbf{x}_Q^T]^T$ the vector of all players' choices, except the qth one, the competition can be expressed as the following set of maximization problems (one for each player):

$$\begin{array}{rl} \boldsymbol{x}_q^* = \arg\max\limits_{\mathbf{x}_q} & \varphi_q(\mathbf{x}_q, \mathbf{x}_{-q}) \\ \text{subject to} & \mathbf{x}_q \in \mathcal{X}_q, \end{array} \qquad \forall q \in \Omega. \tag{6.157}$$

In words, (6.157) means that each player searches for the strategy $\boldsymbol{x}_q$ that maximizes its own payoff, given the strategies $\boldsymbol{x}_{-q}$ adopted by all other players. Note that, for each q, the maximum in (6.157) is taken over $\mathbf{x}_q$, for a *fixed* $\mathbf{x}_{-q}$. Since the payoff of each player is a function of the strategies of the other players, the strategy profile that optimizes the individual payoff function depends on the strategy of the others. A solution of (6.157), if it exists, is the so-called *Nash equilibrium* (NE), which is formally defined in the following manner:

A Nash Equilibrium (NE) of a strategic noncooperative game $\mathcal{G} = \{\Omega, \{\mathcal{X}_q\}_{q\in\Omega}, \{\varphi_q\}_{q\in\Omega}\}$ is a strategy profile $\mathbf{x}^\star \in \mathcal{X}$ such that, for every $q \in \Omega$, $\varphi_q(\mathbf{x}_q^\star, \mathbf{x}_{-q}^\star) \geq \varphi_q(\mathbf{x}_q, \mathbf{x}_{-q}^\star)$, for all $\mathbf{x}_q \in \mathcal{X}_q$.

In words, at a NE, each player, given the strategies of the other players, does not get any increase of its payoff function by changing its own strategy. Alternatively, at a NE, the strategy profile chosen by each player is the *best response* to the strategies adopted by the other players. This interpretation suggests a definition of NE based on the notion of best response. We define the set $\mathbf{B}_q(\mathbf{x}_{-q}) := \{\mathbf{x}_q \in \mathcal{X}_q : \varphi_q(\mathbf{x}_q, \mathbf{x}_{-q}) \geq \varphi_q(\overline{\mathbf{x}}_q, \mathbf{x}_{-q}),$ for all $\overline{\mathbf{x}}_q \in \mathcal{X}_q\}$ as the set of the best responses of player q to the strategies adopted by the other players, given by the vector $\boldsymbol{x}_{-q}$. A NE can thus be formulated as a strategy profile where the strategy of each player is the best response to the strategies adopted by all other players. In formulas, $\mathbf{x}_q^*$ is such that $\mathbf{x}_q^* \in \mathbf{B}_q(\mathbf{x}_{-q}^*)$ for all $q \in \Omega$. This interpretation is useful as it allows us to cast a pure strategy NE $\mathbf{x}^*$ as a *fixed point*, under the best response mapping[14]. We can thus see a Nash equilibrium as a generalization of the fixed point concept that we had already seen in Chapter 4, about power control.

It is important to remark that not every strategic game admits a NE, as defined above. The following theorems give the sufficient conditions for the *existence* and *uniqueness* of a NE [34]

Existence theorem: The strategic game $\mathscr{G} = \{\Omega, \{\mathcal{X}_q\}_{q\in\Omega}, \{\varphi_q\}_{q\in\Omega}\}$ admits at least one NE if, for all $q \in \Omega$:

1. The set $\mathcal{X}_q$ is a nonempty compact convex subset of a Euclidean space;
2. Each payoff function $\varphi_q(\mathbf{x})$ is continuous on $\mathcal{X}$ and quasiconcave on $\mathcal{X}_q$.

To establish the conditions for uniqueness, it is necessary to introduce the notion of diagonally strictly concave functions. Let us consider a weighted nonnegative sum of the functions $\varphi_q(\mathbf{x})$:

$$\sigma(\boldsymbol{x}, \boldsymbol{r}) = \sum_{q=1}^{Q} r_q \varphi_q(\boldsymbol{x}), \quad r_q \geq 0, \tag{6.158}$$

for each nonnegative vector $\boldsymbol{r} := (r_1, \ldots, r_Q)$. For any given vector $\boldsymbol{r}$, let us introduce the mapping $g(\boldsymbol{x}, \boldsymbol{r})$ as

$$g(\boldsymbol{x}, \boldsymbol{r}) : \begin{bmatrix} r_1 \nabla_1 \varphi_1(\boldsymbol{x}) \\ r_2 \nabla_2 \varphi_2(\boldsymbol{x}) \\ \vdots \\ r_Q \nabla_Q \varphi_Q(\boldsymbol{x}) \end{bmatrix}, \tag{6.159}$$

where $\nabla_q \varphi_i(\boldsymbol{x})$ denotes the gradient of $\varphi_i(\boldsymbol{x})$ with respect to the vector $\boldsymbol{x}_q$. Let us also denote by $\mathcal{R}$ the set of vectors $\boldsymbol{x}$ satisfying the problem constraints. The

14 This definition of fixed point is a generalization of the definition adopted in (4.58). In fact, (4.58) referred to fixed points defined for functions (i.e., point-to-point mappings) whereas here we refer to fixed points defined for correspondences (i.e., point-to-set mappings).

function $\sigma(\boldsymbol{x}, \boldsymbol{r})$ introduced in (6.158) is *diagonally strictly concave* for $\boldsymbol{x} \in \mathcal{R}$, and fixed $\boldsymbol{r}$, if, for every $\boldsymbol{x}_0$ and $\boldsymbol{x}_1 \in \mathcal{R}$, we have

$$(\mathbf{x}_1 - \mathbf{x}_0)^T(\mathbf{g}(\mathbf{x}_0, \boldsymbol{r}) - \mathbf{g}(\mathbf{x}_1, \boldsymbol{r})) > 0$$

for some $\boldsymbol{r} \in \mathbb{R}_+^Q$. Alternatively, introducing the matrix $\boldsymbol{G}(\boldsymbol{x}, \boldsymbol{r})$ as the Jacobian of $\boldsymbol{g}(\boldsymbol{x}, \boldsymbol{r})$ with respect to $\boldsymbol{x}$, $\sigma(\boldsymbol{x}, \boldsymbol{r})$ is diagonally strictly concave if the symmetric matrix $\boldsymbol{G}(\boldsymbol{x}, \boldsymbol{r}) + \boldsymbol{G}^H(\boldsymbol{x}, \boldsymbol{r})$ is negative definite, for some $\boldsymbol{r} \in \mathbb{R}_+^Q$.

Given the previous definitions, we can state the following [34]

Uniqueness theorem: The NE is *unique* if, in addition to satisfying the existence conditions, $\varphi_q(\mathbf{x})$ is concave on $\mathcal{X}_q$ and, for some $\boldsymbol{r} \geq \mathbf{0}$, the function $\sigma(\boldsymbol{x}, \boldsymbol{r})$, defined in (6.158) is diagonally strictly concave.

6.9.3 Maximum Information Rate Coding Based on Game Theory

We can now formulate the rate maximization problem (6.156) as a game $\mathscr{G}$, where each source (player) competes against the others to maximize its own transmission rate. We review here the basic results found by Scutari [35] (see also [36]). Each source searches for the power allocation (player's strategy), over the available bandwidth that maximizes its own rate (payoff), under a given limited power budget that puts a constraint on the admissible strategies. The structure of the game is then the set $\mathscr{G} = \{\Omega, \{\mathscr{P}_q\}_{q\in\Omega}, \{R_q\}_{q\in\Omega}\}$, where

$\Omega \equiv \{1, 2, \ldots, Q\}$ is the set of *active links*;

$\mathscr{P}_q = \left\{\mathbf{p}_q := [p_q(0), \ldots, p_q(N-1)]^T \in \mathbb{R}_+^N : 1/N \sum_{k=0}^{N-1} p_q(k) \leq P_q\right\}$ is the set of admissible power allocations for each user;

R_q, as given by (6.155), is the information rate of the qth player.

It is now important to analyze under which conditions there exists a NE and check if the equilibrium is unique. Using the existence theorem recalled before, it is easy to verify that the game always admits at least one NE. In fact, denoting with the N-size vector $\boldsymbol{p}_q$ the power profile adopted by user q, and with $\boldsymbol{p}_{-q}$ the set of power profiles adopted by all other users, except the qth one, it is easy to check that the rate R_q, as given in (6.155), for any given vector $\boldsymbol{p}_{-q}$, is quasi-concave with respect to $\boldsymbol{p}_q$.

The proof of uniqueness is more complicated. Nevertheless, it was proved in [35] that the NE is unique if

$$\Gamma \sum_{r=1, r\neq q}^{Q} \frac{|\bar{H}_{rq}(k)|^2}{|\bar{H}_{qq}(k)|^2} \frac{d_{qq}^{\alpha}}{d_{rq}^{\alpha}} \frac{P_r}{P_q} < 1, \qquad \forall q \in \Omega, \forall k \in [0, N-1]. \qquad (6.160)$$

where we have introduced the normalized channel transfer functions $\bar{H}_{rq}(k) := H_{rq}(k)d_{rq}^{\alpha}$, having indicated with d_{rq}^{α} the attenuation factor due to distance; the exponent α is equal to 2 in free-space propagation and it is typically between 2 and 5 in urban propagations. In (6.160), d_{qq} is the distance between the qth transmitter and the corresponding receiver, whereas d_{rq} is the distance between the rth interfering transmitter and the qth receiver. The ratio d_{qq}/d_{rq} is then a measure of the separation between the links: The greater is d_{rq}/d_{qq}, for $r \neq q$, the more separated are the links and then the lesser is the interference.

Expression (6.160) has a straightforward interesting physical interpretation: The uniqueness condition holds true if the links are sufficiently far apart from each other. This is not surprising as, in such a case, the interference is low enough to make the interaction (interference) among users negligible. And, in the absence of interference, we know from Section 6.3.3 that the maximum information rate problem admits a unique solution.

It is important to remark that condition (6.160) is only a sufficient condition. This means that, if (6.160) is not satisfied, we are simply unable to judge upon the uniqueness.

After having established the conditions for the existence and uniqueness of a Nash equilibrium, it is important to provide algorithms able to reach the equilibrium. One possibility, suggested independently in [37] and [38], is given by the iterative water-filling algorithm adapted to the multipoint-to-multipoint communication scenario. From the definition of NE, in fact, the optimal power allocation strategy for every player must maximize the user rate, given the transmission strategy chosen by the other users. The solution of this problem, for each user, given the other user choices, is precisely the water-filling power distribution that treats the interference coming from the other players as additive (colored) noise. Hence, at any NE, the power allocation must satisfy the following water filling distribution:

$$p_q(k) = \left(\mu_q - \frac{\sigma_w^2 + \sum_{r\neq q, r=1}^{Q} |H_{rq}(k)|^2 p_r(k)}{|H_{qq}(k)|^2} \right)^+, \forall q \in \Omega, \forall k \in [0,\ N-1] \qquad (6.161)$$

where the water-level μ_q is chosen so that the power constraint is met with equality. Since the game admits at least one stable NE, the existence of a simultaneous water-filling solution (6.161) is guaranteed. Then, given an iterative procedure such that, at each step, every player adopts the single-user water-filling power distribution (6.161) that considers the other users as interference, if it converges, it has to converge to one of the Nash equilibria, from any starting point. The algorithm is summarized in the following algorithm.

Iterative water-filling algorithm (IWFA):

1) Start with any power distribution $\mathbf{p}^{(0)} := [\mathbf{p}_1, \ldots, \mathbf{p}_Q]$, such that $\mathbf{p}_q \in \mathscr{P}_q$, $\forall q \in \Omega$;

2) repeat

for $q = 1$ to Q

$$i_q^{(n)}(k) := \sigma_n^2 + \textstyle\sum_{r \neq q, r=1}^{Q} |H_{rq}(k)|^2 p_r^{(n)}(k), \quad \forall k \in [0,\ N-1];$$

$$\mathbf{p}_q^{(n)} = \arg\max \left\{ \frac{1}{N} \sum_{k=0}^{N-1} \log \left(1 + \frac{1}{\Gamma} \frac{|H_{qq}(k)|^2 p_q^{(n)}(k)}{i_q^{(n)}(k)} \right), \text{s.t. } \mathbf{p}_q \in \mathscr{P}_q \right\};$$

end

until the desired accuracy is reached.

The symbol $i_q^{(n)}(k)$ denotes the noise plus interference power received by the qth user, on the kth subcarrier, at the nth iteration. The power vector $\boldsymbol{p}_q^{(n)}$, at the nth iteration, is computed using the water-filling algorithm described in Section 6.3.3. It is important to remark that, for the problem at hand, the optimal power allocation $\mathbf{p}_q$ is computed without any eigen-decomposition (as the channel eigenvectors are the complex sinusoids and the channel eigenvalues are the channel transfer functions). Most important, each radio node (player) computes its own optimal power allocation, as a function of frequency, requiring only the knowledge of its own channel and the covariance matrix of the interference received by its own receiver. In other words, no coordination among the users is required to reach the NE. This is why the iterative water-filling algorithm is suited for decentralized solutions.

6.9.4 Pareto Optimality Versus Nash Equilibria

In general, a Nash equilibrium, even if unique, does not necessarily coincide with some of the optimal Pareto solutions. In fact, a NE is the result of a distributed self-optimization, obtained through a decentralized strategy. Conversely, a Pareto optimum can be reached only through a centralized control node that needs to know all channels. For this reason, Pareto optimization is not suitable for decentralized strategies. However, it is useful as a benchmark term. It is then important to assess the loss resulting from the use of a decentralized game theoretic approach with respect to a centralized Pareto-optimal strategy.

In Figure 6.22, we compare the rate regions achieved with the optimal centralized Pareto optimization and with Nash equilibria, in a scenario composed of two transmitters and two receivers. The two axes represent the rates, in bit/symbol, for the two links. The achievable rates are all the points lying between the two axes and the curves drawn in the figure. The two pairs of nodes are placed at different distances, to test situations with different level of interference. The solid line represents the Pareto-optimal trade-off curve, whereas the dashed line shows the Nash equilibria achieved using different transmit powers for the two users, under a global sum-power constraint. The Pareto-optimal curve has been obtained by using the normal boundary intersection algorithm [39].

From Figure 6.22, we can see that the Nash equilibria approach the optimal Pareto curve as the interference level decreases (i.e., the ratio d_{rq}/d_{qq} increases). But it is also important to notice that the loss resulting from using the decentralized, game-theoretic approach is quite limited also in the case where the links are very close to each other. This suggests that the decentralized approach can be a suitable solution, as it does not require any centralized algorithm, as in the Pareto case, and it yields the solutions using a rather simple algorithm, the iterative water-filling algorithm.

Let us consider now a multiuser system where each user has multiple antennas. As an example, in Figure 6.23 we report the rate region achievable with the game theoretical approach (Nash equilibria), in case of two users, having one antenna (dashed line) or a double antenna (solid line) transceiver. We can see how the use of multiple antennas provides an increase of the admissible rate region.

6.9.5 Examples of Solutions of the Cocktail Party Problem

We can now go back to the cocktail party problem and check which are the coding schemes resulting from the multiobjective maximization problem. In Figures 6.24 and 6.25 we report some examples of power spectral densities achieved as Nash equilibria. The scenario is composed of three pairs of transmitter/receiver. Each transceiver has a single antenna. The channels are FIR filters of order $L = 6$. Each transmitter sends a block of data of size $N = 128$. The two figures refer to two extreme situations: Figure 6.24 reports the results achieved in a low interference scenario where the ratio between the inter-pair distance and the intra-pair distance

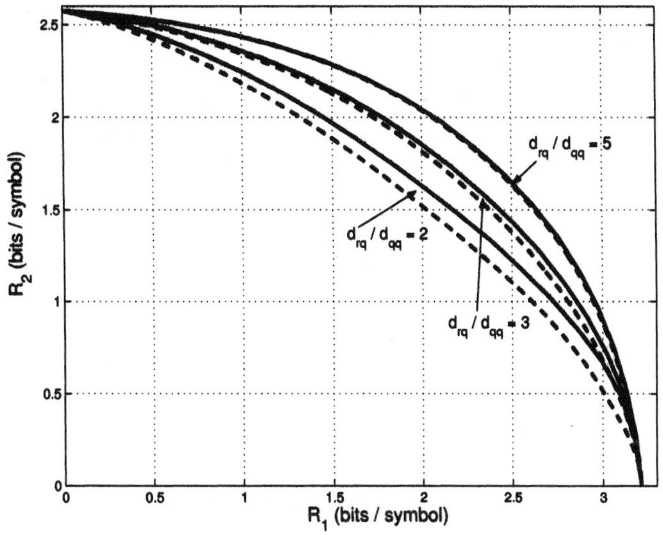

Figure 6.22 Rate region achievable with Pareto optimization (solid curves) and Nash equilibria (dashed lines).

d_{rq}/d_{qq} is five; Figure 6.25 refers to a high interference situation, where the ratio d_{rq}/d_{qq} is equal to 0.9.

We can see that, in the low interference case, all users tend to use the whole available bandwidth. In fact, when the interference is low, it is better to tolerate some interference and use as many channels as possible to maximize the information rate. Conversely, from Figure 6.25 we notice that in the high interference scenario, the users tend to transmit over non-overlapping bands. In such a case, in fact the major problem comes from the interference and then the users tend to avoid the interference by transmitting over orthogonal channels. In other words, when the interference is low, the game evolves in order to assign all the bandwidth to all the users, thus emulating the basic principles of spread spectrum CDMA. On the contrary, when the interference is high, the Nash equilibrium provides a coding strategy that emulates OFDMA, where different users get nonoverlapping portions of the spectrum.

If, instead of having stationary, frequency selective channels, we would have had flat-fading, time-varying channels, invoking duality arguments, the optimal coding strategy would have consisted in assigning distinct time slots to different users, thus

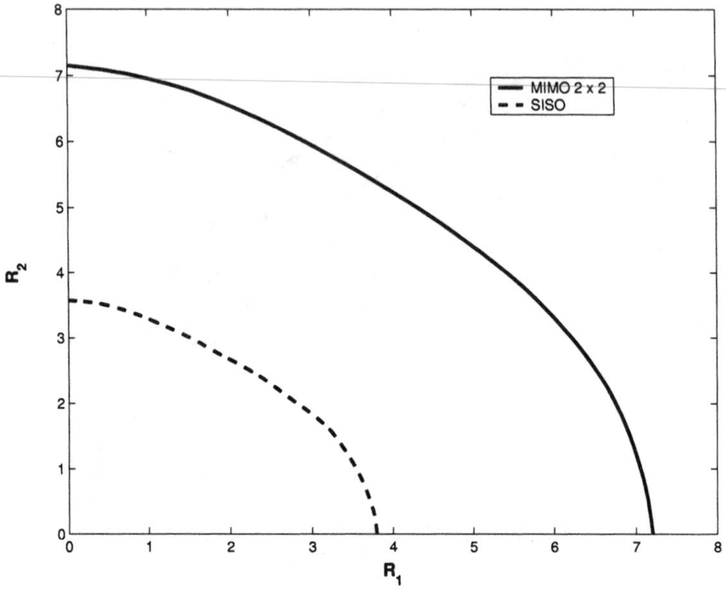

Figure 6.23 Rate region achievable with Nash equilibria in a two-user system with single antenna (dashed line) and double antenna (solid line) transceiver.

leading to TDMA.

Thus, in summary, *DS-CDMA, TDMA, and OFDMA can all be seen as particular solutions to which a general game-theoretic algorithm may converge*, depending on the channel model (frequency-selective versus flat-fading) and on the level of interference. When the interference is very high, each transmitter prefers to transmit over channels orthogonal to the other users. On the contrary, when the interference is very low, each transmitter prefers to use the whole available bandwidth, as in such a case nulling the interference would not be worth the bandwidth reduction. However, in all intermediate situations, the transmitters prefer to share portion of the bandwidth, in the effort to find the best trade-off between interference and rate.

Let us see now an example of what happens when the transceivers have multiple antennas. We report in Figure 6.26 the power spectral densities found again as a Nash equilibrium. The scenario is composed of two pairs of transmitter/receiver, each one with a double antenna transceiver. Figure 6.26a) reports the spectra transmitted by user 1, from its two antennas, and Figure 6.26b) reports the spectra transmitted by user 2, from its antennas. The two pairs are very close to each other

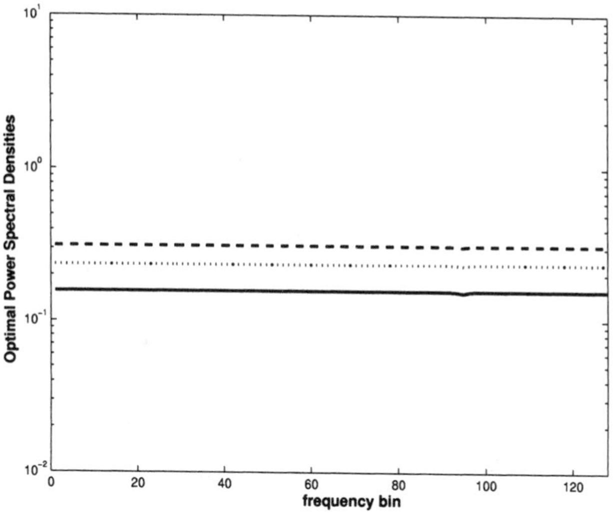

Figure 6.24 Power spectral densities of transmitted signals, corresponding to a Nash equilibrium, in case of low interference ($d_{rq}/d_{qq} = 5$).

so that the interference is high, as in Figure 6.25. Interestingly, we may observe that now, even though the interference is high, the spectra are partially overlapping. This happens because the extra degrees of freedom offered by the multiantenna transceivers allow different users to share the same spectra, without necessarily interfere because they can be still separated exploiting the spatial domain.

6.10 SUMMARY

In this chapter we have shown alternative techniques to synthesize the transmission strategy, when the transmitter knows the channel. In the first part of the chapter, we have considered the single-user context and we have provided the general solution according to alternative criteria. We have then specialized the results to SISO, MISO, and, MIMO systems to provide some physical insight into the optimal solutions. Among other alternatives, the minimum BER approach has been shown to be a good candidate as it has limited losses in terms of information rate and it exhibits a good behavior in terms of BER in many different situations. We have initially assumed perfect channel knowledge at both the transmit and receive sides. Then we have tested the resilience of the optimal systems to imperfect channel knowledge. We have shown how to optimize the transmission strategy in the more practical case where the transmitter has only partial knowledge about the channel. An excellent

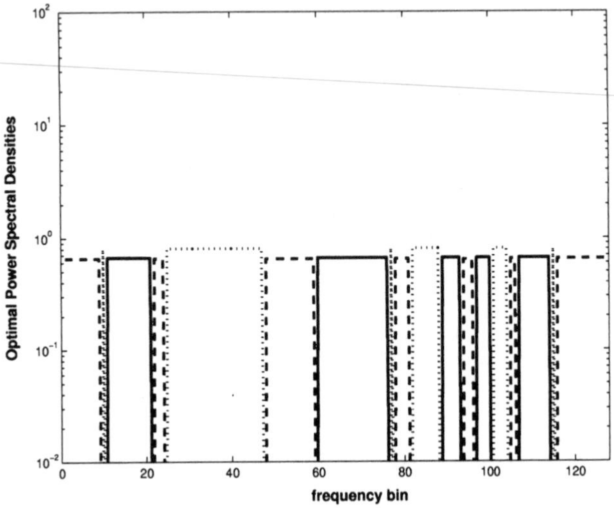

Figure 6.25 Power spectral densities of transmitted signals, corresponding to a Nash equilibrium, in case of high interference ($d_{rq}/d_{qq} = 0.9$).

recent reference on the capacity limits of MIMO channels, under different degrees of channel knowledge at the transmit and receive sides is [40].

In the second part, we have considered the multi-user case. We have analyzed, in particular, the multiple access and the broadcast channels. A very interesting approach to the optimal design of multiantenna multiple access and broadcasting systems is based on the duality between the multiple access and the broadcast channel established in [41, 42, 43]. Finally, we have considered the so called cocktail party problem, where a series of sources transmit to as many corresponding destinations. Instead of approaching the problem by assuming a priori a specific multiplexing strategy, we have searched for the transmission strategy that maximizes the information rate on each channel. We have shown that a game-theoretic formulation of the problem provides the most appropriate multiplexing strategy. In particular, we have shown how the conventional multiplexing schemes, namely TDMA, FDMA, and CDMA, can be seen as specific Nash equilibria. But the optimal solution consists, in general, in allowing partial superposition of the users, depending on the level of interference, rather than assigning orthogonal channels to distinct links.

References

[1] Palomar, D. P., "A unified framework for communications through MIMO channels," *Ph.D.*

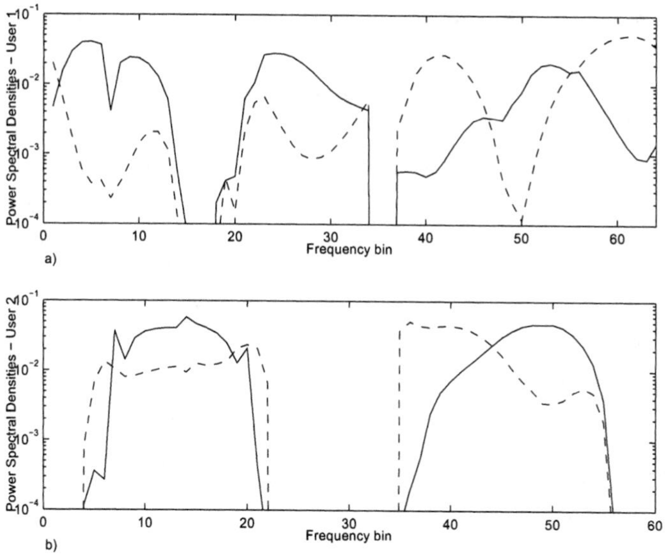

Figure 6.26 Power spectral densities of the transmitted signals: a) user 1 and b) user 2.

Dissertation, Universitat Politécnica de Catalunya, Spain, May 2003.

[2] Palomar, D. P., Cioffi, J.M., Lagunas, M.A., "Joint Tx-Rx beamforming design for multicarrier MIMO channels: A unified framework for convex optimization," *IEEE Trans. on Signal Processing*, vol. 51, Sept. 2003, pp. 2381–2401.

[3] Boyd, S., Vandenberghe, L., *Convex Optimization*, Cambridge, MA: Cambridge University Press, March 2004.

[4] Marshall, A. W., Olkin, I., *Inequalities: Theory of majorization and its applications*, New York: Academic Press, 1979.

[5] Scaglione, A., Giannakis, G.B., Barbarossa, S. , "Redundant filterbank precoders and equalizers - Part I: Unification and optimal design," *IEEE Trans. on Signal Processing*, vol. 47, July 1999, pp. 1988–2006.

[6] Scaglione, A., Stoica, P., Barbarossa, S. , Giannakis, G.B., Sampath, H., "Optimal design for space-time linear precoders and decoders," *IEEE Trans. on Signal Processing*, vol. 50, May 2002, pp. 1051–1064.

[7] Cover, T. M., Thomas, J. A., *Elements of Information Theory*, New York, NY: John Wiley & Sons, 1991.

[8] Proakis, J. , *Digital Communications*, (4th edition), New York, NY: McGraw Hill, 2000.

[9] Ding, Y., Davidson, T.N., Zhang J.K., Luo, Z.-Q., Wong, M., "Minimum BER precoders for zero-forcing equalization" *Proc. of ICASSP '02*, Orlando, FL, May 2002, pp. III.2261–2264.

[10] Scutari, G. , Barbarossa, S., "Generalized Water-Filling for Multiple Transmit Antenna Systems," *Proc. of ICC 2003*, Anchorage, Alaska, May 2003, pp. III.2261–2264.

[11] Caire, G., Shamai, S., "On the capacity of some channels with channel side information," *IEEE Transactions on Information Theory*, Vol. 45, Sept. 1999, pp. 2007–2019.

[12] Visotsky, E., Madhow, U. "Space-time transmit precoding with imperfect feedback," *IEEE Transactions on Information Theory*, Vol. 47, Sept. 2001, pp. 2632– 2639.

[13] Jorswieck, E., A., Boche, H., "Optimal transmission with imperfect channel state information at the transmit antenna array," *Wireless Personal Communications*, Vol. 27, 2003, pp. 33–56.

[14] Jafar, S.A., Goldsmith, A., "Transmitter optimization and optimality of beamforming for multiple antenna systems," *IEEE Transactions on Wireless Communications*, Vol. 3, July 2004, pp.1165–1175.

[15] Jongren, G., Skoglund, M., Ottersten, B., "Combining beamforming and orthogonal space-time block coding," *IEEE Transactions on Information Theory*, Vol. 48, March 2002, pp. 611–627.

[16] Skoglund, M., Jongren, G., "On the capacity of a multiple-antenna communication link with channel side information," *IEEE Journal on Selected Areas in Communications*, Vol. 21, April 2003, pp. 395–405.

[17] Zhou, S., Giannakis, G.B., "Optimal transmitter eigen-beamforming and space-time block coding based on channel correlations," *IEEE Transactions on Information Theory*, Vol. 49, July 2003, pp. 1673–1690.

[18] Zhou, S., Giannakis, G.B., "Optimal transmitter eigen-beamforming and space-time block coding based on channel mean feedback," *IEEE Transactions on Signal Processing*, Vol. 50, Oct. 2003, pp. 2599–2613.

[19] Onggosanusi, E.N., Sayeed, A.M., Van Veen, B.D., "Efficient signaling schemes for wideband space-time wireless channels using channel state information," *IEEE Trans. on Vehicular Technology*, vol. 52, Jan. 2003, pp. 1–13.

[20] Yu, W., Rhee, W., Cioffi, J. M., "Iterative water-filling for Gaussian vector multiple-access channels," *IEEE Transactions on Information Theory*, Vol. 50, Jan. 2004, pp. 145–152.

[21] Yu, W., "Competition and Cooperation in Multi-User Communication Environments", Ph.D. thesis, Stanford Univ., June 2002.

[22] Barbarossa, S., Scutari, G., "Optimal rate allocation for multiple access wideband systems with rate profile constraint and adaptive power control", *Proc. of IEEE Signal Processing Advances in Wireless Communications, SPAWC 2003*, Rome, Italy, June 2003, pp. 115–119.

[23] Cheng, R. S., Verdu, S., "Gaussian Multiaccess Channels with ISI: Capacity Region and Multiuser Water-Filling", *IEEE Trans. on Inform. Theory*, May 1993, pp. 773–785.

[24] ETSI Standard "Digital Video Broadcasting (DVB): Framing structure, channel coding and modulation for digital terrestrial television (DVB-T)", ETS 300 744, March 1997.

[25] Costa, M., "Writing on dirty paper", *IEEE Trans. on Information Theory*, May 1983, pp. 439–441.

[26] Caire, G., Shamai, S., "On achievable rates in a multi-antenna Gaussian channels," *Proc. of* 38th *Allerton Conf. Communications, Control and Computing*, Monticello, IL, Oct. 2000.

[27] Caire, G., Shamai, S., "On the achievable throughput of a multiantenna Gaussian broadcast channel," *IEEE Trans. on Information Theory*, July 2003, pp. 1691–1706.

[28] Ginis, G., Cioffi, J., "A multi-user precoding scheme achieving crosstalk cancellation with aplicayion to DSL systems," *Proc. of* 34th *Asilomar Conf. Signals, Systems and Computers*, Pacific Grove, CA, Nov. 2000.

[29] Fudenberg, D., Tirole, J., *Game Theory*, Cambridge, MA: MIT Press, 1991.

[30] Osborne, M. J. , Rubinstein, A., *A Course in Game Theory*, Cambridge, MA: MIT Press, 1994.

[31] Myerson, R., B., *Game Theory - Analysis of Conflict*, Cambridge, MA: Harvard University Press (fifth printing), 2002.

[32] Basar, T., Olsder, G. J., *Dynamic Noncooperative Game Theory*, Academic Press, 2nd Edition, 1995.

[33] Von Neumann, J., Morgenstern, O., *Theory of Games and Economic Behavior*, Princeton: Princeton University Press, Second Ed., 1947.

[34] Rosen, J., "Existence and Uniqueness of Equilibrium Points for Concave n-Person Games", *Econometrica*, Vol. 33, July 1965, pp. 520–534.

[35] Scutari, G., "Competition and Cooperation in Wireless Communication Networks", *PhD Dissertation*, University of Rome, "La Sapienza", December 2004.

[36] Scutari, G., Barbarossa, S., Ludovici, D., "On the maximum achievable rates in wireless meshed networks: Centralized versus decentralized solutions", *Proc. of ICASSP 2004*, Montreal, Canada, June 2004, pp. iv-574–iv-576.

[37] Scutari, G., Barbarossa, S., Ludovici, D., "Cooperation diversity in multihop wireless networks using opportunistic driven multiple access", *Proc. of IEEE Signal Processing Advances in Wireless Communications, SPAWC 2003*, Rome, Italy, June 2003.

[38] Yu, W., Cioffi, J. M.,"Competitive Equilibrium in the Gaussian Interference Channel", *Proc. of ISIT 2000*, June 2000.

[39] Das, I., Dennis, J., "Normal-Boundary intersection: a new method for generating the Pareto surface in nonlinear multicriteria optimization problems", *SIAM J. on Optimization*, Aug. 1998, pp. 631–657.

[40] Goldsmith, A., Jafar, S.A., Jindal, N., Vishwanath, S., "Capacity limits of MIMO channels," *IEEE Journal on Selected Areas in Communications*, Vol. 21, June 2003, pp. 684–702.

[41] Vishwanath, S., Jindal, N., Goldsmith, A., "Duality, achievable rates, and sum-rate capacity of Gaussian MIMO broadcast channels," *IEEE Trans. on Information Theory*, Oct. 2003, pp.2658–2668.

[42] Viswanathan, H., Venkatesan, S., Huang, H., "Downlink capacity evaluation of cellular networks with known-interference cancellation", *IEEE Journal on Selected Areas in Communications,* June 2003, pp. 802–811.

[43] Jindal, N., Vishwanath, S., Goldsmith, A., "On the duality of Gaussian multiple-access and broadcast channels", *IEEE Trans. on Information Theory*, May 2004, pp. 768–783.

Chapter 7

Space-Time Coding for Flat-Fading Channels

7.1 INTRODUCTION

In Chapter 5 we studied the fundamental limits/performance of MIMO systems, in the situation where the transmitter has no knowledge about the channel. In particular, we have seen that MISO or MIMO systems are capable of achieving full diversity gain, provided that the data is transmitted using codewords that span the space-time domain, jointly. Furthermore, we have observed that a MIMO system has the potential to increase capacity well beyond the value achievable with a SISO system, provided that the transmitter encodes the data properly. We have drawn these conclusions without referring to any specific transmission strategy. We had only anticipated that to achieve the promised benefits, it is necessary to adopt some kind of *joint* space-time coding. In this chapter, we describe and analyze alternative space-time coding strategies that allow us to achieve the full diversity gain and/or the maximum rate gain for systems having no information about the channel. We assume that the channels are flat-fading and the system is single-user. The extension to frequency-selective channels as well as to multiuser systems will be studied in Chapters 8 and 9.

The choice of the specific space-time encoder depends on which performance parameter we wish to improve, for a given set of constraints, such as transmit power and MIMO configuration. As performance parameter, we can use, for example, average error probability or information rate. One more important parameter, albeit not easy to be quantified, is the receiver complexity. Indeed, alternative space-time coding techniques should always be compared taking into account three fundamental aspects: bit error rate, information rate, and receiver complexity.

In recent years, space-time coding has been the subject of a huge number of journal papers, as well as research monographies. We recommend the interested reader to refer to the recent books [1, 2, 3, 4], for a general survey of space-time coding, and the two recent special issues [5, 6], for the state of the art on the research on space-time systems. In this chapter, we will review the fundamental properties of space-time coding and present some novel coding techniques that offer great flexibility in the design of codes with simple receiver structures.

The general scheme of a block space-time coding system, valid for a transceiver with n_T transmit and n_R receive antennas, is depicted in Figure 7.1. In such a

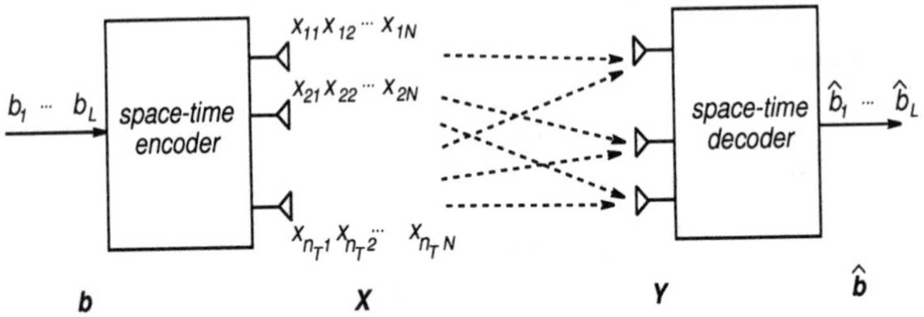

Figure 7.1 Space-time block coding scheme.

system, each block of incoming bits, say $\boldsymbol{b} := (b_1, \ldots, b_L)$, passes through the space-time encoder, which maps the set of bits onto an $n_T \times N$ matrix of, typically complex, symbols $\boldsymbol{X}(\boldsymbol{b})$; N is the number of time instants composing the code. The matrix $\boldsymbol{X}(\boldsymbol{b})$ is the *space-time code* associated to $\boldsymbol{b}$. Different rows of $\boldsymbol{X}(\boldsymbol{b})$ refer to different transmit antennas (and thus space), whereas different columns refer to different time instants.

In this chapter we concentrate on flat-fading channels, as the frequency-selective case will be treated in the ensuing chapter. In case of flat-fading, at each time instant, every receive antenna observes a linear combination of the symbols sent, at the same instant, from the n_T transmit antennas. Denoting with $\boldsymbol{H}$ the $n_R \times n_T$ matrix containing the channel coefficients, the collection of N consecutive samples from the n_R receive antennas gives rise to the $n_R \times N$ matrix $\boldsymbol{Y}$, with

$$\boldsymbol{Y} = \boldsymbol{H}\boldsymbol{X} + \boldsymbol{V}, \tag{7.1}$$

where $\boldsymbol{V}$ is the noise matrix.

Space-time coding can be seen as a *generalization of QAM*: QAM maps a set of bits onto a complex *scalar* symbol; space-time coding maps a set of bits onto a complex *matrix*. Space-time coding can thus be seen as a matrix, as opposed to scalar, coding.

The mapping rule used by the space-time encoder should be designed in order to optimize some performance parameter e.g., the diversity gain or the multiplexing gain. If we concentrate on diversity gain, according to what we have seen in Chapter 4, given any two different bit vectors $\boldsymbol{b}_k$ and $\boldsymbol{b}_l$, the mapping strategy must guarantee that the corresponding difference matrix $\boldsymbol{X}(\boldsymbol{b}_k) - \boldsymbol{X}(\boldsymbol{b}_l)$ be full rank. But the design has an impact on the coding gain and on the rate gain as well. In general, the design is not a simple task. Hence, it is useful to introduce some constraints on the coding rule to simplify the design itself and, possibly, the receiver detection scheme.

The most common simplification consists of assuming a *linear* strategy. For this reason, this chapter is entirely devoted to linear space-time coding systems. More specifically, after introducing the principle of linear coding in Section 7.2, we will describe orthogonal space-time block coding (OSTBC) in Section 7.3, as a tool for achieving full diversity with a very simple receiving structure. Although optimal from the points of view of diversity gain and receiver complexity, OSTBC is sub-optimal in terms of rate. In an effort to improve the rate of OSTBC, without complicating the receiver structure excessively, in Section 7.4 we will study quasi-orthogonal coding (QOD). Then, we will concentrate on layered space-time (LST) coding techniques in Section 7.5, as a general tool to maximize the rate and reach the capacity limit. We will then show how to build codes that guarantee full diversity and full rate, at the same time. Finally, in Section 7.6 we will concentrate on trace-orthogonal design (TOD), as a very general framework to construct space-time codes optimal from the point of view of capacity, that lead to very simple, albeit suboptimal, receiving schemes.

7.2 LINEAR SPACE-TIME BLOCK CODING

Let us consider a set of bits $\boldsymbol{b}$ to be transmitted. The set is first converted onto a set of n_s symbols $\boldsymbol{s} = (s_1, \ldots, s_{n_s})$, belonging, for example, to a QAM constellation. A *linear* space-time encoder maps the vector $\boldsymbol{s}$ onto a space-time matrix $\boldsymbol{X} = \boldsymbol{X}(\boldsymbol{b})$, built as a linear combination of the symbols s_k and their conjugates s_k^*, as follows

$$\boldsymbol{X} = \sum_{k=1}^{n_s} \left(s_k \boldsymbol{C}_k + s_k^* \boldsymbol{D}_k \right), \tag{7.2}$$

where $\{\boldsymbol{C}_k\}$ and $\{\boldsymbol{D}_k\}$ are two sets of (complex) $n_T \times N$ matrices, independent of the transmitted symbols. Alternatively, instead of using the symbols s_k and their conjugated, we can express $\boldsymbol{X}$ in terms of the real and imaginary parts of s_k.

Specifically, setting $s_k = s_{rk} + js_{ik}$, we can write

$$X = \sum_{k=1}^{n_s} (s_{rk}\boldsymbol{A}_k + js_{ik}\boldsymbol{B}_k), \tag{7.3}$$

where $\{\boldsymbol{A}_k\}$ and $\{\boldsymbol{B}_k\}$ are two sets of (complex) $n_T \times N$ matrices, independent of the transmitted symbols, related to the previous sets $\{\boldsymbol{C}_k\}$ and $\{\boldsymbol{D}_k\}$ by the following identities

$$\boldsymbol{C}_k = \frac{\boldsymbol{A}_k + \boldsymbol{B}_k}{2}, \quad \boldsymbol{D}_k = \frac{\boldsymbol{A}_k - \boldsymbol{B}_k}{2}. \tag{7.4}$$

In a linear space-time system, substituting (7.3) in (7.1), the input/output relationship becomes

$$\boldsymbol{Y} = \sum_{k=1}^{n_s} (s_{rk}\boldsymbol{H}\boldsymbol{A}_k + js_{ik}\boldsymbol{H}\boldsymbol{B}_k) + \boldsymbol{V}. \tag{7.5}$$

Since we transmit n_s symbols using N time slots (the number of columns of the matrices $\boldsymbol{A}_k$ and $\boldsymbol{B}_k$), the transmission symbol rate of such a system is

$$R_s = \frac{n_s}{N}. \tag{7.6}$$

Using a constellation of order M, the bit rate is, consequently,

$$R_b = \frac{n_s}{N}\log_2(M). \tag{7.7}$$

Most of the work on linear space-time coding is devoted to the search of the family of coding matrices $\boldsymbol{A}_k$ and $\boldsymbol{B}_k$ that provide the best performance. Some examples are the so-called linear dispersion codes [7], where the matrices $\boldsymbol{A}_k$ and $\boldsymbol{B}_k$ are designed in order to maximize the mutual information between transmitter and receiver. However, this is a non-convex problem and the the solution suggested in [7] is only numerical with no guarantee of convergence to the optimal. Alternatively, in [8] it was proposed a method to derive the matrices $\boldsymbol{A}_k$ and $\boldsymbol{B}_k$ that minimize the error probability, but still the solution is only numerical. In this chapter, we describe alternative approaches capable to strike different trade-offs between the three fundamental aspects of diversity, multiplexing gain, and complexity.

7.3 ORTHOGONAL SPACE-TIME BLOCK CODING

Orthogonal space-time block coding (OSTBC) is a very useful block coding method for achieving full diversity gain with a very simple scalar decoder. To explain the properties of OSTBC, we start with Alamouti's coding [9].

7.3.1 Alamouti Space-Time Coding

The Alamouti space-time encoder is valid for two transmit antennas and any number of receive antennas. The encoder maps every pair of incoming symbols, let us say s_1 and s_2, into the following space-time matrix

$$\boldsymbol{X} = \begin{pmatrix} s_1 & s_2^* \\ s_2 & -s_1^* \end{pmatrix}. \tag{7.8}$$

We adopt the convention of associating different columns to different time instants and different rows to different transmit antennas. The matrix $\boldsymbol{X}$ means that, given the two symbols s_1 and s_2 to be transmitted, in the first time slot we transmit s_1 from the first antenna and s_2 from the second antenna; in the successive time slot, we transmit s_2^* from the first antenna and $-s_1^*$ from the second antenna. The matrix $\boldsymbol{X}$ is a particular case of (7.3), with

$$\boldsymbol{A}_1 = \begin{bmatrix} 1 & 0 \\ 0 & -1 \end{bmatrix}, \boldsymbol{A}_2 = \begin{bmatrix} 0 & 1 \\ 1 & 0 \end{bmatrix}, \boldsymbol{B}_1 = \begin{bmatrix} 1 & 0 \\ 0 & 1 \end{bmatrix}, \boldsymbol{B}_2 = \begin{bmatrix} 0 & -1 \\ 1 & 0 \end{bmatrix}. \tag{7.9}$$

In this case, we have $n_s = 2$, $N = 2$ and the symbol rate is then $R_s = 1$.

Let us consider initially a single antenna receiver and then we will extend the analysis to multiple receive antenna systems. Denoting with y_1 and y_2 the samples received by a single receive antenna in two successive slots, we have the following I/O relationship

$$\begin{aligned} y_1 &= h_1 s_1 + h_2 s_2 + v_1, \\ y_2 &= h_1 s_2^* - h_2 s_1^* + v_2, \end{aligned} \tag{7.10}$$

where v_1 and v_2 are the noise samples. Taking the conjugate of y_2, (7.10) can be rewritten in matrix form as

$$\begin{pmatrix} y_1 \\ y_2^* \end{pmatrix} = \begin{pmatrix} h_1 & h_2 \\ -h_2^* & h_1^* \end{pmatrix} \begin{pmatrix} s_1 \\ s_2 \end{pmatrix} + \begin{pmatrix} v_1 \\ v_2^* \end{pmatrix} \tag{7.11}$$

or, in matrix form, as

$$\boldsymbol{y} = \boldsymbol{\mathcal{H}} \boldsymbol{s} + \boldsymbol{w}. \tag{7.12}$$

The matrix

$$\boldsymbol{\mathcal{H}} := \begin{pmatrix} h_1 & h_2 \\ -h_2^* & h_1^* \end{pmatrix} \tag{7.13}$$

is the *equivalent* channel matrix, whose structure depends on the transmission strategy. The interesting property of Alamouti's coding is that *both* $\boldsymbol{X}$ and $\boldsymbol{\mathcal{H}}$ are *orthogonal* matrices. In fact, we can immediately check that

$$\boldsymbol{X}^H\boldsymbol{X} = (|s_1|^2 + |s_2|^2)\boldsymbol{I} \tag{7.14}$$

and

$$\boldsymbol{\mathcal{H}}^H\boldsymbol{\mathcal{H}} = (|h_1|^2 + |h_2|^2)\boldsymbol{I} = \|\boldsymbol{h}\|^2\boldsymbol{I}, \tag{7.15}$$

where $\|\boldsymbol{h}\|^2 := |h_1|^2 + |h_2|^2$ is the square norm of the vector $\boldsymbol{h} := (h_1, h_2)^T$. The orthogonality of $\boldsymbol{\mathcal{H}}$ is very important, as it leads to a very simple implementation of the optimal maximum likelihood (ML) detection scheme. ML detection will be considered in detail in Section 7.3.3. Here, we anticipate some concept by referring to the ML estimator, assuming no specific properties about the transmitted symbols.

If the additive noise $\boldsymbol{w}$ is white Gaussian noise with zero mean and covariance matrix $\boldsymbol{C}_w = \sigma_n^2\boldsymbol{I}$, the ML estimator amounts to finding the vector $\hat{\boldsymbol{s}}$ that minimizes the square norm

$$\|\boldsymbol{y} - \boldsymbol{\mathcal{H}}\hat{\boldsymbol{s}}\|^2 = \|\boldsymbol{y}\|^2 + \|\boldsymbol{\mathcal{H}}\hat{\boldsymbol{s}}\|^2 - 2\Re\{\boldsymbol{y}^H\boldsymbol{\mathcal{H}}\hat{\boldsymbol{s}}\}. \tag{7.16}$$

Equating the gradient of this expression, with respect to $\hat{\boldsymbol{s}}$, to zero, we get

$$\hat{\boldsymbol{s}} = \left(\boldsymbol{\mathcal{H}}^H\boldsymbol{\mathcal{H}}\right)^{-1}\boldsymbol{\mathcal{H}}^H\boldsymbol{y} = \frac{1}{\|\boldsymbol{h}\|^2}\boldsymbol{\mathcal{H}}^H\boldsymbol{y}. \tag{7.17}$$

Using (7.15), (7.17) simplifies into

$$\hat{\boldsymbol{s}} = \frac{1}{\|\boldsymbol{h}\|^2}\boldsymbol{\mathcal{H}}^H\boldsymbol{\mathcal{H}}\boldsymbol{s} + \frac{1}{\|\boldsymbol{h}\|^2}\boldsymbol{\mathcal{H}}^H\boldsymbol{w} = \boldsymbol{s} + \boldsymbol{w}', \tag{7.18}$$

where $\boldsymbol{w}' := \boldsymbol{\mathcal{H}}^H\boldsymbol{w}/\|\boldsymbol{h}\|^2$. This means that the structure of the ML estimator is very simple, as it can be implemented by simply multiplying the received vector $\boldsymbol{y}$ by $\boldsymbol{\mathcal{H}}^H$. Interestingly enough, this multiplication *diagonalizes the channel without coloring the noise*. The final noise vector $\boldsymbol{w}' := \boldsymbol{\mathcal{H}}^H\boldsymbol{w}/\|\boldsymbol{h}\|^2$ has in fact a covariance matrix

$$\boldsymbol{C}_{w'} = E\{\boldsymbol{w}'\boldsymbol{w}'^H\} = \frac{\sigma_n^2}{\|\boldsymbol{h}\|^4}\boldsymbol{\mathcal{H}}^H\boldsymbol{\mathcal{H}} = \frac{\sigma_n^2}{\|\boldsymbol{h}\|^2}\boldsymbol{I}, \tag{7.19}$$

which is diagonal. In other words, if $\boldsymbol{w}$ is white, $\boldsymbol{w}'$ is also white. Equations (7.18) and (7.19) show that there is no loss of optimality in implementing the optimal ML estimator as two decoupled scalar estimators.

Let us now see what is the diversity gain of the Alamouti scheme. Combining (7.18) and (7.19), the SNR on each estimated symbol is

$$\mathrm{SNR} = \frac{\sigma_s^2 \|\boldsymbol{h}\|^2}{\sigma_n^2} = (|h_1|^2 + |h_2|^2)\frac{\sigma_s^2}{\sigma_n^2}, \tag{7.20}$$

where σ_s^2 denotes the variance of the transmitted symbols. Recalling the results from Section 5.5.1 on the diversity of slowly fading zero-mean MISO systems, (7.20) shows that Alamouti coding yields also the maximum diversity gain, which is equal to two.

Let us compute now the capacity of the Alamouti system. The energy necessary to transmit the Alamouti code is

$$\mathcal{E} = E\{\mathrm{tr}(\boldsymbol{X}\boldsymbol{X}^H)\} = 2n_T\sigma_s^2. \tag{7.21}$$

Consequently, the average power is

$$p_T = \frac{\mathcal{E}}{2} = n_T\sigma_s^2. \tag{7.22}$$

The capacity of the channel corresponding to (7.12) is

$$C = \frac{1}{2}\log_2\left|\boldsymbol{I} + \frac{\sigma_s^2}{\sigma_n^2}\boldsymbol{\mathcal{H}}\boldsymbol{\mathcal{H}}^H\right|, \tag{7.23}$$

where the factor $1/2$ takes into account that Alamouti's code spans over two symbol periods. Using (7.15) and (7.22), (7.23) becomes

$$C = \frac{1}{2}\log_2\left|\boldsymbol{I} + \frac{p_T}{n_T\sigma_n^2}\|\boldsymbol{h}\|^2\boldsymbol{I}\right| = \log_2\left(1 + \frac{p_T}{n_T\sigma_n^2}\|\boldsymbol{h}\|^2\right). \tag{7.24}$$

Let us compare this value with the capacity of the 2×1 MISO channel having the input vector $\boldsymbol{s} = (s_1, s_2)^T$ and the channel row vector $\boldsymbol{h} = (h_1, h_2)$. The capacity of this channel is

$$C = \log_2\left(1 + \frac{p_T}{n_T\sigma_n^2}\|\boldsymbol{h}\|^2\right). \tag{7.25}$$

This expression coincides with (7.24). This proves that Alamouti coding, for a single receive antenna system, is information lossless.

In summary, the Alamouti coding scheme, for a system with two transmit antennas and one receive antenna, is optimal from all points of view: capacity, diversity gain, and receiver simplicity.

Since in most engineering problems, advantages are often paid with some drawbacks, it is rather natural to ask ourselves what is the price to be paid for all these nice features of Alamouti coding. Indeed, whereas Alamouti is optimal under all points of view for 2×1 MIMO configurations, as soon as we depart from such a configuration, Alamouti loses its optimality in terms of information rate. This property will be proved in Section 7.3.6. However, Alamouti remains optimal from the points of view of simplicity and diversity gain, for any number of receive antennas. It is then natural to ask whether there exist more general schemes, valid for systems with more than two transmit antennas, that yield the maximum diversity gain and lead to simple receiver structures. The solution is provided by orthogonal space-time codes.

7.3.2 General Structure of Orthogonal Space-Time Coding

A linear coding strategy is said to be *orthogonal* if, for any set of transmitted symbols, we have

$$\boldsymbol{X}\boldsymbol{X}^H = \sum_{k=1}^{n_s} |s_k|^2 \boldsymbol{I} = \|\boldsymbol{s}\|^2 \boldsymbol{I}. \tag{7.26}$$

This condition holds true *if and only if* the sets of matrices $\{\boldsymbol{A}_k\}$ and $\{\boldsymbol{B}_k\}$ in (7.3) respect the following conditions:

$$\begin{aligned} \boldsymbol{A}_k\boldsymbol{A}_k^H &= \boldsymbol{B}_k\boldsymbol{B}_k^H = \boldsymbol{I}, \; k \in \{1, 2, \ldots, n_s\} \\ \boldsymbol{A}_k\boldsymbol{A}_l^H &= -\boldsymbol{A}_l\boldsymbol{A}_k^H; \; \boldsymbol{B}_k\boldsymbol{B}_l^H = -\boldsymbol{B}_l\boldsymbol{B}_k^H, \; k, l \in \{1, 2, \ldots, n_s\}, \; k \neq l, \\ \boldsymbol{A}_k\boldsymbol{B}_l^H &= \boldsymbol{B}_k\boldsymbol{A}_l^H, \; k, l \in \{1, 2, \ldots, n_s\}, \; k \neq l. \end{aligned} \tag{7.27}$$

A set of matrices $\{\boldsymbol{A}_k, \boldsymbol{B}_k\}$ satisfying the above conditions is called *amicable orthogonal design*[1] [3].

1 Interestingly, we observe from (7.27) that the coding matrices must satisfy an anti-commuting property. Clearly, this property cannot be satisfied by scalar numbers. This is one more argument to prove the usefulness of matrix, as opposed to scalar, modulation for multiple transmit antenna systems.

The sufficient condition can be proved by direct substitution. Using (7.3), we have in fact

$$
\begin{aligned}
\boldsymbol{X}\boldsymbol{X}^H &= \sum_{k=1}^{n_s}\sum_{l=1}^{n_s}(s_{rk}\boldsymbol{A}_k + js_{ik}\boldsymbol{B}_k)\left(s_{rl}\boldsymbol{A}_l^H - js_{il}\boldsymbol{B}_l^H\right) \\
&= \sum_{k=1}^{n_s}\left(s_{rk}^2\boldsymbol{A}_k\boldsymbol{A}_k^H + s_{ik}^2\boldsymbol{B}_k\boldsymbol{B}_k^H\right) + \\
&+ \sum_{k=1}^{n_s}\sum_{l=k+1}^{n_s}\left[s_{rk}s_{rl}\left(\boldsymbol{A}_k\boldsymbol{A}_l^H + \boldsymbol{A}_l\boldsymbol{A}_k^H\right) + s_{ik}s_{il}\left(\boldsymbol{B}_k\boldsymbol{B}_l^H + \boldsymbol{B}_l\boldsymbol{B}_k^H\right)\right] \\
&+ j\sum_{k=1}^{n_s}\sum_{l=1}^{n_s}s_{rk}s_{il}\left(\boldsymbol{B}_l\boldsymbol{A}_k^H - \boldsymbol{A}_k\boldsymbol{B}_l^H\right).
\end{aligned}
\qquad (7.28)
$$

From this expression, we can immediately check that, if all relationships in (7.27) hold true, then (7.28) leads to (7.26). The necessary condition is less easy to prove and the interested reader is invited to check [10] for details.

To prove the basic properties of orthogonal codes, it is useful to introduce the following (real) vector

$$\boldsymbol{s} := \left[s_{r1}, \ldots, s_{rn_s}, s_{i1}, \ldots, s_{in_s}\right]^T, \qquad (7.29)$$

and the matrix

$$\boldsymbol{F} = \begin{pmatrix} \mathrm{vec}^T(\boldsymbol{H}\boldsymbol{A}_1) \\ \cdots \\ \mathrm{vec}^T(\boldsymbol{H}\boldsymbol{A}_{n_s}) \\ j\,\mathrm{vec}^T(\boldsymbol{H}\boldsymbol{B}_1) \\ \cdots \\ j\,\mathrm{vec}^T(\boldsymbol{H}\boldsymbol{B}_{n_s}) \end{pmatrix}^T. \qquad (7.30)$$

If we stack all the columns of the received matrix $\boldsymbol{Y}$ into a vector $\boldsymbol{y} := \mathrm{vec}(\boldsymbol{Y})$, we can write the input/output relationship (7.5) as

$$\boldsymbol{y} := \mathrm{vec}(\boldsymbol{Y}) = \mathrm{vec}(\boldsymbol{H}\boldsymbol{X}) + \mathrm{vec}(\boldsymbol{V}) = \boldsymbol{F}\boldsymbol{s} + \boldsymbol{v}, \qquad (7.31)$$

with $\boldsymbol{v} := \mathrm{vec}(\boldsymbol{V})$. The main properties of OSTBC are a direct consequence of the following property, proved in [11]:

A set of matrices $\{\boldsymbol{A}_k\}$ and $\{\boldsymbol{B}_k\}$ constitutes an orthogonal code *if and only if* the matrix $\boldsymbol{F}$ built as in (7.30) satisfies the following identity

$$\Re\{\boldsymbol{F}^H\boldsymbol{F}\} = \|\boldsymbol{H}\|^2\boldsymbol{I}, \qquad (7.32)$$

for all $\boldsymbol{H}$ [2]. We start proving that an orthogonal code satisfies (7.32). In fact, by construction of $\boldsymbol{F}$, since $\text{vec}(\boldsymbol{HX}) = \boldsymbol{Fs}$, we have

$$\|\boldsymbol{Fs}\|^2 = \boldsymbol{s}^T \Re\{\boldsymbol{F}^H \boldsymbol{F}\} \boldsymbol{s} = \|\boldsymbol{HX}\|^2. \tag{7.33}$$

On the other hand, if $\boldsymbol{X}$ comes from an orthogonal design, then

$$\|\boldsymbol{HX}\|^2 = \text{tr}(\boldsymbol{HXX}^H \boldsymbol{H}^H) = \text{tr}(\boldsymbol{HH}^H)\|\boldsymbol{s}\|^2 = \|\boldsymbol{H}\|^2 \|\boldsymbol{s}\|^2. \tag{7.34}$$

Hence, combining (7.33) with (7.34), we get

$$\boldsymbol{s}^T \Re\{\boldsymbol{F}^H \boldsymbol{F}\} \boldsymbol{s} = \|\boldsymbol{H}\|^2 \|\boldsymbol{s}\|^2. \tag{7.35}$$

Since this equation must hold true for all $\boldsymbol{s}$ and $\Re\{\boldsymbol{F}^H \boldsymbol{F}\}$ is symmetric by construction, $\boldsymbol{F}$ must necessarily satisfy (7.32).

The converse of this statement is also true: If (7.32) holds true, then $\boldsymbol{X}$ must be an orthogonal code. In fact, if (7.32) is true, from (7.33) we can write

$$\text{tr}\left[\boldsymbol{H}(\boldsymbol{XX}^H - \|\boldsymbol{s}\|^2 \boldsymbol{I})\boldsymbol{H}^H\right] = 0. \tag{7.36}$$

This equality must hold for all $\boldsymbol{H}$. Again, the need for (7.36) to be true for all $\boldsymbol{H}$ implies that $\boldsymbol{XX}^H - \|\boldsymbol{s}\|^2 \boldsymbol{I} = \boldsymbol{0}$, i.e., the code is orthogonal.

7.3.3 ML Decoding

Using (7.32), we can immediately prove the basic properties of the ML decoder, in case of orthogonal coding, namely:

P.1: In the presence of AWGN, i.e., $\boldsymbol{v} \sim \mathcal{N}(\boldsymbol{0}, \sigma_n^2 \boldsymbol{I})$, the ML decoder is implemented as a symbol-by-symbol decoder applied to $\hat{\mathbf{s}}$, defined as

$$\hat{\mathbf{s}} := \frac{\Re\{\boldsymbol{F}^H \boldsymbol{y}\}}{\|\boldsymbol{H}\|^2}; \tag{7.37}$$

P.2: The symbol-by-symbol decoder achieves the maximum diversity gain.

Let us prove these statements. In the presence of white Gaussian noise, the ML decoder is

$$\hat{\mathbf{s}} = \arg\min_{\boldsymbol{s} \in \mathcal{S}} \|\boldsymbol{y} - \boldsymbol{Fs}\|^2, \tag{7.38}$$

2 The symbol $\|\boldsymbol{H}\|^2$ in (7.32) has to be intended as the Frobenius norm of $\boldsymbol{H}$, i.e. $\|\boldsymbol{H}\|^2 := \sum_{i=1}^{n_R} \sum_{j=1}^{n_T} |H_{ij}|^2 = \text{tr}(\boldsymbol{H}^H \boldsymbol{H})$.

where $\mathcal{S}$ indicates the symbol constellation. This optimization problem is non-convex (over the symbol constellation set) because the set $\mathcal{S}$ is not convex. For this reason, ML is complicated to implement and computationally intensive. An exhaustive search would be too complicated to implement and too slow to run. Sphere decoding can simplify the detection, but it is still rather complicated to implement. Interestingly, OSTBC leads to a very simple ML decoder. In fact, exploiting (7.32), we can rewrite the square norm in (7.38) as follows

$$\begin{aligned} \|\boldsymbol{y} - \boldsymbol{F}\boldsymbol{s}\|^2 &= \|\boldsymbol{y}\|^2 - 2\Re\{\boldsymbol{y}^H\boldsymbol{F}\boldsymbol{s}\} + \|\boldsymbol{F}\boldsymbol{s}\|^2 \\ &= \|\boldsymbol{y}\|^2 - 2\Re\{\boldsymbol{y}^H\boldsymbol{F}\boldsymbol{s}\} + \|\boldsymbol{H}\|^2\|\boldsymbol{s}\|^2. \end{aligned} \tag{7.39}$$

Introducing the vector $\hat{\boldsymbol{s}}$ defined in (7.37) and recalling that $\boldsymbol{s}$ is real, we have $2\Re\{\boldsymbol{y}^H\boldsymbol{F}\boldsymbol{s}\} = 2\boldsymbol{s}^T\hat{\boldsymbol{s}}\,\|\boldsymbol{H}\|^2$, so that we can rewrite (7.39) as

$$\|\boldsymbol{y} - \boldsymbol{F}\boldsymbol{s}\|^2 = \|\boldsymbol{H}\|^2\|\boldsymbol{s} - \hat{\boldsymbol{s}}\|^2 + \text{constant terms}, \tag{7.40}$$

where the constant terms include all components that do not depend on $\boldsymbol{s}$. From (7.40), it is clear that the ML decoder (7.38) can be implemented by using (7.37) and then taking a scalar decision over each entry of $\hat{\boldsymbol{s}}$.

Furthermore, after pre-multiplying the received vector $\boldsymbol{y}$ by $\boldsymbol{F}^H/\|\boldsymbol{H}\|^2$ and taking the real part, we get

$$\boldsymbol{z} := \frac{\Re\{\boldsymbol{F}^H\boldsymbol{y}\}}{\|\boldsymbol{H}\|^2} = \boldsymbol{s} + \boldsymbol{w}', \tag{7.41}$$

where $\boldsymbol{w}' := \Re\{\boldsymbol{F}^H\boldsymbol{w}\}/\|\boldsymbol{H}\|^2$ is a Gaussian noise vector, with zero mean and covariance matrix

$$\boldsymbol{C}_{w'} = \frac{1}{2}\frac{\Re\{\boldsymbol{F}^H\boldsymbol{F}\}\sigma_n^2}{\|\boldsymbol{H}\|^4} = \frac{1}{2}\frac{\sigma_n^2}{\|\boldsymbol{H}\|^2}\boldsymbol{I}. \tag{7.42}$$

It is not surprising that, if the receiver noise is white, it remains white also after multiplying the received noise by $\boldsymbol{F}^H$ and taking the real part. This is another consequence of the orthogonality of $\boldsymbol{F}$, in the sense of (7.32).

In summary, after the multiplication for $\boldsymbol{F}^H$, the overall system becomes equivalent to a set of n_s parallel scalar, independent sub-channels. On each sub-channel, the SNR is[3]

$$\text{SNR} = \frac{\sigma_s^2}{\sigma_n^2}\|\boldsymbol{H}\|^2. \tag{7.43}$$

3 Please consider that each component of the vector $\boldsymbol{s}$ is real and has variance $\sigma_s^2/2$.

Since $\|\boldsymbol{H}\|^2$ is the summation of the square modulus of $n_T \cdot n_R$ independent complex Gaussian random variables, the diversity gain on each sub-channel is

$$G_D = n_T \cdot n_R. \tag{7.44}$$

Comparing this value with the bound derived in Chapter 5, we can see that *orthogonal space-time coding yields maximum diversity gain.*

An alternative way of deriving the ML detector is useful to grasp some important aspects of the optimal receiver for space-time coded systems. Starting from (7.1), the ML metric to be minimized in the presence of white additive Gaussian noise is the mean square error

$$\text{MSE} = \|\boldsymbol{Y} - \boldsymbol{H}\boldsymbol{X}\|^2 = \|\boldsymbol{Y}\|^2 - 2\Re\{\text{tr}(\boldsymbol{X}^H\boldsymbol{H}^H\boldsymbol{Y})\} + \|\boldsymbol{H}\boldsymbol{X}\|^2. \tag{7.45}$$

Using (7.3) and exploiting the orthogonality of $\boldsymbol{X}$, (7.45) becomes

$$\begin{aligned}\text{MSE} &= \|\boldsymbol{Y}\|^2 - 2\sum_{k=1}^{n_s}\Re\{\text{tr}(\boldsymbol{A}_k^H\boldsymbol{H}^H\boldsymbol{Y})\}s_{rk} - 2\sum_{k=1}^{n_s}\Im\{\text{tr}(\boldsymbol{B}_k^H\boldsymbol{H}^H\boldsymbol{Y})\}s_{ik}\\ &\quad + \|\boldsymbol{H}\|^2\|\boldsymbol{s}\|^2\\ &= \sum_{k=1}^{n_s}\left[-2\Re\{\text{tr}(\boldsymbol{A}_k^H\boldsymbol{H}^H\boldsymbol{Y})\}s_{rk} - 2\Im\{\text{tr}(\boldsymbol{B}_k^H\boldsymbol{H}^H\boldsymbol{Y})\}s_{ik} + \|\boldsymbol{H}\|^2|s_k|^2\right]\\ &\quad + \text{constant terms}\\ &= \|\boldsymbol{H}\|^2\sum_{k=1}^{n_s}\left|s_k - \frac{\Re\{\text{tr}(\boldsymbol{A}_k^H\boldsymbol{H}^H\boldsymbol{Y})\}}{\|\boldsymbol{H}\|^2} - j\frac{\Im\{\text{tr}(\boldsymbol{B}_k^H\boldsymbol{H}^H\boldsymbol{Y})\}}{\|\boldsymbol{H}\|^2}\right|^2\\ &\quad + \text{constant terms,}\end{aligned} \tag{7.46}$$

where the constant terms contain contributions independent of s_k. From (7.46), the optimal (matrix) detector decouples into n_s symbol-by-symbol detectors, where the estimate of the real and imaginary part of the transmitted symbols are

$$\begin{aligned}\hat{s}_{rk} &= \frac{\Re\{\text{tr}(\boldsymbol{A}_k^H\boldsymbol{H}^H\boldsymbol{Y})\}}{\|\boldsymbol{H}\|^2}\\ \hat{s}_{ik} &= \frac{\Im\{\text{tr}(\boldsymbol{B}_k^H\boldsymbol{H}^H\boldsymbol{Y})\}}{\|\boldsymbol{H}\|^2}.\end{aligned} \tag{7.47}$$

Besides the normalization by the scalar $\|\boldsymbol{H}\|^2$, each detector can be seen as a (matrix) filter matched to the symbol of interest. In fact, with linear OSTBC, each symbol is carried, equivalently by a matrix $\boldsymbol{A}_k$ (or $\boldsymbol{B}_k$). The channel modifies this kind of *matrix carrier* into $\boldsymbol{H}\boldsymbol{A}_k$ (or $\boldsymbol{H}\boldsymbol{B}_k$). From this point of view, (7.47) represent a sort of matched filter. The orthogonality of the codes is fundamental to cancel the interference from all other symbols in an easy manner.

In summary, orthogonal coding provides:

1. Maximum SNR, or minimum MSE, on each decoded symbol;
2. Easy ISI cancelation, due to the orthogonality of codes associated to different symbols;
3. Easy implementation of ML decoding, as the matched filter does not color the noise;
4. Full diversity gain.

Given all the positive properties of OSTBC, it is natural to ask ourselves whether there are negative features. Indeed, there are two fundamental questions that arise in the use of OSTBC:

1. Do there exist orthogonal codes for any system configuration (like, e.g., number of transmit antennas, number of transmitted symbols, ...) ?
2. Is OSTBC optimal from the capacity point of view?

We address now these questions in detail.

7.3.4 Existence of Orthogonal Codes

From (7.27), the existence of orthogonal codes is related to the existence of sets of matrices satisfying the relationships (7.27). The existence of such a family of matrices depends on the field of definition of the transmitted symbols. We have to distinguish between real and complex symbols.

Real symbols

If we transmit real symbols, it makes sense to use only the set of matrices $\{\boldsymbol{A}_k,\ k = 1,\ldots,n_s\}$ in (7.3) and, furthermore, to assume that the matrices $\boldsymbol{A}_k$ are real. We also assume that the matrices $\boldsymbol{A}_k$ are square, i.e., $N = n_T$. In such a case, the existence of orthogonal codes boils down to find a set of n_s *real* matrices $\{\boldsymbol{A}_k\}$ satisfying the following relationships:

$$\begin{aligned} \boldsymbol{A}_k\boldsymbol{A}_k^T &= \boldsymbol{I},\quad k=1,\ldots,n_s, \\ \boldsymbol{A}_k\boldsymbol{A}_l^T &= -\boldsymbol{A}_l\boldsymbol{A}_k^T,\quad k,l=1,\ldots,n_s, k\neq l. \end{aligned} \tag{7.48}$$

Given the set $\{\boldsymbol{A}_k,\ k = 1, \ldots, n_s\}$, we can introduce the equivalent set $\{\boldsymbol{V}_k,\ k = 1, \ldots, n_s\}$, with $\boldsymbol{V}_k = \boldsymbol{A}_k \boldsymbol{A}_1^T$, for $k = 2, \ldots, n_s$, and $\boldsymbol{V}_1 = \boldsymbol{I}$. It is straightforward to check that the new set satisfies the conditions

$$\begin{aligned} \boldsymbol{V}_k^T \boldsymbol{V}_k &= \boldsymbol{I}, \quad k = 2, \ldots, n_s, \\ \boldsymbol{V}_k^T &= -\boldsymbol{V}_k, \quad k = 2, \ldots, n_s, \\ \boldsymbol{V}_k^T \boldsymbol{V}_l &= -\boldsymbol{V}_l^T \boldsymbol{V}_k, \quad k, l = 2, \ldots, n_s, k \neq l. \end{aligned} \tag{7.49}$$

A set of matrices $\{\boldsymbol{V}_k\}$ satisfying (7.49) constitutes a *Hurwitz-Radon* (HR) family of matrices. Given the equivalence between the two sets of matrices, the existence of the set of square coding matrices $\{\boldsymbol{A}_k\}$ is equivalent to the existence of an HR family of order N. The conditions for the existence of a HR family were derived by Radon, who proved the following result (see [12] for an in-depth analysis of orthogonal design): For any N, introducing the integer numbers a, b, c, and d, defined implicitly as

$$N = 2^a b, \quad a = 4c + d, \quad 0 \leq d \leq 3, \quad c \geq 0, \tag{7.50}$$

where b is an odd number, and defining the function $\rho(N)$ as

$$\rho(N) = 8c + 2^d, \tag{7.51}$$

then there exists a family of HR matrices $\{\boldsymbol{V}_k\}$ of size $N \times N$ having exactly $\rho(N) - 1$ members.

Clearly, we would like $\rho(N)$ to be as large as possible, so as to have as many matrices $\boldsymbol{A}_k$ as possible. This would allow us to maximize the transmission rate. Since in the above formulation $N = n_T$, the maximum transmission rate for an orthogonal design is one, and it is achievable taking $n_s = N = n_T$. An orthogonal code with rate one is a *full-rate* orthogonal code[4]. From (7.51), it turns out that there exists a full-rate orthogonal only if $\rho(N) = N$. Combining (7.50) and (7.51), this requires that the following equation must be satisfied

$$2^{4c+d} b = 8c + 2^d, \tag{7.52}$$

with a, b, c, and d defined as in (7.50). It is easy to check that (7.52) is satisfied if and only if $c = 0, b = 1$, and $d = 0, 1, 2$, or 3. This means that the only possible

4 It is important to remark that rate one is the maximum rate for orthogonal codes, but it is not the maximum rate in general. We know in fact that a system with n_T transmit and n_R receive antennas can transmit with a symbol rate $R_s = \min(n_T, n_R)$. However, since OSTBC can operate with any number of receive antennas, it is also valid for the worst case i.e., $n_R = 1$. In such a case, the maximum rate is $R_s = 1$.

values of N are $N = 1, 2, 4$, and 8. Neglecting the trivial solution $N = 1$, we see that *a full-rate orthogonal design with real symbols and real square matrices* $\boldsymbol{A}_k$ *exists if and only if the number of transmit antennas is* $n_T = 2, 4$, *or* 8.

It is important to emphasize that the previous results hold true under the assumption that the matrices $\boldsymbol{A}_k$ are square. If we remove this constraint, it is possible to construct full-rate orthogonal codes, *for any* n_T. Some examples of space-time codes will be shown in Section 7.3.5.

Complex symbols

The transmission of real symbols is inefficient when the transmission band is located around a nonnull carrier, i.e., in practice for all wireless communication systems. In such a case, a better use of the available bandwidth entails the transmission of complex symbols, like PSK or QAM symbols. The design of orthogonal codes consists in finding a set of complex *amicable* matrices $\{\boldsymbol{A}_k, \boldsymbol{B}_k\}$ satisfying (7.27). However, in such a case, the solution is less appealing than in the real case, because it has been proven in [13] that *a full-rate, minimal delay, code design exists only for the case* $n_T = 2$.

In case of $n_T = 2$, the full-rate code is the Alamouti code (7.8). If we can tolerate a rate $R = 1/2$, it is quite easy to build orthogonal codes for transmitting complex symbols for systems with $n_T = 4$ or 8. In such a case, in fact, setting $n_s = N/2$, it is sufficient to select a family of real N-size matrices and transmit the real part of the $N/2$ symbols using (for example) the first $N/2$ matrices and the imaginary parts of the symbols using the second $N/2$ matrices. Clearly, this solution is not the best in terms of efficiency. The basic theory concerning the existence of amicable code families was treated in detail in [12]. More recently, the design of complex codes assuming square coding matrices, or a restriction of such case obtained by deleting columns of the coding matrices, has been provided in [14]. In such a case, it was proved in [14] that the maximal achievable rate with n_T transmit antennas is

$$R = \frac{\lceil \log_2 n_T \rceil + 1}{2^{\lceil \log_2 n_T \rceil}}. \tag{7.53}$$

For convenience, the maximal rate, together with the delay and the number of transmitted symbols achievable with the design rule given in [14] is reported in Table 1.

Table 7.1

Table 1 - Maximal Rates for Square Coding Matrices

Tx antennas	delay	symbols	rate
1	1	1	1
2	2	2	1
3 to 4	4	3	3/4
5 to 8	8	4	1/2
2^{K-2} to 2^{K-1}	2^{K-1}	K	$K/2^{K-1}$

7.3.5 Examples

A few examples of codes are useful to better understand OSTBC.

Real symbols

Examples of orthogonal designs in the cases $n_T = 2, 4$, and 8, valid for real symbols, were initially suggested in [13] and they are reported here below:

$n_T = 2$

$$X = \begin{pmatrix} s_1 & s_2 \\ -s_2 & s_1 \end{pmatrix}; \tag{7.54}$$

$n_T = 4$

$$X = \begin{pmatrix} s_1 & s_2 & s_3 & s_4 \\ -s_2 & s_1 & -s_4 & s_3 \\ -s_3 & s_4 & s_1 & -s_2 \\ -s_4 & -s_3 & s_2 & s_1 \end{pmatrix}; \tag{7.55}$$

$n_T = 8$

$$X = \begin{pmatrix} s_1 & s_2 & s_3 & s_4 & s_5 & s_6 & s_7 & s_8 \\ -s_2 & s_1 & s_4 & -s_3 & s_6 & -s_5 & -s_8 & s_7 \\ -s_3 & -s_4 & s_1 & s_2 & s_7 & s_8 & -s_5 & -s_6 \\ -s_4 & s_3 & -s_2 & s_1 & s_8 & -s_7 & s_6 & -s_5 \\ -s_5 & -s_6 & -s_7 & -s_8 & s_1 & s_2 & s_3 & s_4 \\ -s_6 & s_5 & -s_8 & s_7 & -s_2 & s_1 & -s_4 & s_3 \\ -s_7 & s_8 & s_5 & -s_6 & -s_3 & s_4 & s_1 & -s_2 \\ -s_8 & -s_7 & s_6 & s_5 & -s_4 & -s_3 & s_2 & s_1 \end{pmatrix}. \tag{7.56}$$

All previous schemes consist of square matrices. If this assumption is removed, we can build full-rate codes also n_T different from two, four, or eight. An example of (non-minimum) delay code, for $n_T = 3$ is

$$X = \begin{pmatrix} s_1 & s_2 & s_3 & s_4 \\ -s_2 & s_1 & -s_4 & s_3 \\ -s_3 & s_4 & s_1 & -s_2 \end{pmatrix}. \tag{7.57}$$

This matrix has been obtained by simply removing the last row of $\boldsymbol{X}$ in (7.55). This code is orthogonal, as $\boldsymbol{X}\boldsymbol{X}^H = \|\boldsymbol{s}\|^2$, and full-rate, as it allows the transmission of $n_s = 4$ symbols in $N = 4$ time slots.

Similarly, to build full-rate codes for systems with $n_T = 5, 6$, or 7 transmit antennas, it is sufficient to use (7.56) and delete, correspondingly, three, two, or one rows. The loss in such cases is in the decoding delay. In fact, if we have, for example, five transmit antennas and we transmit using an orthogonal code, we need to wait eight time slots to recover the transmitted symbols. The situation where we have an orthogonal code with $n_T = N$ is optimal in the sense that it minimizes the decoding delay and the resulting coding is denoted as *minimal delay design*. From the theory reported above, we know that there exist full-rate, minimum delay code designs, for real symbols, only for $N = 2, 4$, or 8.

Complex symbols

Two possible orthogonal codes for complex symbols when $n_T = 3$, $n_s = 3$, and $N = 4$ i.e., with rate $R = 3/4$, are (see, e.g., [3] and the references therein)

$$\boldsymbol{X} = \begin{bmatrix} s_1 & 0 & s_2 & -s_3 \\ 0 & s_1 & s_3^* & s_2^* \\ -s_2^* & -s_3 & s_1^* & 0 \end{bmatrix} \tag{7.58}$$

and

$$\boldsymbol{X} = \begin{bmatrix} s_1 & -s_2^* & s_3^* & 0 \\ s_2 & s_1^* & 0 & -s_3^* \\ s_3 & 0 & -s_1^* & s_2^* \end{bmatrix}. \tag{7.59}$$

We can write the I/O relationship corresponding to (7.59) as

$$\boldsymbol{y} = \boldsymbol{X}^T\boldsymbol{h} = \boldsymbol{\mathcal{H}}\boldsymbol{s}, \tag{7.60}$$

where $\boldsymbol{y} = (y_1, y_2^*, y_3^*, y_4^*)^T$, $\boldsymbol{s} = (s_1, s_2, s_3)^T$, and

$$\mathcal{H} = \begin{bmatrix} h_1 & h_2 & h_3 \\ h_2^* & -h_1^* & 0 \\ -h_3^* & 0 & h_1^* \\ 0 & h_3^* & -h_2^* \end{bmatrix}. \tag{7.61}$$

We can verify that, even though $\boldsymbol{X}$ and $\mathcal{H}$ are rectangular matrices, they are pseudo-orthogonal matrices. In particular, we have

$$\mathcal{H}^H \mathcal{H} = (|h_1|^2 + |h_2|^2 + |h_3|^2)\boldsymbol{I} \tag{7.62}$$

and

$$\boldsymbol{X}\boldsymbol{X}^H = (|s_1|^2 + |s_2|^2 + |s_3|^2)\boldsymbol{I}. \tag{7.63}$$

$n_T = 4$

A possible orthogonal code, valid for $n_T = 4$, $n_s = 3$, and $N = 4$, i.e., again with symbol rate $R_s = 3/4$, is

$$\boldsymbol{X} = \begin{bmatrix} s_1 & 0 & s_2 & s_3^* \\ 0 & s_1 & s_3 & -s_2^* \\ -s_2^* & -s_3^* & s_1^* & 0 \\ s_3 & s_2 & 0 & s_1^* \end{bmatrix}. \tag{7.64}$$

This matrix has an interesting structure that can be revealed by writing (7.64) as

$$\boldsymbol{X} = \begin{bmatrix} s_1\boldsymbol{I}_2 & \boldsymbol{A}(s_2, s_3) \\ -\boldsymbol{A}^H(s_2, s_3) & s_1^*\boldsymbol{I}_2 \end{bmatrix}, \tag{7.65}$$

where $\boldsymbol{A}(s_2, s_3)$ is the 2×2 Alamouti code (7.8) built with symbols s_2 and s_3. Exploiting the orthogonality of $\boldsymbol{A}(s_2, s_3)$ it is easy to check that $\boldsymbol{X}$ in (7.65) is orthogonal and, in particular, $\boldsymbol{X}\boldsymbol{X}^H = \|\boldsymbol{s}\|^2\boldsymbol{I}$.

$n_T = 8$

Generalizing the structure of (7.65), we can easily build an orthogonal code for $n_T = 8$ antennas:

$$\boldsymbol{X} = \begin{bmatrix} s_1\boldsymbol{I}_4 & \boldsymbol{A}(s_2, s_3, s_4) \\ -\boldsymbol{A}^H(s_2, s_3, s_4) & s_1^*\boldsymbol{I}_4 \end{bmatrix}, \tag{7.66}$$

where $\boldsymbol{A}(s_2, s_3, s_4)$ is the orthogonal design of (7.64), applied to the set of symbols s_2, s_3, and s_4. In this case, we have $n_s = 4$, $N = 8$, and the symbol rate is then $R_s = 1/2$.

7.3.6 Capacity of Orthogonal Coding Schemes

In Section 7.3.1, we have seen that Alamouti design reaches the capacity limit for a 2×1 MIMO system. It is now time to see how orthogonal design, in general, behaves in terms of capacity.

As it has been originally pointed out in [15], OSTBC is capable of achieving the maximum information rate only when the receiver has only one antenna. This implies that the capacity of a truly MIMO system (i.e., a system with a number of both transmit *and* receive antennas strictly greater than one), OSTBC can *never* reach the capacity. This property can be proved as follows (the result is the same as in [15], even though we follow a different proof).

Since OSTBC converts the overall MIMO channel into a set of n_s independent channels, the capacity of the OSTBC system is equal to the sum of the capacities of each sub-channel. As shown before, each sub-channel is an additive Gaussian channel. If we denote by σ_s^2 the variance of the transmitted symbols, the overall energy necessary to transmit the space-time code $\boldsymbol{X}$ is

$$\mathcal{E} = E\{\mathrm{tr}(\boldsymbol{X}\boldsymbol{X}^H)\} = n_T E\{\|\boldsymbol{s}\|^2\} = n_T n_s \sigma_s^2. \tag{7.67}$$

If the space-time code spans N symbol periods, the average power is then

$$p_T = \frac{n_T n_s \sigma_s^2}{N}. \tag{7.68}$$

Taking into account that, to transmit n_s symbols, OSTBC employs N time slots, the capacity, conditioned to the channel $\boldsymbol{H}$, is

$$C_{\mathrm{OSTBC}} = \frac{n_s}{N} \log_2 \left(1 + \frac{N p_T}{n_s n_T \sigma_n^2} \|\boldsymbol{H}\|^2\right). \tag{7.69}$$

On the other hand, the capacity of the equivalent MIMO channel is

$$C_{\mathrm{MIMO}} = \log_2 \left| \boldsymbol{I} + \frac{p_T}{n_T \sigma_n^2} \boldsymbol{H}\boldsymbol{H}^H \right|. \tag{7.70}$$

Using the eigen-decomposition $\boldsymbol{H}\boldsymbol{H}^H = \boldsymbol{U}\boldsymbol{\Lambda}\boldsymbol{U}^H$, we can rewrite the two capacity values in (7.69) and (7.70) as

$$C_{\text{OSTBC}} = \frac{n_s}{N}\log_2\left(1 + \frac{Np_T}{n_s n_T \sigma_n^2}\sum_{i=1}^{q}\lambda_i\right) \tag{7.71}$$

and

$$C_{\text{MIMO}} = \log_2\left|\boldsymbol{I} + \frac{p_T}{n_T\sigma_n^2}\boldsymbol{\Lambda}\right| = \sum_{i=1}^{q}\log_2\left(1 + \frac{p_T}{n_T\sigma_n^2}\lambda_i\right), \tag{7.72}$$

where q is the rank of $\boldsymbol{H}$. Exploiting the property that

$$\alpha\log\left(1 + \frac{x}{\alpha}\right) \leq \log(1+x), \quad \text{for } 0 < \alpha \leq 1 \text{ and } x > 0, \tag{7.73}$$

with the equality sign holding if and only if $\alpha = 1$, and the Jensen's inequality

$$\log(1 + \sum_i x_i) \leq \sum_i \log(1 + x_i), \quad \text{for } x_i \geq 0, \tag{7.74}$$

both valid for any base of the logarithm, we can write

$$\begin{aligned} C_{\text{OSTBC}} - C_{\text{MIMO}} &= \frac{n_s}{N}\log_2\left(1 + \frac{Np_T}{n_s n_T\sigma_n^2}\sum_{i=1}^{q}\lambda_i\right) \\ &- \sum_{i=1}^{q}\log_2\left(1 + \frac{p_T}{n_T\sigma_n^2}\lambda_i\right) \\ &\leq \log_2\left(1 + \frac{p_T}{n_T\sigma_n^2}\sum_{i=1}^{q}\lambda_i\right) - \sum_{i=1}^{q}\log_2\left(1 + \frac{p_T}{n_T\sigma_n^2}\lambda_i\right) \\ &\leq 0, \end{aligned} \tag{7.75}$$

where the first inequality exploits (7.73), taking into account that $n_s \leq N$, whereas the second inequality uses (7.74). The equality sign in (7.75) holds true *if and only if* $n_s = N$ and the channel rank is one ($q = 1$) i.e., the MIMO system is degenerate. Hence, from (7.75), we deduce that, except for the degenerate case where the MIMO channel has rank one and $n_s = N$, the orthogonal design is a lossy coding system i.e., it does not reach capacity.

7.3.7 Performance

In Figures 7.2 and 7.3, we compare three orthogonal coding systems, obtained with $n_T = 2, 4$, and 8 transmit antennas and one receive antenna. The three systems use the same transmit power and the same constellation (QPSK). The symbol rate is then $R_s = 1$ for the the case $n_T = 2$ and it is $R_s = 1/2$ for the other two cases. From

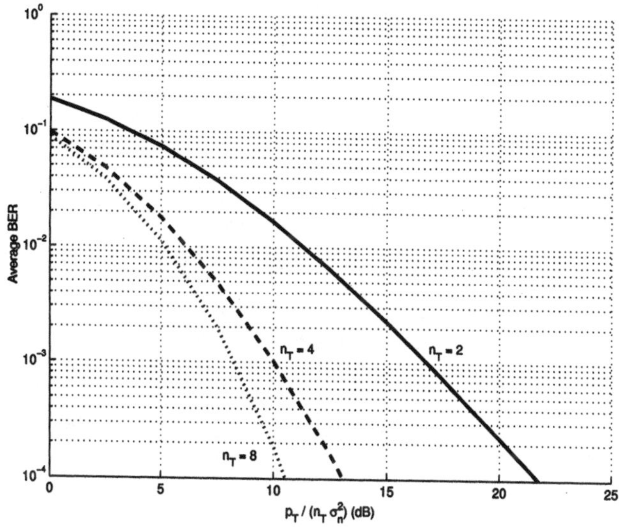

Figure 7.2 Average BER vs. $p_T/(n_T\sigma_n^2)$ for systems with $n_T = 2, 4$ and 8 transmit antennas; all systems transmit a QPSK constellation and use the same overall power.

Figure 7.2, we can clearly see how the diversity gain (slope of the average BER versus SNR, at high SNR) increases as the number of transmit antennas increases.

In Figure 7.2, the comparison is performed enforcing the same transmit power, but not the same bit rate. To make a more fair comparison, we enforced the same bit rate to all systems, by using a $4 - QAM$ constellation when $n_T = 2$ and a $16 - QAM$ constellation for the cases of $n_T = 4$ and 8. Gray encoding is used on each QAM modulator. The result is shown in Figure 7.3. Here, we see again the different diversity gains. However, now at a low SNR, the system with $n_T = 2$ offers slightly better results than the others, because of the higher coding gain.

In summary, OSTBC offers the possibility of achieving the maximum diversity gain with a very simple receiver. However, OSTBC suffers from not being optimal from

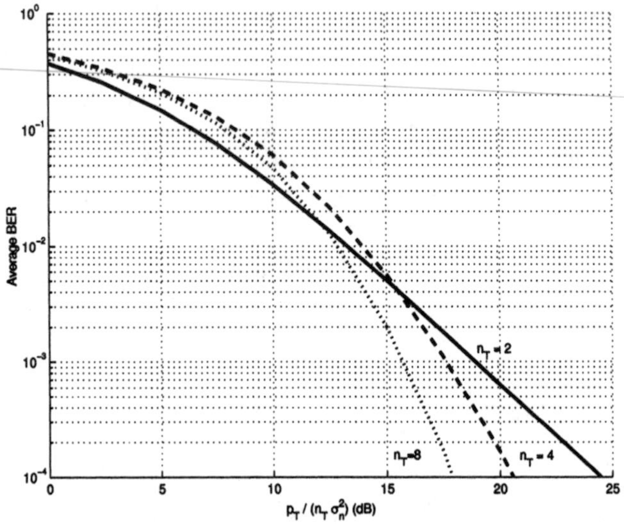

Figure 7.3 Average BER versus $p_T/(n_T\sigma_n^2)$ for systems with $n_T = 2, 4$, and 8 transmit antennas; all systems transmit at the same bit rate and with the same overall power.

the rate point of view. This is particularly evident for truly MIMO systems, with more than one receive antenna.

7.4 QUASI-ORTHOGONAL SPACE-TIME CODING

The main reason for which orthogonal design is not optimal from the rate point of view is the orthogonality constraint, which poses a strong limit on the number of orthogonal space-time matrices of a given size. Relaxing the orthogonality constraint, it is possible to enlarge the family of space-time codes and thus increase the rate. This idea has led to the development of the so called Quasi-Orthogonal Design (QOD) methods. Different QOD strategies have been proposed in [1, 16, 17]. As an example of QOD valid for $n_T = 4$ transmit antennas, we have [16]

$$\boldsymbol{X} = \begin{bmatrix} \boldsymbol{A}(s_1, s_2) & \boldsymbol{A}(s_3, s_4) \\ -\boldsymbol{A}^*(s_3, s_4) & \boldsymbol{A}^*(s_1, s_2) \end{bmatrix} = \begin{pmatrix} s_1 & -s_2^* & s_3 & -s_4^* \\ s_2 & s_1^* & s_4 & s_3^* \\ -s_3^* & s_4 & s_1^* & -s_2 \\ -s_4^* & -s_3 & s_2^* & s_1 \end{pmatrix}, \quad (7.76)$$

where $\boldsymbol{A}(s_1, s_2)$ and $\boldsymbol{A}(s_3, s_4)$ denote two 2×2 Alamouti codes built with the symbols (s_1, s_2) and (s_3, s_4), respectively. This kind of coding allows the transmission with symbol rate $R_s = 1$. The price paid with respect to OSTBC is that the matrix $\boldsymbol{X}$ in (7.76) is not orthogonal. However, it is easy to check that the subspace created by the columns $\boldsymbol{x}_1$ and $\boldsymbol{x}_4$ of $\boldsymbol{X}$ is orthogonal to the subspace spanned by the vectors $\boldsymbol{x}_2$ and $\boldsymbol{x}_3$. This means that ML decoding can be simplified: Instead of taking a joint decision over four symbols, one can take two decisions, each one over two symbols. The idea is then to build families of codes such that $\boldsymbol{X}\boldsymbol{X}^H$ is not diagonal, as with OSTBC, but it is still a *sparse* matrix. QOD offers then some advantage with respect to OSTBC, with limited additional complexity, but still is sub-optimal in terms of rate. In the next sections, we will show how to maximize the rate.

7.5 LAYERED SPACE-TIME CODING

The main tool for increasing the transmission rate with multiple transmit antenna systems consists of transmitting more independent streams, or *layers* of data, from all available transmit antennas, *simultaneously*. This idea was originally suggested by Foschini in [18], who led the group that developed the so called Bell Laboratories layered space-time (BLAST) architecture. BLAST was shown to yield a spectral efficiency up to 40 bit/sec/Hz for indoor communications using a system with $n_T = 8$ transmit and $n_R = 12$ receive antennas [19]. Put in the most simplistic terms, if we have a system with n_T transmit antennas, we can transmit, simultaneously, n_T independent symbols, one from each antenna. At the receiver, we get, at any time instant, n_R observations, one from each receive antenna. Therefore, at any time instant, we have a system of n_R observations in n_T unknowns. If we wish to use a linear estimator[5], the solution of the system of equations requires that the number of independent symbols transmitted at any time instant is equal to $\min(n_T, n_R)$. However, according to what we have proven in Chapter 5, such a system can have, at most, diversity gain equal to n_R, i.e., the system has no transmit diversity gain, whatever its decoding strategy. This is a considerable loss if compared to the achievable gain of $n_T \cdot n_R$ offered by an $n_T \times n_R$ MIMO system. The intuitive reason for such a loss is that, transmitting independent symbols from each antenna, each symbol is observed at the receiver only through n_R replicas of itself. The maximum diversity gain is achieved by combining these replicas and thus a gain factor of n_R is possible, but no more than that.

This argument, in its extreme simplicity, suggests that the way to obtain higher diversity consists in concatenating the streams transmitted from different antennas and

5 In principle, since the symbols belong to a finite alphabet, the symbol recovery does not require $n_R \geq n_T$, necessarily. Of course, this possibility is not available if we restrain the symbol estimator to be linear.

transmitting each symbol, possibly, from *all* transmit antennas. To avoid a sacrifice in terms of rate, it is then necessary to transmit combinations of symbols from each antenna, at the same time. On the other hand, as explained in Chapter 5, to achieve

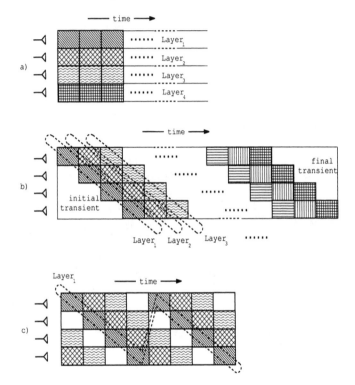

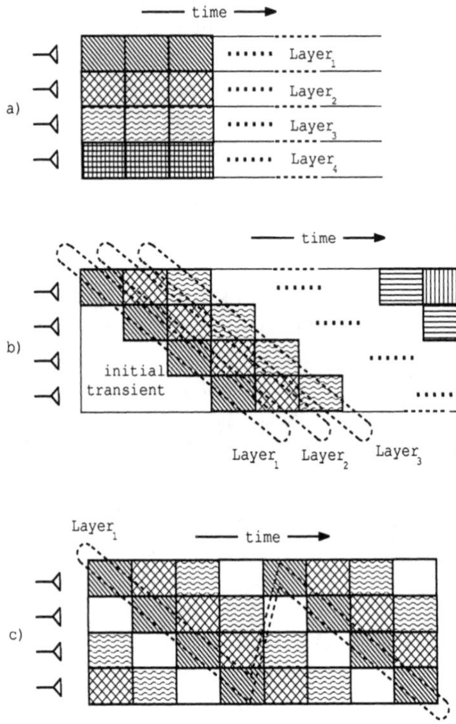

Figure 7.4 Alternative layered space-time coding scheme: a) horizontal layering; b) diagonal layering; and c) threaded layering.

full diversity gain, it is necessary to encode the symbols to be transmitted with a space-time encoder that spans at least n_T time instants. Clearly, this complicates the decoding strategy, at the receiver. In general, ML decoding is the only strategy allowing for maximum diversity gain, but it comes at the expense of computational cost that increases exponentially with the number of antennas and the constellation size. Sphere decoding (SD) can help to simplify the ML decoder, but still it is quite complex to implement.

Given this context, the concept of layering comes out as a way to achieve, possibly, full diversity and full rate gain, but with less computational complexity than ML (or SD). The main features of layering schemes are the following:

- Each layer encodes a certain number of symbols;
- Different layers encode independent sets of symbols, which are successively decoded independently of each other.

Depending on the strategy used to transmit different layers, we can have, for example: *horizontal layering, diagonal layering, threaded layering*, as sketched in Figure 7.4. With horizontal layering, different layers are transmitted from different antennas. Such a scheme can have, at most, diversity gain equal to the number of receive antennas. Using diagonal layering, each layer is transmitted from all antennas. In this way, in principle it is possible to achieve full diversity gain. The schemes in Figure 7.4 (a) and (b) are also known as vertical-BLAST (V-BLAST) and diagonal BLAST (D-BLAST). From Figure 7.4 (b) we can also observe that D-BLAST incurs in a waste of resources because of the two (initial and final) transients. To avoid such a loss, it is possible to *wrap* layers in a cyclic manner, as shown in Figure 7.4 (c). A more general design, always based on cyclically wrapped layers was proposed in [20], where the layers are denoted as *threads* and the resulting scheme is then known as *threaded* space-time coding (TST). TST is more general than D-BLAST, as each thread can last more than n_T symbol periods.

From the point of view of coding, it is useful to distinguish between horizontal, vertical, and diagonal coding, as shown in Figure 7.5. With horizontal coding (HC), the incoming stream is demultiplexed into n_T streams that are coded separately. It is the simplest scheme, as it creates n_T independent streams that can be decoded independently, but it cannot have a diversity gain higher than n_R. Vertical coding (VC) has the potential to be optimal from both diversity and rate point of view, as all the transmitted symbols are virtually related to each other by the initial coding. At the same time, VC requires the highest decoding complexity. Finally, diagonal coding (DC) comes out somehow as a trade-off between performance and complexity. In fact, DC creates n_T independent streams, as with HC. However, the substreams are interleaved through a space-time encoder, before transmission. This creates the possibility for full diversity gain, as opposed to HC. But, differently from VC, it still allows for independent decoding of n_T substreams.

We start our analysis with BLAST techniques, as BLAST played a fundamental role to sparkle interest on MIMO systems.

7.5.1 BLAST

The BLAST techniques proposed by Foschini and his group [18] were designed with the goal of achieving the maximum rate gain offered by the MIMO architecture, with a relatively simple receiver. In the original proposal of BLAST techniques, the number of receive antennas n_R was required to be at least equal to the number of transmit antennas n_T. The distinguishing feature of the BLAST receiving structure is that the two-dimensional (2D) processing necessary to encode and decode the received 2D signal is split in a series of simpler one-dimensional (1D) processing steps. At the transmitter side, the overall encoder is built as the superposition on n_T independent 1D encoders that produce independent *layers*. At the receiver, the layers are separated by properly combining interference suppression and cancelation.

We analyze now in detail the decoding strategies for the V-BLAST system. The transmitter, in its simplest version, is depicted in Figure 7.6. The incoming sequence of information bits is split into n_T separate sequences that are encoded and modulated independently of each other and transmitted from each antenna. The sequence transmitted from the ith antenna is the ith layer and is denoted as $x_i(n)$. At each time instant, the receive array collects the vector

$$\boldsymbol{y}(n) = \boldsymbol{H}\boldsymbol{x}(n) + \boldsymbol{v}(n), \tag{7.77}$$

where $\boldsymbol{x}(n) = [x_1(n), \ldots, x_{n_T}(n)]^T$, $\boldsymbol{H}$ is the $n_R \times n_T$ channel matrix and $\boldsymbol{v}(n)$ is the noise vector, assumed to be $\mathcal{N}(\boldsymbol{0}, \sigma_n^2 \boldsymbol{I})$. The main idea underlying the decoding structure is depicted pictorially in Figure 7.7, where each layer is thought of as immersed in the interference from other layers. The decoding proceeds layer-by-layer, as follows: For each layer, the previous layers are suppressed, whereas the successive layers are canceled through an iterative decision rule. In this way, the decoding structure is a series of 1D decoding steps. We will explain the difference between cancelation and suppression by describing two possible implementations of the decoding rule.

7.5.1.1 QRD interference suppression and interference cancelation

The simplest way to implement the decoding rule based on successive interference suppression and cancelation is based on the QR decomposition (QRD). Having required n_R to be equal or greater than n_T, $\boldsymbol{H}$ can always be decomposed, using the QRD, as

$$\boldsymbol{H} = \bar{\boldsymbol{U}}_R \bar{\boldsymbol{R}}, \tag{7.78}$$

where $\bar{\boldsymbol{U}}_R$ is an $n_R \times n_R$ unitary matrix and $\bar{\boldsymbol{R}}$ is an $n_R \times n_T$ upper triangular matrix (with its lower $n_R - n_T$ rows equal to zero, if $n_R > n_T$). Denoting with $\boldsymbol{U}_R$

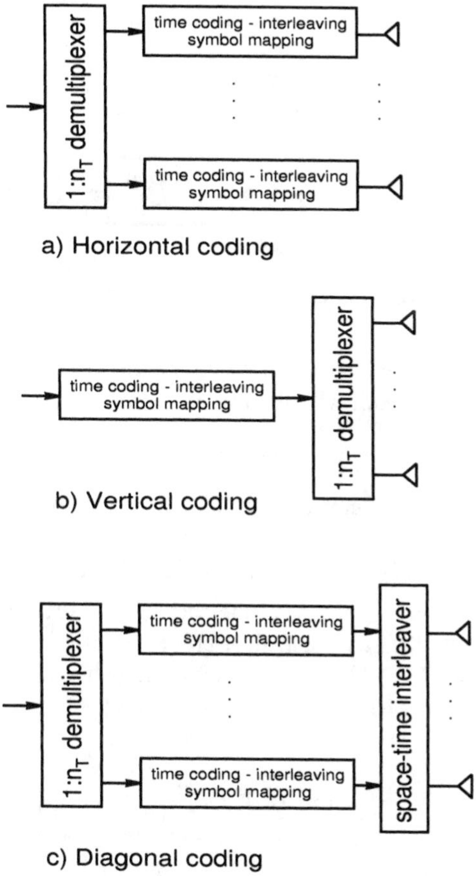

Figure 7.5 Alternative layer encoding schemes.

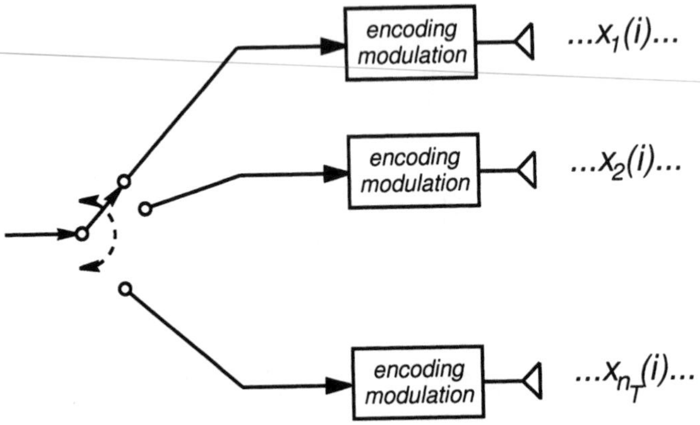

Figure 7.6 V-BLAST encoding scheme.

the $n_R \times n_T$ para-unitary matrix, built with the first n_T columns of $\bar{\boldsymbol{U}}_R$, and with $\boldsymbol{R}$ the $n_T \times n_T$ matrix built with the first n_T rows of $\bar{\boldsymbol{R}}$, (7.78) can be rewritten, equivalently, as

$$\boldsymbol{H} = \boldsymbol{U}_R \boldsymbol{R}. \tag{7.79}$$

Multiplying the received vector $\boldsymbol{y}(n)$ by $\boldsymbol{U}_R^H$ from the left side, we obtain the n_T-size vector

$$\boldsymbol{z}(n) := \boldsymbol{U}_R^H \boldsymbol{y}(n) = \boldsymbol{R}\boldsymbol{x}(n) + \boldsymbol{w}(n), \tag{7.80}$$

where $\boldsymbol{w}(n) := \boldsymbol{U}_R^H \boldsymbol{v}(n)$, because of the para-unitary structure of $\boldsymbol{U}_R$, is a vector noise having exactly the same statistical properties as $\boldsymbol{v}(n)$. The system (7.80) is easier to invert than (7.77), because the coefficient matrix $\boldsymbol{R}$ is triangular. Exploiting the upper triangularity of $\boldsymbol{R}$, we can write (we drop the time index for simplicity of notation)

$$\begin{aligned} z_{n_T} &= R_{n_T,n_T} x_{n_T} + w_{n_T} \\ z_{n_T-1} &= R_{n_T-1,n_T-1} x_{n_T-1} + R_{n_T-1,n_T} x_{n_T} + w_{n_T-1} \\ &\cdots \\ z_1 &= R_{1,1} x_1 + \sum_{k=2}^{n_T} R_{1,k} x_k + w_1 \end{aligned} \tag{7.81}$$

From (7.81), we can set up an iterative decoding and interference cancelation structure that works as follows. We decode x_{n_T} first by simply taking

$$\hat{x}_{n_T} = D\left[\frac{z_{n_T}}{R_{n_T,n_T}}\right], \tag{7.82}$$

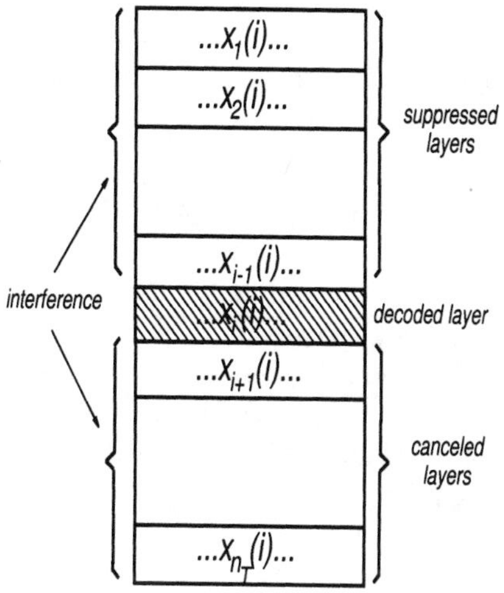

Figure 7.7 V-BLAST decoding principle.

where $D[a]$ denotes the decision rule that yields the constellation symbol closest to a. The contribution of the decoded symbol x_{n_T} can then be canceled from z_{n_T-1} and the symbol x_{n_T-1} can be decoded as

$$\hat{x}_{n_T-1} = D\left[\frac{z_{n_T-1} - R_{n_T-1,n_T}\hat{x}_{n_T}}{R_{n_T-1,n_T-1}}\right]. \tag{7.83}$$

The iterations proceed through successive decoding and cancelations, until x_1 is estimated. In the ith step, we have

$$\hat{x}_i = D\left[\frac{z_i - \sum_{k=i+1}^{n_T} R_{i,k}\hat{x}_k}{R_{i,i}}\right]. \tag{7.84}$$

In words, at, let us say, the ith layer, the interference from the previous layers is suppressed thanks to the QR decomposition. The interference from the successive layers is then canceled, using iterative decisions and cancelations.

Since this method is based on successive decoding and cancelation, the decoding order plays a critical role in the performance of the method itself. To improve the performance, the layers should be ordered so as to decode the layer with the largest SNR first, cancel its contribution from the layer with the next largest SNR, and so on.

What is the diversity offered by this V-BLAST technique? The diversity differs from layer to layer. Using the ZF strategy to remove the interference, and denoting with G_d the desired diversity order, at the ith layer it is possible to remove [21]

$$n_i = n_R - G_d \tag{7.85}$$

interferers and still guarantee a (potential) diversity order G_d. Therefore, at layer n_T, where it is necessary to remove $n_T - 1$ interferers, the maximum diversity gain is $G_d = n_R - n_i = n_R - n_T + 1$. At layer $n_T - 1$, the diversity order is $n_R - n_T + 2$, and so on. This implies, for example, that if $n_R = n_T$, which is the minimum number of receive antennas still allowing the QR decoding, the diversity gain in decoding the layer n_T is 1 (i.e., there is no diversity gain at all). Clearly, even if the other layers have, potentially, a higher diversity gain, the performance of the overall method is bounded by the worst layer, so that the overall diversity is 1.

If compared to OSTBC, V-BLAST offers then a completely opposite performance: V-BLAST has the maximum rate, but minimum diversity, whereas OSTBC yields maximum diversity, but with low rate.

7.5.1.2 MMSE interference suppression and interference cancelation

As an alternative to QR decomposition, we can adopt an MMSE detector. Indeed, it can be proved that successive interference suppression using an MMSE decoder, followed by interference cancelation is optimal from the point of view of capacity.

From Chapter 5, the capacity of a MIMO system is

$$C = \log_2 \left| \boldsymbol{I} + \gamma \boldsymbol{H}\boldsymbol{H}^H \right|, \tag{7.86}$$

where $\gamma = \sigma_s^2/(n_T\sigma_n^2)$. Writing the channel matrix $\boldsymbol{H}$ as

$$\boldsymbol{H} = (\boldsymbol{h}_1, \boldsymbol{H}_{-1}), \tag{7.87}$$

where $\boldsymbol{h}_1$ is the first column of $\boldsymbol{H}$ and $\boldsymbol{H}_{-1}$ is $\boldsymbol{H}$ deprived of its first column, we can always rewrite (7.86) as [22]

$$\begin{aligned}
C &= \log_2 \left| \boldsymbol{I} + \gamma(\boldsymbol{h}_1, \boldsymbol{H}_{-1})(\boldsymbol{h}_1, \boldsymbol{H}_{-1})^H \right| \\
&= \log_2 \left| \boldsymbol{I} + \gamma(\boldsymbol{h}_1\boldsymbol{h}_1^H + \boldsymbol{H}_{-1}\boldsymbol{H}_{-1}^H) \right| \\
&= \log_2 \left| (\boldsymbol{I} + \gamma\boldsymbol{H}_{-1}\boldsymbol{H}_{-1}^H)(\boldsymbol{I} + \gamma(\boldsymbol{I} + \gamma\boldsymbol{H}_{-1}\boldsymbol{H}_{-1}^H)^{-1}\boldsymbol{h}_1\boldsymbol{h}_1^H) \right| \\
&= \log_2 \left| \boldsymbol{I} + \gamma\boldsymbol{H}_{-1}\boldsymbol{H}_{-1}^H \right| + \log_2 \left| 1 + \gamma(\boldsymbol{I} + \gamma\boldsymbol{H}_{-1}\boldsymbol{H}_{-1}^H)^{-1}\boldsymbol{h}_1\boldsymbol{h}_1^H \right| \\
&= \log_2 \left| \boldsymbol{I} + \gamma\boldsymbol{H}_{-1}\boldsymbol{H}_{-1}^H \right| + \log_2 \left| 1 + \gamma\boldsymbol{h}_1^H(1 + \gamma\boldsymbol{H}_{-1}\boldsymbol{H}_{-1}^H)^{-1}\boldsymbol{h}_1 \right| \\
&:= C_{-1} + C_1.
\end{aligned} \tag{7.88}$$

In this expression, C_1 denotes the capacity of the channel where the first symbol is estimated by applying an MMSE estimator that assumes all other symbols as interferers, whereas C_{-1} is the capacity of the system whose channel matrix is $\boldsymbol{H}_{-1}$ (assuming σ_x^2 to be independent of the number of columns of $\boldsymbol{H}$). In fact, writing the received vector $\boldsymbol{y}$ as

$$\boldsymbol{y} = \boldsymbol{h}_1 x_1 + \boldsymbol{H}_{-1}\boldsymbol{x}_{-1} + \boldsymbol{v}, \tag{7.89}$$

where $\boldsymbol{x}_{-1}$ is the $(n_T - 1)$-size vector given by $\boldsymbol{x}$ deprived of its first entry, the MMSE estimate of x_1, assuming that the symbols are zero-mean, uncorrelated and the noise is white, is

$$\hat{x}_1 = \boldsymbol{g}_{\text{mmse}}^H \boldsymbol{y} \tag{7.90}$$

where

$$\boldsymbol{g}_{\text{mmse}}^H = \gamma\boldsymbol{h}_1^H(\boldsymbol{I} + \gamma\boldsymbol{H}\boldsymbol{H}^H)^{-1}. \tag{7.91}$$

Let us evaluate the capacity of system (7.90), with $\boldsymbol{y}$ given by (7.89). The capacity of such a channel is equal to C_1, as defined in (7.88). In fact, the signal-to-interference-plus-noise ratio (SINR) of (7.90) is

$$\text{SINR} = \frac{\left[\gamma\boldsymbol{h}_1^H\left(\boldsymbol{I} + \gamma\boldsymbol{H}\boldsymbol{H}^H\right)^{-1}\boldsymbol{h}_1\right]^2}{\gamma\boldsymbol{h}_1^H\left(\boldsymbol{I} + \gamma\boldsymbol{H}\boldsymbol{H}^H\right)^{-1}\left(\gamma\boldsymbol{H}_{-1}\boldsymbol{H}_{-1}^H + \boldsymbol{I}\right)\left(\boldsymbol{I} + \gamma\boldsymbol{H}\boldsymbol{H}^H\right)^{-1}\boldsymbol{h}}. \tag{7.92}$$

After a few algebraic manipulations, (7.92) can be simplified into

$$\text{SINR} = \frac{\gamma\boldsymbol{h}_1^H\left(\boldsymbol{I} + \gamma\boldsymbol{H}\boldsymbol{H}^H\right)^{-1}\boldsymbol{h}_1}{1 - \gamma\boldsymbol{h}_1^H\left(\boldsymbol{I} + \gamma\boldsymbol{H}\boldsymbol{H}^H\right)^{-1}\boldsymbol{h}_1}. \tag{7.93}$$

Applying the Sherman-Morrison-Woodbury formula[6], (7.93) can be further simplified into

$$\mathrm{SINR} = \gamma \boldsymbol{h}_1^H \left(\boldsymbol{I} + \gamma \boldsymbol{H}_{-1} \boldsymbol{H}_{-1}^H \right)^{-1} \boldsymbol{h}_1. \tag{7.94}$$

Given this SINR expression, it is easy to check that the capacity of system (7.90) is equal to the term C_1, defined in (7.88). On the other hand, the term C_{-1} in (7.88) is the capacity of the channel where the symbol x_1 has been successfully decoded and canceled. In formulas, C_{-1} is the capacity of the channel

$$\boldsymbol{y}_1 = \boldsymbol{y} - \boldsymbol{h}_1 x_1 = \boldsymbol{H}_{-1} \boldsymbol{x}_{-1} + \boldsymbol{v}. \tag{7.95}$$

Since C_{-1} has the same expression as C, except for the structure of $\boldsymbol{H}$, we can iterate the same derivations as in (7.88) and prove that the capacity can be written as

$$C = \sum_{i=1}^{n_T} C_i, \tag{7.96}$$

where C_i is the capacity of the channel obtained using MMSE decoding on the ith iteration, after having canceled the previous $(i-1)$ MMSE estimates, supposing that there are no decision errors.

In summary, the iterative procedure based on successive MMSE estimation and cancelation is information lossless, at least within the (idealistic) assumption that are no intermediate decoding errors.

The performance of the MMSE-BLAST decoder, in terms of average bit error rate, is better than the ZF-BLAST decoder, but both methods exhibit a low diversity gain. Indeed, this is not surprising, as we know that the maximum diversity gain of a MIMO system where each symbol is transmitted by only one antenna (and received by n_R antennas) can be at most n_R. This simple argument suggests that, to increase the diversity, it is necessary to transmit each symbol not only from one antenna, but possibly from all the available antennas. More specifically, from Chapter 5, we know that maximum diversity can be achieved only by spanning each transmitted symbol over both time and space, through a proper space-time encoder. Clearly, the use of space-time encoder per se does not guarantee that the maximum diversity is really achieved. In the next section, we show how to construct space-time codes that guarantee, at the same time, both maximum rate and maximum diversity.

6 The Sherman-Morrison-Woodbury formula proves that, given any n-size vector $\boldsymbol{x}$ and $n \times n$ invertible matrix $\boldsymbol{A}$, the following identity holds true:

$$\frac{\boldsymbol{x}^H (\boldsymbol{A} + \boldsymbol{x}\boldsymbol{x}^H)^{-1} \boldsymbol{x}}{1 - \boldsymbol{x}^H (\boldsymbol{A} + \boldsymbol{x}\boldsymbol{x}^H)^{-1} \boldsymbol{x}} = \boldsymbol{x}^H \boldsymbol{A}^{-1} \boldsymbol{x}.$$

7.5.2 Full-Diversity Full-Rate Design

Quite recently, it has been shown how layering and linear precoding can be combined in order to *guarantee full diversity and full rate* in [23] and [24] (see also the references therein). The possibility of achieving full diversity and full rate, at the same time, could appear in contrast with the fundamental trade-off between information rate and diversity gain, established by Zheng and Tse [25]. Indeed, there is no real contradiction as the information rate used in [25] is different from the transmission rate used in [23] and [24]. In [25], the information rate is directly related to the outage probability over fading channels and thus, for each diversity gain, it gives an upper bound on the achievable rates, that is the rates that, under appropriate channel coding, yield arbitrarily low bit error rates. Conversely, in [23] and [24] the rate is simply the number of symbols transmitted per channel use, without any guarantee about arbitrarily low error probability, as the outage event is not taken into account. What is really important about the methods proposed in [23] and [24] is that they prove that there are full (transmission) rate schemes that are capable of providing full diversity gain. It is not surprising that, since [23] and [24] relax the condition on the error event, the corresponding transmission rate is higher than the information rate of [25].

We review now in detail the full-diversity full-rate (FDFR) design proposed by Ma and Giannakis in [24]. Conceptually, FDFR is strictly related to the universal space-time (UST) coding proposed by El Gamal and Damen [23]. The transmission

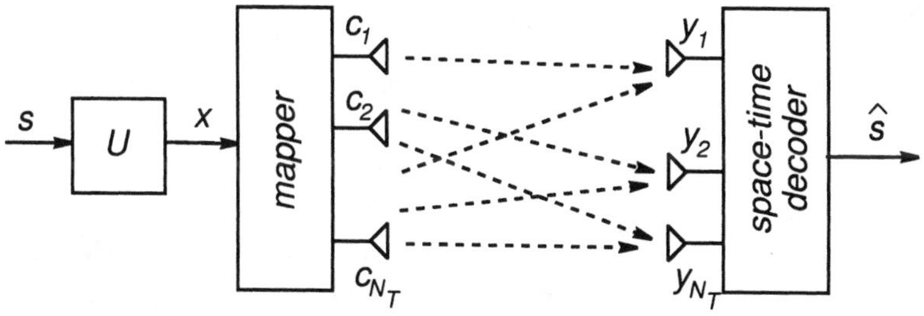

Figure 7.8 Full-rate full-diversity block coding scheme.

strategy is illustrated in Figure 7.8. With respect to the completely general block diagram of Figure 7.1, we notice here the presence of a rotation matrix $\boldsymbol{U}$ applied to the vector of symbols $\boldsymbol{s}$, followed by a mapper onto a space-time code. As we will show next, the initial rotation matrix plays a fundamental role in guaranteeing the full diversity of the overall system. The scheme of Figure 7.8 works as follows. The information symbols, drawn from a finite constellation, are grouped into (column) vectors of size N. For simplicity, we assume that $N = n_T^2$. The vector $\boldsymbol{s}$ is then split into n_T vectors $\boldsymbol{s}_i$, $i = 1, \ldots, n_T$. The role of the initial coding block of Figure 7.8 is to apply a rotation to each vector $\boldsymbol{s}_i$, so as to produce the n_T vectors

$$\boldsymbol{x}_i = \boldsymbol{U}_i \boldsymbol{s}_i = \beta^{i-1} \boldsymbol{U} \boldsymbol{s}_i, \quad i = 1, \ldots, n_T, \tag{7.97}$$

where the $n_T \times n_T$ matrices $\boldsymbol{U}_i = \beta^{i-1} \boldsymbol{U}$ are all unitary matrices built as scaled versions of a primitive unitary matrix $\boldsymbol{U}$, with β such that $|\beta| = 1$. The role of the coefficient β is crucial to guarantee full diversity, as it will be explained later on. The mapper following the initial encoder takes the vectors $\boldsymbol{x}_i$ and maps them onto the matrix

$$\boldsymbol{\mathcal{X}} = \begin{pmatrix} x_1(1) & x_{n_T}(2) & x_{n_T-1}(3) & \cdots & x_2(n_T) \\ x_2(1) & x_1(2) & x_{n_T}(3) & \cdots & x_3(n_T) \\ x_3(1) & x_2(2) & x_1(3) & \cdots & x_4(n_T) \\ \vdots & \vdots & \ddots & \ddots & \vdots \\ x_{n_T}(1) & x_{n_T-1}(2) & \cdots & \cdots & x_1(n_T) \end{pmatrix}, \tag{7.98}$$

where $x_k(l)$ denotes the lth element of the vector $\boldsymbol{x}_k$. The matrix $\boldsymbol{\mathcal{X}}$ is transmitted column-by-column by the transmit array, using n_T consecutive time instants. More specifically, the element $\boldsymbol{\mathcal{X}}_{kl}$ is transmitted from antenna k at time l. In analogy with the layering terminology, each vector $\boldsymbol{x}_i$ is called *layer*, as it plays the same role as the layers described in the previous section. In fact, the transmission strategy specified by the matrix $\boldsymbol{\mathcal{X}}$ means that the layers $\boldsymbol{x}_i$ are wrapped cyclically, exactly as with layered space-time encoding.

Collecting n_T consecutive samples at the receive array, at any time instant n, we get the vector

$$\bar{\boldsymbol{y}}(n) = \boldsymbol{H}\boldsymbol{c}(n) + \bar{\boldsymbol{v}}(n), \tag{7.99}$$

where $\boldsymbol{H}$ is the $n_R \times n_T$ space-time matrix, $\boldsymbol{c}(n)$ is the nth column of $\boldsymbol{\mathcal{X}}$ and $\bar{\boldsymbol{v}}(n)$ is the receive noise. Collecting n_T consecutive blocks, i.e., stacking the n_T vectors $\boldsymbol{y}(n)$ one after the other, we get the $n_R \times n_T$ matrix

$$\boldsymbol{Y} = \boldsymbol{H}\boldsymbol{\mathcal{X}} + \boldsymbol{V}. \tag{7.100}$$

Equivalently, stacking the vectors $\boldsymbol{y}(n)$ into one column vector $\boldsymbol{y} = \text{vec}(\boldsymbol{Y})$, we can rewrite the input-output relationship as

$$\boldsymbol{y} = (\boldsymbol{I} \otimes \boldsymbol{H}) \begin{bmatrix} \boldsymbol{c}(1) \\ \vdots \\ \boldsymbol{c}(n_T) \end{bmatrix} + \boldsymbol{v}. \tag{7.101}$$

To make explicit the relationship between the vectors $\boldsymbol{y}(n)$ and the input vector $\boldsymbol{s}$, it is useful to express $\boldsymbol{c}(n)$ directly as a function of $\boldsymbol{s}$. Using (7.98) and (7.97), we can write

$$\boldsymbol{c}(n) = \left[(\boldsymbol{P}_n \boldsymbol{D}_\beta) \otimes \boldsymbol{u}_n^T\right] \boldsymbol{s}, \tag{7.102}$$

where $\boldsymbol{u}_n^T$ indicates the nth row of $\boldsymbol{U}$, $\boldsymbol{P}_n$ is the permutation matrix obtained by cyclically shifting the rows of the identity matrix $\boldsymbol{I}$ by $n-1$ positions (so that $\boldsymbol{P}_1$ coincides with the identity matrix, $\boldsymbol{P}_2$ is built by cyclically shifting $\boldsymbol{I}$ by one row, and so on), and $\boldsymbol{D}_\beta$ is the diagonal matrix $\boldsymbol{D}_\beta := diag(1, \beta, \beta^2, \ldots, \beta^{n_T-1})$. Introducing the matrices $\boldsymbol{\mathcal{H}} := \boldsymbol{I} \otimes \boldsymbol{H}$ and

$$\boldsymbol{\Phi} := \begin{bmatrix} (\boldsymbol{P}_1 \boldsymbol{D}_\beta) \otimes \boldsymbol{u}_1^T \\ \vdots \\ (\boldsymbol{P}_{n_T} \boldsymbol{D}_\beta) \otimes \boldsymbol{u}_{n_T}^T \end{bmatrix}, \tag{7.103}$$

(7.101) can be rewritten in compact form as

$$\boldsymbol{y} = \boldsymbol{\mathcal{H}} \boldsymbol{\Phi} \boldsymbol{s} + \boldsymbol{v}. \tag{7.104}$$

The symbol vector $\boldsymbol{s}$ can be recovered from (7.104) using ML or sphere decoding, to reduce the complexity. It is important to remark that $\boldsymbol{\Phi}$ is a unitary matrix. In fact, the matrix $\boldsymbol{\Phi}$ is a block matrix, with the block of indices (k, l) of $\boldsymbol{\Phi}\boldsymbol{\Phi}^H$ equal to

$$\begin{aligned} \{\boldsymbol{\Phi}\boldsymbol{\Phi}^H\}_{k,l} &= [(\boldsymbol{P}_k \boldsymbol{D}_\beta) \otimes \boldsymbol{u}_k^T][(\boldsymbol{D}_\beta^* \boldsymbol{P}_l^T) \otimes \boldsymbol{u}_l^*] \\ &= (\boldsymbol{P}_k \boldsymbol{D}_\beta \boldsymbol{D}_\beta^* \boldsymbol{P}_l^T) \otimes (\boldsymbol{u}_k^T \boldsymbol{u}_l^*) \\ &= (\boldsymbol{P}_k \boldsymbol{P}_l^T) \otimes \delta_{kl} = \boldsymbol{I}, \end{aligned} \tag{7.105}$$

where we exploited the orthogonality of the vectors $\boldsymbol{u}_k$ and the property that $(\boldsymbol{A} \otimes \boldsymbol{B})(\boldsymbol{C} \otimes \boldsymbol{D}) = (\boldsymbol{A}\boldsymbol{C}) \otimes (\boldsymbol{B}\boldsymbol{D})$.

It is important to remark, as pointed out in [24], that this transmission strategy is information lossless. In fact, the capacity in the case where the channel is known at the receiver side and unknown at the transmitter side, under a constraint of a maximum transmission power and in the presence of Gaussian noise, is

$$C = \frac{1}{n_T} \log_2 \left| \boldsymbol{I} + \gamma \boldsymbol{\mathcal{H}} \boldsymbol{\Phi} \boldsymbol{\Phi}^H \boldsymbol{\mathcal{H}}^H \right|, \tag{7.106}$$

where $\gamma = p_T / (n_T \sigma_n^2)$. Exploiting (7.105), we obtain

$$C = \frac{1}{n_T} \log_2 \left| \boldsymbol{I} + \gamma \boldsymbol{\mathcal{H}}^H \boldsymbol{\mathcal{H}} \right| = \log_2 \left| \boldsymbol{I} + \gamma \boldsymbol{H}^H \boldsymbol{H} \right|, \tag{7.107}$$

where in the last equality, we have used the property

$$\mathcal{H}^H\mathcal{H} = (\boldsymbol{I}^H \otimes \boldsymbol{H}^H)(\boldsymbol{I} \otimes \boldsymbol{H}) = \boldsymbol{I} \otimes \boldsymbol{H}^H\boldsymbol{H}.$$

Hence, differently from orthogonal coding, *FDFR is information lossless* as it achieves capacity.

The FDFR strategy clearly achieves the maximum transmission rate as it transmits n_T^2 symbols using n_T time slots, so that its symbol rate is

$$R_s = n_T.$$

The main question about the FDFR design is then how to synthesize the coding matrix $\boldsymbol{U}$ in order to achieve also the maximum diversity gain. We recall from Chapter 5 that the maximum diversity is achieved if the error matrix $\boldsymbol{\mathcal{X}} - \boldsymbol{\mathcal{X}}'$, built using any two coding matrices having the structure of (7.98), corresponding to any pair of (different) encoded vectors, is full rank. It was proved in [24] that, for information symbols carved from a finite constellation, there exists at least a unitary matrix $\boldsymbol{U}$ and a scalar β such that $\boldsymbol{\mathcal{X}} - \boldsymbol{\mathcal{X}}'$ is full-rank, for *every* pair of encoded vectors.

We omit the details of the proof here, but we remind two basic steps of the proof, as they have a direct impact on the design of the pair $(\boldsymbol{U}, \beta)$. First, the matrix $\boldsymbol{U}$ must be such that, for any difference vector $\boldsymbol{s}_i - \boldsymbol{s}_j$, the corresponding vector $\boldsymbol{x}_i - \boldsymbol{x}_j = \boldsymbol{U}(\boldsymbol{s}_i - \boldsymbol{s}_j)$ has no zero entry. This problem was solved by Belfiore *et al.* in [26, 27, 28] in the context of SISO transmission schemes. Specifically, it was proven in [26] that, given vectors of symbols $\boldsymbol{s}_i$ and $\boldsymbol{s}_j$, with $\boldsymbol{s}_i \neq \boldsymbol{s}_j$, belonging to a PAM or QAM constellation, it is always possible to find a rotation matrix $\boldsymbol{U}$ such that the vectors $\boldsymbol{U}(\boldsymbol{s}_i - \boldsymbol{s}_j)$ has no null entries. The scope of the rotation matrix $\boldsymbol{U}$ in Figure 7.8 is then precisely to guarantee that each difference vector $\boldsymbol{U}(\boldsymbol{s}_i - \boldsymbol{s}_j)$ has no null entries, i.e., that each layer gets maximum diversity. As explained in [24], the role of β is to guarantee full diversity also *across* the layers. Systematic ways to design FDFR codes were given in [24], where it was shown that the rotation matrix $\boldsymbol{U}$ can assume, for example, the form

$$\boldsymbol{U} = \boldsymbol{W}_{n_T}\mathrm{diag}(1, \alpha, \ldots, \alpha^{n_T-1}), \tag{7.108}$$

where $\boldsymbol{W}_{n_T}$ is the DFT unitary matrix of order n_T, i.e. the matrix with elements $W(k,l) = e^{j2\pi kl/n_T}/\sqrt{n_T}$, and α is a complex number that has to be chosen together with β. One possibility, for example, consists of taking $\alpha = e^{j/2}$ and $\beta = \alpha^{1/n_T}$. The use of the transcendental number $e^{j/2}$ had been also suggested in [29].

It is important to emphasize that the *full-rate full-diversity property is achieved without putting any constraint on the number of transmit antennas.*

What is the price paid for having the full-rate and full-diversity capabilities at the same time? The major drawback is certainly the receiver complexity. In fact, the use of ML decoding is prohibitive when the number of antennas or the constellation cardinality are high. Sphere decoding is better than plain ML, but still it is rather complex to implement. For this reason, in the next section, we examine coding strategies that have a complexity comparable to orthogonal space-time coding, but with a higher rate gain and no constraints on the number of transmit antennas.

Examples

We show now a few examples of FDFR schemes, taken from [23].

Ex. 1: $n_T = 2, n_R = 2$

A 2×2 MIMO system can use the following code [29]

$$\boldsymbol{X} = \begin{pmatrix} s_1 + \phi s_2 & \theta(s_3 + \phi s_4) \\ \theta(s_3 - \phi s_4) & s_1 - \phi s_2 \end{pmatrix}. \tag{7.109}$$

In case of 4-QAM, the two coefficients θ and ϕ that guarantee full diversity and coding gain, are $\theta = e^{j/4}$ and $\phi = e^{j/2}$. From (7.109), we notice that there are two layers in this case: The first one is on the main diagonal and it carries combinations of the symbols s_1 and s_2; the second layer is on the extra-diagonal entries and it carries combinations of the symbols s_3 and s_4. The symbol rate is $R_s = 2$ and the bit rate is $R_b = 4$.

Ex. 2: $n_T = 3, n_R = 3$

A code valid for a 3×3 MIMO system is, for example,

$$\boldsymbol{X} = \begin{pmatrix} x_1(1) & \phi^{2/3}x_3(2) & \phi^{1/3}x_2(3) \\ \phi^{1/3}x_2(1) & x_1(2) & \phi^{2/3}x_3(3) \\ \phi^{2/3}x_3(1) & \phi^{1/3}x_2(2) & x_1(3) \end{pmatrix}, \tag{7.110}$$

where each vector (layer) $\boldsymbol{x}_i := (x_i(1), x_i(2), x_i(3))^T$ is obtained from a correspondent symbol vector $\boldsymbol{s}_i$, as in (7.97) and $\phi = e^{j\pi/12}$.

7.6 TRACE-ORTHOGONAL DESIGN

In this section we consider a very general approach to the design of space-time codes, named trace-orthogonal design (TOD), that was proposed in [30] and further studied in [31]. The most interesting features of TOD are the following:

1. Necessary and sufficient condition for having an information lossless space-time coding system, for systems with perfect channel knowledge at the receiver but no channel knowledge at the transmitter, is to use TOD;
2. Adopting a linear MMSE receiver, TOD with unitary matrices insures minimum BER;
3. The complexity of TOD MMSE decoding is similar to that of orthogonal space-time decoding.

In this section, we review these properties in detail. We start with the so called *unitary error matrices*. Unitary error matrices have been studied extensively for applications in quantum information and dense coding [32, 33]. There are two fundamentally different constructions of unitary bases, namely: the so called *nice error bases* [33] and the *shift-and-multiply* bases [32]. We consider here only the shift-and-multiply bases, as they have many traits in common with layered space-time coding.

In general, a set $\mathcal{M} := \{\boldsymbol{A}_k, k = 1, \ldots, N^2\}$ of N^2 unitary $N \times N$ matrices $\boldsymbol{A}_k$ is called a *unitary error basis* if and only if, defining the inner product between matrices as

$$\langle \boldsymbol{A}_k, \boldsymbol{A}_l \rangle := \operatorname{tr}(\boldsymbol{A}_k^H \boldsymbol{A}_l), \tag{7.111}$$

any two pairs of matrices in the set are orthogonal, i.e.[7]

$$\langle \boldsymbol{A}_k, \boldsymbol{A}_l \rangle := \operatorname{tr}(\boldsymbol{A}_k^H \boldsymbol{A}_l) = \delta_{kl}, \tag{7.112}$$

and

$$\boldsymbol{A}_k^H \boldsymbol{A}_k = \frac{1}{N} \boldsymbol{I}, \quad k = 1, \ldots, N. \tag{7.113}$$

Shift-and-multiply bases

A concise definition of unitary error matrices exploits the Latin square symbol. We recall that a *Latin square* of order N is an $N \times N$ matrix $\boldsymbol{L}_N$ such that each element in the integer set $\{1, \ldots, N\}$ is contained only once in each row and in each column of $\boldsymbol{L}_N$. For each order, there are many different ways to build a Latin square. As an example, a Latin square of order N can be built as $L_N(k, l) = (l - k) \bmod N + 1$.

7 δ_{kl} denotes the Kronecker symbol, equal to one if $k = l$ and zero otherwise.

In such a case, a possible Latin square of order 4 is, for example

$$\boldsymbol{L}_4 = \begin{pmatrix} 1 & 2 & 3 & 4 \\ 4 & 1 & 2 & 3 \\ 3 & 4 & 1 & 2 \\ 2 & 3 & 4 & 1 \end{pmatrix} \tag{7.114}$$

Given a set of N unitary matrices $\mathcal{U} := \{\boldsymbol{U}_i, i = 1, \ldots, N\}$, of size $N \times N$, and the Latin square $\boldsymbol{L}_N$, the shift-and-multiply basis associated to the family $(\mathcal{U}, \boldsymbol{L}_N)$ is given by the N^2 unitary matrices

$$\boldsymbol{A}_{kl} = \boldsymbol{P}_l \operatorname{diag}(\boldsymbol{U}_l(k, :)), \quad k = 1, \ldots, N, l = 1, \ldots, N, \tag{7.115}$$

where $\boldsymbol{P}_l$ is the permutation matrix whose entries are $P_l(L_N(l, m), m) = 1$, for $1 \leq m \leq N$ and 0, otherwise, $\operatorname{diag}(\boldsymbol{x})$ is the diagonal matrix whose main diagonal is $\boldsymbol{x}$, and $\boldsymbol{U}_l(k, :)$, using Matlab notation, denotes the k-th row of $\boldsymbol{U}_l$. In words, the matrix $\boldsymbol{A}_{kl}$ is a cyclic shift of a diagonal matrix having on the diagonal a row of a unitary matrix. As an example, setting all matrices $\boldsymbol{U}_l$ equal to the DFT matrix $\boldsymbol{W}$ of order N, where $W(n, m) = \rho^{mn}/\sqrt{N}$, with $\rho = e^{j2\pi/N}$, we have

$$\boldsymbol{A}_{1,2} = \frac{1}{2} \begin{pmatrix} 0 & 1 & 0 & 0 \\ 0 & 0 & 1 & 0 \\ 0 & 0 & 0 & 1 \\ 1 & 0 & 0 & 0 \end{pmatrix}, \quad \boldsymbol{A}_{2,3} = \frac{1}{2} \begin{pmatrix} 0 & 0 & \rho^2 & 0 \\ 0 & 0 & 0 & \rho^3 \\ 1 & 0 & 0 & 0 \\ 0 & \rho & 0 & 0 \end{pmatrix}. \tag{7.116}$$

In words, the coding matrices are built as follows. For any integer number $1 \leq k \leq N$, we pick up from $\boldsymbol{L}_N$ the matrix positions where we find k, i.e., all pairs of indices (m, n) where $\boldsymbol{L}_N(m, n) = k$. Then we build N matrices having all zeros, except in the previous pairs of indices, where we write one of the N orthogonal vectors. Thus, for every k, that is for every matrix structure, we have N different matrices, each one corresponding to one of the N orthogonal vectors. Repeating the same construction, for every integer k, we build N^2 matrices. It is easy to prove that *a shift-and-multiply basis is a unitary error basis*. In fact, multiplying any two matrices corresponding to different integers k and m, the product is null over the main diagonal and then its trace is zero. Conversely, if we multiply two matrices referring to the same integer k, the product has a diagonal different from zero, but the trace is equal to the scalar product between two orthogonal vectors, and then the result is, again, equal to zero.

TOD is a linear space-time encoder that maps a vector of n_s (complex) symbols $\boldsymbol{s} = (s_1 \; s_2 \; \cdots \; s_{n_s})^T$ onto a space-time code matrix $\boldsymbol{X}$, of dimension $n_T \times Q$, according to the rule[8]

$$\boldsymbol{X} = \sum_{k=1}^{n_s} \boldsymbol{A}_k s_k, \tag{7.117}$$

where $\boldsymbol{A}_k$, with $k = 1, \ldots, n_s$, are complex $n_T \times Q$ matrices that satisfy (7.112). We say that the TOD design is *unitary* if the matrices $\boldsymbol{A}_k$ satisfy both (7.112) and (7.113). The set of matrices $\{\boldsymbol{A}_k\}$, with $k = 1, \ldots, N^2$, defined in (7.115) is only a particular example of *unitary TOD* (UTOD).

Applying the vec$(\cdot)$ operator to (7.117), we obtain

$$\boldsymbol{x} = \text{vec}(\boldsymbol{X}) = \sum_{k=1}^{n_s} \text{vec}(\boldsymbol{A}_k) s_k := \boldsymbol{F}\boldsymbol{s}, \tag{7.118}$$

where $\boldsymbol{F}$ is the matrix whose k-th column is vec$(\boldsymbol{A}_k)$. The minimal assumption about the encoder is injectivity, which requires that, if $\boldsymbol{s}_1$ is mapped onto $\boldsymbol{X}_1$ and $\boldsymbol{s}_2$ is mapped onto $\boldsymbol{X}_2$, then $\boldsymbol{s}_1 \neq \boldsymbol{s}_2$ implies $\boldsymbol{X}_1 \neq \boldsymbol{X}_2$. We are in fact able to recover all symbols s_k from the code matrix $\boldsymbol{X}$ if and only if this requirement is met. We introduce then the following

Definition: A linear space-time code is *nonsingular* if the mapping described by (7.117) is injective.

Non-singularity implies a constraint on the maximum number of symbols that can be transmitted using $n_T \times Q$ code matrices. In fact, since the code mapping is completely described by $\boldsymbol{F}$, the mapping is injective if and only if $\ker(\boldsymbol{F}) = \{\boldsymbol{0}\}$, which is equivalent to say that $\boldsymbol{F}$ is full-rank and, moreover, $\text{rank}(\boldsymbol{F}) = n_s$. This means that in order to have *nonsingular* codes, the following inequality must be satisfied

$$n_s \leq Q \cdot n_T. \tag{7.119}$$

7.6.1 Information Lossless Space-Time Coding

In general, imposing a structure on the transmitted data, like introducing space-time coding, may induce a capacity loss. We have seen that this happens, for example, using orthogonal space-time coding. It is then of interest to see under what conditions a space-time encoder is information lossless. We provide now necessary and sufficient conditions for insuring such a property.

8 From now on, we adopt the single index notation for the coding matrices $\boldsymbol{A}_k$, for simplicity of notation. This requires only a renumbering of the set $\{\boldsymbol{A}_{kl}\}$.

Let us consider a flat fading MIMO systems with n_T transmit and n_R receive antennas, described by the input/output relationship

$$\tilde{\boldsymbol{y}} = \boldsymbol{H}\tilde{\boldsymbol{x}} + \tilde{\boldsymbol{v}}, \tag{7.120}$$

where $\boldsymbol{H}$ is the $n_R \times n_T$ channel matrix, $\tilde{\boldsymbol{x}}$ is the vector of transmitted symbols and $\tilde{\boldsymbol{v}}$ is the noise vector assumed to be zero mean circularly symmetric complex Gaussian with covariance matrix $\sigma_v^2\boldsymbol{I}$. We assume that: i) the receiver has perfect channel knowledge; ii) the transmitter has no channel knowledge, and iii) the transmit power is upper bounded by P_T. Under these conditions, the capacity of the uncoded channel (7.120) is

$$C_H^{\text{unc}} = \log\left|\boldsymbol{I} + \gamma\boldsymbol{H}\boldsymbol{H}^H\right| = \log\left|\boldsymbol{I} + \gamma\boldsymbol{H}^H\boldsymbol{H}\right|, \tag{7.121}$$

where $\gamma = P_T/(n_T\sigma_v^2)$ is the average SNR per receive antenna.

Let us consider now a linear space-time encoded system transmitting the matrix codewords $\boldsymbol{X}$ over the same channel $\boldsymbol{H}$, with $\boldsymbol{X}$ having the structure (7.117). Assuming that the channel is constant over Q successive channel uses (quasi-static fading), and stacking Q consecutive received vectors into a matrix, the received matrix corresponding to $\boldsymbol{X}$ is

$$\boldsymbol{Y} = \boldsymbol{H}\boldsymbol{X} + \boldsymbol{V}, \tag{7.122}$$

where $\boldsymbol{V}$ is the receive noise matrix.

Applying the vec$(\cdot)$ operator[9] to (7.122), and using (7.117) and (7.118), (7.122) can be rewritten in vector form as

$$\boldsymbol{y} \;=\; \text{vec}(\boldsymbol{H}\boldsymbol{X}) + \text{vec}(\boldsymbol{V}) = (\boldsymbol{I}_Q \otimes \boldsymbol{H})\boldsymbol{F}\boldsymbol{s} + \boldsymbol{v}, \tag{7.123}$$

where all symbols have been previously defined.

We prove now that the capacity of the uncoded system (7.120) is equal to the capacity of the coded system (7.122) (or (7.123)), for any channel realization, *if and only if* the encoding matrices $\boldsymbol{A}_k$ satisfy (7.112). Proceeding as in [31], we prove the following important result:

9 If $\boldsymbol{A}(r \times t)$, $\boldsymbol{X}(t \times p)$, and $\boldsymbol{B}(p \times s)$ are matrices, we recall that $\text{vec}(\boldsymbol{A}\boldsymbol{X}\boldsymbol{B}) = (\boldsymbol{B}^T \otimes \boldsymbol{A})\,\text{vec}(\boldsymbol{X})$, where $\otimes$ is the Kronecker product.

If the transmitted symbols are zero mean, uncorrelated, with variance σ_s^2, the coded system (7.122) has the same capacity as the uncoded system (7.120), for any channel realization $\boldsymbol{H}$, *if and only if* $\boldsymbol{F}\boldsymbol{F}^H = \boldsymbol{I}$.

We prove the sufficient condition first. If $\boldsymbol{F}\boldsymbol{F}^H = \boldsymbol{I}$, the capacity of the coded system (7.122) or, equivalently, (7.123), for any given realization of the channel $\boldsymbol{H}$, is

$$C_H^{\text{cod}} = \frac{1}{Q}\log\left|\boldsymbol{I} + \gamma(\boldsymbol{I}\otimes\boldsymbol{H})\boldsymbol{F}\boldsymbol{F}^H(\boldsymbol{I}\otimes\boldsymbol{H}^H)\right| \tag{7.124}$$

$$= \frac{1}{Q}\log\left|\boldsymbol{I} + \gamma(\boldsymbol{I}\otimes\boldsymbol{H})(\boldsymbol{I}\otimes\boldsymbol{H}^H)\right| \tag{7.125}$$

where $\gamma = P_T/(n_T\sigma_v^2)$; the factor $1/Q$ accounts for the Q uses of the channel. Using the property $(\boldsymbol{A}\otimes\boldsymbol{B})(\boldsymbol{C}\otimes\boldsymbol{D}) = (\boldsymbol{A}\boldsymbol{C}\otimes\boldsymbol{B}\boldsymbol{D})$, (7.125) becomes

$$\begin{aligned} C_H^{\text{cod}} &= \frac{1}{Q}\log\left|\boldsymbol{I} + \gamma(\boldsymbol{I}\otimes\boldsymbol{H}\boldsymbol{H}^H)\right| \\ &= \frac{1}{Q}\log\left|\boldsymbol{I}\otimes(\boldsymbol{I} + \gamma\boldsymbol{H}\boldsymbol{H}^H)\right| \\ &= \log\left|\boldsymbol{I} + \gamma\boldsymbol{H}\boldsymbol{H}^H\right| = C_H^{\text{unc}}, \end{aligned} \tag{7.126}$$

with C_H^{unc} given in (7.121), having used the identity $|\boldsymbol{I}_n\otimes\boldsymbol{M}| = |\boldsymbol{M}|^n$. Note that (7.126) holds true for *any* realization $\boldsymbol{H}$. This proves sufficiency.

Let us analyze now the necessity condition. We consider the proof of necessity only for the case $n_R \geq n_T$, but the result applies to any choice of n_T and n_R. To prove necessity, we start from the following equality

$$C_H^{\text{cod}} = C_H^{\text{unc}}, \qquad \forall\boldsymbol{H}\in\mathbb{C}^{n_R\times n_T} \tag{7.127}$$

Using (7.124) and the identity $|\boldsymbol{I}+\boldsymbol{A}\boldsymbol{B}| = |\boldsymbol{I}+\boldsymbol{B}\boldsymbol{A}|$, (7.127) can be rewritten as

$$\left|\boldsymbol{I}_{Q\cdot n_T} + \gamma(\boldsymbol{I}_Q\otimes\boldsymbol{H}^H\boldsymbol{H})\boldsymbol{F}\boldsymbol{F}^H\right| = \left|\boldsymbol{I}_{n_T} + \gamma\boldsymbol{H}^H\boldsymbol{H}\right|^Q, \ \forall\boldsymbol{H}\in\mathbb{C}^{n_R\times n_T}, \tag{7.128}$$

where the dimensions of the identity matrices involved are made explicit. Since (7.128) holds for *any* channel realization, it is satisfied, in particular, for channel matrices $\hat{\boldsymbol{H}}_\lambda$ such that

$$\gamma\hat{\boldsymbol{H}}_\lambda^H\hat{\boldsymbol{H}}_\lambda = \lambda\boldsymbol{I}_{n_T} \qquad \lambda\in\mathbb{R}^+ \tag{7.129}$$

where $\mathbb{R}^+ \equiv [0, +\infty)$. Note that such channel realizations always exist, since we assumed $n_R \geq n_T$. As an example, $\hat{\boldsymbol{H}}_\lambda$ can be built from any set of n_T orthonormal column vectors $\boldsymbol{h}_i \in \mathbb{C}^{n_R}$, as follows

$$\hat{\boldsymbol{H}}_\lambda = \sqrt{\frac{\lambda}{\gamma}}\, [\boldsymbol{h}_1\, \boldsymbol{h}_2\, \cdots\, \boldsymbol{h}_{n_T}], \qquad \lambda \in \mathbb{R}^+. \tag{7.130}$$

Substituting $\hat{\boldsymbol{H}}_\lambda$ in (7.128), we get

$$\left|\boldsymbol{I}_{Q\cdot n_T} + \lambda \boldsymbol{F}\boldsymbol{F}^H\right| = \left|\boldsymbol{I}_{n_T} + \lambda \boldsymbol{I}_{n_T}\right|^Q, \qquad \lambda \in \mathbb{R}^+. \tag{7.131}$$

Since $\boldsymbol{F}\boldsymbol{F}^H$ is Hermitian positive semidefinite, let us indicate with φ_k ($k = 1, \ldots, Q \cdot n_T$) its eigenvalues. Note that $\varphi_k \in \mathbb{R}^+$, as this property will be useful later on. Hence, (7.131) can be written as

$$\prod_{k=1}^{Q\cdot n_T} (1 + \lambda \varphi_k) = (1 + \lambda)^{Q\cdot n_T}, \qquad \lambda \in \mathbb{R}^+. \tag{7.132}$$

Equation (7.132) is an identity between polynomials of degree $Q \cdot n_T$, in the indeterminate λ (it holds for any $\lambda \in \mathbb{R}^+$). But two polynomials coincide if and only if the coefficients of the corresponding powers of λ are equal. Hence, equating, in particular, the coefficients of $\lambda^{Q\cdot n_T}$ and λ, from (7.132) we get

$$\prod_{k=1}^{Q\cdot n_T} \varphi_k = 1, \tag{7.133}$$

and

$$\sum_{k=1}^{Q\cdot n_T} \varphi_k = Q \cdot n_T. \tag{7.134}$$

Since $\varphi_k \in \mathbb{R}^+$, it is easy to verify[10] that (7.133) and (7.134) jointly imply

$$\left(\prod_{k=1}^{Q\cdot n_T} \varphi_k\right)^{\frac{1}{Q\cdot n_T}} = \frac{1}{Q \cdot n_T} \sum_{k=1}^{Q\cdot n_T} \varphi_k. \tag{7.135}$$

The well know relation between geometric and arithmetic means implies that (7.135) holds if and only if all the terms φ_k are equal to each other. This in turn

10 If $x \in \mathbb{R}^+$, then $x = 1$ implies $x^{1/n} = 1$, for any $n \in \mathbb{N}$.

implies, using (7.133) or (7.134), that

$$\varphi_1 = \varphi_2 = \cdots = \varphi_{Q \cdot n_T} = 1, \tag{7.136}$$

which allows us to fully characterize the product $\boldsymbol{F}\boldsymbol{F}^H$. In fact, we use the property that, if $\boldsymbol{G} \in \mathbb{C}^{n \times n}$ is a Hermitian matrix with all eigenvalues equal to 1, then $\boldsymbol{G} = \boldsymbol{I}_n$ [31]. Thanks to this property, we can deduce from (7.136) that

$$\boldsymbol{F}\boldsymbol{F}^H = \boldsymbol{I}_{Q \cdot n_T}. \tag{7.137}$$

This proves the necessity condition for the subclass of channels satisfying (7.129). But since (7.137) is also the condition that guarantees (7.127) for *any* channel matrix, as it follows from the proof of sufficiency, we can conclude a fortiori that (7.137) is also a necessary condition for all channel matrices.

Stated differently, the sufficiency condition proves the existence of a solution to the problem of capacity invariance, with respect to the channel realization, and the necessity condition guarantees that the solution is unique showing that for a subclass of channel matrices, condition (7.137) should necessarily hold. The proof is thus complete.

It is important to remark that since we did not make any assumption about the channel statistics, the previous result guarantees invariance of the corresponding ergodic capacity, regardless of the channel statistics.

The condition $\boldsymbol{F}\boldsymbol{F}^H = \boldsymbol{I}$ holds true only if $\boldsymbol{F}$ is full-rank, that is only if

$$n_s \geq Q \cdot n_T . \tag{7.138}$$

This is equivalent to say that the symbol rate associated to the code, defined as

$$R_s = \frac{n_s}{Q} , \tag{7.139}$$

should be at least n_T. Combining (7.138) with (7.119), we arrive at the following equality

$$n_s = Q \cdot n_T. \tag{7.140}$$

Now, since a code with symbol rate $R_s = n_T$ is called *full-rate*, (7.140) can be recast saying that a necessary condition for *nonsingular* codes to be information lossless is to be *full-rate*.

Moreover, condition (7.140) forces $\boldsymbol{F}$ to be square and this, together with $\boldsymbol{F}\boldsymbol{F}^H = \boldsymbol{I}$, implies that $\boldsymbol{F}$ is unitary. This property leads directly to TOD. In fact, the product $\boldsymbol{F}^H\boldsymbol{F}$ can be rewritten as

$$\begin{aligned}\boldsymbol{F}^H\boldsymbol{F} &= \begin{bmatrix}\mathrm{vec}^H(\boldsymbol{A}_1)\\ \vdots \\ \mathrm{vec}^H(\boldsymbol{A}_{n_s})\end{bmatrix}[\mathrm{vec}(\boldsymbol{A}_1)\ \cdots\ \mathrm{vec}(\boldsymbol{A}_{n_s})]\\ &= \begin{bmatrix}\mathrm{tr}(\boldsymbol{A}_1^H\boldsymbol{A}_1) & \dots & \mathrm{tr}(\boldsymbol{A}_1^H\boldsymbol{A}_{n_s})\\ \vdots & \vdots & \vdots\\ \mathrm{tr}(\boldsymbol{A}_{n_s}^H\boldsymbol{A}_1) & \dots & \mathrm{tr}(\boldsymbol{A}_{n_s}^H\boldsymbol{A}_{n_s})\end{bmatrix} = \boldsymbol{I},\end{aligned} \tag{7.141}$$

where we used the property $\mathrm{vec}^H(\boldsymbol{A}_k)\mathrm{vec}(\boldsymbol{A}_l) = \mathrm{tr}(\boldsymbol{A}_k^H\boldsymbol{A}_l)$. Combining the last equality in (7.141) with (7.140), we can summarize the results of this section saying that a space-time encoder is information lossless if and only if it is TOD (i.e., it respects (7.112)) and full rate.

7.6.2 MMSE Decoding

Maximum Likelihood decoding is the optimal decoder, but it suffers from exponential complexity. Sphere-Decoding, albeit suboptimal, could be used instead of ML, but its complexity is still quite high. It is then important to analyze the performance of sub-optimal decoders, more appealing from the implementation point of view. We consider the minimum mean square error (MMSE) receiver, which derives first a soft estimate of the transmitted symbols and then it takes a hard decision, based on those estimates. In the sequel, we show that: 1) TOD leads to very simple scalar MMSE decoders; and 2) within the class of MMSE encoders/decoders, unitary-TOD yields the minimum BER.

We derive the MMSE estimator for the vector model (7.123), assuming that the symbols are i.i.d., zero mean, with variance σ_s^2 and the noise samples are zero mean, i.i.d., circularly symmetric complex Gaussian random variables, with variance σ_n^2, and statistically independent of the symbols.

The linear MMSE estimate of the kth symbol assumes the following expression

$$\hat{s}_k = \mathrm{tr}(\boldsymbol{A}_k^H\boldsymbol{W}\boldsymbol{Y}),\ k = 1,\dots,n_s, \tag{7.142}$$

where, setting $\gamma = \sigma_s^2/\sigma_v^2$,

$$\boldsymbol{W} = (\boldsymbol{H}^H\boldsymbol{H} + \frac{1}{\gamma}\boldsymbol{I}_{n_T})^{-1}\boldsymbol{H}^H. \tag{7.143}$$

In fact, the MMSE estimate of $\boldsymbol{s}$, denoted by $\hat{\boldsymbol{s}} = (\hat{s_1}\ \hat{s_2}\ \cdots\ \hat{s}_{n_s})^T$, is linear in the observation vector $\boldsymbol{y}$ and it takes the form

$$\hat{\boldsymbol{s}} = \boldsymbol{K}_{sy}\boldsymbol{K}_{yy}^{-1}\boldsymbol{y}, \tag{7.144}$$

where

$$\boldsymbol{K}_{sy} = E\{\boldsymbol{s}\boldsymbol{y}^H\} = \sigma_s^2\boldsymbol{F}^H(\boldsymbol{I}_Q \otimes \boldsymbol{H}^H) \tag{7.145}$$

and

$$\boldsymbol{K}_{yy} = E\{\boldsymbol{y}\boldsymbol{y}^H\} = \sigma_s^2\left[\boldsymbol{I}_Q \otimes \left(\boldsymbol{H}\boldsymbol{H}^H + \frac{1}{\gamma}\boldsymbol{I}_{n_R}\right)\right]. \tag{7.146}$$

In deriving (7.146), we exploited the property $\boldsymbol{F}\boldsymbol{F}^H = \boldsymbol{I}$. Substituting (7.145) and (7.146) in (7.144), we get

$$\hat{\boldsymbol{s}} = \boldsymbol{F}^H \text{vec}\left[\left(\boldsymbol{H}^H\boldsymbol{H} + \frac{1}{\gamma}\boldsymbol{I}_{n_T}\right)^{-1}\boldsymbol{H}^H\boldsymbol{Y}\right], \tag{7.147}$$

where in (7.147) we used the property $\text{vec}(\boldsymbol{W}\boldsymbol{Y}) = (\boldsymbol{I} \otimes \boldsymbol{W})\text{vec}(\boldsymbol{Y})$. Now, using (7.143) and recalling that $\boldsymbol{F} = [\text{vec}(\boldsymbol{A}_1)\ \text{vec}(\boldsymbol{A}_2)\ \cdots\ \text{vec}(\boldsymbol{A}_{n_s})]$, we are able to write

$$\tilde{\boldsymbol{s}} = \boldsymbol{F}^H\text{vec}(\boldsymbol{W}\boldsymbol{Y}) = \begin{bmatrix} \text{tr}(\boldsymbol{A}_1^H\boldsymbol{W}\boldsymbol{Y}) \\ \text{tr}(\boldsymbol{A}_2^H\boldsymbol{W}\boldsymbol{Y}) \\ \vdots \\ \text{tr}(\boldsymbol{A}_{n_s}^H\boldsymbol{W}\boldsymbol{Y}) \end{bmatrix}, \tag{7.148}$$

from which (7.142) follows.

Let us analyze now the main features of the proposed estimator. The first appealing property of (7.142) is low complexity. Since the observation vector $\boldsymbol{y}$ is linearly related to the symbol vector $\boldsymbol{s}$ by a $Qn_R \times n_s$ coefficient matrix, to recover the symbols through a linear decoder it is necessary, in general, to (pseudo) invert a matrix of dimension $Q \cdot n_R \times n_s$. For example, in the minimum delay full-rate case, where $Q = n_T$ and $n_s = n_T^2$, it is necessary to pseudo-invert an $n_T \cdot n_R \times n_T^2$ matrix. Conversely, with TOD, the MMSE estimate requires the inversion of an $n_T \times n_T$ matrix, to compute $\boldsymbol{W}$ in (7.143). Furthermore, this inversion has to be computed at a rate that depends only on the channel coherence time, but not on the symbol rate. To this regard, this detector is comparable in terms of complexity to the ML detector used for orthogonal space-time coding [13]. However, TOD has the advantage of TOD of being full-rate. The price paid for having this low complexity is that MMSE is a sub-optimal detector for TOD, whereas it is optimal for orthogonal coding.

Let us now analyze the BER of the MMSE detector. In [34], it was proved that the linear space-time encoding strategy that maximizes the signal-to-interference-plus-noise ratio (SINR), where signal, interference and noise powers are averaged with respect to the channel, leads to trace-orthogonal encoding matrices. Based on this result, in [34] it was given a sufficient condition for minimizing the BER, for BPSK transmission. Here, we prove that TOD with unitary matrices minimizes the BER, for any channel realization.

The MMSE estimate is not immune from inter-symbol interference (ISI). However, invoking the central limit theorem, when n_s is sufficiently large one can get a fairly good approximation of the final BER by modeling ISI as additive complex Gaussian noise. Within the limit of such an approximation, the error probability for the k-th symbol can be expressed as

$$P_{e_k} = \frac{1}{2}\mathrm{erfc}\left(\sqrt{\frac{\mathrm{SINR}_k}{2}}\right), \tag{7.149}$$

where SINR_k is the SINR on the k-th symbol, defined as

$$\mathrm{SINR}_k = \frac{\sigma^2_{s_k}}{\sigma^2_{i_k+n_k}}. \tag{7.150}$$

where $\sigma^2_{s_k}$ is the variance of the useful component in $\tilde{s}_k$, whereas $\sigma^2_{i_k+n_k}$ is the variance of the ISI and noise contained in $\tilde{s}_k$. Considering now the average probability of error

$$\bar{P}_\epsilon = \frac{1}{2n_s}\sum_{k=1}^{n_s}\mathrm{erfc}\left(\sqrt{\frac{\mathrm{SINR}_k}{2}}\right), \tag{7.151}$$

it is interesting to investigate if there is any subclass of TOD decoders that minimize (7.151). With this objective in mind, let us consider the covariance matrix of the estimation errors, that is

$$\boldsymbol{K}_\epsilon = \sigma_s^2\left[\boldsymbol{I}_{n_s} - \boldsymbol{F}^H(\boldsymbol{I}_Q \otimes \boldsymbol{W}\boldsymbol{H})\boldsymbol{F}\right], \tag{7.152}$$

where $\boldsymbol{W}$ is defined in (7.143). The diagonal entries of $\boldsymbol{K}_\epsilon$ are the n_s mean square errors $\sigma^2_{\epsilon,k}$ and are given by

$$\sigma^2_{\epsilon,k} = \sigma_s^2 - \sigma_s^2\,\mathrm{tr}(\boldsymbol{A}_k^H\boldsymbol{W}\boldsymbol{H}\boldsymbol{A}_k). \tag{7.153}$$

Note that $\sigma^2_{\epsilon,k}$ depends only on the encoding matrix with the same index k, but not on the other matrices.

It is possible to prove that $\sigma^2_{\epsilon,k}$ is related to SINR_k through the following relation

$$\mathrm{SINR}_k = \frac{\sigma_s^2}{\sigma^2_{\epsilon,k}} - 1 \,. \tag{7.154}$$

Consider now that, from (7.152), the sum of all MSE's is

$$\sum_{k=1}^{n_s} \sigma^2_{\epsilon,k} = \mathrm{tr}(\boldsymbol{K}_\epsilon) = \sigma_s^2 \left[n_s - Q \cdot \mathrm{tr}(\boldsymbol{WH}) \right], \tag{7.155}$$

where we have exploited the equality $\boldsymbol{F}\boldsymbol{F}^H = \boldsymbol{I}$. The last term in (7.155) does not depend on the encoding matrices. Thus, (7.155) shows the invariance of the sum of all MSE's with respect to the choice of the matrices $\boldsymbol{A}_k$, provided that TOD is used. This result is important since allows us to characterize the minimizers for (7.151). In fact, substituting (7.154) in (7.151), we get

$$\bar{P}_\epsilon = \frac{1}{2n_s} \sum_{k=1}^{n_s} \mathrm{erfc} \left[\sqrt{\frac{1}{2} \left(\frac{\sigma_s^2}{\sigma^2_{\epsilon,k}} - 1 \right)} \right] . \tag{7.156}$$

Let us consider now the function $\mathrm{erfc}(\sqrt{\frac{\sigma_s^2}{2x} - 1/2})$ that appears in (7.156). This is a convex function of x. Hence, applying Jensen's inequality to (7.156) leads to

$$\bar{P}_\epsilon \geq \frac{1}{2} \mathrm{erfc} \left(\sqrt{\frac{\sigma_s^2}{\frac{1}{2\,n_s} \sum_{k=1}^{n_s} \sigma^2_{\epsilon,k}} - \frac{1}{2}} \right), \tag{7.157}$$

with equality if and only if all the $\sigma^2_{\epsilon,k}$ are equal. But, because of (7.155), the term in the denominator does not depend on the encoding matrices $\boldsymbol{A}_k$. This means that, for each channel realization, (7.157) is the minimum achievable average BER. The minimum of (7.157) is reached if and only if $\sigma^2_{\epsilon,1} = \cdots = \sigma^2_{\epsilon,n_s}$ that is, if and only if (see (7.153))

$$\mathrm{tr}(\boldsymbol{WHA}_1\boldsymbol{A}_1^H) = \cdots = \mathrm{tr}(\boldsymbol{WHA}_{n_s}\boldsymbol{A}_{n_s}^H) \tag{7.158}$$

for any channel realization. A sufficient condition for (7.158) to be true is

$$\boldsymbol{A}_1\boldsymbol{A}_1^H = \cdots = \boldsymbol{A}_{n_s}\boldsymbol{A}_{n_s}^H \,. \tag{7.159}$$

In particular, (7.159) is satisfied if the matrices $\boldsymbol{A}_k$ $(k = 1, \ldots, n_s)$ are unitary matrices, like, for example the shift-and-multiply basis, described before.

This concludes the proof that TOD, with the additional constraint of using uni-

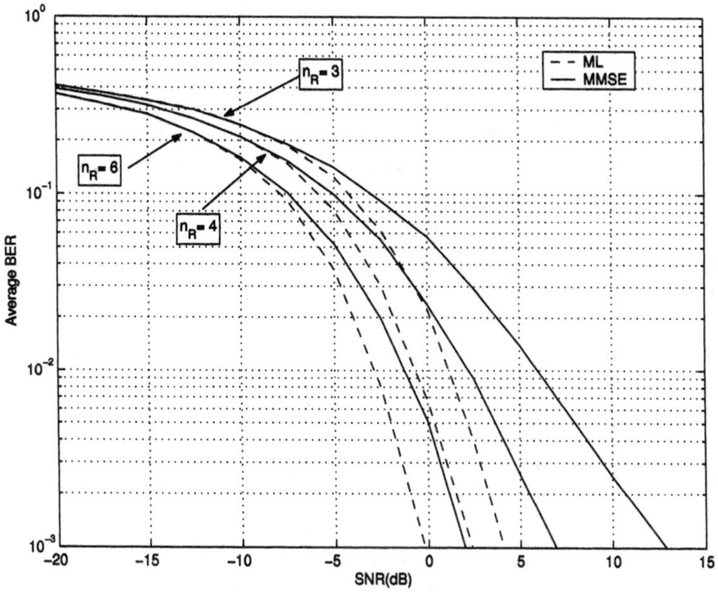

Figure 7.9 Average BER obtained with full-rate unitary TOD, using shift and multiply bases and ML (dashed line) or MMSE (solid line) decoding.

tary matrices, yields minimum error probability, provided that the receiver uses an MMSE symbol estimator.

Since MMSE is obviously sub-optimal, it is necessary to compare its performance with the optimal ML decoder. For this reason, in Fig.7.9 we compare the BER, averaged over $10,000$ independent channel realizations, obtained with full-rate unitary-TOD, using the shift-and-multiply basis (7.115). We used the following set of parameters: $n_T = 3$ and $n_R = 3, 4$ and 6. We can see that, for $n_R = n_T$, the MMSE decoder has a considerable loss with respect to ML, but, as soon as n_R increases with respect to n_T, the loss becomes smaller and smaller.

In summary, if the number of receive antennas is greater than the number of transmit antennas, TOD is an interesting choice, as it provides performance not too distant from the optimal ones, even though it is full rate and it requires only a simple scalar decoder.

7.7 SUMMARY

In summary, in this chapter we have seen how to build space-time codes that target different aspects of wireless links: diversity gain, multiplexing gain, and receiver simplicity. Viewed over an imaginary three-dimensional (3D) reference system, each space-time code design can be represented as a point, whose coordinates represent the diversity gain, the rate gain and the receiver simplicity. A pictorial view is sketched in Figure 7.10, where we compare OSTBC, FDFR and the Quasi-Orthogonal Design (QOD). In summary, OSTBC guarantees the highest diversity

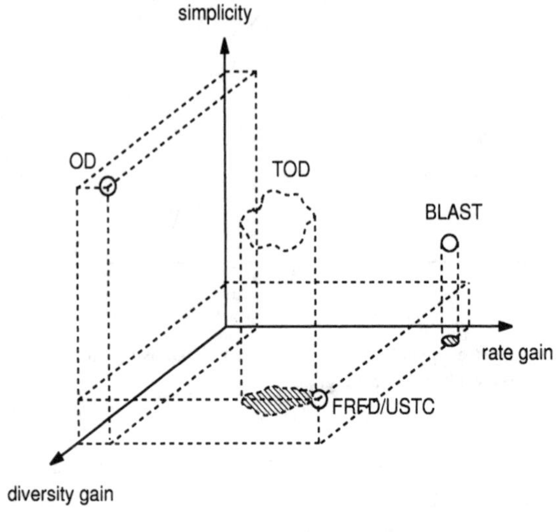

Figure 7.10 Pictorial comparison of different space-time designs in terms of rate gain, diversity gain, and simplicity.

gain and decoding simplicity, but it has a low rate; FDFR or UST have high diversity and rate gains, but the decoding is rather complex (simplicity is low). BLAST-like techniques are optimal from the point of view of capacity, they do not guarantee full diversity and have intermediate complexity. Finally, TOD offers great flexibility in designing systems capable of achieving a complexity comparable with OSTBC, but with maximum rate gain, and it is valid for any number of transmit antennas. TOD is especially valuable for systems with a high number of antennas, requiring high rate, where there are no alternatives with comparable simplicity.

The reader interested in space-time coding is invited to check books entirely devoted to the subject, like e.g. [1, 2, 3, 4]. In particular, the book by Hottinen *et al.* is especially useful for the quasi-orthogonal design. The book by Vucetic *et al.* covers nonlinear space-time coding techniques, like, e.g., space-time trellis coding, and iterative decoding in great detail. The book by Larsson and Stoica is especially devoted to linear block space-time coding techniques. Important papers that contributed to draw most of the attention on the space-time coding are [35] and [36]. An interesting recent work shedding new light in the relationship between space-time codes and error-correction codes is [37]. The two recent special issues [5, 6], are also highly recommended to get an idea about the state of the art of the research on space-time systems.

References

[1] Hottinen, A., Tirkkonen, O., Wichman, R., *Multi-antenna - Transceiver techniques for 3G and beyond*, West Sussex, UK: John Wiley & Sons Ltd., 2003.

[2] Paulraj, A., Nabar, R., Gore, D., *Introduction to space-time wireless communications*, Cambridge, UK: Cambridge University Press, 2003.

[3] Larsson, E., G., Stoica, P., *Space-time block coding for wireless communications*, Cambridge, UK: Cambridge University Press, 2003.

[4] Vucetic, B., Yuan, J., *Space-time coding*, New York, NY: Wiley, 2003.

[5] *IEEE Transaction on Information Theory*, Special issue on "Space-time methods", Oct. 2003.

[6] *IEEE Transaction on Signal Processing*, Special issue on "MIMO wireless communications", Nov. 2003.

[7] Hassibi, B., Hochwald, B.M., "High-rate codes that are linear in space and time", *IEEE Transactions on Information Theory*, Vol. 48, July 2002, pp. 1804–1824.

[8] Heath, R., W., Paulraj, A. J., "Linear dispersion codes for MIMO systems based on frame theory", *IEEE Transactions on Signal Processing*, Vol. 50, Oct. 2002, pp. 2429–2441.

[9] Alamouti, S.,M., "A simple transmitter diversity scheme for wireless communications," *IEEE Journal on Selected Areas in Communications*, Vol. 16, Oct. 1998, pp. 1451–1458.

[10] Geramita, A.,V., Geramita, J. M., "Complex orthogonal design", *Journal of Combinatorial Theory*, Series A, pp.211–225, 1978.

[11] Ganesan, G., Stoica, P., "Space-time block codes: a maximum SNR approach", *IEEE Transactions on Information Theory*, Vol. 47, May 2001, pp. 1650–1656.

[12] Geramita, A.,V., Seberry, J., *Orthogonal designs, quadratics forms and Hadamards matrices - Lecture notes in pure and applied mathemathics*, New York, NY: Marcel Dekker, 1979.

[13] Tarokh, V., Jafarkhani, H., Calderbank, A.R., "Space-time block codes from orthogonal designs", *IEEE Transactions on Information Theory*, Vol. 45, July 1999, pp. 1456–1467.

[14] Tirkkonen, O., Hottinen, A., "Square-matrix embeddable space-time block codes for complex signal constellations", *IEEE Transactions on Information Theory*, Vol. 48, Febr. 2002, pp. 384–395.

[15] Sandhu, S., Paulraj, A., "Space-time block codes: a capacity perspective", *IEEE Communications Letters*, Vol. 4, Dec. 2000, pp. 384–386.

[16] Jafarkhani, H., "A quasi-orthogonal space-time block code", *IEEE Transactions on Communications*, Vol. 49, Jan. 2001, pp. 1–4.

[17] Sharma, N., Papadias, C.B., "Improved quasi-orthogonal codes through constellation rotation," *IEEE Transactions on Communications*, Vol. 51, March 2003, pp-332–335.

[18] Foschini, G., J., "Layered space-time architecture for wireless communication in a fading environment when using mutiple antennas", *Bell Lab. Tech. J.*, Vol. 1, 1996, pp. 41–59.

[19] Golden, G., D., Foschini, G., J., Valenzuela, R. A., Wolniansky, P.W., "Detection algorithm and initial laboratory results using V-BLAST space-time communication architecture", *Electronics Letters*, Vol. 35, Jan. 1999, pp. 14–16.

[20] El Gamal, H., Hammons, A.R.,"A new approach to layered space-time coding and signal processing", *IEEE Transactions on Information Theory*, Vol. 47, Sept. 2001, pp. 2321–2334.

[21] Winters, J. H., Salz, J., Gitlin, R.D., "The impact of antenna diversity on the capacity of of wireless communication systems", *IEEE TRansactions on Communications*, Vol. 42, Apr. 1994, pp. 1740–1751.

[22] Varanasi, M., Guess, T., "Optimum decision feedback multiuser equalization with successive decoding achieves the total capacity of the Gaussian multiple access channel," *Proc. of Asilomar Conf. on Signals, Systems and Computers*, Pacific Grove, CA, November 1997, pp. 1405–1409.

[23] El Gamal, H., Damen, M., O., "Universal space-time coding", *IEEE Transactions on Information Theory*, Vol. 49, May 2003, pp. 1097–1119.

[24] Ma, X., Giannakis, G.B., "Full-diversity full-rate complex-field space-time coding", *IEEE Transactions on Signal Processing*, Vol. 49, Nov. 2003, pp. 2917–2930.

[25] Zheng, L., Tse, D.N.C., "Diversity and multiplexing: A fundamental tradeoff in multiple antenna channels,", *IEEE Trans. on Information Theory*, Vol. 49, May 2003, pp. 1073–1096.

[26] Giraud, X., Belfiore, J.C., "Constellations matched to the Rayleigh fading channel", *IEEE Transactions on Information Theory*, Vol. 42, Jan. 1996, pp. 106–115.

[27] Boutros, J., Viterbo, E., Rastello, C., Belfiore, J.C., "Good lattice constellations for both Rayleigh and Gaussian channels", *IEEE Transactions on Information Theory*, Vol. 42, Mar. 1996, pp. 502–518.

[28] Giraud, X., Boutillon, E., Belfiore, J.C., "Algebraic tools to build modulation schemes for fading channels", *IEEE Transactions on Information Theory*, Vol. 43, May 1997, pp. 938–952.

[29] Damen, M. O., Tewfik, A., Belfiore, J.-C., "A construction of a space-time code based on number theory", *IEEE Transactions on Information Theory*, Vol. 48, Mar. 2002, pp. 753–760.

[30] Barbarossa, S., "Trace-orthogonal design of MIMO systems with simple scalar detectors, full diversity and (almost) full rate", *Proc. of the V IEEE Signal Proc. Workshop on Signal Proc. Advances in Wireless Commun., SPAWC '2004*, Lisbon, Portugal, July 11–14, 2004.

[31] Barbarossa, S., Fasano, A., "Trace-orthogonal space-time coding, " submitted to *ICC '05*.

[32] Werner, R., "All teleportation and dense coding schemes", *J. Phys. A*, Vol. 34, 1994, pp. 7081-7094.

[33] Knill, E., "Group representations, error bases and quantum codes", *Los Alamos National Lab. Rreport* LAUR-96-2807, 1996.

[34] Liu, J., Zhang, J. K. , Wong, K. M., "Design of Optimal Orthogonal Linear Codes in MIMO Systems for MMSE Receiver", *ICASPP 2004*, 2004.

[35] Naguib, A.F., Tarokh, V., Seshadri, N., Calderbank, A.R., "A space-time coding modem for high-data-rate wireless communications," *IEEE Journal on Selected Areas in Communications*, Vol. 16, Oct. 1998, pp. 1459–1478.

[36] Tarokh, V., Jafarkhani, H., Calderbank, A.R., "Space-time block coding for wireless communications: performance results", *IEEE Journal on Selected Areas in Communications*, Vol. 17, March 1999, pp. 451–460.

[37] Biglieri, E., Taricco, G., Tulino, A., "Performance of space-time codes for a large number of antennas", *IEEE Transactions on Information Theory*, Vol. 48, July 2002, pp. 1794–1803.

Chapter 8

Space-Time Coding for Frequency-Selective Channels

8.1 INTRODUCTION

Given the attractive capabilities of space-time coding to increase the information rate, for a given bandwidth, or reduce the transmission power to guarantee a required BER, it is not surprising that STC is an appealing candidate for applications in third generation cellular systems as well as for broadband wireless LANs. In both cases, the incorporation of space-time coding techniques has to face compatibility with currently standardized systems or systems under standardization. Clearly, the simpler is the compatibility with current systems, the most likely is the possibility of including space-time coding techniques. From this perspective, the first basic step to making space-time coding an appealing candidate is its extension to frequency-selective channels and multiuser systems. As mentioned in Chapter 3, the current standardization for broadband wireless LAN, see IEEE 802.11a for example, tends to use OFDM as a powerful tool to counteract channel frequency selectivity and still allow for simple receiving structures. For these reasons, this chapter is devoted to the generalization of space-time coding to frequency-selective systems. We will pay special attention to the application of space-time coding to block transmission systems using cyclic prefixes, as a way to have the benefits of space-time coding, with affordable complexity at the receiver side. The multiuser case will be then specifically analyzed in Chapter 9.

As already observed in Chapter 7, the choice of the space-time encoder depends on the performance parameter that is considered to be the most important to optimize, for the application at hand (i.e., diversity gain, multiplexing gain, and/or decoder simplicity). Each choice achieves a certain trade-off among the previous three fundamental parameters. The same happens to the generalizations of space-time coding to frequency-selective channels. In this chapter we will combine the space-time coding techniques of Chapter 7 with the block transmission schemes of Chapter 3. We will start with orthogonal-STBC combined with CP-block transmission schemes and then we will consider full-rate transmission systems.

8.2 ORTHOGONAL STBC FOR CP-BLOCK TRANSMISSIONS

In Chapter 3 we showed that block transmissions over stationary FIR channels of order L lead to very simple decoding schemes if a cyclic prefix (CP) of length not smaller than L is appended at the beginning of each block. We will denote this kind of transmission as CP-block transmission. We also observed, in Section 7, that orthogonal STC yields maximum spatial diversity $G_d = n_T n_R$ for transmissions over flat-fading channels, using a very simple scalar decoder. In this section, we merge the basic ideas of CP-block transmissions with orthogonal STC. We will show how, thanks to appropriate coding, frequency selectivity can be turned into a source of multipath diversity.

Among the first works that considered the combination of STBC with OFDM or CDMA, we can mention [2, 3, 4, 6, 13]. In this section, we will follow the formulation of [15].

8.2.1 Block Alamouti

To simplify the explanation of orthogonal STBC combined with time encoding, we start with the simple case of two transmit antennas. In such a case, orthogonal STBC reduces to the *block* Alamouti's scheme. In its most general formulation, block Alamouti proceeds as follows. The information bits are mapped into pairs of consecutive symbol vectors $\boldsymbol{s}(n)$ and $\boldsymbol{s}(n+1)$, of size N. The two vectors are then premultiplied by the precoding $N \times N$ matrices $\boldsymbol{F}$ (or $\boldsymbol{G}$). A CP of length L is then appended at the beginning of each block, with L greater than the channel order.

Denoting with $\boldsymbol{F}_{CP}$ (or $\boldsymbol{G}_{CP}$) the $(N+L) \times N$ matrix built adding on top of $\boldsymbol{F}$ (alternatively $\boldsymbol{G}$) its bottom L rows, block Alamouti proceeds as follows. In the nth time slot, the block $\boldsymbol{x}_1(n) = \boldsymbol{F}_{CP}\boldsymbol{s}(n)$ is transmitted, serially, from the first antenna, whereas $\boldsymbol{x}_2(n) = \boldsymbol{F}_{CP}\boldsymbol{s}(n+1)$ is transmitted from the second antenna. In the successive $(n+1)$-th time slot, the block $\boldsymbol{x}_1(n+1) = \boldsymbol{G}_{CP}\boldsymbol{s}^*(n+1)$ is transmitted from the first antenna whereas $\boldsymbol{x}_2(n+1) = -\boldsymbol{G}_{CP}\boldsymbol{s}^*(n)$ is transmitted from the second antenna. For the moment, we leave the choice of the two matrices $\boldsymbol{F}$ and $\boldsymbol{G}$ completely open, except for the property that they must be full column rank. We will show later on which relationship between $\boldsymbol{F}$ and $\boldsymbol{G}$ must hold true to insure maximum spatial diversity gain.

This formulation is quite general and encompasses a series of cases of practical interest. The matrices $\boldsymbol{F}$ and $\boldsymbol{G}$ could be in fact composed of the user-codes of a multiuser system. In such a case, the entries of the vectors $\boldsymbol{s}(n)$ and $\boldsymbol{s}(n+1)$ would be the symbols of different users. Alternatively, the matrices $\boldsymbol{F}$ and $\boldsymbol{G}$ could be the FFT matrices of an OFDM system. In any case, using block-Alamouti, the transmission of two blocks of N symbols each requires $2(N+L)T_s$ seconds, where T_s is the symbol duration. Hence the symbol rate is $R_s = N/(N+L)$, which is exactly as with any single antenna CP-block transmission scheme.

The vectors received in any pair of two consecutive time slots, from a single receive antenna system, after discarding the guard intervals of length L between consecutive blocks, are

$$\begin{array}{lclcl} \boldsymbol{y}(n) & = & \boldsymbol{H}_1\boldsymbol{F}\boldsymbol{s}(n) & + & \boldsymbol{H}_2\boldsymbol{F}\boldsymbol{s}(n+1) + \boldsymbol{v}(n) \\ \boldsymbol{y}(n+1) & = & \boldsymbol{H}_1\boldsymbol{G}\boldsymbol{s}^*(n+1) & - & \boldsymbol{H}_2\boldsymbol{G}\boldsymbol{s}^*(n) + \boldsymbol{v}(n+1), \end{array} \tag{8.1}$$

where $\boldsymbol{H}_1$ and $\boldsymbol{H}_2$ are circulant, Toeplitz matrices and $\boldsymbol{v}(n), \boldsymbol{v}(n+1)$ denote the noise vectors, supposed to be AWGN. We denote with $\boldsymbol{W}$ the unitary $N \times N$ DFT matrix, whose entries are $\{\boldsymbol{W}\}_{kl} := \exp(j2\pi kl/N)/\sqrt{N}$, with $k, l = 0, \ldots, N-1$. The structure of the channel matrices $\boldsymbol{H}_1$ and $\boldsymbol{H}_2$ insures that they are diagonalized as $\boldsymbol{H}_i = \boldsymbol{W}\boldsymbol{\Lambda}_i\boldsymbol{W}^H$, where $\boldsymbol{\Lambda}_i$ is the diagonal matrix whose diagonal entries are the samples of the channel transfer function corresponding to the ith channel [i.e., $\{\boldsymbol{\Lambda}_i\}_{kk} = H_i(k) = \sum_{l=0}^{L} h_i(l)\exp(-j2\pi kl/N)$]. Using this diagonalization, multiplying in (8.1) $\boldsymbol{y}(n)$ by $\boldsymbol{W}^H$ and $\boldsymbol{y}^*(n+1)$ by $\boldsymbol{W}^T$, both from their left side, we get

$$\begin{array}{lclcl} \boldsymbol{W}^H\boldsymbol{y}(n) & = & \boldsymbol{\Lambda}_1\boldsymbol{W}^H\boldsymbol{F}\boldsymbol{s}(n) & + & \boldsymbol{\Lambda}_2\boldsymbol{W}^H\boldsymbol{F}\boldsymbol{s}(n+1) \\ \boldsymbol{W}^T\boldsymbol{y}^*(n+1) & = & \boldsymbol{\Lambda}_1^*\boldsymbol{W}^T\boldsymbol{G}^*\boldsymbol{s}(n+1) & - & \boldsymbol{\Lambda}_2^*\boldsymbol{W}^T\boldsymbol{G}^*\boldsymbol{s}(n). \end{array} \tag{8.2}$$

To obtain the spatial diversity gain, it is sufficient to choose $\boldsymbol{G}$ so that $\boldsymbol{W}^H\boldsymbol{F} = \boldsymbol{W}^T\boldsymbol{G}^*$. This amounts to take $\boldsymbol{G}^* = \boldsymbol{J}\boldsymbol{F}$, where $\boldsymbol{J} := \boldsymbol{W}^*\boldsymbol{W}^H$ is a permutation matrix whose structure depends on N. For N even, we have

$$\boldsymbol{J} = \begin{pmatrix} 1 & 0 & \cdots & \cdots & 0 \\ 0 & \iddots & \iddots & \iddots & 1 \\ \vdots & \iddots & \iddots & \iddots & 0 \\ \vdots & \iddots & \iddots & \iddots & \vdots \\ 0 & 1 & 0 & \cdots & 0 \end{pmatrix}, \tag{8.3}$$

otherwise, for N odd, we have

$$\boldsymbol{J} = \begin{pmatrix} 0 & 0 & \cdots & \cdots & 1 \\ 0 & \iddots & \iddots & \iddots & 0 \\ \vdots & \iddots & \iddots & \iddots & 0 \\ \vdots & \iddots & \iddots & \iddots & \vdots \\ 1 & 0 & 0 & \cdots & 0 \end{pmatrix}. \tag{8.4}$$

This implies that the columns of $\boldsymbol{G}$ are obtained by simply reverting the order of the corresponding columns of $\boldsymbol{F}$ and cyclically shifting them by one sample[1]. Because of the relationship between $\boldsymbol{F}$ and $\boldsymbol{G}$, we have that $\boldsymbol{G}\boldsymbol{s}^*(n) = \boldsymbol{J}[\boldsymbol{F}\boldsymbol{s}(n)]^*$. As a

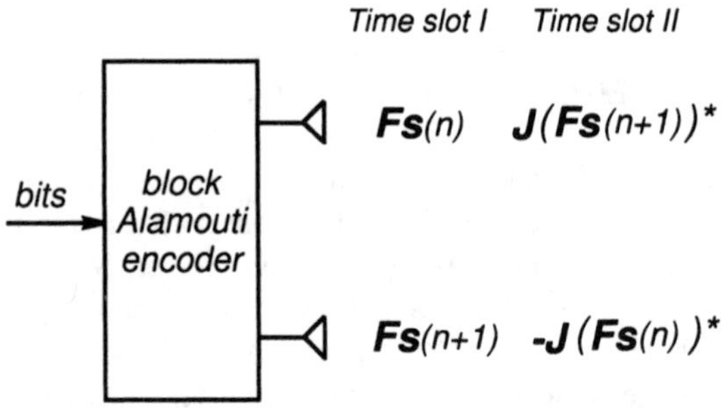

Figure 8.1 Block Alamouti transmission scheme.

consequence, the structure of the block Alamouti encoder is as reported in Figure 8.1, valid for a two-transmit antennas and any number of receive antennas.

Exploiting the equality $\boldsymbol{W}^H\boldsymbol{F} = \boldsymbol{W}^T\boldsymbol{G}^*$ and introducing the matrix $\tilde{\boldsymbol{F}} := \boldsymbol{W}^H\boldsymbol{F}$, (8.2) can be rewritten in matrix form as

$$\begin{bmatrix} \tilde{\boldsymbol{y}}(n) \\ \tilde{\boldsymbol{y}}(n+1) \end{bmatrix} := \begin{bmatrix} \boldsymbol{W}^H\boldsymbol{y}(n) \\ \boldsymbol{W}^T\boldsymbol{y}^*(n+1) \end{bmatrix} = \begin{pmatrix} \boldsymbol{\Lambda}_1 & \boldsymbol{\Lambda}_2 \\ -\boldsymbol{\Lambda}_2^* & \boldsymbol{\Lambda}_1^* \end{pmatrix} \cdot \begin{bmatrix} \tilde{\boldsymbol{F}}\boldsymbol{s}(n) \\ \tilde{\boldsymbol{F}}\boldsymbol{s}(n+1) \end{bmatrix}. \tag{8.5}$$

It is straightforward to check that the matrix

$$\boldsymbol{\Lambda} := \begin{pmatrix} \boldsymbol{\Lambda}_1 & \boldsymbol{\Lambda}_2 \\ -\boldsymbol{\Lambda}_2^* & \boldsymbol{\Lambda}_1^* \end{pmatrix} \tag{8.6}$$

1 In case of N even, there is a cyclic shift by one sample, instead of pure time reversal.

is an orthogonal matrix. In fact

$$\mathbf{\Lambda}^H \mathbf{\Lambda} = \begin{pmatrix} |\mathbf{\Lambda}_1|^2 + |\mathbf{\Lambda}_2|^2 & \mathbf{0} \\ \mathbf{0} & |\mathbf{\Lambda}_1|^2 + |\mathbf{\Lambda}_2|^2 \end{pmatrix}. \tag{8.7}$$

Given $\mathbf{\Lambda}$, we can always build the unitary matrix $\boldsymbol{U}_\Lambda$ as

$$\boldsymbol{U}_\Lambda := \begin{pmatrix} \mathbf{\Lambda}_1^* & -\mathbf{\Lambda}_2 \\ \mathbf{\Lambda}_2^* & \mathbf{\Lambda}_1 \end{pmatrix} \begin{pmatrix} (|\mathbf{\Lambda}_1|^2 + |\mathbf{\Lambda}_2|^2)^{-1/2} & \mathbf{0} \\ \mathbf{0} & (|\mathbf{\Lambda}_1|^2 + |\mathbf{\Lambda}_2|^2)^{-1/2} \end{pmatrix}. \tag{8.8}$$

In the presence of additive Gaussian noise, we can multiply the vector $[\tilde{\boldsymbol{y}}^T(n), \tilde{\boldsymbol{y}}^T(n+1)]^T$ in (8.5) by $\boldsymbol{U}_\Lambda$ without any loss of optimality and without altering the noise properties. In particular, introducing the matrix

$$\mathbf{\Lambda}_{12} := (|\mathbf{\Lambda}_1|^2 + |\mathbf{\Lambda}_2|^2)^{1/2}, \tag{8.9}$$

we obtain

$$\begin{aligned} \begin{bmatrix} \boldsymbol{z}(n) \\ \boldsymbol{z}(n+1) \end{bmatrix} &:= \boldsymbol{U}_\Lambda \begin{bmatrix} \tilde{\boldsymbol{y}}(n) \\ \tilde{\boldsymbol{y}}(n+1) \end{bmatrix} \\ &= \begin{pmatrix} \mathbf{\Lambda}_{12} & \mathbf{0} \\ \mathbf{0} & \mathbf{\Lambda}_{12} \end{pmatrix} \begin{bmatrix} \tilde{\boldsymbol{F}}\boldsymbol{s}(n) \\ \tilde{\boldsymbol{F}}\boldsymbol{s}(n+1) \end{bmatrix} + \begin{bmatrix} \boldsymbol{w}(n) \\ \boldsymbol{w}(n+1) \end{bmatrix} \end{aligned} \tag{8.10}$$

where the noise vectors $\boldsymbol{w}(n)$ and $\boldsymbol{w}(n+1)$ are still AWGN vectors with zero mean and variance σ_n^2. We see from (8.10) that the $2N$ equations system in (8.10) can be decoupled, without any loss of optimality, into two systems of N equations each. We can then concentrate on the solution of the two systems of N equations

$$\boldsymbol{z}(l) = \mathbf{\Lambda}_{12}\tilde{\boldsymbol{F}}\boldsymbol{s}(l) + \boldsymbol{w}(l), \; l = n, n+1, \tag{8.11}$$

separately. If the two blocks $\boldsymbol{s}(n)$ and $\boldsymbol{s}(n+1)$ have been coded independently of each other, from (8.11) there is no loss in decoding each block separately. Otherwise, it is necessary to look at the systems in (8.11), jointly.

8.2.2 Orthogonal STC/OFDM

If we have an option in the selection of the precoding matrix $\tilde{\boldsymbol{F}}$, the choice that leads to the simplest solution of system (8.11) is uncoded OFDM (i.e., $\tilde{\boldsymbol{F}} = \boldsymbol{I}$ or, equivalently, $\boldsymbol{F} = \boldsymbol{W}$). The combination Alamouti/OFDM leads to the simplest decoding structure, as it decouples the $2N$-equation system (8.11) into $2N$ *independent scalar* equations.

Indicating with $s_k(n)$ (or $z_k(n)$) the kth entry of the vector $\boldsymbol{s}(n)$ ($\boldsymbol{z}(n)$), (8.11) becomes

$$z_k(n) = [|H_1(k)|^2 + |H_2(k)|^2]^{1/2} s_k(n) + w_k(n), \quad k = 0, \ldots, N-1. \quad (8.12)$$

In such a case, there is no loss of optimality in applying a scalar detector. In particular, we may use the ZF detector

$$\hat{s}_k(n) = [|H_1(k)|^2 + |H_2(k)|^2]^{-1/2} z_k(n), \quad k = 0, \ldots, N-1, \quad (8.13)$$

or the MMSE detector

$$\hat{s}_k(n) = \frac{[|H_1(k)|^2 + |H_2(k)|^2]^{1/2}}{|H_1(k)|^2 + |H_2(k)|^2 + \sigma_n^2/\sigma_s^2} z_k(n), \quad k = 0, \ldots, N-1. \quad (8.14)$$

Let us consider now the most general orthogonal STBC scheme, valid for a number of transmit antennas greater than two, in conjunction with OFDM. Applying the same steps shown above to the more general situation of n_T transmit and n_R receive antennas[2], it is straightforward to show that we obtain a received vector of the form

$$z_k(n) = \left(\sum_{l=1}^{n_T} \sum_{m=1}^{n_R} |H_{ml}(k)|^2 \right)^{1/2} s_k(n) + w_k(n), \quad k = 0, \ldots, N-1, \quad (8.15)$$

where $H_{ml}(k)$ is the transfer function of the channel between the lth transmit and the mth receive antennas, evaluated at the kth sub-carrier.

Let us compute now the performance of OSTBC/OFDM for Rayleigh fading, frequency-selective, channels. In such a case, the coefficients of the channel impulse responses $h_{ml}(k)$ are i.i.d. complex Gaussian random variables, with zero mean and variance σ_h^2. Let us consider, for simplicity the ZF decoder. Since ZF is not affected by ISI, the bit error rate is rather easy to compute. In particular, using QAM constellations, the bit error rate on the kth symbol is [23]

$$P_e(k) = c\, Q\left(\sqrt{g \frac{\mathcal{E}_b}{\sigma_n^2} \alpha(k)} \right) \leq c\, e^{-g\mathcal{E}_b \alpha^2(k)/\sigma_n^2}. \quad (8.16)$$

where

$$\alpha(k) = \sum_{l=1}^{n_T} \sum_{m=1}^{n_R} |H_{ml}(k)|^2, \quad (8.17)$$

2 In the orthogonal case, the value of n_T is such that the corresponding orthogonal encoder exists.

$\mathcal{E}_b$ is the energy per bit, σ_n^2 is the noise variance, and c and g are two coefficients that depend on the order M of the QAM constellation [23]:

$$\begin{aligned} c &= 4\frac{\sqrt{M}-1}{\sqrt{M}\cdot\log_2 M}, \\ g &= \frac{3}{M-1}\log_2 M. \end{aligned} \tag{8.18}$$

As a consequence of the Rayleigh fading model, the variables $H_{ml}(k)$ are themselves complex Gaussian random variables, with zero mean and variance $(L+1)\sigma_h^2$. Hence, $\alpha(k)$ has the same kind of pdf as α in (5.82), except that n_R in (5.82) is substituted by $n_T n_R$. Hence, the average BER has the same form as (5.83), substituting n_T with $n_T n_R$. This shows that uncoded OSTBC/OFDM has diversity gain

$$G_d = n_T n_R. \tag{8.19}$$

In summary, *uncoded OSTBC/OFDM has full spatial diversity, but it does not have any multipath diversity.* OSTBC/OFDM is certainly amenable from the complexity point of view, as it leads to the simplest detection scheme, but it is not capable of gathering all potential diversity.

Before concluding this section, it is useful to remark that the BER averaged over all the subchannels can be lower-bounded, at high SNR, as follows[3]

$$\frac{1}{N}\sum_{k=0}^{N-1} c\,Q\left(\sqrt{g\frac{\mathcal{E}_b}{\sigma_n^2}\alpha(k)}\right) \geq c\,Q\left(\sqrt{g\frac{\mathcal{E}_b}{\sigma_n^2}\frac{1}{N}\sum_{k=0}^{N-1}\alpha(k)}\right). \tag{8.20}$$

The lower bound in this inequality is reached if and only if all channels give rise to the same value of $\alpha(k)$. With SISO systems, there is typically a great variability of $H(k)$, and then $\alpha(k)$, as a function of k. As a consequence, the average BER may be well above the minimum bound in (8.20). However, with multiantenna systems, as the product $n_R n_T$ increases, the coefficients $\alpha(k)$ tend to assume a more stable value[4]. As a result, the loss in average BER with respect to the minimum bound is lower than in the SISO case. Intuitively speaking, multiantenna systems give rise to an equivalent channel that is more stable (or less fluctuating) than SISO channels, and this improves the system reliability.

3 This bound is a consequence of the convexity of the function $Q(\sqrt{x})$, as already pointed out in Chapter 6.

4 Asymptotically speaking, as the number of transmit/receive antennas tend to infinity, $\alpha(k)$ tends to the expected value of $|H_{ml}(k)|^2$

8.2.3 Diversity Gain

In this section, we study the diversity achievable with different coding schemes. We recall the analysis of [16], starting again, for simplicity, from the case with two transmit and one receive antennas. We assume that the two symbol blocks $\boldsymbol{s}(n)$ and $\boldsymbol{s}(n+1)$ are independent. As shown in Chapter 4, for a given channel realization, the pairwise error probability (PEP) $\Pr\{\boldsymbol{s} \rightarrow \boldsymbol{s}'/\boldsymbol{h}_1, \boldsymbol{h}_2\}$ of transmitting the block $\boldsymbol{s}$ and erroneously deciding for the vector $\boldsymbol{s}'$, for a given realization of the two channels $\boldsymbol{h}_1$ and $\boldsymbol{h}_2$, is upper-bounded by

$$\Pr\{\boldsymbol{s} \rightarrow \boldsymbol{s}'/\boldsymbol{h}_1, \boldsymbol{h}_2\} \leq e^{-d^2(\boldsymbol{z}, \boldsymbol{z}'/\boldsymbol{h}_1, \boldsymbol{h}_2)E_s/4N_0}, \tag{8.21}$$

where $d^2(\boldsymbol{z}, \boldsymbol{z}'/\boldsymbol{h}_1, \boldsymbol{h}_2)$ is the Euclidean distance between $\boldsymbol{z}$ and $\boldsymbol{z}'$, conditioned to the channels $\boldsymbol{h}_1$ and $\boldsymbol{h}_2$. The PEP averaged with respect to the channel realizations is then upper-bounded as

$$\Pr\{\boldsymbol{s} \rightarrow \boldsymbol{s}'\} \leq E_{\boldsymbol{h}_1, \boldsymbol{h}_2}\left\{e^{-d^2(\boldsymbol{z}, \boldsymbol{z}'/\boldsymbol{h}_1, \boldsymbol{h}_2)E_s/4N_0}\right\}, \tag{8.22}$$

where $E_{\boldsymbol{h}_1, \boldsymbol{h}_2}\{f(\boldsymbol{h}_1, \boldsymbol{h}_2)\}$ denotes expected value with respect to the vector random variables $\boldsymbol{h}_1$ and $\boldsymbol{h}_2$. From (8.11), introducing the symbol error vector $\boldsymbol{e} := \boldsymbol{s} - \boldsymbol{s}'$ [5], we can write the distance $d^2(\boldsymbol{z}, \boldsymbol{z}'/\boldsymbol{h}_1, \boldsymbol{h}_2)$ as

$$d^2(\boldsymbol{z}, \boldsymbol{z}'/\boldsymbol{h}_1, \boldsymbol{h}_2) = \|\boldsymbol{\Lambda}_{12}\tilde{\boldsymbol{F}}\boldsymbol{e}\|^2 = \boldsymbol{e}^H\tilde{\boldsymbol{F}}^H\boldsymbol{\Lambda}_{12}^2\tilde{\boldsymbol{F}}\boldsymbol{e}. \tag{8.23}$$

Using (8.9), we can rewrite (8.23) as

$$d^2(\boldsymbol{z}, \boldsymbol{z}'/\boldsymbol{h}_1, \boldsymbol{h}_2) = \|\boldsymbol{\Lambda}_1\tilde{\boldsymbol{F}}\boldsymbol{e}\|^2 + \|\boldsymbol{\Lambda}_2\tilde{\boldsymbol{F}}\boldsymbol{e}\|^2. \tag{8.24}$$

It is useful to rewrite the diagonal matrices $\boldsymbol{\Lambda}_m$, $m = 1, 2$, as $\boldsymbol{\Lambda}_m = \mathrm{diag}(\boldsymbol{V}\boldsymbol{h}_m)$, where $\boldsymbol{V}$ is the $N \times (L+1)$ Vandermonde matrix

$$\boldsymbol{V} := \begin{pmatrix} 1 & \cdots & \cdots & 1 \\ 1 & \rho & \cdots & \rho^L \\ \vdots & \vdots & \vdots & \vdots \\ 1 & \rho^{N-1} & \cdots & \rho^{(N-1)L} \end{pmatrix}, \tag{8.25}$$

where $\rho := e^{-j2\pi/N}$. This is nothing more than writing the transfer function of the mth channel in matrix form, as the premultiplication of $\boldsymbol{h}_m$ by $\boldsymbol{V}$ is the N-point

5 We drop the symbol index, for simplicity of notation, whenever the index does not affects the derivations.

DFT of $\boldsymbol{h}_m$. Hence, introducing the diagonal matrix $\boldsymbol{D}_e := \text{diag}(\tilde{\boldsymbol{F}}\boldsymbol{e})$, we can rewrite (8.24) as[6]

$$d^2(\boldsymbol{z}, \boldsymbol{z}') = \|\boldsymbol{D}_e \boldsymbol{V} \boldsymbol{h}_1\|^2 + \|\boldsymbol{D}_e \boldsymbol{V} \boldsymbol{h}_2\|^2 := d_1^2(\boldsymbol{z}, \boldsymbol{z}'/\boldsymbol{h}_1) + d_2^2(\boldsymbol{z}, \boldsymbol{z}'/\boldsymbol{h}_2). \quad (8.26)$$

From (8.22), if the two channels $\boldsymbol{h}_1$ and $\boldsymbol{h}_2$ are statistically independent, the bound on the average PEP can be expressed as the product

$$\begin{aligned} \Pr\{\boldsymbol{s} \to \boldsymbol{s}'\} &\leq E_{\boldsymbol{h}_1,\boldsymbol{h}_2}\left\{e^{-d^2(\boldsymbol{z},\boldsymbol{z}'/\boldsymbol{h}_1,\boldsymbol{h}_2)E_s/4N_0}\right\} \\ &= E_{\boldsymbol{h}_1}\left\{e^{-d_1^2(\boldsymbol{z},\boldsymbol{z}'/\boldsymbol{h}_1)E_s/4N_0}\right\} E_{\boldsymbol{h}_2}\left\{e^{-d_2^2(\boldsymbol{z},\boldsymbol{z}'/\boldsymbol{h}_2)E_s/4N_0}\right\}. \end{aligned} \quad (8.27)$$

Hence, in case of independent channels, we can concentrate on the two factors of (8.27), separately.

Let us denote with $\boldsymbol{C}_m := E\{\boldsymbol{h}_m \boldsymbol{h}_m^H\}$ the covariance matrix of the channel vector $\boldsymbol{h}_m$. We say that the "effective" channel length $L_m + 1$ is equal to the rank of $\boldsymbol{C}_m$, or that the effective channel order is $L_m = \text{rank}(\boldsymbol{C}_m) - 1$. Since $\boldsymbol{C}_m$ is square Hermitian, it always admits the eigen-decomposition

$$\boldsymbol{C}_m = \boldsymbol{U}_m \boldsymbol{\Gamma}_m \boldsymbol{U}_m^H, \quad (8.28)$$

where $\boldsymbol{\Gamma}_m$ is an $(L+1) \times (L+1)$ diagonal matrix, whose diagonal entries are the eigenvalues of $\boldsymbol{C}_m$, and $\boldsymbol{U}_m$ is a unitary matrix whose columns are the eigenvectors of $\boldsymbol{C}_m$. We can then introduce the equivalent channel vector

$$\bar{\boldsymbol{h}}_m := \boldsymbol{\Gamma}^{-1/2} \boldsymbol{U}_m^H \boldsymbol{h}_m \quad (8.29)$$

that has, by construction, an identity covariance matrix. Expressing the distances in (8.26) as a function of $\bar{\boldsymbol{h}}_m$, we can write

$$d_m^2(\boldsymbol{z}, \boldsymbol{z}'/\boldsymbol{h}_m) = \|\boldsymbol{D}_e \boldsymbol{V} \boldsymbol{h}_m\|^2 = \bar{\boldsymbol{h}}_m^H \boldsymbol{A}_m \bar{\boldsymbol{h}}_m, \quad m = 1, 2, \quad (8.30)$$

where

$$\boldsymbol{A}_m := \boldsymbol{\Gamma}_m^{H/2} \boldsymbol{U}_m^H \boldsymbol{V}^H |\boldsymbol{D}_e|^2 \boldsymbol{V} \boldsymbol{U}_m \boldsymbol{\Gamma}_m^{1/2} \quad (8.31)$$

and $\bar{\boldsymbol{h}}_m$ is a vector composed of i.i.d. complex Gaussian random variables with zero mean and unit variance. Hence, we can use (5.59) to compute the bound on the average PEP

$$E\left\{e^{-\bar{\boldsymbol{h}}_m^H \boldsymbol{A}_m \bar{\boldsymbol{h}}_m}\right\} = \frac{1}{|\boldsymbol{I} + \boldsymbol{A}_m E_s/(4N_0)|}. \quad (8.32)$$

6 In (8.26) we exploit the obvious property that, given two equal size vectors $\boldsymbol{a}$ and $\boldsymbol{b}$, then $\text{diag}(\boldsymbol{a})\boldsymbol{b} = \text{diag}(\boldsymbol{b})\boldsymbol{a}$.

Denoting with $\lambda_{m,i}$ the eigenvalues of $\boldsymbol{A}_m$, we can write (8.32) as

$$E\left\{e^{-\bar{\boldsymbol{h}}_m^H \boldsymbol{A}_m \bar{\boldsymbol{h}}_m}\right\} = \prod_{i=1}^{L+1} \frac{1}{1+\lambda_{m,i} E_s/(4N_0)}. \tag{8.33}$$

At high SNR (i.e., when $\lambda_{m,i} E_s \gg 4N_0$, for all i), if g_{dm} is the rank of $\boldsymbol{A}_m$, we have the following asymptotic bevavior

$$E\left\{e^{-\bar{\boldsymbol{h}}_m^H \boldsymbol{A}_m \bar{\boldsymbol{h}}_m}\right\} \approx \frac{1}{(g_{cm}\mathrm{SNR})^{g_{dm}}}, \tag{8.34}$$

where $g_{cm} := \left(\prod_{i=1}^{L+1} \lambda_{m,i}\right)^{1/g_{dm}}$ and $\mathrm{SNR} = E_s/4N_0$. Inserting this expression in (8.27), we get the final bound

$$\begin{aligned} \Pr\{\boldsymbol{s} \to \boldsymbol{s}'\} &\le \prod_{i=1}^{g_{d1}} \frac{1}{1+\lambda_{1,i} E_s/(4N_0)} \prod_{i=1}^{g_{d2}} \frac{1}{1+\lambda_{2,i} E_s/(4N_0)} \\ &\approx \frac{1}{(g_{c1} g_{c2} \mathrm{SNR})^{g_{d1}+g_{d2}}}, \end{aligned} \tag{8.35}$$

where the last approximation holds true asymptotically, at high SNR.

The previous derivations can be easily extended to all other orthogonal space-time coding schemes. The result is that the asymptotic behavior of the error bound is

$$\Pr\{\boldsymbol{s} \to \boldsymbol{s}'\} \approx \frac{1}{\left(\prod_{i=1}^{n_T} \prod_{j=1}^{n_R} g_{c,ij} \mathrm{SNR}\right)^{\sum_{i=1}^{n_T} \sum_{i=1}^{n_R} g_{d,ij}}}, \tag{8.36}$$

where $g_{c,ij}$ and $g_{d,ij}$ are the coding and diversity gains of the channel between the ith transmit and the jth receive antenna, respectively.

If, for all pairs of symbol vectors, the matrix $\boldsymbol{A}_{ij}$ relative to channel $\boldsymbol{h}_{ij}$ is full-rank, that is $\mathrm{rank}(\boldsymbol{A}_{ij}) = L+1$, for all i and j, the error bound decreases, asymptotically, as

$$\Pr\{\boldsymbol{s} \to \boldsymbol{s}'\} \approx \frac{1}{(G_c SNR)^{n_T n_R (L+1)}}. \tag{8.37}$$

This shows that the system is capable of achieving full diversity

$$G_d = n_T n_R (L+1),$$

incorporating both space and multipath diversity.

8.2.4 Full-Diversity Schemes

The theoretical analysis shows that full diversity can be achieved if the encoder and the channels are such that the matrices $\boldsymbol{A}_{ij}$ are all full-rank and the decoder is the maximum likelihood decoder. In this section, we show how to design the precoders in order to achieve full diversity. From the definition of $\boldsymbol{A}_{ij}$ in (8.31), a major role in the determination of the diversity gain is played by the rank of $\boldsymbol{D}_e$ (i.e., by the number of nonnull entries of the vector $\tilde{\boldsymbol{F}}\boldsymbol{e}$). Generalizing the concept of Hamming distance commonly used in coding theory[7], we call such a number the symbol-wise Hamming distance between the complex vectors $\tilde{\boldsymbol{F}}\boldsymbol{s}$ and $\tilde{\boldsymbol{F}}\boldsymbol{s}'$. Considering all possible pairs of vectors $\boldsymbol{s}$ and $\boldsymbol{s}'$, the performance is ultimately determined by the minimum symbol-wise Hamming distance of the set of codewords $\{\boldsymbol{u} : \boldsymbol{u} = \tilde{\boldsymbol{F}}\boldsymbol{s}, \forall \boldsymbol{s}\}$.

As an important example, if we use OFDM without precoding (i.e., $\tilde{\boldsymbol{F}} = \boldsymbol{I}$) we need to check the minimum complex Hamming distance of the set $\{\boldsymbol{u} = \boldsymbol{s}\}$. Clearly, among all the error vectors $\boldsymbol{e} = \boldsymbol{s} - \boldsymbol{s}'$, there are vectors with all entries equal to zero except one (occurring when the symbol vectors $\boldsymbol{s}$ and $\boldsymbol{s}'$ differ only by one element). In such a case, the rank of $\boldsymbol{D}_e$ is one and then the rank of $\boldsymbol{A}_m$ is also one. Consequently, the maximum diversity for OFDM is $n_T n_R$. We find again, even though following different argumentations, what we had already stated in Section 8.2: Uncoded OFDM is uncapable of achieving multipath diversity.

Intuitively speaking, OFDM converts the MIMO frequency-selective channel into N parallel MIMO flat-fading channels. If a symbol is transmitted over one of these MIMO flat-fading systems, the receiver can achieve, on that symbol, a diversity order of $n_T n_R$, but no more than that. This intuitive argument suggests that, if we want to increase the diversity, we need to encode the information bits so that each bit is transmitted over more than one flat-fading MIMO channel. To avoid a rate loss, we should then transmit, on each MIMO subchannel, a combination of the information bits. More specifically, according to the above theoretical analysis, to increase diversity, it is necessary to encode the symbols in order to generate codewords such that the corresponding error vector $\boldsymbol{e}$ had at least $L + 1$ nonzero entries.

The previous analysis was carried out assuming linear precoding. However, the same results apply equally well if the coded vector set $\{\boldsymbol{u}\}$ is generated by any mapping from the finite set $\{\boldsymbol{s}\}$ to the set $\{\boldsymbol{u}\}$. Within the nonlinear class, nonredundant

7 We recall that, given two binary codewords $\boldsymbol{a} = [a_1, \ldots, a_N]$ and $\boldsymbol{b} = [b_1, \ldots, b_N]$, their Hamming distance $d_H(\boldsymbol{a}, \boldsymbol{b})$ is the number of distinct bits between $\boldsymbol{a}$ and $\boldsymbol{b}$, i.e. $d_H(\boldsymbol{a}, \boldsymbol{b}) = |\{i : a(i) \neq b(i)\}|$.

coding is still a rather unexplored field, because of its difficulties. Conversely, many algorithms are available to generate codewords that differ by a fixed minimum Hamming distance, if the encoder is redundant (i.e., the number n of coded bits is greater than the number k of information bits [24]). For example, according to the previous analysis, a *necessary condition for achieving maximum diversity is to use a coding strategy that provides a minimum Hamming distance of, at least,* $L+1$ *bits*. This condition is only necessary. Once a code has been chosen, one should check that, for each pair of different codes, the matrices $\boldsymbol{A}_{ij}$ are all full-rank. The disadvantage of this approach is the redundancy and thus the consequent rate loss that increases as L increases. The advantage is the great availability of coding strategies that provide the desired Hamming distance.

Linear coding can also be used to enforce the desired minimum complex Hamming distance, and thus the maximum diversity, as suggested for example in [17]. In particular, linear encoding admits nonredundant coding. In such a case, it is in fact possible to show that, for *any* finite signal constellation, there exists a rotation matrix $\boldsymbol{\Theta}$ ensuring that *each* entry of the vector $\boldsymbol{\Theta}(\boldsymbol{s}-\boldsymbol{s}')$ is different from zero. This approach guarantees the maximum diversity gain [21, 22]. It is important to remark that, in case of nonredundant encoding, the rotation matrix $\boldsymbol{\Theta}$ depends on the symbol constellation. Conversely, using redundant linear encoding, the encoder can be made independent of the symbol constellation. An excellent unified approach for building rotation matrices that yield maximum diversity is [17]. Here, we recall an example of nonredundant rotation precoding that provides maximum diversity.

A possible structure of $\boldsymbol{\Theta}$ is the $N \times N$ Vandermonde matrix

$$\boldsymbol{\Theta} = \frac{1}{K}\begin{pmatrix} 1 & \rho_1 & \cdots & \rho_1^{N-1} \\ 1 & \rho_2 & \cdots & \rho_2^{N-1} \\ \vdots & \vdots & \vdots & \vdots \\ 1 & \rho_N & \cdots & \rho_N^{N-1} \end{pmatrix}, \tag{8.38}$$

where K is a normalization coefficient used to enforce the property $\mathrm{tr}(\boldsymbol{\Theta}^H\boldsymbol{\Theta}) = N$, whereas the complex numbers $\rho_1, \ldots, \rho_N$ have to be chosen properly, as detailed next.

Before showing how to select the numbers $\rho_1, \ldots, \rho_N$, it is necessary to recall a few definitions (see also [17, 18, 21]).

Euler numbers: Given an integer number P, its corresponding Euler number, denoted as $\phi(P)$, is the cardinality of the set $\{p : \gcd(p, P) = 1, p \in [1, P]\}$, where $\gcd(p, P)$ denotes the greatest common divisor between p and P. In words, the

Euler number of P is the number of positive integers less than P that are prime to P.

For a given block size N, the coefficients $\rho_1, \ldots, \rho_N$ in (8.38) can be found as follows. First of all, one has to select an integer multiple P of N, such that $\phi(P) = 2mN$, with m positive integer. Then the coefficients ρ_n are chosen as

$$\rho_n = e^{j2\pi(P(n-1)/N+p)/P}, \tag{8.39}$$

where p is selected from $[1, P/N)$, such that $\gcd[P(n-1)/N + p, P] = 1, \forall n$.

In such a case, the normalization factor K in (8.38) is equal to N. This choice of the rotation matrix $\boldsymbol{\Theta}$ meets a series of important properties [18]: 1) $\boldsymbol{\Theta}$ exists for any size N; and 2) $\boldsymbol{\Theta}$ enables full diversity over QAM and PAM constellations.

To maximize the coding gain, P should be selected in order to have the minimum value of m.

As a final remark for this section, it is important to point out that, even choosing the right precoder, capable of, potentially, full diversity, the whole diversity gain is achieved if an ML decoder is employed. In applications where the receiver complexity cannot be increased excessively, the use of the ML decoder can be prohibitive. This is especially true when the block length is high, because ML decoding has a complexity that increases exponentially with the block length. It is then useful to search for alternative suboptimal decoding schemes.

8.2.5 Suboptimum MMSE Estimation

One of the possible alternatives to ML is the MMSE solution. Given (8.11), the soft MMSE estimate of the symbol vector is

$$\hat{\boldsymbol{s}}_{\text{mmse}}(n) = \left(\tilde{\boldsymbol{F}}^H \boldsymbol{\Lambda}_{12}^2 \tilde{\boldsymbol{F}} + \frac{\sigma_n^2}{\sigma_s^2}\boldsymbol{I}\right)^{-1} \tilde{\boldsymbol{F}}^H \boldsymbol{\Lambda}_{12}\boldsymbol{z}(n). \tag{8.40}$$

The decoding is then performed by taking a hard decision on each entry of $\hat{\boldsymbol{s}}_{\text{mmse}}(n)$. This approach is clearly suboptimum, but it is much simpler to implement than ML. The difficulty resides primarily in the matrix inversion in (8.40). However, this operation has to be repeated with a period that is on the order of magnitude of the channel coherence time.

Following the same arguments as in Section 3.3.3, applied to system (8.11) instead of (3.21) [8], we can state that the MMSE estimator is also minimum BER if we use a precoding matrix such that the entries of $\tilde{\boldsymbol{F}}$ have constant modulus. For example, we can take $\boldsymbol{F}$, and thus $\tilde{\boldsymbol{F}}$ as in (3.42) or (3.43).

Of course, the minimality of the BER holds true only within the class of MMSE decoders. ML decoding would provide better performance, but clearly ML could be too complex to implement, especially in broadband real-time applications.

We show now a few examples of performance.

Example 1: Effect of channel order

In Figure 8.2, we report the average BER as a function of the SNR for different channel orders ($L = 1$, 3 or 15) of a block-Alamouti scheme, with linear precoding. Dotted lines refer to a precoder that is pure OFDM ($\boldsymbol{F} = \boldsymbol{W}$), whereas solid lines

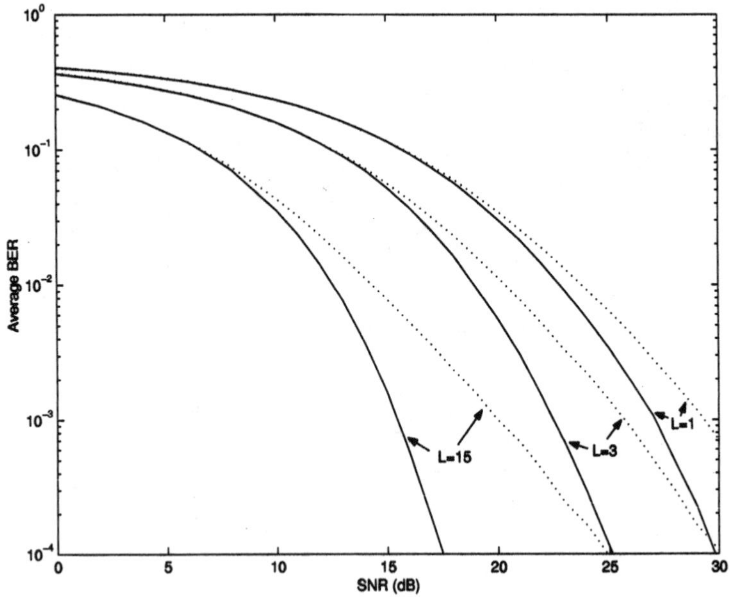

Figure 8.2 Average BER versus SNR for a block CP Alamouti scheme with MMSE decoding: OFDM precoder (dotted line) and identity precoder (solid line).

refer to the identity precoder ($\boldsymbol{F} = \boldsymbol{I}$) that simply adds the CP to the block to

8 This requires only the substitution of $\boldsymbol{\Lambda}$ in (8.11) with $\boldsymbol{\Lambda}_{12}$ in (3.21).

be transmitted. In both cases the decoder is the MMSE decoder. The channels are FIR of various order, with i.i.d. complex Gaussian coefficients, having zero mean and unit variance. We can see from Figure 8.2 that the OFDM/Alamouti scheme provides the spatial diversity gain $G_d = 2$, but there is no multipath gain. In fact, increasing the channel order, we observe a shift the curves due to a higher coding gain, but there is no change of slope. Conversely, the identity precoder is capable of providing additional multipath diversity gain, even though the MMSE decoder is suboptimal.

Example 2: Effect of number of transmit antennas

In Figure 8.3 we report the average BER as a function of the SNR, for different numbers of transmit antennas. The channel order is fixed and equal to $L = 3$. As in Figure 8.1, in Figure 8.3 we report two cases of precoders: OFDM and identity precoder. The decoder is MMSE, as in Figure 8.1. We can see that, as the number

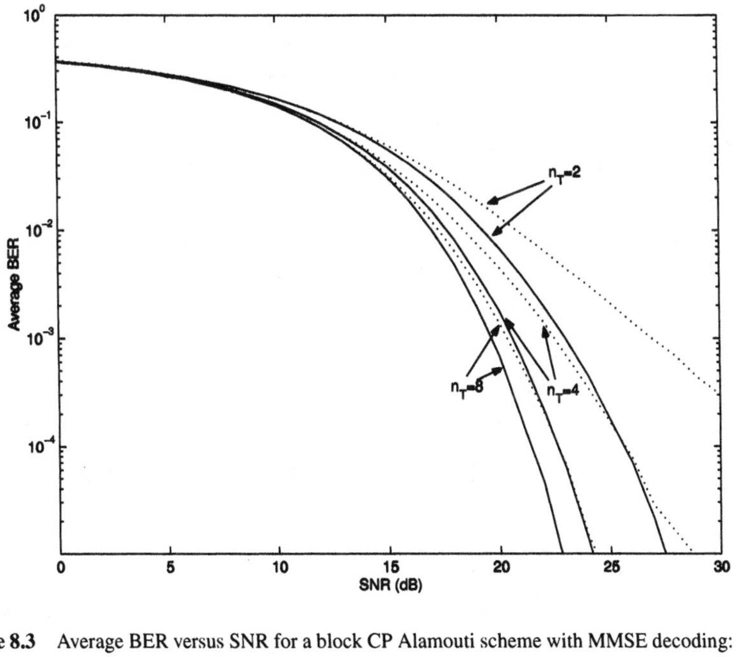

Figure 8.3 Average BER versus SNR for a block CP Alamouti scheme with MMSE decoding: OFDM precoder (dotted line) and identity precoder (solid line).

of transmit antennas increases, both schemes increase their diversity gain and the relative advantage of the identity precoder with respect to the OFDM precoder reduces.

8.3 FULL-RATE SYSTEMS

The orthogonal space-time coding techniques analyzed in the previous sections are primarily intended for systems capable of maximum diversity and variable complexity. However, orthogonal design is known for not being optimal from the rate point of view, especially when the number of receive antennas is high. On the other hand, orthogonal design is important in many applications as it does not impose any constraint on the number of receive antennas.

In this section, we consider schemes capable of, potentially, maximum rate gain. To check this possibility, we assume that the number of receive antennas is at least equal to the number of transmit antennas. Within this context, in Chapter 7 we had shown that trace-orthogonal design (TOD) provides great flexibility in designing space-time encoders capable of reaching a good trade-off between rate, diversity and complexity. In this section, we extend the TOD strategy to the frequency-selective channel case.

Following the same approach used in the previous section, we consider a block transmission, with cyclic prefixes appended to each block, exactly as in the block orthogonal case.

8.3.1 TOD for Frequency-Selective Channels

Let us consider a system that maps the information bits into $N \cdot n_s$ symbols $s_n(k)$, $n = 1, \ldots, n_s$, $k = 0, \ldots, N-1$. We group the symbols into N blocks of n_s symbols $\boldsymbol{s}(k) = [s_1(k)\ s_2(k)\ \cdots\ s_{n_s}(k)]^T$ $(k = 0, \ldots, N-1)$. The encoder proceeds through the following steps:

1. Form the N matrices $\boldsymbol{X}(k)$ of size $n_T \times n_T$, as

$$\boldsymbol{X}(k) = \sum_{n=1}^{n_s} s_n(k)\boldsymbol{A}_n(k), \quad k = 0, \ldots, N-1, \tag{8.41}$$

   where the $n_T \times n_T$ matrices $\boldsymbol{A}_n(k)$ are built as in Section 7.6[9];

2. Form the n_T^2 column vectors $\boldsymbol{x}_{i,j}$, for $i, j = 1, \ldots, n_T$, of size N, composed of the (i, j) entry of the matrices $\boldsymbol{X}(k)$, for $k = 0, \ldots, N-1$;

3. Apply an N-size IFFT to each vector $\boldsymbol{x}_{i,j}$, thus obtaining the vectors $\boldsymbol{u}_{ij} = \boldsymbol{W}\boldsymbol{x}_{i,j}$;

9 We use the symbol $\boldsymbol{A}_n(k)$ here to distinguish between the subcarrier index k and the symbol index n, for each subcarrier.

4. Transmit the blocks $\boldsymbol{u}_{ij}$, serially, from the ith transmit antenna, during the jth time slot (each time slot has a duration of N samples).

As a consequence of this transmission strategy, the receiver collects $n_R \cdot n_T$ vectors $\boldsymbol{y}_{i,j}$, of size N, from the n_R receive antennas. We can group together these vectors to form the block matrix (for simplicity of notation, the receiver noise will be added later on)

$$\begin{pmatrix} \boldsymbol{y}_{1,1} & \cdot\cdot & \boldsymbol{y}_{1,n_T} \\ \cdots & \cdot\cdot & \cdots \\ \boldsymbol{y}_{n_R,1} & \cdot\cdot & \boldsymbol{y}_{n_R,n_T} \end{pmatrix} = \begin{pmatrix} \boldsymbol{H}_{1,1} & \cdot\cdot & \boldsymbol{H}_{1,n_T} \\ \cdots & \cdot\cdot & \cdots \\ \boldsymbol{H}_{n_R,1} & \cdot\cdot & \boldsymbol{H}_{n_R,n_T} \end{pmatrix} \begin{pmatrix} \boldsymbol{u}_{1,1} & \cdot\cdot & \boldsymbol{u}_{1,n_T} \\ \cdots & \cdot\cdot & \cdots \\ \boldsymbol{u}_{n_R,1} & \cdot\cdot & \boldsymbol{u}_{n_R,n_T} \end{pmatrix} \tag{8.42}$$

Thanks to the insertion of the CP, each matrix $\boldsymbol{H}_{i,j}$ is Toeplitz and circulant and then it can be factorized as $\boldsymbol{H}_{i,j} = \boldsymbol{W}\boldsymbol{\Lambda}_{i,j}\boldsymbol{W}^H$. Applying an FFT to each vector $\boldsymbol{y}_{i,j}$, we obtain the matrix

$$\begin{pmatrix} \boldsymbol{z}_{1,1} & \cdot\cdot & \boldsymbol{z}_{1,n_T} \\ \cdots & \cdot\cdot & \cdots \\ \boldsymbol{z}_{n_R,1} & \cdot\cdot & \boldsymbol{z}_{n_R,n_T} \end{pmatrix} = \begin{pmatrix} \boldsymbol{\Lambda}_{1,1} & \cdot\cdot & \boldsymbol{\Lambda}_{1,n_T} \\ \cdots & \cdot\cdot & \cdots \\ \boldsymbol{\Lambda}_{n_R,1} & \cdot\cdot & \boldsymbol{\Lambda}_{n_R,n_T} \end{pmatrix} \begin{pmatrix} \boldsymbol{x}_{1,1} & \cdot\cdot & \boldsymbol{x}_{1,n_T} \\ \cdots & \cdot\cdot & \cdots \\ \boldsymbol{x}_{n_R,1} & \cdot\cdot & \boldsymbol{x}_{n_R,n_T} \end{pmatrix} \tag{8.43}$$

where $\boldsymbol{z}_{mn} := \boldsymbol{W}^H \boldsymbol{y}_{mn}$. The channel matrix in (8.43) is composed of diagonal blocks. Each block $\boldsymbol{\Lambda}_{mn}$ has, on its diagonal, the values assumed by the transfer function of the channel between the nth transmit and the mth receive antenna, on N frequency bins. Taking the kth component of each vector $\boldsymbol{z}_{i,j}$, the above system is equivalent to the following set of N systems

$$\boldsymbol{Z}(k) = \boldsymbol{H}(k)\boldsymbol{X}(k) + \boldsymbol{V}(k), \quad k = 0, \ldots, N-1 \tag{8.44}$$

where $\boldsymbol{Z}(k)$ is $n_R \times n_T$, $\boldsymbol{H}(k)$ is $n_R \times n_T$ and $\boldsymbol{X}(k)$ is $n_T \times n_T$. The matrix $\boldsymbol{V}(k)$ is $n_R \times n_T$ and it is the receiver additive noise.

Comparing system (8.44) with the analogous relation in Section 7.6, we see that block coding with CP transforms the overall system into N parallel TOD systems transmitting over flat fading channels. If we set $n_s = n_T^2$ and we use all matrices $\boldsymbol{A}_n(k)$, $n = 1, \ldots, n_s$, over each subcarrier, the maximum symbol rate is

$$R_s = \frac{n_T N}{N + L}. \tag{8.45}$$

Hence, the rate exhibits a loss with respect to the flat-fading case, equal to $\eta = N/(N+L)$, which is the common loss factor present in all block systems employing a CP.

The decoding process depends, of course, on the relationship between symbols transmitted over different subcarriers. If the symbols over different subcarriers are encoded independently of each other, the overall decoder splits into N parallel independent decoders over each subcarrier, where each decoder has the same form as TOD decoders for flat fading channels. In particular, assuming $n_s = n_T^2$ (full-rate systems), we can use for each sub-carrier, the simple MMSE decoder

$$\hat{s}_n(k) = \mathrm{tr}\left[\boldsymbol{A}_n^H(k)\boldsymbol{W}^H(k)\boldsymbol{Z}(k)\right], \tag{8.46}$$

where $\hat{s}_n(k)$, with $n = 1, \ldots, n_s$, and $k = 0, \ldots, N-1$, denotes the nth estimate in the kth sub-carrier, and

$$\boldsymbol{W}(k) = \left[\boldsymbol{H}^H(k)\boldsymbol{H}(k) + \frac{1}{\gamma}\boldsymbol{I}_{n_T}\right]^{-1}\boldsymbol{H}^H(k), \tag{8.47}$$

with $\gamma = \sigma_s^2/\sigma_v^2$, where σ_s^2 and σ_n^2 are the variances of the transmitted symbols and the noise, respectively.

8.3.2 Full-Diversity Schemes

The properties of the decoder seen in the previous section are exactly the same as the TOD decoder for flat fading channels examined in the previous chapter, as the N sub-channels are totally decoupled. Hence, that scheme is capable of spatial diversity[10], but it does not yield any multipath diversity. To obtain multipath diversity, it is necessary to combine the symbols transmitted through different subcarriers. Proceeding as in Section 8.2.4, to get multipath diversity, we must apply a rotation to the symbols before space-time encoding. More specifically, let us organize the symbols $s_n(k)$ into n_s vectors $\boldsymbol{s}_n$ of size N, where $\boldsymbol{s}_n = [s_n(0), \cdots, s_n(N-1)]^T$.

The encoding proceeds through the following steps:

1. Each vector $\boldsymbol{s}_n$ is first rotated, through the unitary matrix $\boldsymbol{\Theta}$, thus obtaining the vectors
$$\boldsymbol{r}_n = \boldsymbol{\Theta}\boldsymbol{s}_n, \tag{8.48}$$
for $n = 1, \ldots, n_s$;

2. The N matrices $\boldsymbol{X}(k)$ of size $n_T \times n_T$ are built as
$$\boldsymbol{X}(k) = \sum_{n=1}^{n_s} r_n(k)\boldsymbol{A}_n(k), \quad k = 0, \ldots, N-1, \tag{8.49}$$

10 Full diversity if ML decoding is used.

where the matrices $\boldsymbol{A}_n(k)$ are the same as used in Section (7.6), and $r_n(k)$ is the kth component of $\boldsymbol{r}_n$;

3. The n_T^2 column vectors $\boldsymbol{x}_{i,j}$, for $i, j = 1, \ldots, n_T$, of size N, are formed by taking the (i, j) entry of all the matrices $\boldsymbol{X}(k)$, for $k = 0, \ldots, N-1$;

4. An N-size IFFT is applied to each vector $\boldsymbol{x}_{i,j}$, thus obtaining the vectors $\boldsymbol{u}_{ij} = \boldsymbol{W}\boldsymbol{x}_{i,j}$;

5. The blocks $\boldsymbol{u}_{ij}$ are transmitted, serially, from the ith transmit antenna, during the jth time slot (each time slot has a duration of N samples).

In this way, each symbol is transmitted, equivalently, through *all* sub-channels. As a consequence, the MMSE decoder for $s_n(k)$ cannot be decoupled, in general, into N decoders. However, since the information symbols $s_n(k)$ are related to the rotated ones $r_n(k)$ through a unitary transformation, the MMSE estimates $\hat{s}_n(k)$ can be obtained following a two-step procedure: 1) we compute first the MMSE estimates $\hat{r}_n(k)$ of the rotated symbols; then 2) we recover the MMSE estimates $\hat{s}_n(k)$ by applying the inverse rotation $\boldsymbol{\Theta}^H$ to the estimates $\hat{r}_n(k)$. To prove the validity of this procedure, let us introduce the following vectors

$$\boldsymbol{s}(k) = [s_1(k)\; s_2(k) \;\cdots\; s_{n_s}(k)]^T, \tag{8.50}$$

and

$$\boldsymbol{r}(k) = [r_1(k)\; r_2(k) \;\cdots\; r_{n_s}(k)]^T, \tag{8.51}$$

for $k = 0, \ldots, N-1$, where $r_j(k)$ is the kth component of $\boldsymbol{r}_j$. Relation (8.48), for $n = 1, \ldots, n_s$, can be expressed in compact form as

$$[\boldsymbol{r}_1\; \boldsymbol{r}_2 \;\cdots\; \boldsymbol{r}_{n_s}] = \boldsymbol{\Theta}[\boldsymbol{s}_1\; \boldsymbol{s}_2 \;\cdots\; \boldsymbol{s}_{n_s}]. \tag{8.52}$$

Introducing the $n_s \times N$ matrices $\boldsymbol{\mathcal{R}} = [\boldsymbol{r}_1\; \boldsymbol{r}_2 \;\cdots\; \boldsymbol{r}_{n_s}]^T$, and $\boldsymbol{\mathcal{S}} = [\boldsymbol{s}_1\; \boldsymbol{s}_2 \;\cdots\; \boldsymbol{s}_{n_s}]^T$, it is easy to recognize that the vectors (8.50) and (8.51) are the columns of $\boldsymbol{\mathcal{R}}$ and $\boldsymbol{\mathcal{S}}$, respectively. Hence, (8.52) can be recast, equivalently, as

$$[\boldsymbol{r}(0)\; \boldsymbol{r}(1) \;\cdots\; \boldsymbol{r}(N-1)] = [\boldsymbol{s}(0)\; \boldsymbol{s}(1) \;\cdots\; \boldsymbol{s}(N-1)]\boldsymbol{\Theta}^T, \tag{8.53}$$

or

$$\boldsymbol{\mathcal{R}} = \boldsymbol{\mathcal{S}}\boldsymbol{\Theta}^T. \tag{8.54}$$

Now, applying the vec$(\cdot)$ operator[11] to (8.53), and defining $\boldsymbol{r} = \text{vec}(\boldsymbol{\mathcal{R}})$ and $\boldsymbol{s} = \text{vec}(\boldsymbol{\mathcal{S}})$, we obtain, eventually

$$\boldsymbol{r} = (\boldsymbol{\Theta} \otimes \boldsymbol{I}_{n_s})\boldsymbol{s}, \tag{8.55}$$

11 We use the property $\text{vec}(\boldsymbol{A}\boldsymbol{X}\boldsymbol{C}^T) = (\boldsymbol{C} \otimes \boldsymbol{A})\text{vec}(\boldsymbol{X})$.

which relates all the information symbols to the rotated ones. Since $\boldsymbol{\Theta}$ is unitary, this relationship is invertible as

$$\boldsymbol{s} = (\boldsymbol{\Theta}^H \otimes \boldsymbol{I}_{n_s})\,\boldsymbol{r}. \tag{8.56}$$

The relation between transmitted (rotated) symbols and received samples, over each subcarrier, is given by (8.44). Applying the vec$(\cdot)$ operator to (8.44), we get

$$\boldsymbol{z}(k) = [\boldsymbol{I}_{n_T} \otimes \boldsymbol{H}(k)]\boldsymbol{F}(k)\boldsymbol{r}(k) + \boldsymbol{v}(k)\,, \tag{8.57}$$

where $\boldsymbol{z}(k) = \text{vec}[\boldsymbol{Z}(k)]$, $\boldsymbol{v}(k) = \text{vec}[\boldsymbol{V}(k)]$, and $\boldsymbol{F}(k)$ is the $n_T^2 \times n_s$ matrix whose columns are vec$[\boldsymbol{A}_j(k)]$, for $j = 1, \ldots, n_s$. Stacking $\boldsymbol{z}(k)$ in (8.57), for $k = 0, \ldots, N-1$, we obtain

$$\boldsymbol{z} = \begin{bmatrix} [\boldsymbol{I}_{n_T} \otimes \boldsymbol{H}(0)]\boldsymbol{F}(0) & \cdots & \boldsymbol{0} \\ \vdots & \ddots & \vdots \\ \boldsymbol{0} & \cdots & [\boldsymbol{I}_{n_T} \otimes \boldsymbol{H}(N-1)]\boldsymbol{F}(N-1) \end{bmatrix} \boldsymbol{r} + \boldsymbol{v}, \tag{8.58}$$

where $\boldsymbol{z} = [\boldsymbol{z}(0)^T \, \cdots \, \boldsymbol{z}(N-1)^T]^T$, $\boldsymbol{v} = [\boldsymbol{v}(0)^T \, \cdots \, \boldsymbol{v}(N-1)^T]^T$. The vector $\boldsymbol{r}$ depends on the information symbols through (8.55).

To derive the (sub-optimal) linear MMSE estimator for $\boldsymbol{s}$, we assume that the symbols $s_n(k)$ are independent with zero mean and variance σ_s^2, and that $\boldsymbol{v} \sim \mathcal{CN}(\boldsymbol{0}, \sigma_v^2 \boldsymbol{I})$ and it is independent of the symbols. Denoting by $\hat{\boldsymbol{s}}$ and $\hat{\boldsymbol{r}}$ the linear MMSE estimate of $\boldsymbol{s}$ and $\boldsymbol{r}$, respectively, using (8.56), the following chain of equalities holds true

$$\begin{aligned} \hat{\boldsymbol{s}} &= \mathbb{E}\left\{\boldsymbol{s}\,\boldsymbol{z}^H\right\}\left\{\mathbb{E}\left\{\boldsymbol{z}\boldsymbol{z}^H\right\}\right\}^{-1}\boldsymbol{z} \\ &= \mathbb{E}\left\{(\boldsymbol{\Theta}^H \otimes \boldsymbol{I}_{n_s})\boldsymbol{r}\,\boldsymbol{z}^H\right\}\left\{\mathbb{E}\left\{\boldsymbol{z}\boldsymbol{z}^H\right\}\right\}^{-1}\boldsymbol{z} \\ &= (\boldsymbol{\Theta}^H \otimes \boldsymbol{I}_{n_s})\mathbb{E}\left\{\boldsymbol{r}\,\boldsymbol{z}^H\right\}\left\{\mathbb{E}\left\{\boldsymbol{z}\boldsymbol{z}^H\right\}\right\}^{-1}\boldsymbol{z} \\ &= (\boldsymbol{\Theta}^H \otimes \boldsymbol{I}_{n_s})\hat{\boldsymbol{r}}\,. \end{aligned} \tag{8.59}$$

The last equality in (8.59) states that relation (8.55) holds true also between the vectors $\hat{\boldsymbol{s}}$ and $\hat{\boldsymbol{r}}$ obtained through MMSE estimation. As a consequence, $\hat{\boldsymbol{s}}$ can be simply obtained by applying an inverse rotation to the MMSE vector estimate $\hat{\boldsymbol{r}}$. To derive $\hat{\boldsymbol{r}}$, it is useful to observe that, because of the independence of the information symbols and of the unitarity of relation (8.55), the components of $\boldsymbol{r}$ are also uncorrelated. This implies, together with the fact that the noise vector $\boldsymbol{v}$ has independent entries, that the linear MMSE estimate $\hat{\boldsymbol{r}}$ decouples into the linear

MMSE estimate of N vectors, namely $\boldsymbol{r}(k)$, with $k = 0, \ldots, N-1$, exactly as pointed out in the previous section. This means that the same formulas can be used. In particular when $n_s = n_T^2$ (full-rate systems) the MMSE estimate of the nth component of $\boldsymbol{r}(k)$, that is the estimate of the nth (rotated) symbol transmitted trough the kth sub-carrier, is given by

$$\hat{r}_n(k) = \text{tr}\left[\boldsymbol{A}_n^H(k)\boldsymbol{W}^H(k)\boldsymbol{Z}(k)\right], \tag{8.60}$$

where

$$\boldsymbol{W}(k) = \left[\boldsymbol{H}^H(k)\boldsymbol{H}(k) + \frac{1}{\gamma}\boldsymbol{I}_{n_T}\right]^{-1}\boldsymbol{H}^H(k), \tag{8.61}$$

with $\gamma = \sigma_s^2/\sigma_v^2$. Subsequently, for any $n = 1, \ldots, n_s$, we collect the N estimates $\hat{r}_n(0), \cdots, \hat{r}_n(N-1)$ in the vector $\hat{\boldsymbol{r}}_n = [\hat{r}_n(0) \;\; \hat{r}_n(1) \;\; \cdots \;\; \hat{r}_n(N-1)]^T$. Finally, the MMSE estimate $\hat{s}_n(k)$ is obtained as the kth component of vector $\hat{\boldsymbol{s}}_n$ given by

$$\hat{\boldsymbol{s}}_n = \boldsymbol{\Theta}^H\hat{\boldsymbol{r}}_n. \tag{8.62}$$

To achieve full diversity and full rate, it is necessary to choose the rotation parameters properly. Since TOD subsumes, as a particular case, the full diversity/full rate scheme of [19], the parameters yielding both full diversity and full rate, over a frequency-selective channel, are as suggested in [19]. However, it is important to remark that, even choosing the optimal parameters, full diversity is guaranteed only if the receiver adopts the optimal ML decoder.

We show next a numerical example, useful to grasp the main properties of the (sub-optimal) MMSE decoder described before. In Fig. 8.4 we report the average BER obtained with a 4×4 MIMO system, composed of independent FIR channels of order $L = 3$ or $L = 5$. The channel coefficients are generated as i.i.d. circularly symmetric Complex Gaussian random variables, with zero mean and unit variance. The block-length is $N = 6$. The constellation is BPSK. The system is full-rate, which means that it transmits at a rate of 4 symbols per channel use. We can see, from Figure 8.4, that the rotation introduces a clear multipath diversity gain. This is evidenced by the greater slope of the average BER of the rotated TOD with respect to the non-rotated TOD. Increasing the channel order L, we notice also a coding gain, testified by the shift of the average BER curves.

8.4 SUMMARY

In this chapter we have shown how to extend the space-time coding techniques to transmissions over wideband multipath channels. We have seen how to convert the

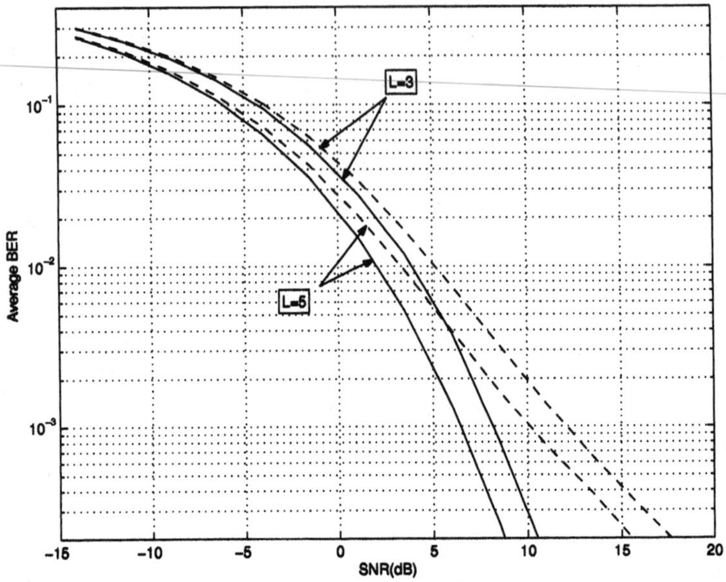

Figure 8.4 Average BER versus SNR (dB) for TOD over frequency selective channels using an MMSE decoder with symbol rotation (solid line) or without symbol rotation (dashed line).

channel frequency-selectivity into a useful source of diversity, by applying proper coding to the information symbols. More specifically, we have seen how a block transmission scheme with cyclic prefix can easily incorporate space-time coding. We have shown that the direct combination of uncoded OFDM with space-time coding yields full spatial diversity and minimum receiver complexity. To reach full diversity, including multipath diversity, it is necessary to encode the data before applying OFDM. The precoding can be either linear or nonlinear, operating over a finite field. In the first case, there are rotation matrices that, for a given QAM constellation, guarantee full diversity, even without requiring any additional redundancy. In the second case, one has to resort to conventional error correction coding, with minimum Hamming distance related to the channel order. The higher the channel order (i.e., delay dispersion), the greater the required redundancy.

We have considered separately the case of orthogonal design, valid for any number of receive antenna, and leading to the simplest decoding scheme, and the case of full-rate design. In the second case, we have studied in detail only the trace-orthogonal design as a rather general framework.

The reader interested in a more in-depth analysis of space-time coding for frequency selective channels may read, for example, [17, 18, 19], where the conditions for insuring the maximum diversity gains are treated in great detail. For a more in-depth analysis MIMO OFDM principles applied to broadband wireless systems, recent surveys are given in [26, 27].

References

[1] Diggavi, S., N., et al., "Great expectations: The value of spatial diversity in wireless networks," *Proceedings of the IEEE*, Feb. 2004, pp. 219–270.

[2] Agrawal, D. , et al., "Space-time coded OFDM for high data-rate wireless communication over wideband channels," *VTC 98*, vol. 3, 1998, Ottawa, ON, Canada, 1998, pp. 2232–2236.

[3] Kim, J., Cimini, L., Chuang, J., "Coding strategies for OFDM with antenna diversity high bit-rate mobile data applications," *Proc. of VTC '98*, Ottawa, Canada, May 1998, pp. 763–767.

[4] Lang, L., Cimini, L., Chuang, J., "Turbo codes for OFDM with antenna diversity," *Proc. of VTC '99*, Houston, TX, May 1999, pp. 1664–1668.

[5] Papadias, C., "On the spectral efficiency of space-time spreading schemes for multiple antenna CDMA systems," *Proc. of Asilomar Conf. on Signals, Systems and Computers*, Pacific Grove, CA, October 1999, pp. 639–643.

[6] Bölcskei, H., Paulraj, A., "Space-frequency coded broadband OFDM systems," *Proc. IEEE WCNC '00*, Chicago, IL, Sept. 2000, pp. 1–6.

[7] Choi, W.-J., Cioffi, J. M., "Space-Time Block Codes over Frequency Selective Rayleigh Fading Channels," *Proc. of IEEE VTC '99*, pp. 2541–2545.

[8] Al-Dhahir, N., "Single-Carrier Frequency-Domain Equalization for Space-Time Block-Coded Transmissions Over Frequency-Selective Fading Channels," *IEEE Commu. Letters*, Vol. 5, July 2001, pp. 304–306.

[9] Zhou, S., Giannakis, G. B.,"Space-Time Coded Transmissions with Maximum Diversity Gains over Frequency-Selective Multipath Fading Channels," *IEEE Sig. Proc. Letters*,vol. 8, no. 10, October 2001, pp. 269-272.

[10] Blum, R., et al., "Improved space-time coding for MIMO-OFDM wireless communications," *IEEE Trans. on Commun.*, Vol. 49, 2001, pp. 1873–1878.

[11] Alamouti, S. M., "A simple transmit diversity technique for wireless communications," *IEEE Journal on Selected Areas in Communications*, Oct. 1998, pp. 1451–1458.

[12] G. Klang, A. Naguib, "Transmit diversity based on space-time block codes in frequency-selective Rayleigh fading DS-CDMA channels," *VTC 00*, 2000, pp. 264–268.

[13] Liu, Z., Giannakis, G.B., Barbarossa, S., Scaglione, A., "Block Precoding and Transmit-Antenna Diversity for Decoding and Equalization of Unknown Multipath Channels," *Proc. of* 33rd *Asilomar Conf. on Signals, Systems, and Computers*, vol. 2, Pacific Grove, CA, Nov. 1-4, 1999, pp. 1557-1561.

[14] Lindskog, E., Paulaj, A., "A transmit diversity scheme for channels with intersymbol interference," *Proc. of ICC '00*, New Orleans, LA, 2000, pp. 307–311.

[15] Barbarossa, S. , Cerquetti, F. , "Simple Space-Time Coded SS-CDMA Systems Capable of Perfect MUI/ISI Elimination," *IEEE Communications Letters*, Vol. 5, Dec. 2001, pp. 471–473.

[16] Zhou, S., Giannakis, G. B., "Single-carrier space-time block-coded transmissions over frequency-selective fading channels," *IEEE Trans. on Information Theory*, Jan. 2003, pp. 164–179.

[17] Xin, Y., Wamg, Z., Giannakis, G. B., "Space-time diversity systems based on linear constellation precoding," *IEEE Trans. on Wireless Commun.*, March 2003, pp. 294–309.

[18] Ma, X., Giannakis, G. B., "Complex field coded MIMO systems: Performance, rate, and trade-offs," *Wireless Communications and mobile computing*, Febr. 2002, pp. 693–717.

[19] Ma, X., Giannakis, G.B., "Full-diversity full-rate complex-field space-time coding," *IEEE Transactions on Signal Processing*, Vol. 49, Nov. 2003, pp. 2917–2930.

[20] El Gamal, H., Damen, M., O., "Universal space-time coding," *IEEE Transactions on Information Theory*, Vol. 49, May 2003, pp. 1097–1119.

[21] Giraud, X., Boutillon E., Belfiore, J.C., "Algebraic tools to build modulation schemes for fading channels," *IEEE Trans. on Information Theory*, May 1997, pp. 938–952.

[22] Boutros, J., Viterbo, E., "Signal space diversity: A power and bandwidth efficient diversity technique for the Rayleigh fading channel," *IEEE Trans. on Information Theory*, July 1998, pp. 1453–1467.

[23] Simon, M. K., Alouini, M.-S., *Digital communications over fading channels: A unified approach to performance analysis*, New York: John Wiley & Sons, 2000.

[24] Proakis, J., *Digital Communications*, (4th edition), New York: McGraw Hill, 2000.

[25] Barbarossa, S., Fasano, A., "Trace-orthogonal space-time coding," *IEEE Wireless Communications*, submitted (November 2004).

[26] Stuber, G.L., Barry, J.R., McLaughlin, S.W., Ye Li, Ingram, M.A., Pratt, T.G., "Broadband MIMO-OFDM wireless communications," *Proceedings of the IEEE*, Vol. 92, Feb. 2004, pp.2 71–294.

[27] Van Zelst, A., Schenk, T.C.W., "Implementation of a MIMO OFDM-based wireless LAN system," *IEEE Trans. on Signal Processing*, Vol. 52, Feb. 2004, pp. 483–494.

Chapter 9

Space-Time Coding for Multiuser Systems

9.1 INTRODUCTION

Multiantenna transceivers may bring considerable performance gains in multiuser systems with respect to single antenna systems [1]. The availability of multiantenna receivers allows in fact for interference cancelation. Given a system with N users, each with one transmit antenna, it was proved in [2] that, in a Rayleigh flat-fading environment, a linear receiver with $K + N$ antennas is capable of nulling $N - 1$ interferers and, at the time, yielding diversity gain $K+1$ for every remaining user. In other words, the cancelation of $N-1$ interferers, through linear combining, involves the use of $N - 1$ degrees of freedom out of the overall $N + K$ available degrees. These degrees of freedom are then lost from the point of view of diversity gain. This is why the diversity gain on each user passes from the potential maximum value of $N+K$ to $K+1$. The remarkable property is that, after linear cancelation, *every user* is seen free of interference and with diversity gain $K + 1$. It is important to remark that this trade-off between interference cancelation and diversity gain is valid under the assumption of using a *linear* canceler. Using joint ML decoding, this trade-off is no longer valid. But joint ML detection can be quite complex to implement.

In this chapter we consider the situation where each user has multiple transmit antennas and uses space-time coding. In the previous chapter, we have shown how to combine space-time coding with block transmissions in order to make space-time coding suitable for implementations on broadband systems. The approach followed in the previous chapter was rather general and then able to incorporate different situations, acting on the precoding (or multiplexing) strategies. However, the systems analyzed in the previous chapter were, essentially, single-user systems. In this chapter, we concentrate on multiuser systems where each radio terminal has multiple antennas and different users share the same physical resources (channels). We will show that space-time coding offers indeed a rather general framework that allows the design of multiuser systems capable to strike the desired trade-off between performance, rate, and complexity. As in the previous chapter, we consider, successively, two major cases: 1) orthogonal and 2) trace-orthogonal space time coding.

9.2 ORTHOGONAL CODING

Orthogonal space-time coding is the method to choose when one wishes to reach the best trade-off between diversity gain and receiver complexity. However, orthogonal coding is suboptimal from the rate point of view. The rate loss of orthogonal coding can be intuitively justified thinking that the orthogonality condition puts a strong constraint on the encoder design that prevents the system from having all the potential degrees of freedom necessary to reach the maximum rate. In other words, orthogonal coding is a sort of redundant coding, as it does not allow the system to operate at full capacity (unless in a very specific example where the transmitter has two antennas and the receiver has only one antenna). Looking at the problem from a different perspective, this remark suggests that the implicit redundancy of orthogonal coding, in case of multiple receive antennas, can be exploited to cancel the interference from cochannel users.

9.2.1 A Short Digression on Quaternions

Before starting the analysis, it is useful to recall a few interesting properties of the set of 2×2 matrices having the structure

$$\mathcal{A} := \begin{pmatrix} \alpha & \beta \\ -\beta^* & \alpha^* \end{pmatrix}, \tag{9.1}$$

where the coefficients α and β are complex numbers. The interesting property of this set of matrices is that the set (9.1) is *closed under addition, multiplication, inversion, and multiplication with a real-valued scalar coefficient.* This means that any combination of the operations mentioned above over any set of matrices having the structure (9.9) gives rise to another matrix that preserves the same structure.

Note that this property reveals something well known in algebra. The structure (9.1) is in fact isomorphic to the field of *quaternions*. Quaternions are a generalization of complex numbers. While the complex numbers are obtained by adding the element $i = \sqrt{-1}$ to the real numbers, the quaternions are obtained by adding the elements i, j and k to the real numbers, with i, j, and k satisfying the following relations:

$$i^2 = j^2 = k^2 = ijk = -1.$$

Every quaternion is a real linear combination of the *unit quaternions* $1, i, j$, and k (i.e., every quaternion is uniquely expressible in the form $a+bi+cj+dk$ with a, b, c, and d real). Addition of quaternions is accomplished by adding corresponding coefficients, as with complex numbers. Multiplication of quaternions is completely determined by the multiplication table shown below

$\cdot$	1	i	j	k
1	1	i	j	k
i	i	-1	k	$-j$
j	j	$-k$	-1	i
k	k	j	$-i$	-1

The conjugate of the quaternion $z = a + bi + cj + dk$ is defined as $z^* = a - bi - cj - dk$ and the absolute value of z is the nonnegative real number defined by

$$|z| = \sqrt{z \cdot z^*} = \sqrt{a^2 + b^2 + c^2 + d^2}. \tag{9.2}$$

Note that $(wz)^* = z^*w^*$ is not equal, in general, to w^*z^*. The multiplicative inverse of the nonzero quaternion z can be computed as

$$z^{-1} = z^*/|z|^2. \tag{9.3}$$

The set of quaternions with the operations defined above constitutes a *division ring*, that is, an algebraic structure similar to a field, except for commutativity of multiplication. In particular, multiplication is still associative and every nonzero element has a unique inverse. However, unlike real or complex numbers, multiplication of quaternions is not commutative (e.g., $ij = k$, but $ji = -k$).

The field of complex numbers is isomorphic to the set of 2×2 real matrices having a specific structure. More specifically, every complex number $a + jb$, with a and b real, has an equivalent member in the set of the 2 real matrices having the structure

$$\begin{pmatrix} a & b \\ -b & a \end{pmatrix}. \tag{9.4}$$

The equivalence is sketched in Figure 9.1. This means that all operations like addition, multiplication, and inversion over the complex numbers have an equivalent operation among the corresponding members of the set of real matrices having the structure (9.4).

As a generalization of the concept of complex numbers, the field (more precisely, the division ring) of quaternions is isomorphic to the set of 2×2 *complex* matrices.

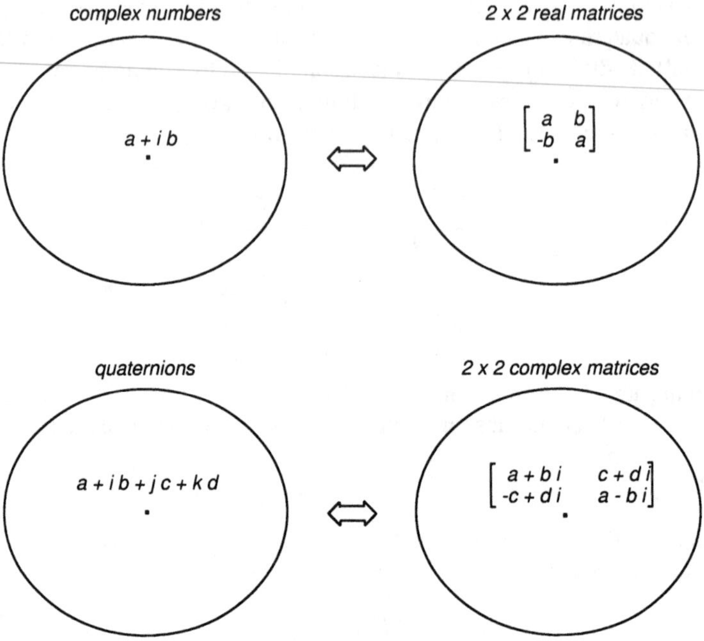

Figure 9.1 Equivalence between: a) complex numbers and 2×2 real matrices; and b) quaternions and 2×2 complex matrices.

In particular, every quaternion $a + bi + cj + dk$ is in a unique correspondence with the complex matrix

$$\begin{pmatrix} a+bi & c+di \\ -c+di & a-bi \end{pmatrix} \equiv \begin{pmatrix} \alpha & \beta \\ -\beta^* & \alpha^* \end{pmatrix}, \tag{9.5}$$

having introduced the complex numbers $\alpha := a + bi$ and $\beta := c + di$.

Interestingly, *the matrix structure in (9.5) is exactly the same as the Alamouti coding scheme*, where α and β are the two transmitted symbols. This equivalence between Alamouti coding and quaternions will be useful in analyzing the performance of multiuser systems having two transmit/recive antennas.

9.2.2 Multiuser Suppression in Orthogonal STC Systems

Let us consider now a multiple access system with two users, where each terminal (including both mobile nodes and access point) has two antennas. Each user transmits using an Alamouti coding strategy. Therefore, recalling the results of Section 7.2, denoting by $[s_k(n), s_k(n+1)]$ the pair of symbols transmitted by user k over two consecutive time slots, the corresponding pair of samples collected by the mth receive antenna from user k, with $k = 1, 2$, is (we omit the noise contribution here, for notation simplicity)

$$\begin{bmatrix} y_m(n) \\ y_m^*(n+1) \end{bmatrix} = \begin{bmatrix} h_{m1}^{(k)} & h_{m2}^{(k)} \\ -h_{m2}^{(k)*} & h_{m1}^{(k)*} \end{bmatrix} \begin{bmatrix} s_k(n) \\ s_k(n+1) \end{bmatrix}, \quad m = 1, 2, \tag{9.6}$$

where $h_{mn}^{(k)}$ denotes the channel from the nth antenna of user k to the mth receive antenna. In matrix form, (9.6) can be written as

$$\boldsymbol{y}_m = \boldsymbol{\mathcal{H}}_{mk}\boldsymbol{s}_k. \tag{9.7}$$

Collecting the contributions from both antennas and both users, we have

$$\begin{bmatrix} \boldsymbol{y}_1 \\ \boldsymbol{y}_2 \end{bmatrix} = \begin{pmatrix} \boldsymbol{\mathcal{H}}_{11} & \boldsymbol{\mathcal{H}}_{12} \\ \boldsymbol{\mathcal{H}}_{21} & \boldsymbol{\mathcal{H}}_{22} \end{pmatrix} \begin{bmatrix} \boldsymbol{s}_1 \\ \boldsymbol{s}_2 \end{bmatrix}, \tag{9.8}$$

where

$$\boldsymbol{\mathcal{H}}_{mk} := \begin{bmatrix} h_{m1}^{(k)} & h_{m2}^{(k)} \\ -h_{m2}^{(k)*} & h_{m1}^{(k)*} \end{bmatrix}. \tag{9.9}$$

Given the vector $\boldsymbol{y} = [\boldsymbol{y}_1^T, \boldsymbol{y}_2^T]^T$ collected at the receiver, we can separate the two users by premultiplying $\boldsymbol{y}$ by the matrix

$$\boldsymbol{G} := \begin{pmatrix} \boldsymbol{I} & -\boldsymbol{\mathcal{H}}_{12}\boldsymbol{\mathcal{H}}_{22}^{-1} \\ -\boldsymbol{\mathcal{H}}_{21}\boldsymbol{\mathcal{H}}_{11}^{-1} & \boldsymbol{I} \end{pmatrix}. \tag{9.10}$$

In fact, the multiplication $\boldsymbol{G}\boldsymbol{y}$ gives rise to the system

$$\boldsymbol{G}\begin{bmatrix} \boldsymbol{y}_1 \\ \boldsymbol{y}_2 \end{bmatrix} = \begin{pmatrix} \boldsymbol{\mathcal{H}}_{11} - \boldsymbol{\mathcal{H}}_{12}\boldsymbol{\mathcal{H}}_{22}^{-1}\boldsymbol{\mathcal{H}}_{21} & \boldsymbol{0} \\ \boldsymbol{0} & \boldsymbol{\mathcal{H}}_{22} - \boldsymbol{\mathcal{H}}_{21}\boldsymbol{\mathcal{H}}_{11}^{-1}\boldsymbol{\mathcal{H}}_{12} \end{pmatrix} \begin{bmatrix} \boldsymbol{s}_1 \\ \boldsymbol{s}_2 \end{bmatrix}. \tag{9.11}$$

Introducing the vector $\boldsymbol{z} := (\boldsymbol{z}_1^T, \boldsymbol{z}_2^T)^T := \boldsymbol{G}\boldsymbol{y}$, (9.11) is then decoupled into the pair of systems

$$\begin{aligned} \boldsymbol{z}_1 &= (\boldsymbol{\mathcal{H}}_{11} - \boldsymbol{\mathcal{H}}_{12}\boldsymbol{\mathcal{H}}_{22}^{-1}\boldsymbol{\mathcal{H}}_{21})\boldsymbol{s}_1 \\ \boldsymbol{z}_2 &= (\boldsymbol{\mathcal{H}}_{22} - \boldsymbol{\mathcal{H}}_{21}\boldsymbol{\mathcal{H}}_{11}^{-1}\boldsymbol{\mathcal{H}}_{12})\boldsymbol{s}_2. \end{aligned} \tag{9.12}$$

Hence, after multiplication of the received vector by $\boldsymbol{G}$ the two users are decoupled and they can be decoded independently.

The previous comment holds true only in the absence of noise. In fact, in the presence of noise, even if the noise vectors present at the separate receivers are independent, the noise vectors on the two separate channels, after multiplication by $\boldsymbol{G}$, become statistically dependent. Hence, independent decoding is suboptimal.

Interestingly, thanks to the property of the set of matrices (9.1) to be closed under summation, multiplication, and inversion, the two separate channels in (9.12) are equivalent to two Alamouti systems between each user and the destination, provided that the channel matrix from each user to the access point is substituted by the matrices

$$\tilde{\mathcal{H}}_{12} := \mathcal{H}_{11} - \mathcal{H}_{12}\mathcal{H}_{22}^{-1}\mathcal{H}_{21} \tag{9.13}$$

$$\tilde{\mathcal{H}}_{21} := \mathcal{H}_{22} - \mathcal{H}_{21}\mathcal{H}_{11}^{-1}\mathcal{H}_{12}. \tag{9.14}$$

By virtue of the previous properties, the matrices $\tilde{\mathcal{H}}_{12}$ and $\tilde{\mathcal{H}}_{21}$ are orthogonal.

We can now extend the previous formulation to the general case of M users, where each user has two transmit antennas, and the access point has M antennas. Repeating the same derivations as in the previous case, the set of samples received by the M receive antennas, over two consecutive time slots can be written as

$$\boldsymbol{y} := \left[\begin{array}{c} \boldsymbol{y}_1 \\ \boldsymbol{y}_2 \\ \vdots \\ \hline \boldsymbol{y}_M \end{array}\right] = \left(\begin{array}{cccc|c} \mathcal{H}_{11} & \mathcal{H}_{12} & \cdots & & \mathcal{H}_{1M} \\ \mathcal{H}_{21} & \mathcal{H}_{22} & \cdots & & \mathcal{H}_{2M} \\ \cdots & \cdots & \cdots & & \cdots \\ \hline \mathcal{H}_{M1} & \mathcal{H}_{M2} & \cdots & & \mathcal{H}_{MM} \end{array}\right) \left[\begin{array}{c} \boldsymbol{s}_1 \\ \boldsymbol{s}_2 \\ \vdots \\ \hline \boldsymbol{s}_M \end{array}\right], \tag{9.15}$$

where every matrix $\mathcal{H}_{nm}$ has an Alamouti structure. To single out the Mth user, it is useful to exploit the partition evidenced in (9.15), which we rewrite as

$$\left[\begin{array}{c} \boldsymbol{y}_{-M} \\ \boldsymbol{y}_M \end{array}\right] = \left(\begin{array}{cc} \boldsymbol{A} & \boldsymbol{B} \\ \boldsymbol{C} & \boldsymbol{D} \end{array}\right) \left[\begin{array}{c} \boldsymbol{s}_{-M} \\ \boldsymbol{s}_M \end{array}\right], \tag{9.16}$$

where $\boldsymbol{y}_{-M}$ is the $2(M-1)$-size vector containing all two-element vectors $\boldsymbol{y}_i$, except the Mth one, $\boldsymbol{s}_{-M}$ is the $2(M-1)$-size vector containing all symbol vectors, except the Mth one, and the matrices $\boldsymbol{A}$, $\boldsymbol{B}$, $\boldsymbol{C}$ and $\boldsymbol{D}$, are $2(M-1) \times 2(M-1)$, $2(M-1) \times 2$, $2 \times 2(M-1)$, and 2×2, respectively. Given the structure (9.16), to single out the Mth user, it is sufficient to multiply the vector $\boldsymbol{y}$ by the $2M \times 2M$ matrix

$$\boldsymbol{G} := \left(\begin{array}{cc} \boldsymbol{I}_{2(M-1)} & -\boldsymbol{B}\boldsymbol{D}^{-1} \\ -\boldsymbol{C}\boldsymbol{A}^{-1} & \boldsymbol{I}_2 \end{array}\right), \tag{9.17}$$

where the identity matrices have dimensions congruent with the matrices involved in (9.17).

After multiplication, we obtain the vector

$$\begin{pmatrix} \boldsymbol{r}_{-M} \\ \boldsymbol{r}_M \end{pmatrix} := \boldsymbol{G}\boldsymbol{y} = \begin{bmatrix} (\boldsymbol{A} - \boldsymbol{B}\boldsymbol{D}^{-1}\boldsymbol{C})\boldsymbol{s}_{-M} \\ (\boldsymbol{D} - \boldsymbol{C}\boldsymbol{A}^{-1}\boldsymbol{B})\boldsymbol{s}_M \end{bmatrix}. \tag{9.18}$$

Interestingly, the matrices $\boldsymbol{A} - \boldsymbol{B}\boldsymbol{D}^{-1}\boldsymbol{C}$ and $\boldsymbol{D} - \boldsymbol{C}\boldsymbol{A}^{-1}\boldsymbol{B}$ have the same structure as the matrices $\boldsymbol{A}$ and $\boldsymbol{D}$, respectively. This is a consequence of the equivalence between quaternions and complex matrices having the Alamouti structure, suggested in [3, 4]. To prove this statement, it is sufficient to prove first that the matrix $\boldsymbol{A}^{-1}$ has the same structure as $\boldsymbol{A}$. This can be proved by induction. Let us consider, for example, the 4×4 matrix, assumed to be invertible

$$\boldsymbol{M} := \begin{pmatrix} \boldsymbol{P} & \boldsymbol{R} \\ \boldsymbol{Q} & \boldsymbol{S} \end{pmatrix}, \tag{9.19}$$

where all matrices $\boldsymbol{P}, \boldsymbol{Q}, \boldsymbol{R}$, and $\boldsymbol{S}$ are 2×2 Alamouti complex matrices (and then equivalent to quaternions). The inverse of $\boldsymbol{M}$ is

$$\boldsymbol{M}^{-1} = \begin{pmatrix} (\boldsymbol{P} - \boldsymbol{R}\boldsymbol{S}^{-1}\boldsymbol{Q})^{-1} & -\boldsymbol{P}^{-1}\boldsymbol{R}(\boldsymbol{S} - \boldsymbol{Q}\boldsymbol{P}^{-1}\boldsymbol{R})^{-1} \\ -(\boldsymbol{S} - \boldsymbol{Q}\boldsymbol{P}^{-1}\boldsymbol{R})^{-1}\boldsymbol{Q}\boldsymbol{P}^{-1} & (\boldsymbol{S} - \boldsymbol{Q}\boldsymbol{P}^{-1}\boldsymbol{R})^{-1} \end{pmatrix}. \tag{9.20}$$

Since all the operations appearing in (9.20) are, equivalently inversions, multiplications, and additions of quaternions, thanks to the closeness of the quaternions division ring with respect to these operations, each submatrix in (9.20) is still equivalent to a quaternion (i.e., it is an Alamouti complex matrix). Iterating this argument, the matrix $\boldsymbol{A}^{-1}$ in (9.17) has the same structure as the matrix $\boldsymbol{A}$. Consequently, the matrix $\boldsymbol{D} - \boldsymbol{C}\boldsymbol{A}^{-1}\boldsymbol{B}$ in (9.18) is itself an Alamouti matrix and the matrix $\boldsymbol{A} - \boldsymbol{B}\boldsymbol{D}^{-1}\boldsymbol{C}$ in (9.18) has the same structure as $\boldsymbol{A}$.

The previous derivations show that the interference cancelation, through premultiplication by the matrix $\boldsymbol{G}$ in (9.17), singles out only one user and, interestingly enough, it *preserves the Alamouti structure* of the channel between the remaining user and the access point.

Iterating the same derivations shown above, we can then successively extract one user at the time. After cancelation, every user will still be seen as arriving through an equivalent Alamouti channel.

It is natural to ask ourselves what is then the price paid for suppressing the interference with a simple linear combiner. The price is a loss of diversity. In fact, in case of a $2 \times M$ MIMO system, the potential diversity gain with orthogonal coding is $2M$. However, after linear interference suppression, the diversity gain for each user becomes equal to two.

As an example of performance, in Figure 9.2, we report the average BER obtained with the following configurations: 1) one user, with two transmit and one receive antenna (solid line); 2) one user, with two transmit and two receive antennas (dashed line); 3) one user, with two transmit and three receive antennas (dotted line); 4) two users, with two two transmit and two receive antennas (solid line with stars); and 5) three users, with two transmit and three receive antennas (solid line with circles). In the multiuser case, each user is decoded independently of the others by applying

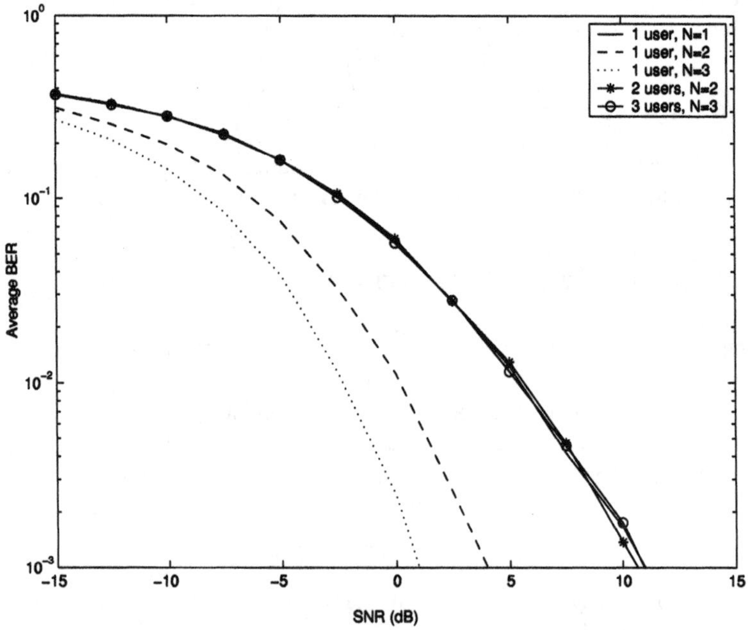

Figure 9.2 Average BER versus SNR; comparison between single user and multiuser systems, with different number of receive antennas.

the iterative cancelation scheme shown before.

We can see, from Figure 9.2, that the multiuser cases where the number of receive antennas is equal to the number of users provide the same results. The system does not gain in diversity as much as it could, but it is able to cancel the interference with a simple linear receiver and still get diversity gain equal to two. The previous result is somehow a generalization of what found in [2, 5] for single transmit antenna systems.

Before concluding this section, it is important to emphasize that the previous statement about the loss of diversity resulting from the cancelation of multiuser interference holds true only when the canceler is *linear*. In case of *joint* optimal multiuser detection, there is no diversity loss because of interference cancelation. The problem, in case of joint optimal (nonlinear) decoding, is complexity.

9.3 FULL-RATE CODING

Orthogonal STC coding leads to very simple receive structures also in the multiuser case. However, orthogonal coding is suboptimal in terms of rate. To avoid the rate losses of orthogonal coding, we can resort to the general scheme of trace-orthogonal design (TOD), developed in Chapter 7. In this section, we generalize the TOD strategy to the multiuser case, as suggested in [7]. We assume flat-fading channels. The extension to frequency-selective channels can be done following the same guidelines reported in Chapter 8.

Each radio node, comprising both mobile nodes and access points, is assumed to be equipped with N transmit/receive antennas. As a consequence of the TOD design, each code spans N symbol periods. The multiuser TOD is particularly appealing as far as rate flexibility is concerned. The multirate capability, so important in current and future wireless communications, is in fact inherently built in the coding strategy. The multiuser version of TOD is in fact obtained by simply assigning a set of matrices to each user. We recall, from Chapter 7, that the number of available matrices is N^2. Let us indicate with $\{\boldsymbol{A}_i^{(k)}\}_{i\in\mathcal{I}_k}$ the set of matrices assigned to user k. Given a maximum number N_u of users, if we denote by N_k the number of matrices assigned to user k, multiuser TOD gives rise to a multirate system, where each user has symbol rate $R_k = N_k/N$. The aggregate rate of the system is

$$R_A = \frac{\sum_{k=1}^{N_u} N_k}{N}. \tag{9.21}$$

If we assign distinct coding matrices to different users, we can get, at most, an aggregate rate equal to N, as in the single-user case. This is what happens in the downlink case. However, as we will see later on, in the uplink channel we can remove this constraint and assign the same matrices to different users. In this way, we may get a maximum aggregate rate equal to $N_u \cdot N$.

9.3.1 Downlink Channel

The downlink channel is a straightforward generalization of the single-user case. Denoting with $\boldsymbol{H}_k$ the MIMO channel between the access point and the kth user, the space-time code matrix transmitted by the AP is

$$\boldsymbol{X} = \sum_{m=1}^{N_u} \sum_{i=1}^{N_m} s_i^{(m)} \boldsymbol{A}_i^{(m)}, \tag{9.22}$$

where $s_i^{(m)}, i = 1, \ldots, N_m$, is the sequence of symbols for user m. The code matrix received by user k is

$$\boldsymbol{Y}_k = \boldsymbol{H}_k \boldsymbol{X} + \boldsymbol{V} = \boldsymbol{H}_k \sum_{m=1}^{N_u} \sum_{i=1}^{N_k} s_i^{(m)} \boldsymbol{A}_i^{(m)} + \boldsymbol{V}. \tag{9.23}$$

This expression is perfectly equivalent to the single-user case examined in Section 7.5. Hence, the performance can be evaluated exactly as in Section 7.5.

9.3.2 Uplink Channel

Let us now turn our attention to the more complicated multiple access case. We review the basic results derived in [7]. We consider a multiple access system composed of N_u users, each with n_T transmit antennas, and an access point (AP), with n_R receive antennas. The generic user k encodes its own n_s (complex) symbols $s_k(j), j = 1, \ldots, n_s$, through the following space-time linear encoder:

$$\boldsymbol{X}_k = \sum_{j=1}^{n_s} \boldsymbol{A}_k(j) s_k(j), \tag{9.24}$$

where $\{\boldsymbol{A}_k(j), j = 1, \ldots, n_s\}$ is the set of $n_T \times Q$ complex matrices assigned to user k. A multi-user (MU) space-time encoder is said to be an MU *Trace-Orthogonal Design* (TOD), if the matrices $\boldsymbol{A}_k(j)$ satisfy[1], for each user k,

$$\langle \boldsymbol{A}_k(j), \boldsymbol{A}_k(m) \rangle := \mathrm{tr}[\boldsymbol{A}_k^H(j) \boldsymbol{A}_k(m)] = \delta_{jm}. \tag{9.25}$$

Furthermore, we say that it is a Unitary TOD (U-TOD) if the following additional condition holds true

$$\boldsymbol{A}_k(j) \boldsymbol{A}_k^H(j) = \frac{1}{n_T} \boldsymbol{I}, \quad j = 1, \ldots, n_s, \tag{9.26}$$

1 δ_{kl} denotes the Kronecker symbol, equal to one if $k = l$ and zero otherwise.

as in the single user case studied in Chapter 7.

Let us consider now a multiple access channel, where all channels $\boldsymbol{H}_k$ between the kth user and the AP are constant over Q successive channel uses (quasi-static fading). If each user transmits the matrix $\boldsymbol{X}_k$, built as in (9.24), the received matrix is

$$\boldsymbol{Y} = \sum_{k=1}^{N_u} \boldsymbol{H}_k \boldsymbol{X}_k + \boldsymbol{V} := \boldsymbol{\mathcal{H}}\boldsymbol{\mathcal{X}} + \boldsymbol{V}, \tag{9.27}$$

where

$$\boldsymbol{\mathcal{H}} := (\boldsymbol{H}_1, \ldots, \boldsymbol{H}_{N_u}) \text{ and } \boldsymbol{\mathcal{X}} := \begin{pmatrix} \boldsymbol{X}_1 \\ \vdots \\ \boldsymbol{X}_{N_u} \end{pmatrix}. \tag{9.28}$$

Let us assume now that the following conditions hold true: 1) the receiver has perfect knowledge of all channels; 2) each transmitter (user) has no channel knowledge; 3) the transmit power of each user is upper bounded by p_T; 4) the users encode their symbols independently of the other users; 5) the transmitted symbols are zero mean, uncorrelated, with variance $\sigma_s^2 = p_T/n_T$; 6) each user uses a space-time code having maximum symbol rate $R_s = n_T$. It was proven in [7] that, under these assumptions, the trace-orthogonal multiuser system (9.27) is information lossless, for any channel realization. This generalizes the result already proven in Chapter 7, about the single user TOD.

It is important to remark that assumption 6) implies that *the aggregate symbol rate of multiuser TOD is equal to the product* $N_u\, n_T$.

The optimal decoding of the multiuser TOD scheme requires a maximum likelihood decoder. However, this can be computationally prohibitive. A suboptimum approach consists in using a linear MMSE decoder [2]. It was proven in [7] that, under the previous assumption, the MMSE decoder for the multiuser case is a direct generalization of the single user case already analyzed in Chapter 7. In particular, the MMSE estimator for user k is

$$\hat{s}_k(j) = \text{tr}\left[\boldsymbol{A}_k^H(j)\, \boldsymbol{P}_k^H\, \boldsymbol{\mathcal{W}}\, \boldsymbol{Y}\right], \tag{9.29}$$

where $\hat{s}_k(j)$ is the estimate of the j-th symbol transmitted by the k-th user. The matrix $\boldsymbol{\mathcal{W}}$ is the same for all the users and is equal to

$$\boldsymbol{\mathcal{W}} = \left(\boldsymbol{\mathcal{H}}^H\boldsymbol{\mathcal{H}} + \frac{1}{\gamma}\boldsymbol{I}_{N_u \cdot n_T}\right)^{-1} \boldsymbol{\mathcal{H}}^H \tag{9.30}$$

2 This possibility is paid in terms of number of receive antennas. To get acceptable performance, it is necessary to select n_R at least equal to $N_u \cdot n_T$.

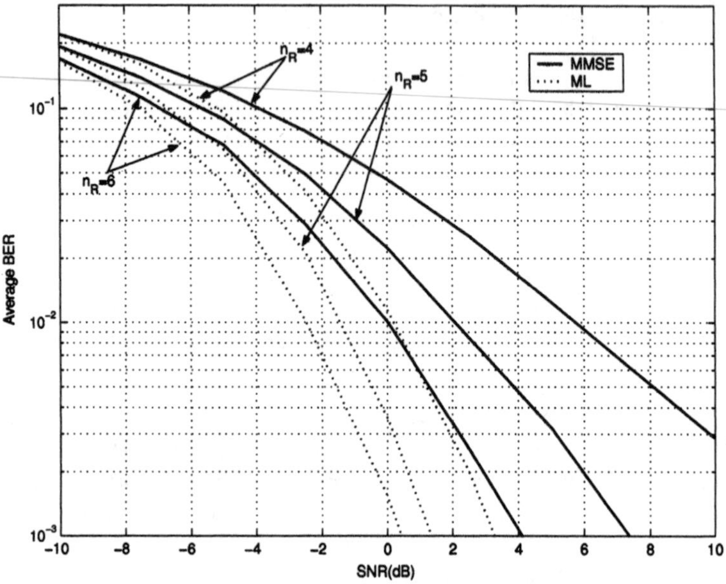

Figure 9.3 Average BER obtained with full rate unitary TOD, using shift and multiply bases and ML (dashed line) or MMSE (solid line) decoding.

with $\gamma = \sigma_s^2/\sigma_v^2$. The matrix $\boldsymbol{P}_k$ is a selection matrix, defined as $\boldsymbol{P}_k := \boldsymbol{e}_k \otimes \boldsymbol{I}_{n_T}$, where $\boldsymbol{e}_k$ is the kth N_u-dimensional unit vector[3], whose role is to extract from $\boldsymbol{\mathcal{W}}$ the subset of rows relevant for the k-th user. Using Matlab notation, $\boldsymbol{P}_k$ is such that $\boldsymbol{P}_k^H \boldsymbol{\mathcal{W}} = \boldsymbol{\mathcal{W}}\left[(k-1)\cdot n_T + 1 : k \cdot n_T, :\right]$.

The main feature of (9.29) is low complexity. The complexity of the estimator is comparable to that of orthogonal space-time coding, but with the advantage of being full rate. The price paid for having this low complexity is that MMSE is sub-optimal. As in the single user case, it was proven in [7] that if we use Unitary-TOD, then the (suboptimal) MMSE decoder provides the minimum BER.

To check the performance of the MMSE receiver, in Figure 9.3 we compare the average BER obtained with full rate Unitary-TOD, using the MMSE and the ML decoders. The encoding matrices are the shift-and-multiply basis described in Chapter 7. The system parameters are: $n_T = 2$ and $n_R = 4, 5,$ and 6. The number of users N_u is two. Every user is transmitting at full-rate, i.e., n_T symbols pcu. The aggregate rate is then $N_u \cdot n_T$. The BER is averaged over $10,000$ independent channel realizations. We can see from Figure 9.3 that, for $n_R = 2n_T$, the MMSE

3 The vector $\boldsymbol{e}_k$ has all elements equal to zero, except the kth one, equal to one.

decoder has a considerable loss with respect to ML, but, as soon as n_R increases with respect to its minimum value $N_u \cdot n_T$, the loss decreases. Considering the huge difference in complexity between the MMSE and the ML decoders, TOD is then an interesting candidate for multi-user systems, provided that the access point has a sufficient number of receive antennas. It is also interesting to check the performance of multiuser TOD in the case where the system is not working full rate. As an example, in Figure 9.4 we report the average error rate resulting from the following scenario: the number of user is $N_u = 8$, $n_T = 2$, and $n_R = 16$. The symbol rate is the same for all users and it assumes the following values: $1/2, 1, 3/4$, or 2. The aggregate rate is then $R_a = 4, 8, 12$, or 16. We can notice, from Figure 9.4, how the

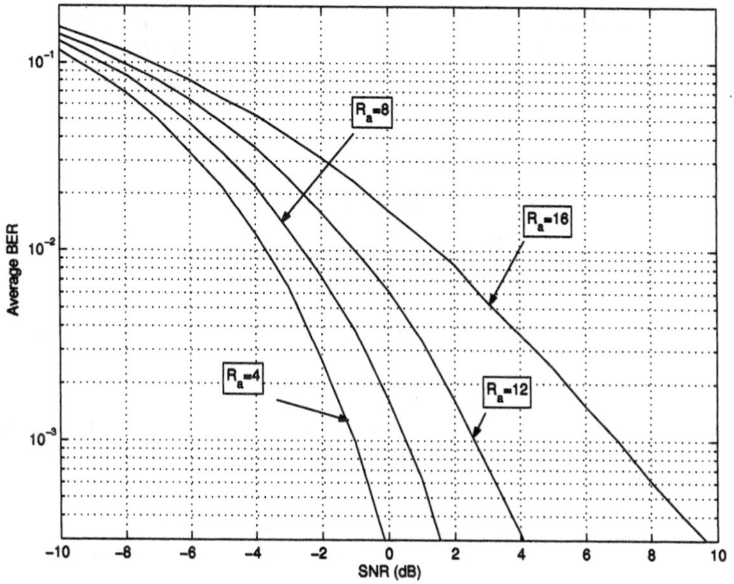

Figure 9.4 Average BER versus SNR; comparison between single user and multiuser systems, with different number of receive antennas.

performance improve dramatically, as the rate decreases, even though the aggregate rate remains quite high.

9.4 SUMMARY

In summary, orthogonal design in a multiuser context makes possible the implementation of fairly simple receiving schemes, capable of nulling the interference with a simple linear combiner and still providing some diversity gain. Compared with the single user case, where the maximum diversity achievable with a $2 \times M$ MIMO system, using a simple scalar detector, is $2M$, in the multiuser case the diversity gain achievable with a scalar detector is only two. In words, diversity is traded with interference suppression. To increase the transmission rate, the trace-orthogonal design offers a very flexible tool capable of achieving very good performance with limited complexity. The price paid for achieving full rate, with minimum complexity, is that the MMSE detector does not guarantee full diversity. As in the single-user case, TOD is shown to be information lossless also in the multiuser case. Finally, it is worth emphasizing that, in the uplink channel, if the access point has a sufficient number of receive antennas, multiuser-TOD can achieve an aggregate rate, with acceptable performance, equal to the product between the number of transmit antennas and the number of users. However, this requires a receive array with a number of elements at least equal to such a product. In practice, the flexibility of multiuser-TOD allows us to trade performance with rate and number of receive antennas.

References

[1] Blum, R. S., Winters, J. H., Sollenberger, N. R., "On the capacity of cellular systems with MIMO," *IEEE Communications Letters*, June 2002, pp. 242–244.

[2] Winters, J. H., Salz, J., Gitlin, R. D., "The impact of antenna diversity on the capacity of wireless communication systems," *IEEE Transactions on Communications*, Febr./March/April 1994, pp. 1740–1751.

[3] Stamoulis, A., Al-Dhahir, N., Calderbank, A.R., "Further results on interference cancelation and space-time block codes," *Proc. of Asilomar Conf. on Signals, Systems and Computers*, Pacific Grove, CA, Oct. 2001, pp. 257–261.

[4] Diggavi, S., N., Al-Dhahir, N., Calderbank, A., R., "Algebraic properties of space-time block codes in intersymbol interference multiple-access channels," *IEEE Trans. on Information Theory*, Oct. 2003, pp. 2403–2414.

[5] Winters, J., H., "On the capacity of radio communication systems with diversity in a Rayleigh fading environment," *IEEE Journal on Selected Areas in Communications*, June 1987, pp. 871–878.

[6] Barbarossa, S., "Trace-orthogonal design of MIMO systems with simple scalar detectors, full diversity and (almost) full rate," *Proc. of the V IEEE Signal Proc. Workshop on Signal Proc. Advances in Wireless Commun., SPAWC '2004*, Lisbon, Portugal, July 11–14, 2004.

[7] Barbarossa, S., Fasano, A., "Trace-orthogonal space-time coding design for multiuser systems," *Proc. of ICASSP 2005*, Philadelphia, March 2005.

Chapter 10

Cooperative Networks

10.1 INTRODUCTION

The information signals flow through telecommunication networks through paths that depend on nodes location, traffic, etc. The routing problem has been typically formulated as the problem of finding the best path from the source to the destination. This formulation is perfectly reasonable, but it is somehow biased by the fact that the initial research in this field assumed the network to be wired. A wireless network has certainly its own problems, due primarily to the randomness of the links, but it offers, at the same time, a greater flexibility in designing the most appropriate routing strategy than wired networks. In wireless networks, in fact, the information packets might go from the source to the destination following more paths, *at the same time*. With a wireless network, we can think of the information emanating from the source as an *information wave* that propagates through a medium composed of many nodes that can, in principle, retransmit the received information at the same time. Clearly, these simultaneous paths create also interference. However, if the cooperating nodes retransmit the information in a coordinated manner, the potential interference can be turned into a *constructive interference* that can improve the link performance.

Multihop radio networking is a field whose study started long time ago (see [1] and the references therein). Some basic theorems on the capacity of relaying networks were established in [2, 3, 4, 5], for example. More recently, the interest about relaying networks, especially in the form of cooperation among nodes, has increased considerably. One result that sparked great interest was that cooperation among users can increase the capacity in an uplink multiuser channel [6]. A thorough analysis of the diversity gain achievable with cooperation was given in [7, 8, 9], where different distributed cooperation protocols were compared. Cooperation was proved to be very useful to combat shadowing effects, as shown in

[10], and it can occur in different forms, as suggested in many recent works, like [8, 11, 12, 13, 14, 15, 16, 17, 18, 19, 20, 21, 22, 23].

One more idea that is currently receiving great attention and it is likely to play a fundamental role in the deploying of future networks is the network *self-organization* capability. This is a feature appealing in the so called *ad hoc* networks and *sensor* networks.

Ad hoc networks are telecommunication networks with no pre-existing infrastructure. After deployment, the nodes tend to organize themselves, assigning different roles to each node, dynamically. In an ad hoc network, there are no access points. A recent standard for ad hoc networks is IEEE 802.15.3 [24], devised for high rate wireless personal area networks (WPANs). IEEE 802.15.3 allows for peer-to-peer communications, at data rates ranging from 11 to 55 Mbps, in the unlicensed 2.4 to 2.4835 GHz band, at distances up to 70m, with assigned quality of service.

Sensor networks are also self-organizing networks, but with totally different characteristics than ad hoc networks. First of all, the network is primarily a sensor that has to reveal the occurrence of some event of interest, like sensing the temperature or the concentration of polluting elements, etc., and communicate this information to a control node, typically identified as a *sink* (of information). Since sensor networks are typically deployed in hard to reach areas, the primary constraint on sensor networks is energy consumption. Each sensor has a very limited energy storage capability. It might recharge its own batteries using small solar cells or piezoelectric cells, wherever applicable. Typically, each sensor is "on" only for a small percentage of time. On the other hand, the data rate required from a sensor network is typically much lower than in a telecommunication network.

The aim of this chapter is to show how the space-time coding idea described in the previous chapters can provide an important tool to devise novel cooperation strategies in wireless networks. We will start deriving the conditions for the connectivity of a wireless network, exploiting basic mathematical tools borrowed from random geometric graphs. We will show how the cooperation among radio nodes can reduce the overall power necessary to deliver the information. Then, we will illustrate some example of cooperation, based on the so called distributed space-time coding.

10.2 RANDOM GEOMETRIC GRAPHS

Given a set of radio nodes distributed over a given territory, how can we assess whether each node can communicate with each other node through the network?

This question can be formulated in terms of *connectivity* of a graph. We review now the basic properties of random graphs. The interested reader may check [25, 26, 27] to dig further into this fascinating field. An in-depth study of the statistics od spatial data is given in [28].

A network may be modeled as an undirected graph $G(V, E)$ composed of a set V of n radio nodes (vertices) and a set E of links (edges). We are interested here only in *random graphs*, where the nodes are placed randomly within a given space. A *geometric random graph* is a random graph where the n nodes are placed independently and uniformly and there exists a link between two nodes if their distance is smaller than a given range r_0. We say that a node u is *neighbor* of a node v if the distance between u and v is less than r_0. An example of geometric graph is reported in Figure 10.1, where each black dot is a radio node and the circles of radius r_0 indicate the coverage of each node. In this example, one node is isolated as its minimum distance from all other nodes is greater than r_0. The *degree* of a

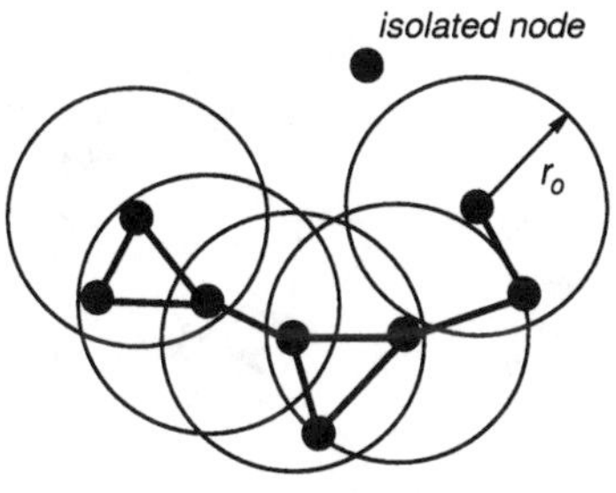

Figure 10.1 Random network with an isolated node.

node v, denoted as $d(v)$, is the number of neighbors of v. A node is isolated (i.e., it has no neighbors), if its degree is zero. The minimum degree or, simply, the degree of a graph is the minimum of the degrees of all the nodes

$$\delta = \min_{\forall v \in G} \{d(v)\}. \tag{10.1}$$

A *path* from a node u to a node v is a sequence of distinct nodes $(x_1 = u, x_2, \cdots, x_{n-1}, x_n = v)$, such that each pair (x_{n-1}, x_n) belongs to E (i.e., any two consecutive nodes are neighbors of each other). Given two nodes u and v, two paths connecting them, namely $(u, x_2, \cdots, x_{n-1}, v)$ and $(u, y_2, \cdots, y_{n-1}, v)$, are *independent* if they have no nodes in common (except of course the end-points). A graph is said to be *k-connected* if, for each node pair, there exist at least k independent paths connecting them. Equivalently, a graph is k-connected if and only if it is not possible to find any set of $(k-1)$ nodes whose removal would make the graph disconnected. The maximum value of k such that a graph is k-connected is the *connectivity* κ of the network. In words, the connectivity is the smallest number of nodes failures necessary to disconnect the network. For any graph, it holds true that $\kappa \leq d_{\min}$.

As an example of connectivity, in Figure 10.2 we report a random graph built using $n = 100$ points distributed uniformly within a circle of radius $R = 18$m. The density is $\rho = 0.1$ and r_0 is one quarter of R. We can see that, in this example, the graph degree is one, but the connectivity is zero, as there exist four separated *graph components* disconnected from each other. In general, in fact, the degree of a network and the connectivity are two distinct parameters. Nevertheless, Penrose

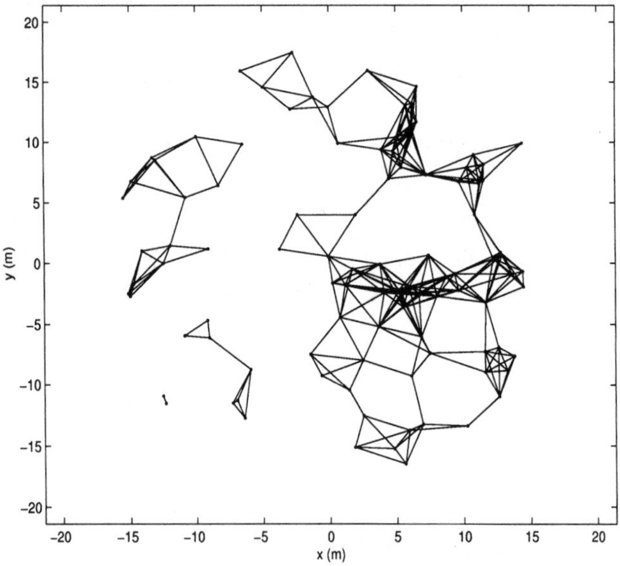

Figure 10.2 Network connectivity.

proved that there is a basic relationship between connectivity and degree of a geometric graph. In particular, Penrose proved that, denoting with $r(\kappa \geq k)$ the minimum distance r_0 such that a random graph is k-connected and with $r(\delta \geq k)$ the minimum distance such that the graph has minimum degree k, then [29]

$$\lim_{n\to\infty} \Pr\{r(\kappa \geq k) = r(\delta \geq k)\} = 1, \ \ \forall k. \tag{10.2}$$

This means that, asymptotically as n tends to infinity, the coverage radius that insures a degree k is the same as the radius that yields connectivity k.

As an example of connectivity, in Figure 10.3 we report three cases of networks obtained generating 100 points distributed uniformly within a circle of radius $R = 18\text{m}$, for different values of r_0, equal to $0.25R$, $0.3R$, and $0.35R$. We can check that, in the three cases mentioned above, the network is unconnected, 1-connected, or 2-connected.

Let us consider now the statistics of the node positions. If we assume that in a given geographical area, the density of the radio nodes is constant, we can model the position of the radio nodes as a two-dimensional (2D) Poisson process [28]. The spatial (or 2D) Poisson point process, similarly to the time (or 1D) Poisson process, can be interpreted as the asymptotic limit of a Bernoulli distribution [30]. Let us consider, as an example, a circle of radius R, centered at the origin of the reference system, where we throw n nodes. We denote with $C(\boldsymbol{x}_0, r)$ the circle of radius r, centered around the point $\boldsymbol{x}_0$. Without any loss of generalization, we assume that the circle of radius R is centered around $\mathbf{0}$. If the dots are uniformly spaced within $C(\mathbf{0}, R)$, the probability of finding k dots within a circle $C(\boldsymbol{x}, r_0)$ of radius r_0, strictly included in $C(\mathbf{0}, R)$, is

$$\begin{aligned}&\Pr\{k \text{ dots out of } n \text{ lie in } C(\boldsymbol{x}, r_0)\}\\ &= \binom{n}{k}\left(\frac{\pi r_0^2}{\pi R^2}\right)^k \left(1-\frac{\pi r_0^2}{\pi R^2}\right)^{n-k} := \binom{n}{k} p^k (1-p)^{n-k},\end{aligned} \tag{10.3}$$

where $p = r_0^2/R^2$ is the probability of finding a dot within a circle of radius r_0. The above formula is independent of the position of $C(\boldsymbol{x}, r_0)$, provided that $C(\boldsymbol{x}, r_0)$ is included in $C(\mathbf{0}, R)$, or, in formulas, $C(\boldsymbol{x}, r_0) \subseteq C(\mathbf{0}, R)$. In the limit, as R and n go to infinity, but the ratio $\rho := n/\pi R^2$ remains constant, if $k \ll n$ and $p \ll 1$ (i.e., $r_0 \ll R$), the Bernoulli distribution (10.3) tends to the Poisson distribution (10.4) (see, e.g. [30] for the equivalent derivation in the time domain):

$$\begin{aligned}&\mathcal{P}\{k \text{ dots out of } n \text{ lie in } C(\boldsymbol{x}, r_0)\}\\ &\approx \frac{n(n-1)\cdots(n-k+1)}{k!}\left(\frac{\rho\pi r_0^2}{n}\right)^k e^{-\rho\pi r_0^2 (n-k)/n} \approx \frac{(\rho\pi r_0^2)^k}{k!} e^{-\rho\pi r_0^2},\end{aligned} \tag{10.4}$$

having set $p = \rho\pi r_0^2/n$ and having used the approximations $1 - x \approx e^{-x}$ and $(n-k)/n \approx 1$.

Using a rigorous definition, denoting with $d\boldsymbol{S}$ an infinitesimal region of the space $\mathcal{R}^2$, with $|d\boldsymbol{S}|$ the area of $d\boldsymbol{S}$, and with $N(d\boldsymbol{S})$ the number of points falling in $d\boldsymbol{S}$, the point process N is a *homogeneous* Poisson point process, with density ρ, if

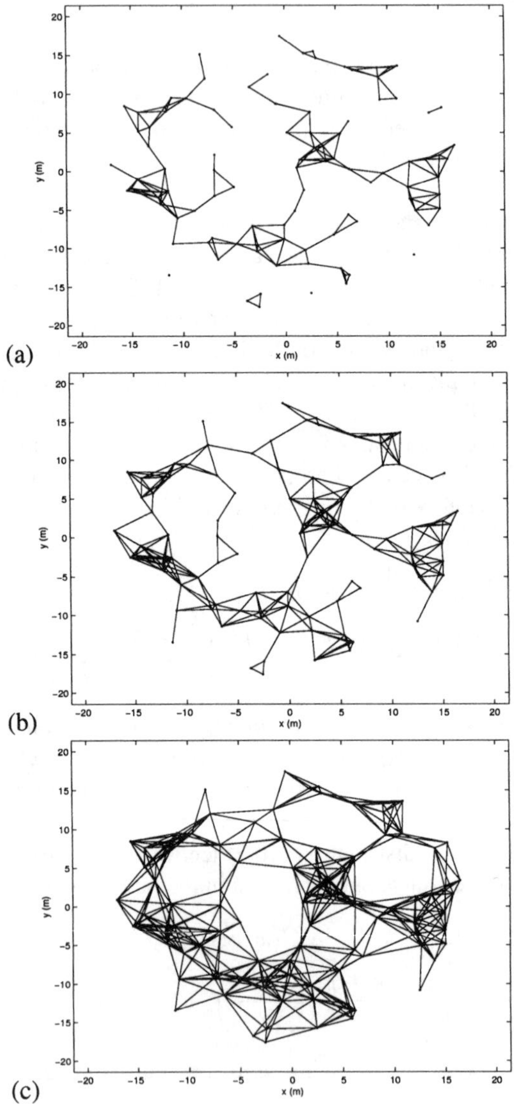

Figure 10.3 Examples of network connectivity: (a) degree 0; (b) degree 1; and (c) degree 2.

1. $\lim_{|ds| \to 0} \frac{1-\Pr(N(d\boldsymbol{S})=0)}{|d\boldsymbol{S}|} = \rho;$

2. $\lim_{|ds| \to 0} \frac{\Pr(N(d\boldsymbol{S})=1)}{|d\boldsymbol{S}|} = \rho;$

3. For any pair of disjoint domains $d\boldsymbol{S}_1$ and $d\boldsymbol{S}_2$, $N(d\boldsymbol{S}_1)$ and $N(d\boldsymbol{S}_2)$ are statistically independent.

If ρ is allowed to be a function of space [i.e., $\rho = \rho(\boldsymbol{S})$] the Poisson process is *inhomogeneous*. In a $2D$ space, the spatial equivalent of "white noise" is *complete spatial randomness*. This property indicates the absence of any spatial structure of the point process. For some authors, complete spatial randomness is synonymous of a *homogeneous* Poisson process.

In the following sections, we will study wireless networks assuming that the network nodes are distributed as a homogeneous Poisson point process.

One of the fundamental questions in a wireless network is what is the amount of power that each node has to transmit to guarantee that the information signals reach the intended receiver, from any point in the network to any other node. The *critical* minimum power guaranteeing connectivity was derived by Gupta and Kumar [3], in asymptotic sense, as the number of nodes tends to infinity. The analysis carried out in [3] assumes a random distribution of n points within a disk of unit area and shows how the coverage radius of each point must vary with n to guarantee the connectivity with probability one. In this chapter, we consider a different approach. We assume a set of points randomly distributed over an infinite region, with assigned density. Each node has fixed coverage radius, independent of n. Furthermore, we require that the connectivity is ensured with a prescribed probability (not necessarily one). We review first some results on the connectivity of geometric random graphs, derived in [31] (see also [32]).

One of the most important properties for our ensuing derivations is that the pdf of the distance between a point and its nearest neighbor in a homogeneous Poisson process is a Rayleigh pdf. In formulas, denoting with Ξ the random variable indicating the distance between one node and its nearest neighbor, the pdf of Ξ is [28]

$$p_{\Xi}(\xi) = 2\pi\rho\xi e^{-\rho\pi\xi^2}. \tag{10.5}$$

Starting from (10.5), the probability that a node is isolated is

$$P_{\text{isolated}} = 1 - P\{\xi \leq r_0\} = e^{-\pi\rho r_0^2}. \tag{10.6}$$

The probability that there are no isolated nodes is, approximately

$$p_{\text{con}} \simeq (1 - e^{-\pi\rho r_0^2})^n. \tag{10.7}$$

Within the validity of the approximation leading to (10.7), inverting (10.7) we find the coverage radius r_0 that ensures that the degree of the network is greater than zero:

$$r_0 = \sqrt{\frac{-\log\left(1 - p_{con}^{1/n}\right)}{\pi\rho}}. \tag{10.8}$$

Of course, this same radius does not guarantee the connectivity, as the network could be composed of isolated clusters, even if with no isolated nodes. However, the result (10.2) guarantees that, asymptotically, as n tends to infinity, the radius r_0 in (10.8) tends to coincide with the radius that guarantees also the connectivity. Hence, the value r_0 in (10.8) denotes, approximately, the minimum coverage that each node has to provide in order to guarantee the connectivity of the whole graph, with a given probability.

A test of validity of the previous approximations is reported in Figure 10.4, that shows p_{con} as a function of r_0, as given in (10.7) (solid line), and the probability that the network is connected (dashed line), estimated over 200 independent realizations of random geometric graphs composed of 1000 points distributed over a toroidal surface, to avoid border effects. We can see a rather good agreement between theory and simulation. The slight difference is due to the fact that the independence assumption is not exactly valid and that (10.2) is valid only for n going to infinity. It is interesting to notice, from Figure 10.4, the rapid change from probability zero to probability one. This is indeed a characteristic of random geometric graphs, that is reminiscent of phase transitions in chemistry.

If we wish to increase the fault tolerance of the graph, we need to increase the connectivity order. It was shown in [31] that the probability that a graph is k-connected is approximately equal to the probability that the minimum degree of the graph is at least k; that is

$$\Pr_k := \Pr\{G \text{ is } k\text{−connected}\} \simeq \Pr\{d_{min} \geq k\} = \left[1 - \sum_{i=0}^{k-1} \frac{(\rho\pi r_0^2)^i}{i!} e^{-\rho\pi r_0^2}\right]^n. \tag{10.9}$$

This expression does not admit an inverse in closed form, but it is certainly invertible as it is a monotonic increasing function of r_0. The value of r_0 providing the desired value $\mathcal{P}_k$, to be found numerically, guarantees the k-connectivity, within the limits of approximations in (10.9).

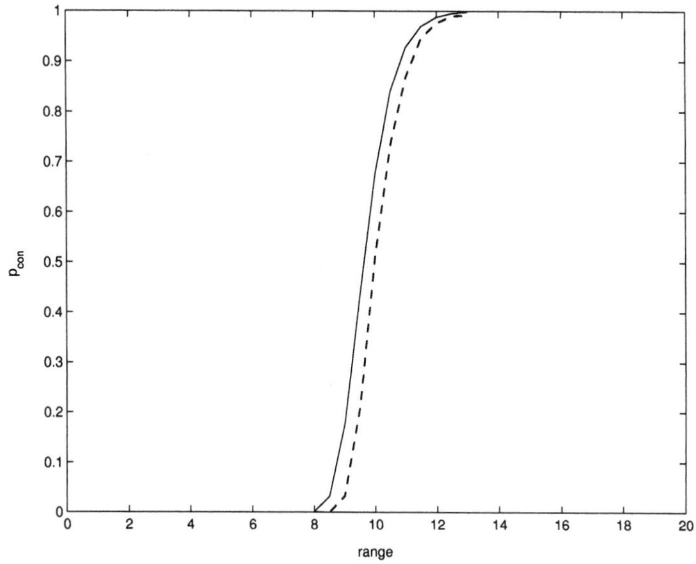

Figure 10.4 Probability that a random geometric graph is connected: theory (solid line) and simulation results (dashed line).

10.3 CONNECTIVITY OF A WIRELESS NETWORK

In a wireless network, besides the position of the radio nodes, there is one more source of randomness: the fading of the radio links. We compute now the connectivity of a wireless network modeling the network as a geometric random graph incorporating the channel statistics. In particular, instead of choosing the coverage radius r_0 arbitrarily, as in the previous section, we choose the value of r_0 that guarantees a certain reliability of each radio link. We start with single antenna transceivers and then we will extend the analysis to multiantenna terminals in the next sections. To simplify the theoretical analysis, we assume that all links are interference-free and that the channels between different pairs of nodes are independent Rayleigh flat-fading channels, with variance σ_h^2. This means that the system is implicitly allocating orthogonal channels to different links, with a consequent rate loss.

10.3.1 Connectivity of a SISO Flat-Fading Network

We say that two nodes are linked to each other if the bit error rate on their link is smaller than a given target value P_{target} with a given out-of-service probability

P_{out}. Using QAM constellations, the bit error rate is

$$P_e = c_M \; Q\left(\sqrt{g_M \frac{\mathcal{E}_b}{\sigma_n^2}|h|^2}\right) \leq c_M \; e^{-g_M \mathcal{E}_b |h|^2/\sigma_n^2}. \tag{10.10}$$

where $\mathcal{E}_b$ is the energy per bit, σ_n^2 is the noise variance, h is the flat-fading coefficient, and c_M and g_M are two coefficients that depend on the order M of the QAM constellation as follows [33]

$$\begin{aligned} c_M &= 4\frac{\sqrt{M}-1}{\sqrt{M}\cdot\log_2 M}, \\ g_M &= \frac{3}{M-1}\log_2 M. \end{aligned} \tag{10.11}$$

Exploiting the upper bound in (10.10), we can upper-bound the out-of-service probability as

$$P_{\text{out}} = \Pr\{P_e > P_{\text{target}}\} \leq \Pr\{c_M e^{-g_M \mathcal{E}_b |h|^2/\sigma_n^2} > P_{\text{target}}\}. \tag{10.12}$$

Since the channel is Rayleigh, $|h|^2$ is an exponential random variable. Hence P_{out} can be upper-bounded as

$$P_{\text{out}} \leq 1 - e^{-\sigma_n^2 \log(c_M/p_{\text{emax}})/(g_M \mathcal{E}_b \sigma_h^2)}. \tag{10.13}$$

We say that a link is reliable, and it is then established, if the out-of-service event occurs with a probability smaller than a given value. We assume that $\sigma_h^2 = 1/r^\alpha$. The exponent α depends on the environment where the propagation takes place. Typically, α is between two and five. Setting $\sigma_h^2 = 1/r^\alpha$ in (10.13) and inverting (10.13), we find the coverage radius

$$r_{\text{cov}} = \left[-\frac{g_M \mathcal{E}_b \log(1-P_{\text{out}})}{\sigma_n^2 \log(c_M/P_{\text{target}})}\right]^{1/\alpha}. \tag{10.14}$$

Since P_{out} is typically small, we can use the approximation $\log(1-P_{\text{out}}) \approx P_{\text{out}}$ to rewrite (10.14) as

$$r_{\text{cov}} \simeq \left[\frac{g_M \mathcal{E}_b P_{\text{out}}}{\sigma_n^2 \log(c_M/P_{\text{target}})}\right]^{1/\alpha}. \tag{10.15}$$

Equating (10.14) to (10.8), we get the relationship between the node density and the transmitted power necessary to insure the network connectivity, for a given number of nodes n. For example, in the case where $\alpha = 2$, we have

$$\frac{\mathcal{E}_b}{\sigma_n^2} \geq \frac{1}{\rho} \frac{\log(1 - p_{\text{con}}^{1/n}) \log(c_M / P_{\text{target}})}{\pi g_M \log(1 - P_{\text{out}})}. \tag{10.16}$$

In a sensor network scenario or in the deployment of an ad hoc network, (10.16) can be rewritten as the minimum node density ρ that guarantees the connectivity, for a given set of sensors having a specified transmitted power

$$\rho \geq \frac{\sigma_n^2}{\mathcal{E}_b} \frac{\log(1 - p_{\text{con}}^{1/n}) \log(c_M / P_{\text{target}})}{\pi g_M \log(1 - P_{\text{out}})}. \tag{10.17}$$

10.3.2 Connectivity of a MIMO Flat-Fading Network

We show now how the connectivity improves if the radio nodes have multiple antennas. In such a case, the network may benefit from the diversity gain. To make a fair comparison with the single antenna case seen before, denoting with n_T the number of transmit/receive antennas, we set the power transmitted by each antenna equal to p_T / n_T, where p_T is the overall transmit power.

Using an $n_T \times n_T$ MIMO system, using a space-time coding technique capable of achieving full diversity, the error probability is

$$P_e = c_M \, Q\left(\sqrt{g_M \frac{\mathcal{E}_b}{\sigma_n^2 n_T} \sum_{i=1}^{n_T^2} |h_i|^2}\right) \leq c_M \, e^{-g_M \mathcal{E}_b z / \sigma_n^2 n_T}, \tag{10.18}$$

having introduced, in the last approximation, the random variable $z := \sum_{i=1}^{n_T^2} |h_i|^2$. The out-of-service probability is then

$$P_{\text{out}} \leq \Pr\left\{ z \leq \frac{\sigma_n^2 n_T \log(c_M / P_{\text{target}})}{g_M \mathcal{E}_b} \right\} = D_Z\left(\frac{\sigma_n^2 n_T \log(c_M / P_{\text{target}})}{g_M \mathcal{E}_b} \right),$$

where $D_Z(z)$ is the cumulative distribution function (CDF) of z. If the channels are Rayleigh fading, independent, and all with the same variance σ_h^2, the CDF of z is

$$D_Z(z) = 1 - e^{-z/\sigma_h^2} \sum_{k=0}^{n_T^2 - 1} \left(\frac{z}{\sigma_h^2} \right)^k \frac{1}{k!}. \tag{10.19}$$

For small values of z, more specifically for $z \ll \sigma_h^2$, $D_Z(z)$ can be approximated as

$$D_Z(z) \approx \left(\frac{z}{\sigma_h^2} \right)^{n_T^2} \frac{1}{n_T^2!}. \tag{10.20}$$

For the sake of finding closed form, albeit approximated, expressions, it is useful to introduce the normalized random variable $x = z/\sigma_h^2$, whose CDF is

$$D_X(x) = 1 - e^{-x} \sum_{k=0}^{n_T^2-1} x^k \frac{1}{k!} \approx \frac{x^{n_T^2}}{n_T^2!}. \tag{10.21}$$

Repeating the same kind of derivations as in the SISO case, the out-of-service probability for the MIMO case can be written as

$$P_{\text{out}} \leq D_X \left(\frac{\sigma_n^2 n_T \log(c_M/p_{\text{target}})}{g_M \mathcal{E}_b \sigma_h^2} \right). \tag{10.22}$$

From (10.22), setting $\sigma_h^2 = 1/r^\alpha$, we can derive the coverage of each node[1]

$$r_{\text{cov}} = \left[\frac{g_M \mathcal{E}_b}{\sigma_n^2 n_T \log(c_M/p_{\text{target}})} D_X^{-1}(P_{\text{out}}) \right]^{1/\alpha}. \tag{10.23}$$

To derive an approximate closed form expression for r_{cov}, since we are interested in small values of the out-of-service probability, we can use the approximation (10.21) to invert $D_X(x)$. The result is

$$r_{\text{cov}} \simeq \left[\frac{g_M \mathcal{E}_b}{\sigma_n^2 n_T \log(c_M/p_{\text{target}})} (n_T^2! P_{\text{out}})^{1/n_T^2} \right]^{1/\alpha}. \tag{10.24}$$

Equating (10.23) to (10.8), we get the minimum transmit power guaranteeing the connectivity of a MIMO network

$$\frac{\mathcal{E}_b}{\sigma_n^2} = \frac{n_T \log(c_M/p_{\text{target}})}{g_M (n_T^2! P_{\text{out}})^{1/n_T^2}} \left[\frac{-\log\left(1 - p_{\text{con}}^{1/n}\right)}{\pi\rho} \right]^{\alpha/2}. \tag{10.25}$$

As a numerical example, in Figure 10.5 we show the density ρ, as a function of the transmitted energy per bit, normalized to the noise power, for different numbers of antennas per terminal. For a fair comparison, all curves refer to the same overall transmitted power. The constellation is QPSK. The overall number of transmit/receive antennas is also the same in all cases. More specifically, we used the following combinations: $n = 100$ and $n_T = 1$ (dotted line), $n = 50$ and $n_T = 2$ (dashed line), and $n = 25$ and $n_T = 4$ (solid line). The connectivity is insured

1 $D_X^{-1}(x)$ denotes the inverse of $D_X(x)$ and it certainly exists because in this case $D_X(x)$ is strictly monotone.

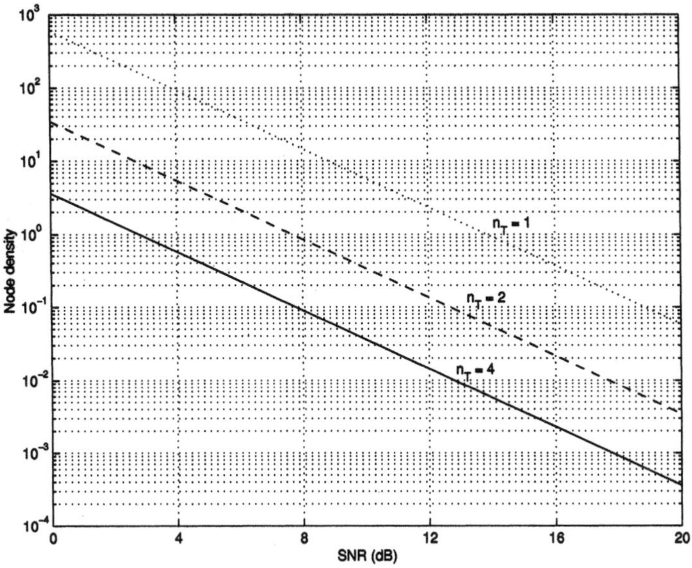

Figure 10.5 Node density (number of nodes per area unit) versus $\mathcal{E}_b/\sigma_n^2$, for different number of antennas per terminal.

with probability $p_{\text{con}} = 0.99$ and the out-of-service event refers to a target BER of 10^{-3} and it is required to occur with a maximum time percentage of $P_{\text{out}} = 10^{-2}$. We can see, from Figure 10.5, the advantage of diversity that makes possible a considerable decrease of the nodes density. Clearly, terminals with multiple antennas provide a larger coverage than single antenna terminals, but at the cost of increased complexity.

The connectivity is also a function of the bit rate. As an example, in Figure 10.6 we show the minimum SNR required for connectivity, assuming an efficiency of 2, 4, and 6 bits/sec/Hz, achieved using 4, 16, or 64-QAM constellations. As expected, an increase of rate requires an increase of node density, for a given power budget. Hence, the bit rate may result as a compromise between the number of antennas, node density, energy per node, and complexity.

10.3.3 Cooperative Communications

The next question is "What can we do to improve the connectivity if we have only single antenna transceivers?". We can resort to cooperation among nearby terminals. The idea is pictorially sketched in Figure 10.7, where we see three nodes, A, B,

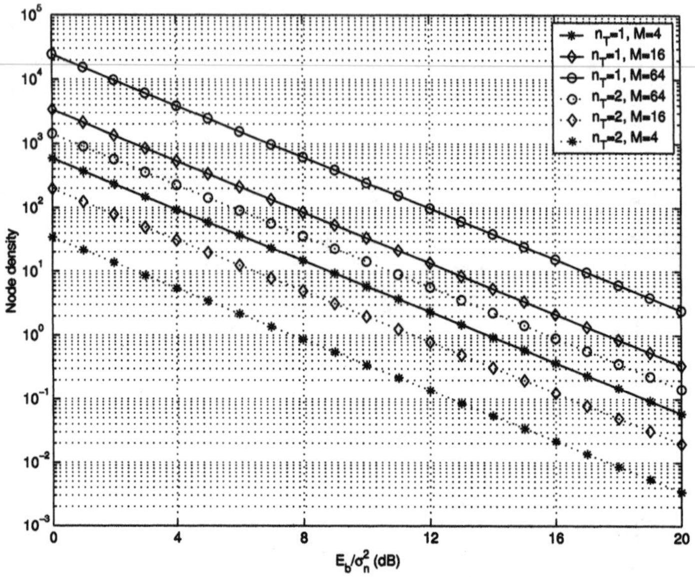

Figure 10.6 Node density (number of nodes per area unit) versus $\mathcal{E}_b/\sigma_n^2$, for different number of antennas per terminal ($n_T = 1$ or 2) and different transmission rates ($M = 4, 16, 64$).

and C. In the absence of any cooperation, each node covers a circle centered on its location, of radius r_0 given by (10.14). For example, in Figure 10.7, A and B are connected, but C is isolated. However, if nodes A and B cooperate, they can give rise to a *virtual* transmit array capable of covering an area larger than the one covered with a single antenna. The idea is represented, pictorially, in Figure 10.7 (b), where the bigger circle is the area covered by a system located in the center of gravity of the nodes A and B, with a bigger radius resulting from the use of a MISO system with two transmit and one receive antenna. Proceeding as in the previous section, denoting with n_{relay} the number of cooperating (relay) nodes, we have the potential of diversity gain $n_{\text{relay}} + 1$ (the relays plus the source itself). The existence of the bigger circle is a result of the cooperation between A and B. Thanks to cooperation, a disconnected network may become connected, as shown in the example of Figure 10.7 (b), using the same overall transmit power.

Increase of connectivity

Let us now quantify how much is the coverage increase, due to cooperation, and how this affects the connectivity. If we have n_{relay} relays cooperating with a source and a destination with one receive antenna, the (maximum) diversity gain

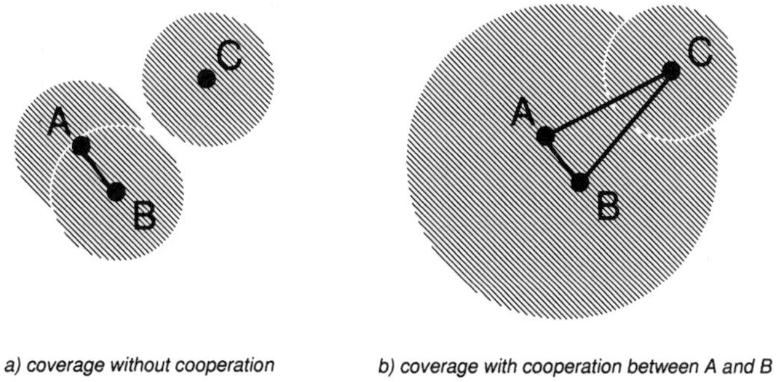

Figure 10.7 Coverage in cooperating networks.

is $n_T := n_{\text{relay}} + 1$. Repeating derivations similar to the ones described in the previous section, with the only exception that now we only have transmit diversity, but no receive diversity because each receiver has a single antenna, the coverage radius in case of cooperation is

$$\begin{aligned} r_{\text{coop}} &\simeq \left[\frac{g_M \mathcal{E}_b}{\sigma_n^2 n_T \log(c_M/p_{\text{target}})} (n_T! P_{\text{out}})^{\frac{1}{n_T}} \right]^{\frac{1}{\alpha}} = \left[\frac{(n_T!)^{\frac{1}{n_T}}}{n_T P_{\text{out}}^{(n_T-1)/n_T}} \right]^{\frac{1}{\alpha}} r_{\text{cov}} \\ &:= \beta r_{\text{cov}}, \end{aligned} \tag{10.26}$$

with r_{cov} given by (10.15). Therefore, the coverage increases by a factor β that depends on the number of cooperating nodes and on the desired out-of-service probability.

The effect of the coverage increase on the network connectivity is illustrated in Figure 10.8, where we report the connection probabilities obtained without cooperation (dashed line) and with cooperation (solid line). In case of cooperation, we considered only the case of no more than two cooperating terminals. The probabilities shown in Figure 10.8 have been estimated over a set of 200 independent network realizations. As a comparison term, we report in Figure 10.8 the connection probability of a non-coperative network, but having a coverage radius βr_{cov} (dotted line). We can see that cooperation between pairs of radio nodes is sufficient to yield an SNR gain of approximately seven dB. Interestingly, the curve obtained without cooperation, but with a coverage radius βr_{cov} has approximately the same connectivity as the cooperative case, where the coverage of each node is r_{cov}. Recalling, from (10.26), that the transmitted power is proportional to r^2, if r is the coverage, we infer that cooperation among pairs of terminals yields an improvement in terms

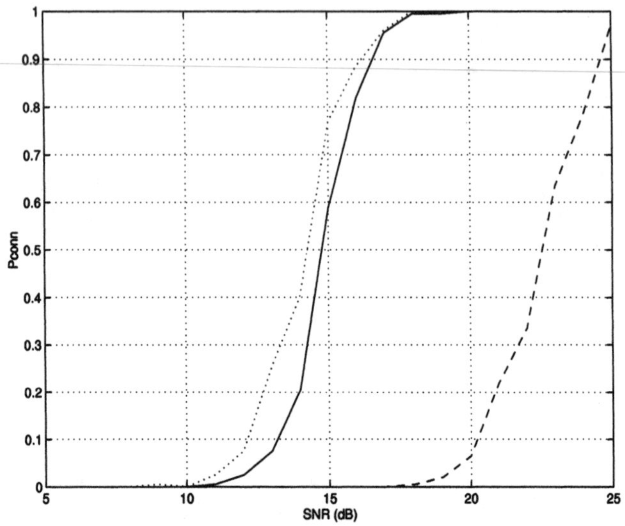

Figure 10.8 Connection probabilities versus SNR (dB): without cooperation (dashed line), with cooperation (solid line), and without cooperation, but using $\beta r_{\rm cov}$ instead of $r_{\rm cov}$ (dotted line).

of transmitted power approximately equal to β^2, when $n_T = 2$.

Average BER

In the previous section, the existence of radio nodes available for relaying was given for granted. In practice, this availability depends on the relay location with respect to the source, on the density of stand-by nodes having no traffic to transmit, and on the power transmitted by the source. In this section, we model the availability of a node to act as a relay as a random event and we compute the average BER, taking into account both the channel randomness and the probability of finding relays.

We consider a source node S wishing to communicate with a node D, using, possibly, relaying. We denote with $p_{r_0}(k)$ the probability for S to find k relays within a distance r_0 and with $P_e(k+1; \boldsymbol{h})$ the error probability corresponding to a multiantenna transmit system having $k+1$ antennas (the k relays' antennas plus the source itself), conditioned to a set of channels $\boldsymbol{h}$. The error probability at the destination can be written as a sum of the error probabilities $P_e(k+1; \boldsymbol{h})$, weighted by the probability of finding k relays $p_{r_0}(k)$

$$P_e(\boldsymbol{h}) \simeq \sum_{k=0}^{\infty} p_{r_0}(k) P_e(k+1; \boldsymbol{h}). \tag{10.27}$$

This formula is approximated as it assumes that there are no errors at the relay nodes. However, the goodness of this approximation can be kept under control by adopting a relay discovery strategy such that a node is chosen as a relay only if its distance from the source S is less than the value r_0 given in (10.14). This guarantees that the error probability of a relay node is, with probability $1 - P_{\text{out}}$, smaller than a given threshold P_{target}. Assuming a spatial Poisson distribution for the relays, the probability $p_{r_0}(k)$ of finding k relays is then given by (10.4), with r_0 given by (10.14). If the final destination has n_R receive antennas, the error probability $P_e(k+1; \boldsymbol{h})$ is[2]

$$P_e(k+1; \boldsymbol{h}) = c\,Q\left(\sqrt{g \frac{\mathcal{E}_b}{\sigma_n^2(k+1)} \sum_{i=1}^{(k+1)n_R} |h_i|^2}\right). \tag{10.28}$$

Assuming that the channels are statistically independent, the expected value of $P_e(k+1) := E_h\{P_e(k+1; \boldsymbol{h})\}$ is

$$\begin{aligned} P_e(k+1) &= \frac{4\sqrt{M}-1}{\sqrt{M}\log_2(M)} \left(\frac{1-\mu}{2}\right)^{(k+1)n_R} \\ &\cdot \sum_{m=0}^{(k+1)n_R-1} \binom{(k+1)n_R + m - 1}{m} \left(\frac{1+\mu}{2}\right)^m, \end{aligned} \tag{10.29}$$

where

$$\mu := \sqrt{\frac{3\mathcal{E}_b \log_2(M)\sigma_h^2}{3\mathcal{E}_b \log_2(M)\sigma_h^2 + 2(M-1)(k+1)\sigma_n^2}}. \tag{10.30}$$

The average error probability, in case of cooperation is then

$$\bar{P}_e = \sum_{k=0}^{\infty} p_{r_0}(k) P_e(k+1), \tag{10.31}$$

with $p_{r_0}(k)$ and $P_e(k+1)$ given by (10.4) and (10.29), respectively.

2 In case of coordinated transmission from $k+1$ nodes, we normalize the transmit power of each node by $k+1$, so that the overall radiated power is independent of k.

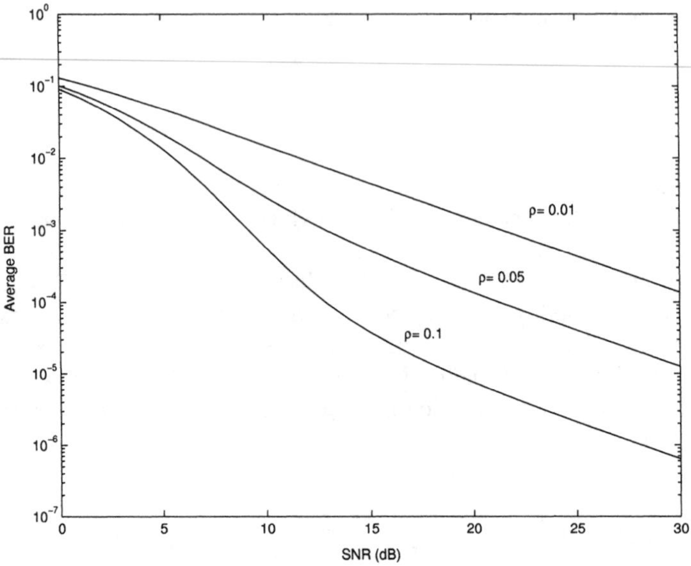

Figure 10.9 Average BER at the final destination in a cooperating network, for three different node densities.

An example of performance is shown in Figure 10.9, where we report the average BER (10.31) obtained with three different density values. Clearly, as SNR tends to infinity, the diversity is given by the worst case (i.e., by the term with $k = 0$ in (10.27), which, at high SNR, it behaves as 1/ SNR). Hence, *there is no diversity gain* when we take into account the probability of finding relays. However, as we can see from Figure 10.9, there is a considerable coding gain.

Cooperation coding gain

The asymptotic behavior of $\bar{P}_e$ (10.31), at high SNR, is determined by the worst case, i.e., the term with $k = 0$ in (10.31). The asymptotic behavior of the average BER is then

$$\bar{P}_e \propto p_{r_0}(0)\frac{1}{\text{SNR}} = \frac{1}{G_c\text{SNR}}. \tag{10.32}$$

The quantity

$$G_c = e^{\rho\pi r_0^2} \tag{10.33}$$

introduced in (10.32) is the *cooperation coding gain*.

This gain can be controlled by acting on the product ρr_0^2. For a given density ρ, from (10.14) we see that we can increase r_0 and then improve the gain by increasing the transmit power or decreasing the bit rate. This implies that, for a given node density and transmit power, one has to decrease the transmission rate. Alternatively, for a given power and rate, one has to increase the node density (if this is a controllable variable, as in the deployment of a sensor network, for example).

To illustrate the behavior of the coding gain with the transmission rate, we report in Figure 10.10 the coding gain as a function of the number of bits n_b per symbol, for an SNR of 30 dB. The out-of-service probability is $P_{\text{out}} = 10^{-2}$ and it is referred to a maximum error rate of $P_{\text{target}} = 10^{-3}$. We can see that the coding gain decreases, as n_b increases.

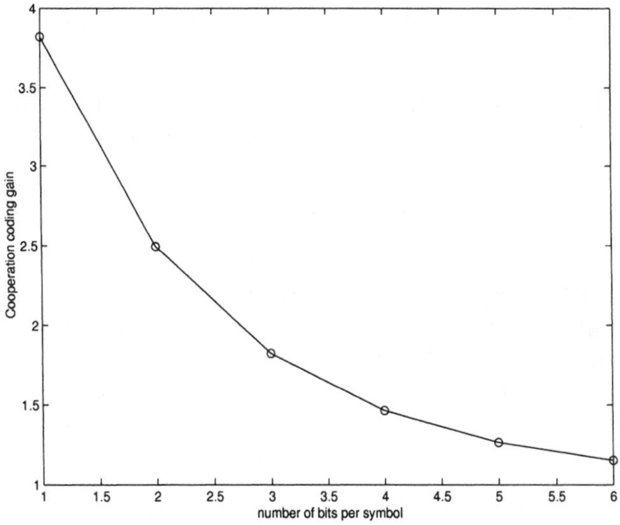

Figure 10.10 Cooperation coding gain versus rate (bits pcu).

10.4 COOPERATION PROTOCOLS

After having derived the potential gains achievable through cooperation, it is now time to describe how these advantages can be obtained. From what we have seen in Chapters 7 and 8, space-time coding provides a powerful framework for achieving the desired diversity and/or multiplexing gain in multiantenna systems. Combining now the cooperative idea with the space-time coding algorithms, we realize that the cooperation gain can be achieved by making the cooperating nodes coordinate their transmission strategies as if they were the antennas of a single system adopting space-time coding. This form of cooperation gives rise to the so called *distributed space-time coding* (DSTC), as space-time coding is distributed among cooperating radio nodes. The main difference of DSTC with respect to STC is that the transmit array, as well as the receive array, are *virtual*, in the sense that they are the result of cooperation. In other words, if two or more nodes are able to exchange their data at sufficiently low BER, they can, subsequently, act as if they were the antennas of a single source having a multiantenna transmitter. The same arguments can be applied at the reception side.

In establishing a cooperative communication, we need to decide first the role of each relay. A relay could work according to the following strategies:

- *Amplify and forward ($A\&F$):* The relay amplifies and forwards the received packets;
- *Decode and forward ($D\&F$):* The relay decodes the symbols in the received packets and retransmits them, irrespective of its own BER;
- *Selective decode and forward ($S/D\&F$):* The relay decodes the symbols in the received packets and retransmits them only if its own BER is below a given threshold.

The $A\&F$ scheme can be very useful to simplify the implementation of the relay, because in an $A\&F$ system the relay needs to have only an antenna and an RF amplifier. However, an $A\&F$ transmits also part of its own noise. $S/D\&F$ is the best scheme, in terms of performance, but it requires the relay to know if its decision on the received symbols is correct or not. This requires the encoding of the packets transmitted by the source towards the relays with error detection coding of sufficient redundancy. This makes the system more robust, but it has a price in terms of bit rate. In this chapter, we will consider only the $D\&F$ strategy. The difference between amplify and forward and decode and forward strategies was considered, for example, in [21].

Within the context of $D\&F$ relaying, the basic differences between DSTC and STC are listed below:

- *Errors in the relay nodes:* In STC systems, of course there are no errors in sending bits to different transmit antennas; conversely, in DSTC systems operating in a D&F mode, there might be decision errors at the relays;
- *Power allocation among different nodes:* In STC systems, all transmit antennas transmit with the same power; conversely, in DSTC, since the antennas (radio

nodes) are not co-located, it is advisable to use different power from different antennas, depending on the nodes location as well as on the nodes reliability (i.e., on the BER);

- *Synchronization:* In STC systems, the symbols transmitted from different antennas arrive at the same time at the receiver[3]. In DSTC systems, since the transmitting antennas are not co-located, even if the transmitters are synchronous, the packets coming from source and relay may arrive at the final destination at different times;
- *Time scheduling:* To allow for cooperative communications, the source has to send its data to the relay first; this requires the introduction of time slots dedicated to source-relay communications.

The introduction of the source-relay time slot yields an inevitable rate loss. To limit such a loss, it is advisable to assign the same time slot to more than one source-relay set.

In general, a reasonable strategy for selecting potential relays is that a source and its relays should be as close as possible. At the same time, a relay associated to a given source should be as far as possible from the relays associated to other sources. This is justified by the following concurring reasons: 1) less power is wasted in the source-relay link; 2) there are less synchronization problems in the final link towards the destination; and 3) there is less interference in the time slot where more sources send their data to their own relays.

In summary, the relay discovery phase follows a strategy that gives rise to many spatially separated microcells where each source acts as a local base station broadcasting to its relays, who may get interference from other microcells (sources). A typical situation is sketched in Figure 10.11, where we see: a destination (D) node, in the center, several relay nodes (black dots), and three sources ('o'). The circle around each source denotes the region containing the potential relays for that source. As shown in Figure 10.11, different sources may find different numbers of relays.

Cooperative transmissions proceed through a series of phases:

- Relay discovery (RD);
- Transmission from source to relays (S2R);

3 To be exact, in principle there is a delay, but it is too small to be detected.

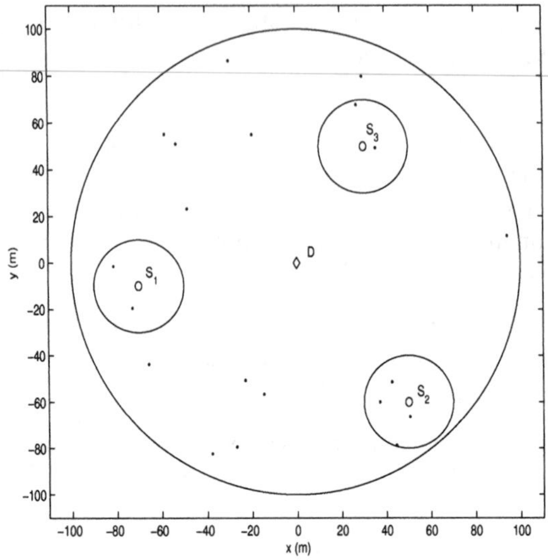

Figure 10.11 Geometry of cooperative network with three source nodes (circles), potential relays (dots) and one destination (D).

- Coordinated transmission from source and relays to the destination (SR2D).

All these steps require a proper resource allocation (RA) strategy. Let us now analyze these steps, starting with the RA phase that encompasses all the others.

10.4.1 Resource Allocation (RA)

We may distinguish between *coordinated* and *uncoordinated* multihop networking. In coordinated multihop networking there are distinct time slots dedicated to different phases (i.e., RD, S2R, and SR2D transmission). Conversely, in an uncoordinated strategy, the different phases occur without having any specific resource allocated for relaying. Here, we will concentrate on coordinated networking. The two basic resources are power and channels. We have to decide how much of the available power can be spent to find out a relay and how much can be used to send data to the relays. Furthermore, we have to decide which is the most appropriate multiple access strategy. To avoid an excessive waste of resources, as proposed in [20], it is useful to reuse the same time slot for the simultaneous exchange of data between each source and its own relays. Then, to avoid interference at the final destination, we can assign orthogonal channels to different links from the source-relays sets to the final destination. The situation is depicted in Figure 10.12, where there is

$S_1 \rightarrow R_{11}, R_{12} \ldots$ $S_2 \rightarrow R_{21}, R_{22} \ldots$ ...	$S_1, R_{11}, R_{12} \ldots \rightarrow D$	$S_2, R_{21}, R_{22} \ldots \rightarrow D$	...

Figure 10.12 Time schedule of a relaying network based on TDD.

one common initial time slot dedicated to the link between each source and its own relays, followed by a series of time slots, each one allocated to a distinct set of source-relays. This is only an example of time-division duplexing (TDD), but clearly one could use other multiple access schemes.

In the next section we will show how to balance the transmit power between S and R in order to minimize the BER at the final destination.

10.4.2 Resource Discovery (RD)

A source wishing to find relays starts sending sounding signals to verify whether there are available neighbors. The sounding signal might be a pseudonoise code identifying the source. A potential relay may receive the sounding signals from more than one source. The radio nodes available to act as relays compute the signal-to-noise plus interference ratio (SNIR) for each source. This step requires the node to be able to separate the signals coming from different sources. This is made possible by the use of orthogonal codes. The potential relays retransmit an acknowledgment signal back only to those sources whose SNIR exceeds a certain threshold. Each source receives the acknowledgments and the estimate of the SNIR from its potential relays and it decides which relays to use. This phase insures that the relay, once chosen, is sufficiently reliable. Given the variability of the wireless channel, this operation has to be repeated at least once every channel coherence time. To avoid excessive complications, it is simpler to assume that a node may act as a relay for no more than one source.

10.4.3 Source to Relays (S2R) Transmission

Once a source has found its own relays, it transmits its information to them in a dedicated S2R slot. To limit the rate reduction resulting from the insertion of this slot, each source would like to use a constellation order as high as possible, in this slot, in order to reduce its duration. On the other hand, as already illustrated in the previous section, increasing the bit rate determines a reduction of the coverage, for a given source power budget and a required out-of-service probability at the relay. Limiting the coverage excessively would reduce considerably the probability of finding a relay and then it would nullify the whole relaying process. Hence, the choice of the constellation order has to result as a trade-off between rate reduction and probability of finding a relay.

10.4.4 Source-Relays to Destination (SR2D) Transmission

Finally, each set of source and relative relays transmit toward the destination using dedicated resources (e.g., time slots, frequency bands, or codes) so that SR2D transmissions from different sets do not interfere with each other. Each SR2D transmission occurs using a distributed space-time coding strategy.

10.5 ALTERNATIVE DISTRIBUTED STC STRATEGIES

In selecting the right distributed space-time coding, we can choose among the many STC techniques illustrated in Chapter 7. DSTC has received great attention through the last year. In particular, Laneman and Wornell showed how to get the maximum diversity gain out of cooperation, comparing different relaying strategies [7, 8]. They showed under which condition it is advantageous to pass through a relay terminal. The decision errors at the relay node were incorporated explicitly in [7], where it was showed that the optimal decision rule consists in combining properly the data from S with the data coming from R, properly *clipped* in amplitude, with a clipping level depending on the error rate on the S-R link. The coding strategy proposed in [7, 8] was essentially a time repetition code, as D was supposed to receive data from S and R separately. Instead of time, one could use the frequency domain to separate the data coming from R and S, but in any case it was supposed in [7, 8] to have two dedicated (orthogonal) channels for the link from S, R, and D. Clearly, this implies a loss in terms of rate. Hunter and Nosratinia proposed a different cooperation strategy in [11, 12], valid for a different set-up, where both S and R are two nodes that have their own data to transmit.

A different approach to achieve the maximum diversity gain, without suffering from the rate loss of repetition coding, consists in merging the idea of cooperation with space-time coding, giving rise to the so called *distributed* space-time coding, where S and R share their antennas to create a *virtual* transmit array. This idea was suggested by several authors [13, 16, 17, 18]. In all these works, the relay was assumed to be error-free, meaning that a radio node acts as a relay only after verifying that there are no errors in the decoded bits. However, this assumption requires the introduction of sufficient redundant coding and it might be too restrictive. In fact, a limited number of decoding errors might be well tolerated by the DSTC system that could still take advantage out of the cooperation. To check this possibility, a different approach was pursued in [22], where the decoding errors at the relay were specifically taken into account. In [22], it was also shown how to improve the system performance by incorporating the knowledge of the BER at the relay in the detector present at the final destination. A comparison of alternative DSTC techniques was carried out in [23].

In general, the choice of the right space-time coding technique depends on several factors. Primarily, from Chapter 7, we can choose among the following classes of techniques: 1) orthogonal STC (OSTC), to maximize the diversity gain and minimize the receiver complexity; 2) full-rate/full-diversity codes (FRFD), to maximize both diversity gain and transmission rate, but with a rather high receiver complexity; 3) BLAST codes, to maximize the rate, sacrificing part of the diversity gain, but with intermediate receiver complexity; and 4) trace-orthogonal design (TOD), as a flexible way to trade complexity, bit rate, and bit error rate.

What it is useful to emphasize here is that the optimal trade-off among these alternatives, in the distributed case, does not coincide, necessarily, with the conventional space-time coding case.

In this section, we start comparing alternative coding strategies, in terms of rate and BER. We will assume here that the channels are all flat-fading. In the next section, we will extend the analysis to frequency-selective channels.

We denote with T_{S2R} and T_{SR2D} the duration of the S2R and SR2D time slots. T_s is the symbol duration in all slots. For a given bit rate, the time slot duration depends on the constellation order used in that slot. We denote with Q and M the constellation orders used in the S2R and in the SR2D slots, respectively. We assume also that the S2R slot is shared among N_p source-relay pairs. The frame containing both S2R and the N_p SR2D links has then a duration $T_F = T_{S2R} + N_p T_{SR2D}$. The rate reduction factor, with respect to the noncooperative case, in a TDMA context, is then

$$\eta = \frac{N_p T_{\mathrm{SR2D}}}{T_{\mathrm{S2R}} + N_p T_{\mathrm{SR2D}}}. \tag{10.34}$$

The rate loss can be reduced by decreasing T_{S2R} (i.e., by increasing Q with respect to M) or by increasing N_p. In the first case, the relay needs a higher SNIR, to decode the higher dimension constellation; in the second case there is a SNIR increase at the relay, because of additional interference. In both cases, it is less likely to discover a relay with sufficient SNIR. Hence, the right choice has to result from a trade-off between rate and performance. We discuss now in detail some alternative DSTC choices. In all cases, $s(n)$ denotes the sequence of symbols sent by S to the relay, whereas $\hat{s}(n)$ indicates the estimate of $s(n)$ performed at the relay. For simplicity, we consider the presence of one relay only.

10.5.1 Distributed Orthogonal STC (D-OSTC)

D-OSTC guarantees maximum receiver simplicity and full diversity and it can be implemented also when the final destination has a single antenna. D-OSTC transmits 2 symbols over two successive time periods, so that $T_{\text{SR2D}} = 2T_s$ and $T_{\text{S2R}} = 2\log_2(M)T_s/\log_2(Q)$. The sequence transmitted by the source-relay pair is

$$\begin{bmatrix} s(n) & -s^*(n+1) \\ \hat{s}(n+1) & \hat{s}^*(n) \end{bmatrix}. \tag{10.35}$$

The first row of this matrix contains the symbols transmitted by the source, whereas the second row refers to the symbols transmitted by the relay; different columns refer to successive time instants. The overall bit rate, incorporating the rate loss due to the insertion of the S2R slot, is

$$R = \frac{2N_p \log_2 M}{2N_p + 2\log_2 M/\log_2 Q} \text{ b/s/Hz}. \tag{10.36}$$

10.5.2 Distributed Full-Rate/Full-Diversity (D-FRFD)

If the final destination has two antennas, the cooperation among the nodes establishes a virtual 2×2 MIMO, which, in principle, is capable of a rate increase. This can be achieved, for example, using distributed-FRFD (D-FRFD) or distributed-BLAST (DiBLAST)[4]. With D-FRFD, the pair S-R transmits four symbols over two consecutive time periods. The transmitted matrix is

$$\begin{bmatrix} s(n) + \varphi s(n+1) & \theta\left(s(n+2) + \varphi s(n+3)\right) \\ \theta\left(\hat{s}(n+2) - \varphi\hat{s}(n+3)\right) & \hat{s}(n) - \varphi\hat{s}(n+1) \end{bmatrix}, \tag{10.37}$$

where $\varphi = e^{j/2}$, $\theta = e^{j/4}$ are two rotation parameters (see, e.g. [34] or [35], for the choice of φ and θ). The bit rate is

$$R = \frac{4N_p \log_2 M}{2N_p + 4\log_2 M/\log_2 Q} \text{ b/s/Hz}. \tag{10.38}$$

4 We use different acronyms here to distinguish distributed-BLAST (DiBLAST) from diagonal BLAST (D-BLAST).

10.5.3 Distributed BLAST (DiBLAST)

We consider here the version of BLAST where two independent streams of data are transmitted from the two antennas. In its distributed version, DiBLAST requires that the relay receives only half of the bits to be transmitted. This implies an advantage with respect to both D-OSTC and D-FRFD, as it allows us to reduce the duration of the S2R time slot. The price paid with respect to D-FRFD is that DiBLAST is not full-diversity. The transmitted matrix in the DiBLAST case, is

$$\begin{bmatrix} s(n) & s(n+2) \\ \hat{s}(n+1) & \hat{s}(n+3) \end{bmatrix} \tag{10.39}$$

and the bit rate is

$$R = \frac{4N_p \log_2 M}{2N_p + 2\log_2 M/\log_2 Q} \text{ b/s/Hz.} \tag{10.40}$$

Comparing the transmission rates of all the distributed schemes, for a given choice of the constellation orders Q and M, we see that DiBLAST has the highest transmission rate (as DiBLAST does not send all the data to the relay).

10.5.4 Performance Comparison

We compare now the performance of the previous distributed coding schemes, incorporating a conventional noncooperative system as a reference benchmark. To make a fair comparison, we enforce all systems to transmit with the same overall energy. More specifically, if $\mathcal{E}$ is the energy radiated by the noncooperative case, we denote with $\alpha\mathcal{E}$ the energy radiated by S in the S2R slot and with $(1-\alpha)\mathcal{E}/2$ the energy radiated by S and R in the SR2D slot, with $\alpha < 1$[5]. The overall radiated energy is then always $\mathcal{E}$. In all the simulations, we assume that the constellation used in the S2R link is 16-QAM (i.e., $Q = 16$). In Figure 10.13, we report the average BER, averaged over $16,000$ independent channel realizations. Solid lines refer to the noncooperative case, whereas dashed lines refer to the cooperative scenario. The number of source-relay pairs is $N_p = 10$. The destination has two real antennas. To have a global view of both BER and information bit rate, in Figure 10.14, we report, for the same cases analyzed in Figure 10.13, the average information rate computed by averaging the capacity of the equivalent binary channel over the channel statistics. We recall that the capacity of a binary symmetric channel,

5 We assume here that S and R are sufficiently close to each other that their channels towards the destination have the same statistics. Later on, we will show how to optimize the power distribution between source and relay when the statistics are different.

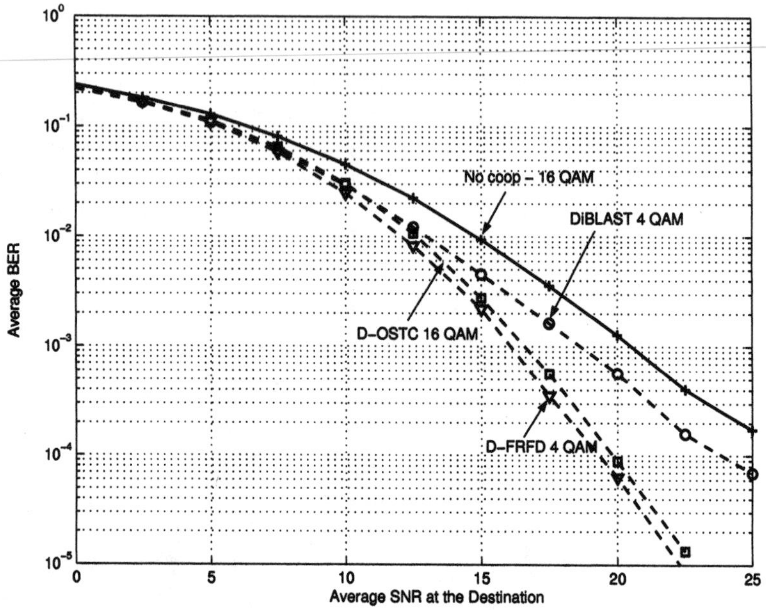

Figure 10.13 Final average BER: comparison of alternative coding strategies.

characterized by bit error rate $P_e(\boldsymbol{h})$, is [36]

$$C(\boldsymbol{h}) = 1 + P_e(\boldsymbol{h}) \log_2[P_e(\boldsymbol{h})] + [1 - P_e(\boldsymbol{h})] \log_2[1 - P_e(\boldsymbol{h})]. \tag{10.41}$$

The information rate, incorporating the rate loss resulting from the insertion of the S2R slot, is then

$$R(\boldsymbol{h}) := \eta C(\boldsymbol{h}), \tag{10.42}$$

with η given by (10.34). The result shown in Figure 10.14 is the average of $R(\boldsymbol{h})$ over 16,000 independent channel realizations. The results shown in Figures 10.13 and 10.14 refer to the situation where the density of the relay nodes is high enough to guarantee that the probability of finding one relay is one. The SNIR at each relay is fixed and equal to 20 dB. The percentage of energy devoted to the S2R link is $\alpha = 0.1$.

Observing Figures 10.13 and 10.14, we see that D-FDFR, with 4-QAM, achieves the best performance in terms of BER. D-OSTC, with 16-QAM, is very close to D-FDFR. DiBLAST is worse than D-FDFR, as it cannot achieve full diversity, however, it is a little better in terms of bit rate, because it suffers from less insertion

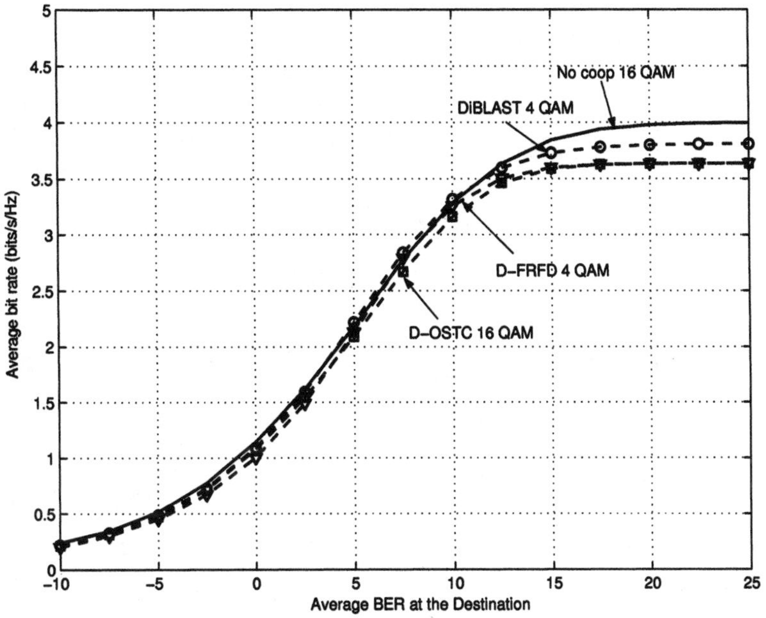

Figure 10.14 Average information rate as a function of SNR.

losses.

The results shown before assumed that the source was always able to find a relay with sufficient SNIR. In a more realistic scenario, the fading in the S2R link and the interference at the relay might prevent the source from finding a relay with sufficient SNIR. In such a case, the source would transmit without relaying. In general, the probability of finding a relay depends on both the relay and source spatial densities. The effect of having a relay discovery probability different from one, in a multiuser context, depending on the required SNIR at the relay, was studied in [23]. We report some results in Figure 10.15, where we show the average BER as a function of the SNR, at the destination. The scenario analyzed in this figure is composed of 10 active sources and 190 potential relays, all scattered uniformly and independently of each other within a circular cell of radius 300m. Requiring, as an example, a SNIR of 12.5 dB, the probability of finding one relay turns out to be $p_{r_0}(1) = 0.72$, whereas at 15 dB, we have $p_{r_0}(1) = 0.65$. We observe from Figure 10.15 the presence of a floor on the average BER, due to the intermediate decision errors at the relay nodes. The floor decreases as the target SNIR, at the relay, increases. We can check from Figure 10.15 that indeed, increasing the SNIR from 12.5 to 15, even

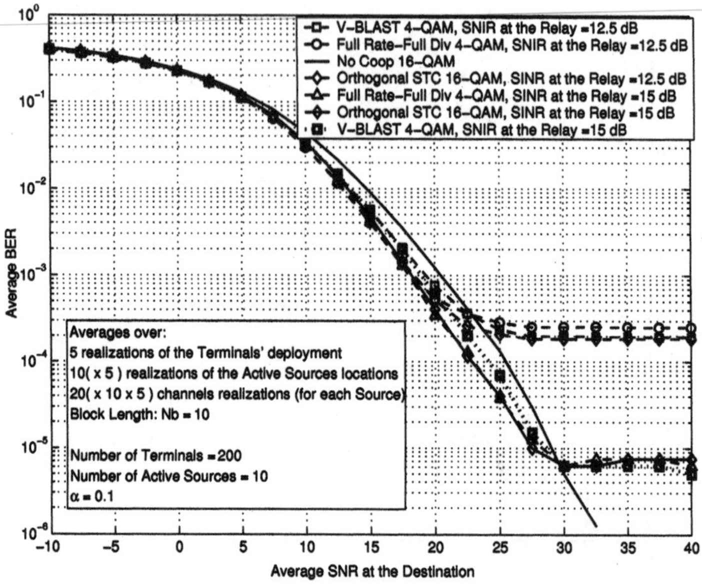

Figure 10.15 Final average BER incorporating the possibility of not finding a relay.

though $p_{r_0}(1)$ decreases, the floor on the BER decreases by more than a decade. However, we cannot choose an excessively high target SNIR because, otherwise, the probability of finding a relay would tend to zero, thus avoiding all the potential advantages of cooperative communications. Finally, in the case of a SNIR of 15 dB, we can observe a gain of approximately 3 dB at $BER = 10^{-4}$.

10.6 DISTRIBUTED ORTHOGONAL SPACE-TIME BLOCK CODING FOR FREQUENCY-SELECTIVE CHANNELS

We consider now in detail a distributed STC based on orthogonal coding, valid in the general case of frequency-selective channels. We incorporate the decision errors at the relay explicitly, both in the performance analysis as well as in the design of the final decision rule. We assume, for simplicity, that the source has only one relay. In such a case, a good choice for the D-OSTC scheme is a block Alamouti scheme, as suggested in [17, 22]. As in Chapter 8, we make the following assumptions: (a1) the channels are FIR of order L_h and time-invariant over at least a pair of consecutive blocks; (a2) the channel coefficients are zero mean complex Gaussian random variables; (a3) the transmission scheme for all terminals is a block strategy, where each block is composed of M symbols, incorporating a cyclic prefix of length L_h, at the beginning of each block, so as to simplify the elimination of interblock interference and channel equalization; (a4) the information symbols are i.i.d. BPSK and each symbol may assume the values A or $-A$ with equal probability[6]; (a5) the received data are degraded by additive white Gaussian noise (AWGN), denoted by the vector $\boldsymbol{w}$; and (a6) all radio nodes are synchronous. We start describing the distributed space-time protocol assuming that each terminal is equipped with one antenna; then we will extend the scheme to the case of multiple receive antennas at the destination. In the following, we will use the following notation. We denote with $h_{sd}(i)$, $h_{sr}(i)$, and $h_{rd}(i)$, the impulse responses between S and D, S and R, and R and D, respectively.

In the S2R time slot, S send blocks of symbols $\boldsymbol{s}(i)$ to the relay, which decodes them and generates the symbol vector $\hat{\boldsymbol{s}}(i)$. Then, in the successive SR2D time slot, S and R transmit simultaneously, using a block Alamouti's strategy. More specifically, in the first half of the SR2D time slot, S transmits $\boldsymbol{x}_s(i+2) = \alpha_1 \boldsymbol{F}\boldsymbol{s}(i)$ and R transmits $\boldsymbol{x}_r(i+2) = \alpha_2 \boldsymbol{F}\hat{\boldsymbol{s}}(i+1)$. In the second half, S transmits $\boldsymbol{x}_s(i+3) = \alpha_1 \boldsymbol{G}\boldsymbol{s}^*(i+1)$ while R transmits $\boldsymbol{x}_r(i+3) = \alpha_2 \boldsymbol{G}\hat{\boldsymbol{s}}^*(i)$. The two real coefficients α_1 and α_2 are related to each other by $\alpha_1^2 + \alpha_2^2 = 1$. They are introduced in order to have a degree of freedom in the power distribution between S and R, under a given total transmit power. In Section 10.6.3, we will show how to choose α_1 (and then α_2) in order to minimize the final average BER. To guarantee maximum spatial diversity, the two matrices $\boldsymbol{G}$ and $\boldsymbol{F}$ are related to each other by $\boldsymbol{G} = \boldsymbol{J}\boldsymbol{F}^*$, as shown in Chapter 8, where $\boldsymbol{J}$ is a time reversal (plus a one chip cyclic shift when N is even) matrix.

The blocks received by D in the two consecutive time-slots, $i+2$ and $i+3$, after discarding the CP, are

$$\begin{aligned} \boldsymbol{y}(i+2) &= \alpha_1 \boldsymbol{H}_{sd}\boldsymbol{F}\boldsymbol{s}(i) + \alpha_2 \boldsymbol{H}_{rd}\boldsymbol{F}\hat{\boldsymbol{s}}(i+1) + \boldsymbol{v}(i+2) \\ \boldsymbol{y}(i+3) &= \alpha_1 \boldsymbol{H}_{sd}\boldsymbol{G}\boldsymbol{s}^*(i+1) - \alpha_2 \boldsymbol{H}_{rd}\boldsymbol{G}\hat{\boldsymbol{s}}^*(i) + \boldsymbol{v}(i+3), \end{aligned} \tag{10.43}$$

where $\boldsymbol{H}_{sd}$ and $\boldsymbol{H}_{rd}$ refer to the channels between S and D and between R and D, respectively. Exploiting, as in Chapters 3 and 8, the diagonalizations $\boldsymbol{H}_{sd} = \boldsymbol{W}\boldsymbol{\Lambda}_{sd}\boldsymbol{W}^H$ and $\boldsymbol{H}_{rd} = \boldsymbol{W}\boldsymbol{\Lambda}_{rd}\boldsymbol{W}^H$, if we pre-multiply $\boldsymbol{y}(i+2)$ by $\boldsymbol{W}^H$ and

6 This assumption is made only for simplifying our derivations, but there is no restriction to use higher order constellations.

$\boldsymbol{y}^*(i+3)$ by $\boldsymbol{W}^T$, we get

$$\begin{aligned}\boldsymbol{W}^H\boldsymbol{y}(i+2) &= \alpha_1\boldsymbol{\Lambda}_{sd}\tilde{\boldsymbol{F}}\boldsymbol{s}(i)+\alpha_2\boldsymbol{\Lambda}_{rd}\tilde{\boldsymbol{F}}\hat{\boldsymbol{s}}(i+1)+\boldsymbol{w}(i+2)\\ \boldsymbol{W}^T\boldsymbol{y}^*(i+3) &= \alpha_1\boldsymbol{\Lambda}_{sd}\tilde{\boldsymbol{G}}^*\boldsymbol{s}(i+1)-\alpha_2\boldsymbol{\Lambda}_{rd}\tilde{\boldsymbol{G}}^*\hat{\boldsymbol{s}}(i)+\boldsymbol{w}^*(i+3),\end{aligned} \tag{10.44}$$

where $\tilde{\boldsymbol{F}} := \boldsymbol{W}^H\boldsymbol{F}$, $\tilde{\boldsymbol{G}} := \boldsymbol{W}^H\boldsymbol{G}$, $\boldsymbol{w}(i+2) := \boldsymbol{W}^H\boldsymbol{v}(i+2)$, and $\boldsymbol{w}^*(i+3) := \boldsymbol{W}^T\boldsymbol{v}^*(i+3)$. For the sake of simplicity, we assume that OFDM is performed at both S and R nodes, so that $N = M$, $\boldsymbol{F} = \boldsymbol{W}$. With this choice, we have $\tilde{\boldsymbol{F}} = \boldsymbol{I}_N$ and $\boldsymbol{G} = \boldsymbol{W}$. We also introduce the orthogonal matrix

$$\boldsymbol{\Lambda} := \begin{pmatrix} \alpha_1\boldsymbol{\Lambda}_{sd} & \alpha_2\boldsymbol{\Lambda}_{rd} \\ -\alpha_2\boldsymbol{\Lambda}_{rd} & \alpha_1\boldsymbol{\Lambda}_{sd}\end{pmatrix} \tag{10.45}$$

such that

$$\boldsymbol{\Lambda}^H\boldsymbol{\Lambda} = \begin{pmatrix} \alpha_1^2|\boldsymbol{\Lambda}_{sd}|^2+\alpha_2^2|\boldsymbol{\Lambda}_{rd}|^2 & \boldsymbol{0} \\ \boldsymbol{0} & \alpha_1^2|\boldsymbol{\Lambda}_{sd}|^2+\alpha_2^2|\boldsymbol{\Lambda}_{rd}|^2\end{pmatrix} := \boldsymbol{I}_2\otimes\bar{\boldsymbol{\Lambda}}^2, \tag{10.46}$$

where $\bar{\boldsymbol{\Lambda}}^2 := \alpha_1^2|\boldsymbol{\Lambda}_{sd}|^2+\alpha_2^2|\boldsymbol{\Lambda}_{rd}|^2$, whereas $\otimes$ denotes the Kronecker product. We introduce also the unitary matrix[7] $\boldsymbol{Q} := \boldsymbol{\Lambda}\,(\boldsymbol{I}_2\otimes\bar{\boldsymbol{\Lambda}}^{-1})$, satisfying the relationships $\boldsymbol{Q}^H\boldsymbol{Q} = \boldsymbol{I}_{2N}$ and $\boldsymbol{Q}^H\boldsymbol{\Lambda} = \boldsymbol{I}_2\otimes\bar{\boldsymbol{\Lambda}}$. Exploiting the above equalities and multiplying the vector

$$\boldsymbol{u} := \begin{bmatrix}\boldsymbol{W}^H\boldsymbol{y}(i+2)\\ \boldsymbol{W}^T\boldsymbol{y}^*(i+3)\end{bmatrix}$$

by the matrix $\boldsymbol{Q}^H$ [8], we get

$$\begin{aligned}\begin{bmatrix}\boldsymbol{r}(i)\\ \boldsymbol{r}(i+1)\end{bmatrix} &:= \boldsymbol{Q}^H\boldsymbol{u}\\ &= \begin{bmatrix}|\tilde{\boldsymbol{\Lambda}}_{sd}|^2 & -\tilde{\boldsymbol{\Lambda}}_{sd}^*\tilde{\boldsymbol{\Lambda}}_{rd}\\ \tilde{\boldsymbol{\Lambda}}_{sd}\tilde{\boldsymbol{\Lambda}}_{rd}^* & |\tilde{\boldsymbol{\Lambda}}_{sd}|^2\end{bmatrix}\begin{bmatrix}\boldsymbol{s}(i)\\ \boldsymbol{s}(i+1)\end{bmatrix}\\ &+ \begin{bmatrix}|\tilde{\boldsymbol{\Lambda}}_{rd}|^2 & \tilde{\boldsymbol{\Lambda}}_{sd}^*\tilde{\boldsymbol{\Lambda}}_{rd}\\ -\tilde{\boldsymbol{\Lambda}}_{sd}\tilde{\boldsymbol{\Lambda}}_{rd}^* & |\tilde{\boldsymbol{\Lambda}}_{rd}|^2\end{bmatrix}\begin{bmatrix}\hat{\boldsymbol{s}}(i)\\ \hat{\boldsymbol{s}}(i+1)\end{bmatrix}+\bar{\boldsymbol{w}},\end{aligned} \tag{10.47}$$

where $\tilde{\boldsymbol{\Lambda}}_{sd} := \alpha_1\boldsymbol{\Lambda}_{sd}\bar{\boldsymbol{\Lambda}}^{-1/2}$, $\tilde{\boldsymbol{\Lambda}}_{rd} := \alpha_1\boldsymbol{\Lambda}_{rd}\bar{\boldsymbol{\Lambda}}^{-1/2}$, $\bar{\boldsymbol{w}} := [\bar{\boldsymbol{w}}^T(i),\,\bar{\boldsymbol{w}}^T(i+1)]^T = \boldsymbol{Q}^H[\boldsymbol{w}^T(i+2),\,\boldsymbol{w}^H(i+3)]^T$.

7 We suppose that the channels do not share common zeros on the grid $z_q = e^{j2\pi q/N}$, with q integer, so that $\bar{\boldsymbol{\Lambda}}$ is invertible.

8 This operation does not affect the optimality of the decision since $\boldsymbol{Q}$ is unitary.

The previous equations reduce to the classical block Alamouti equations (see, e.g., Chapter 8), when $\alpha_1 = \alpha_2$, that is when the two antennas transmit with the same power and when $\hat{\boldsymbol{s}}(n) \equiv \boldsymbol{s}(n), n = i, i+1$, i.e., there are no decision errors at the relay node.

Since $\boldsymbol{Q}^H$ is unitary, if $\boldsymbol{w}$ is white, $\bar{\boldsymbol{w}}$ is also white, with covariance matrix $\boldsymbol{C}_w = \sigma_n^2 \boldsymbol{I}_{2N}$. Furthermore, since all matrices $\boldsymbol{\Lambda}$ appearing in (10.47) are diagonal, the system (10.47) of $2N$ equations can be decoupled into N independent systems of two equations in two unknowns, each equation referring to one sub-carrier. More specifically, introducing the vectors

$$\begin{aligned} \boldsymbol{r}_k &:= \begin{bmatrix} r_k(i) \\ r_k(i+1) \end{bmatrix}, \boldsymbol{s}_k := \begin{bmatrix} s_k(i) \\ s_k(i+1) \end{bmatrix}, \\ \hat{\boldsymbol{s}}_k &:= \begin{bmatrix} \hat{s}_k(i) \\ \hat{s}_k(i+1) \end{bmatrix}, \bar{\boldsymbol{w}}_k := \begin{bmatrix} \bar{w}_k(i) \\ \bar{w}_k(i+1) \end{bmatrix}, \end{aligned} \tag{10.48}$$

referring to the k-th sub-carrier, with $k = 0, \ldots, N-1$ (for simplicity of notation, we drop the block index and we set $\tilde{\Lambda}_{sd} = \tilde{\Lambda}_{sd}(k,k)$ and $\tilde{\Lambda}_{rd} = \tilde{\Lambda}_{rd}(k,k)$), (10.47) is equivalent to the following systems of equations

$$\boldsymbol{r}_k = \begin{pmatrix} |\tilde{\Lambda}_{sd}|^2 & -\tilde{\Lambda}_{sd}^* \tilde{\Lambda}_{rd} \\ \tilde{\Lambda}_{sd} \tilde{\Lambda}_{rd}^* & |\tilde{\Lambda}_{sd}|^2 \end{pmatrix} \boldsymbol{s}_k + \begin{pmatrix} |\tilde{\Lambda}_{rd}|^2 & \tilde{\Lambda}_{sd}^* \tilde{\Lambda}_{rd} \\ -\tilde{\Lambda}_{sd} \tilde{\Lambda}_{rd}^* & |\tilde{\Lambda}_{rd}|^2 \end{pmatrix} \hat{\boldsymbol{s}}_k + \bar{\boldsymbol{w}}_k. \tag{10.49}$$

Since the noise vector $\bar{\boldsymbol{w}}_k$ is also white with covariance matrix $\boldsymbol{C}_w = \sigma_n^2 \boldsymbol{I}_{2N}$, and there is no inter-symbol interference (ISI) between the vectors $\boldsymbol{s}_k$ and $\boldsymbol{r}_k$ corresponding to different sub-carriers, $\boldsymbol{r}_k$ represents a sufficient statistic for the decision on the transmitted symbols vector $\boldsymbol{s}_k$.

Since S and R are not co-located, the blocks transmitted from S and R arrive at D at different times. This is a specific difference of distributed space-time block coding with respect to the classical distributed space-time coding techniques. However, if the difference in arrival times τ_d is incorporated in the CP used from both S and R, D is still able to get N samples from each received block, without inter-block interference (IBI). In such a case, the different arrival time does not cause any trouble to the final receiver. In fact, let us take as the reference starting instant the time of arrival of the i-th block from R. If the block coming from S arrives with a delay of L_d samples, thanks to the block spectral processing operated at the receiver, the only difference with respect to the case of perfect synchronization is that the transfer function $\tilde{\Lambda}_{sd}(k)$ in (10.49) is substituted by $\tilde{\Lambda}_{sd}(k)e^{-j2\pi L_d k/N}$. But, from (10.49), it is clear that such a substitution does not affect the useful term, as it only affects the interfering term. In fact, in the hypothesis of Rayleigh fading channel, $\tilde{\Lambda}_{sd}(k)$ is statistically indistinguishable from $\tilde{\Lambda}_{sd}(k)e^{-j2\pi L_d k/N}$. Hence, the combination of Alamouti (more generally, orthogonal STC) and OFDM is robust with respect to lack of synchronization between the time of arrivals of packets from S and from R, as long as the CP incorporates also the relative delay between the packets arriving at D from S and R. The price paid for this property is an efficiency loss. This loss becomes negligible when S and R are close to each other, because the relative delay diminishes.

10.6.1 ML Detector

We derive now the structure of the maximum likelihood (ML) detector, under (a3). We denote with $\mathscr{S}$ the set of all possible transmitted vectors $\boldsymbol{s}_k$ and with $p_{e1}(k)$ and $p_{e2}(k)$ the conditional (to a given channel realization) error probabilities, at the relay node, on $s_k(1)$ and $s_k(2)$, respectively. After detection, at the node R, we have $\hat{s}_k(l) = s_k(l)$, with probability $(1 - p_{el}(k))$, or $\hat{s}_k(l) = -s_k(l)$, with probability $p_{el}(k), l = 1, 2$. Since the symbols are independent, the probability density function of the received vector $\boldsymbol{r}_k$, conditioned to having transmitted $\boldsymbol{s}_k$, is [22]

$$
\begin{aligned}
f_{\boldsymbol{r}_k|\boldsymbol{s}_k}(\boldsymbol{r}_k|\boldsymbol{s}_k) &= \frac{1}{\pi^2\sigma_n^2}\Big\{[1 - p_{e1}(k)][1 - p_{e2}(k)]\, e^{-|\boldsymbol{r}_k - \boldsymbol{A}_k(1,1)\boldsymbol{s}_k|^2/\sigma_n^2} \\
&+ p_{e1}(k)p_{e2}(k)\, e^{-|\boldsymbol{r}_k - \boldsymbol{A}_k(-1,-1)\boldsymbol{s}_k|^2/\sigma_n^2} \\
&+ [1 - p_{e1}(k)]p_{e2}(k), e^{-|\boldsymbol{r}_k - \boldsymbol{A}_k(1,-1)\boldsymbol{s}_k|^2/\sigma_n^2} \\
&+ p_{e1}(k)[1 - p_{e2}(k)]\, e^{-|\boldsymbol{r}_k - \boldsymbol{A}_k(-1,1)\boldsymbol{s}_k|^2/\sigma_n^2}\Big\}, \qquad (10.50)
\end{aligned}
$$

where $\boldsymbol{A}_k(m, n)$ is defined as follows

$$
\boldsymbol{A}_k(m,n) := \begin{bmatrix} \alpha_1|\tilde{\Lambda}_s|^2 + \alpha_2|\tilde{\Lambda}_r|^2 m, & \alpha_2\tilde{\Lambda}_s^*\tilde{\Lambda}_r n - \alpha_1\tilde{\Lambda}_s^*\tilde{\Lambda}_r \\ \alpha_1\tilde{\Lambda}_s\tilde{\Lambda}_r^* - \alpha_2\tilde{\Lambda}_s\tilde{\Lambda}_r^* m, & \alpha_1|\tilde{\Lambda}_s|^2 + \alpha_2|\tilde{\Lambda}_r|^2 n \end{bmatrix}
$$

with $m, n = \pm 1$. Based on (10.50), the ML detector is

$$
\hat{\boldsymbol{s}}_k = \arg\max_{\boldsymbol{s}_k \in \mathscr{S}} \left\{ f_{\boldsymbol{r}_k|\boldsymbol{s}_k}(\boldsymbol{r}_k|\boldsymbol{s}_k) \right\}. \qquad (10.51)
$$

The ML detector (10.51) assumes that the vector of error probabilities $p_{e_1}(k)$ and $p_{e_2}(k)$, $k = 0, \ldots, N - 1$, occurring at the relay are known at the destination side. This requires an exchange of information between R and D. This information has to be updated with a rate depending on the channel coherence time. We will show later on an alternative (sub-optimum) detection scheme that does not require such a knowledge.

In deriving the optimal receiver (10.51), we assumed that D processes only the vectors received in the SR2D time-slot. If D were in a listening mode also during the

S2R time slot, it was shown in [22] that the system could achive a further diversity gain, in case of fast-varying channel.

10.6.2 Suboptimum Detector

The ML detector described above assumes the knowledge, at the destination node, of the set of error probabilities $p_{e1}(k)$ and $p_{e2}(k)$, with $k = 0, \ldots, N-1$. If this knowledge is not available, a sub-optimum scalar detector can be implemented, instead of the ML detector. More specifically, the decision on the transmitted symbol $s_k(n)$ can be simply obtained as

$$\hat{\boldsymbol{s}}(n) = \text{sign}\left\{\Re[\boldsymbol{r}(n)]\right\}, \quad n = i, i+1, \tag{10.52}$$

where $\boldsymbol{r}(n)$ is given by (10.47). Note that, for high SNR at the relay (i.e., when R makes no decision errors), the symbol-by-symbol decision in D becomes optimal and, thus, the decoding rule (10.52) provides the same performance of the optimal receiver (10.51). However, when the decision errors at the relay side cannot be neglected, the sub-optimal receiver introduces a floor in the bit-error-rate (BER) curve, because the symbol-by-symbol decision (10.52) treats the wrong received symbols as interference. The choice between the decoding rules (10.51) and (10.52) should then result as a trade-off between performance and computational complexity, taking into account the need, for the ML detector, to make available, at the destination node, the error probabilities of the relay node. We will show a comparison between ML and sub-optimum strategies in the next section.

10.6.3 Power Allocation

In this section, we show how to allocate the available power optimally between source and relay. We will provide a closed form analysis in the ideal case where there are no decision errors at the relay and then we will show some performance results concerning the real case where the errors are taken into account.

In case of no decoding errors at the relay, from (10.49) the SNR_k on the k-th bit of the received block is

$$\text{SNR}_k = \frac{A^2}{\sigma_n^2}\left[\alpha|\Lambda_{sd}(k)|^2 + (1-\alpha)|\Lambda_{rd}(k)|^2\right], \tag{10.53}$$

having set $\alpha = \alpha_1^2$ (and thus $1 - \alpha = \alpha_2^2$). The BER averaged over the channel realizations is [36]

$$P_b = \frac{1}{2}\sum_{k=1}^{2}\pi_k\left[1 - \sqrt{\frac{\gamma_k}{1+\gamma_k}}\right] \tag{10.54}$$

with

$$\gamma_1 := \frac{A^2}{\sigma_n^2} \frac{\alpha\tilde{\sigma}_h^2}{d_{sd}^r}, \ \gamma_2 := \frac{A^2}{\sigma_n^2} \frac{(1-\alpha)\tilde{\sigma}_h^2}{d_{rd}^r}.$$

The maximum available diversity gain is two. It is straightforward to show that, if D is equipped with n_R antennas, the maximum diversity gain would be $2n_R$. In [22], it was shown that, in case of fast fading, if D is in a listening mode also in the S2R time slot, the maximum diversity gain is $3n_R$.

We show now how to distribute a given power budget between S and R in order to minimize the BER, depending on the relative distances between S, R and D. As a numerical example, in Figure 10.16 we report the average BER versus α, for different values of the distance d_{rd} (and thus of SNR_R) between R and D (all distances are normalized with respect to the distance d_{sd} between S and D). In Figure 10.16a), we consider the ideal case where there are no errors at the relay node. The SNR_D at the final destination is fixed and equal to 10 dB. We can observe that, when $d_{sd} = d_{rd} = 1$, the value of α that minimizes the average BER is $\alpha = 0.5$, i.e., the two transmitter use the same power. However, as R gets closer to D (i.e., $d_{rd} < 1$), the optimal α tends to increase, i.e., the system allocates more power to S, with respect to R. The reverse happens when S is closer to D, than R (i.e., $d_{rd} > 1$). Thus, as expected, the system tends to put, somehow, S and R in similar conditions with respect to D, to get the maximum diversity gain.

The real case, where there are decision errors at the relay, is reported in Figure 10.16b), where the average BER is again plotted as a function of α, but for different values of the SNR_R at the relay node. We can observe that, as SNR_R decreases, the system tends to allocate less power to the relay node (the optimal value of α is greater than 0.5), as the relay node becomes less and less reliable. We compare now the alternative cooperative approaches described above. We assume a block length $N = 32$ and channel order $L = 6$. To make a fair comparison of the alternative transmission schemes, we enforce all systems to transmit with the same overall power. More specifically, if $\mathcal{P}$ is the total power radiated by the non cooperative case, we denote by $\mathcal{P}_I$ the power radiated by S during the first time-slot and by $\alpha\mathcal{P}_{II}$ and $(1-\alpha)\mathcal{P}_{II}$ the power spent respectively by S and R in the second time-slot. Since the overall radiated power is always $\mathcal{P}$, it must be $\mathcal{P} = \mathcal{P}_I + \mathcal{P}_{II}$. The coefficient α is chosen in order to minimize the final average bit-error probability (10.54). The power $\mathcal{P}_I$ is chosen in order to achieve a required average SNR_R at the relay, defined as $\text{SNR}_R := \mathcal{P}_I/(\sigma_n^2\, d_{sr}^2)$. All distances in the network are normalized with respect to the distance d_{sd} between S and D.

Example - ML versus sub-optimum detector: In Figure 10.17, we compare the average BER obtained using cooperative and non-cooperative schemes, in a slow

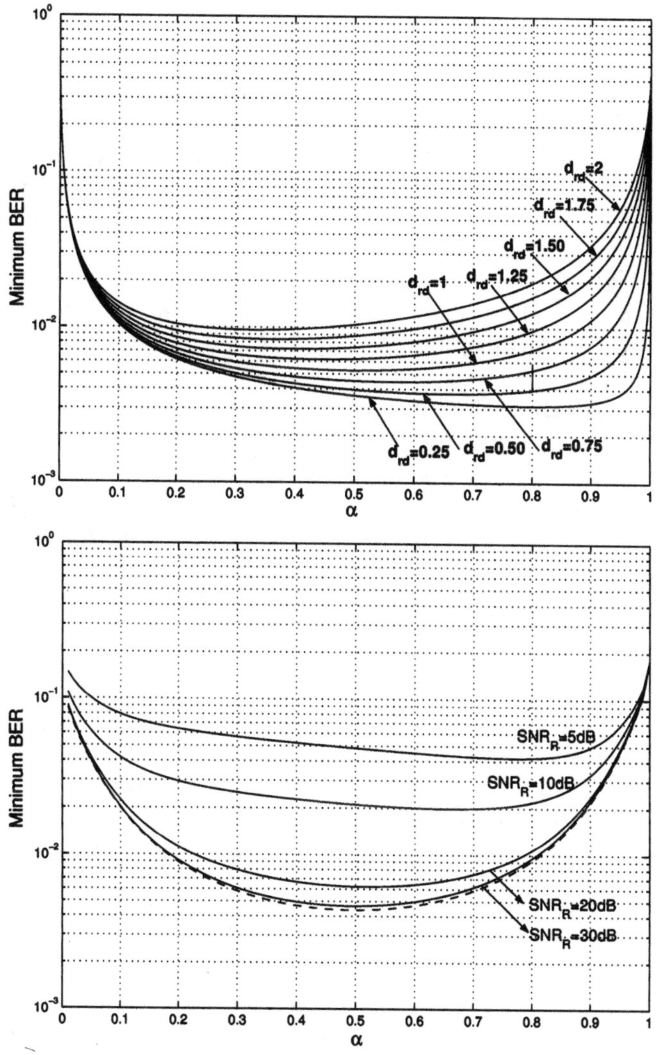

Figure 10.16 Average BER versus α: (a) no errors at the relay; and (b) including errors at the relay node.

fading scenario. The BER is averaged over 2,000 independent channel realizations. All curves are plotted versus the SNR in D, defined as $\text{SNR}_D := \frac{\mathcal{P}}{\sigma_n^2 d_{sd}^2}$. This is also the SNR of the non-cooperative case. The variance of the noise at both R and D is unitary. In this example, we set $d_{sr} = 0.1$ and $d_{rd} = 0.9$. The results shown in Figure 10.17 are achieved transmitting with a power $\mathcal{P}_I$ yielding an average SNR_R at the relay equal to 15 dB, in Figure 10.17a), and 25 dB in Figure 10.17b), for all values of SNR_D reported in the abscissas. Since the noise power and SNR_R are both fixed, increasing SNR_D means that $\mathcal{P}_{II}$ increases. In Figure 10.17 we report, for the sake of comparison, the average BER obtained with the following schemes: 1) the non-cooperative case (dotted line); 2) the ideal ML detector for the DSTC scheme, with no errors at the relay (dashed and dotted line); 3) the ML detector, incorporating the decision errors at the relay (dashed line); 4) the sub-optimum receiver for the DSTC scheme, showing both the theoretical average BER (solid line) and the corresponding simulation results (circles).

The floor on the BER of the sub-optimum receiver is due to the decision errors at the relay node. The floor can decrease only if the SNR_R at the relay increases, as is evident by comparing Figures 10.17 a) and b). Conversely, the ML detector that takes into account the probability of errors at the relay is more robust against the errors. In fact, the ML detector has the possibility to discard (or weight with a vanishing coefficient) the data coming from the relay, when the associated error probability is high. Clearly, this requires some exchange of information between S and R and this is the price to be paid.

It is also interesting to notice, from Figure 10.17, that the sub-optimum DSTC scheme exhibits performance very close to the optimal ML detector, at low SNR_D, i.e., before the BER floor. This indicates that the sub-optimum detector is a good choice whenever the final BER floor is lower than the required BER, because it is much less complicated to implement than the ML detector and, most important, differently from the ML strategy, it does not require any exchange of information between R and D, about the BER in R. The price paid for this simplicity is that the R node must have a sufficiently high SNR to guarantee that the BER of interest be above the floor.

Finally, looking at the slopes of the average BER curves of the ML DSTC detector, shown in Figures 10.17 a) and b), it is worth noticing that, in the absence of errors at the relay, the cooperative scheme achieves full spatial diversity gain. However, in the presence of decoding errors, the cooperative schemes do not achieve the full diversity gain. Nevertheless, there is a considerable coding gain, which justifies the use of cooperation. Indeed, a more attentive look at the results shows that the average BER starts approaching the slope with maximum diversity, as far as

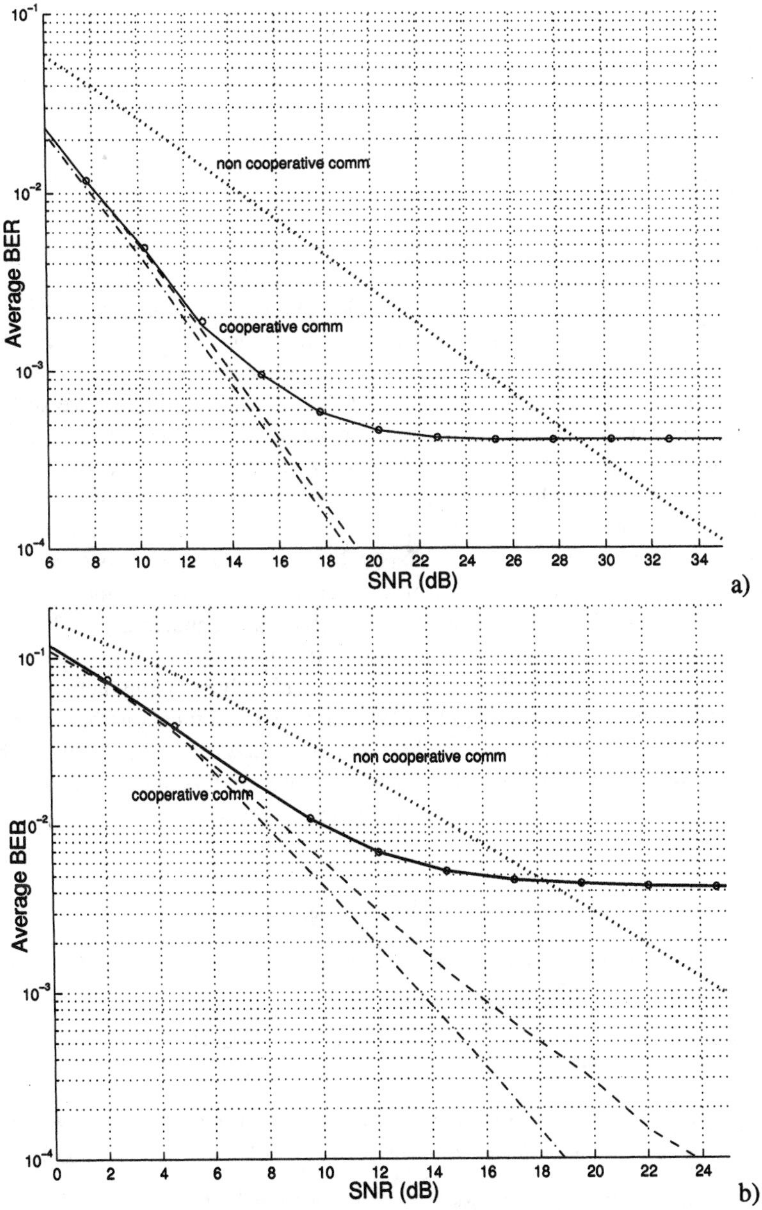

Figure 10.17 Comparison between average BER vs. SNR (dB) achieved with different decision schemes: single S/D link (dotted line); ideal ML detector (dashed and dotted line); real ML detector (dashed line); sub-optimum receiver - theoretical results (solid line) and simulation (circles); a) $\text{SNR}_R = 25$ dB, b) $\text{SNR}_R = 15$ dB; $d_{sr} = 0.1, d_{rd} = 0.9$.

the errors at the relay are negligible with respect to the errors at the destination. Then, when the errors at the relay become dominant, the final BER curve follows a $1/\mathrm{SNR}_D$ behavior.

10.7 SUMMARY

In this chapter we have shown that cooperation among radio nodes can be helpful to reduce the power strictly necessary to insure a desired BER. Cooperation in the form of relaying has been known and used for a long time. In this chapter, we have shown that coordinating the transmission of cooperative radio nodes using the general and powerful framework of distributed space-time coding can provide a considerable gain. This field is still at the beginning, but the preliminary results are sufficiently promising to foresee an increasing interest in the years to come. We have formulated the connectivity problem of a wireless network in terms of the connectivity of random geometric graphs. We have limited the analysis to graphs (networks) where all links have the same statistical properties. A more general network fading model, assuming different statistics for different links might give rise to *small world* or *scale-free* network models, which may be expected to play a fundamental role in the analysis and design of novel wireless network architectures, to decrease the average distance (measured in terms of minimum number of hops) between radio nodes of a wireless network. A very lively and enjoyable reading on the fascinating field of complex networks is [37]. Cooperative communications has been barely touched upon in this chapter, but this is another area where cross-fertilization from other fields, like population biology or self-organization paradigms may produce a substantial leap towards future communication systems.

References

[1] Tobagi, F., "Modeling and performance analysis of multihop radio networks," *Proc. of the IEEE*, Jan. 1987, pp. 135–155.

[2] Cover, T. M., El Gamal, A. A., "Capacity theorems for the relay channel," *IEEE Trans. on Inform. Theory*, Sept. 1979, pp. 572–584.

[3] Gupta, P., Kumar, P. R., "Critical power for asymptotic connectivity in wireless networks," in *Stochastic Analysis, Control, Optimization and Applications*, Birkhäuser, 1998, pp. 547–566.

[4] Gupta, P., Kumar, P.R., "The capacity of wireless networks," *IEEE Trans. on Information Theory*, Vol. 46, March 2000, pp.388–404.

[5] Host-Madsen, A., On the capacity of wireless relaying," *Proc. of VTC 2002-Fall*, Vol. 3, 24–28 Sept. 2002, pp. 1333 -1337.

[6] Sendonaris, A., Erkip, E., Aazhang, B., "Increasing uplink capacity via user cooperation diversity," *Proc. of ISIT 1998*, 16-21 Aug. 1998, p. 156.

[7] Laneman, J. N., Wornell, G. W.,"Energy-efficient antenna sharing and relaying for wireless networks," *Proc. of WCNC 2000*, Vol. 1, 23–28 Sept. 2000, pp. 7–12.

[8] Laneman, J. N., Wornell, G. W., "Distributed space-time coded protocols for exploiting cooperative diversity in wireless networks," *Proc. of GLOBECOM 2002*, Nov 17-21, 2002, pp. 77–81.

[9] Laneman, J. N., Wornell, G. W., "Distributed space-time-coded protocols for exploiting cooperative diversity in wireless networks," *IEEE Trans. on Information Theory*, Oct. 2003, pp. 2415–2425.

[10] Emamian, V., Kaveh, M., "Combating shadowing effects for systems with transmitter diversity by using collaboration among mobile users," *Proc. of Intern. Symp. on Commun.*, Taiwan, Nov. 2001, Vol. 9.4, pp. 105.1–105.4.

[11] Hunter, T. E., Nosratinia, A., "Cooperation diversity through coding," *Proc. of ISIT 2002*, 2002, p. 220.

[12] Hunter, T. E., Nosratinia, A., "Performance analysis of coded cooperation diversity," *Proc. of ICC 2003*, Vol. 4, 2003, pp. 2688 –2692.

[13] M. Dohler, E. Lefranc, H. Aghvami, "Space-time block codes for virtual antenna arrays", *Proc. of the 13th IEEE Int. Symp. on Personal, Indoor and Mobile Radio Commun., PIMRC 2002*, Vol.1, pp. 414–417 , 15-18 Sept., 2002.

[14] Stefanov, A., Erkip, E., "Cooperative coding for wireless networks," *Proc. of the* 4th *Int. Workshop on Mobile and Wireless Communications Network*, 9–11 Sept. 2002, pp. 273–277.

[15] Stefanov, A., Erkip, E., "On the performance analysis of cooperative space-time coded systems," *Proc. of WCNC 2003*, Vol. 2, 16-20 March 2003, pp. 729–734.

[16] Anghel, P. A., Leus, G., Kaveh, M., "Multi-user space-time coding in cooperative networks," *Proc. of ICASSP 2003*, Vol. 4, April 6-10, 2003, pp. IV-73–IV-76.

[17] Anghel, P. A., Leus, G., Kaveh, M., "Relay assisted uplink communication over frequency-selective channels," *Proc. of SPAWC 2003*, June 15-18, 2003.

[18] Hammerstroem, I., et al. "Space-Time Processing for Cooperative Relay Networks," *Proc. of SPAWC 2003*, Rome, June 15-18, 2003.

[19] Barbarossa, S., Scutari, G., "Cooperative diversity through virtual arrays in multihop networks," *Proc. of IEEE ICASSP 2003*, Hong Kong (China), April 2003, pp. 209–212.

[20] Scutari, G., Barbarossa, S., Ludovici, D., "Cooperation diversity in multihop wireless networks using opportunistic driven multiple access," *Proc. of IEEE Signal Processing Advances in Wireless Communications, SPAWC 2003*, Rome, Italy, June 2003, pp. 170–174.

[21] Barbarossa, S., Scutari, G., "Distributed space-time coding strategies for wideband multihop networks: Regenerative vs. non-regenerative relays," *Proc. of ICASSP 2004*, Montreal, Canada, May 2004, pp. 501–504.

[22] Scutari, G., Barbarossa, S., "Distributed space-time coding for regenerative relay networks," to appear on the *IEEE Transactions on Wireless Communications* (2005); see also Barbarossa, S., Scutari, G., "Distributed space-time coding for multihop networks," *Proc. of ICC 2004*, Paris, France, June 2004, pp. 916–920.

[23] Barbarossa, S., et al. "Cooperative wireless networks based on distributed space-time coding," *Proc. of Int. Workshop on Wireless and Ad-Hoc Networks (IWWAN 2004)*, Oulu, Finland, May 31-June 3, 2004.

[24] IEEE Std 802.15.3-2003, "Part 15.3: Wireless Medium Access Control (MAC) and Physical Layer (PHY) Specifications for High Rate Wireless Personal Area Networks (WPANs)", 2003.

[25] Bollobás, B., *Random Graphs*, Academic Press, 1985.

[26] Bollobás, B., *Modern Graph Theory*, Springer, 1998.

[27] Penrose, M. D., *Geometric Random Graphs*, Oxford, UK: Oxford University Press, 2003.

[28] Cressie, N. A. C., *Statistics for Spatial Data*, New York: John Wiley & Sons, 1993.

[29] Penrose, M. D., "On k-conectivity for a geometric random graph," *Wiley Random Structures and Algorithms*, vol. 15, no. 2, 1999, pp.145–164.

[30] Papoulis, A., *Probability, Random Variables and Stochastic Processes,*, (third edition), New York: McGraw Hill, 1995.

[31] Bettstetter, C., "On the minimum node degree and connectivity of a wireless multihop network," *Proc. of MOBIHOC '02*, EPF Lausanne, 16-20 June 2002, pp. 80–91.

[32] Cheng, Y.-C., Robertazzi, T. G., "Critical connectivity phenomena in multihop radio models," *IEEE Trans. on Communications*, vol. 37, July 1989.

[33] Simon, M. K., Alouini, M.-S., *Digital communications over fading channels: A unified approach to performance analysis*, New York: John Wiley & Sons, 2000.

[34] El Gamal, H., Damen, M. O., "Universal space-time coding," *IEEE Transactions on Information Theory*, Vol. 49, May 2003, pp. 1097–1119.

[35] Ma, X., Giannakis, G. B., "Full-diversity full-rate complex-field space-time coding," *IEEE Transactions on Signal Processing*, Vol. 49, Nov. 2003, pp. 2917–2930.

[36] Proakis, J. , *Digital Communications*, (4th edition), New York: McGraw Hill, 2000.

[37] Barabasi, A., L., *Linked: How everything is connected to everything else and what it means*, Plume Books, 2003.

Symbols

$\approx$	approximately equal to
$\otimes$	the Kronecker product
$\mathbf{0}_m$	$m \times m$ all zeros matrix
$\mathbf{0}_{m,n}$	$m \times n$ all zeros matrix
$\lvert a \rvert$	the magnitude of the scalar a
$\boldsymbol{A}^*$	the elementwise conjugate of $\boldsymbol{A}$
$\boldsymbol{A}^\dagger$	the pseudoinverse of $\boldsymbol{A}$
$[\boldsymbol{A}]_{i,j}$	the (i, j)th element of $\boldsymbol{A}$
$\lvert\lvert \boldsymbol{A} \rvert\rvert$	the Frobenius norm of $\boldsymbol{A}$
$\boldsymbol{A}^H$	the conjugate transpose of $\boldsymbol{A}$
$\boldsymbol{A}^T$	the transpose of $\boldsymbol{A}$
$\lvert \boldsymbol{A} \rvert$	the determinant of $\boldsymbol{A}$
$\text{rank}(\boldsymbol{A})$	the rank of the matrix $\boldsymbol{A}$
$\text{tr}(\boldsymbol{A})$	the trace of $\boldsymbol{A}$
$\text{vec}(\boldsymbol{A})$	the vector obtained by stacking the columns of $\boldsymbol{A}$ on top of each other
$\boldsymbol{I}_N$	the identity matrix of dimension $N \times N$
$\lvert \Omega \rvert$	the cardinality of the set Ω
$\delta(x)$	the Dirac delta function
δ_{kl}	the Kronecker symbol (equal to one if $k = l$ and zero otherwise)
$E\{\cdot\}$	the expected value
$f_X(x)$	the pdf of the random variable X
$f_{\mathbf{x}}(x_1, x_2, \ldots, x_N)$	the joint pdf of the random variables $X_1, X_2, \ldots, X_N$
$F_X(x)$	the CDF of the random variable X
$F_{\mathbf{x}}(x_1, x_2, \ldots, x_N)$	the joint CDF of the random variables $X_1, X_2, \ldots, X_N$
$\min(a_1, a_2, \ldots, a_n)$	the minimum of $a_1, a_2, \ldots, a_n$

$Q(x)$	the Gaussian $Q(x)$-function
$\mathbb{C}$	the complex field
$\mathbb{R}$	the real field
$\Re\{a\}, \Im\{a\}$	the real and imaginary parts of scalar a, respectively
$\Re\{\boldsymbol{A}\}, \Im\{\boldsymbol{A}\}$	the elementwise real and imaginary parts of matrix $\boldsymbol{A}$, respecti
$\log_2(x)$	the base-2 logarithm of x
$\propto$	proportional to
$u(x)$	the unit step function
$\Pr\{A\}$	the probability of the event A
$\mathrm{rect}(x)$	the rectangular function: it assumes the value 1 for $\lvert x\rvert < 1/2$, or 0 otherwise
$\mathbf{1}[A]$	the indicator function: it assumes the value 1 if A is true, or 0 if A is false
$\boldsymbol{x} \sim \mathcal{N}(\boldsymbol{\mu}, \boldsymbol{C})$	$\boldsymbol{x}$ is a Gaussian random vector with mean μ and covariance matrix $\boldsymbol{C}$
$\boldsymbol{x} \sim \mathcal{CN}(\boldsymbol{\mu}, \boldsymbol{C})$	$\boldsymbol{x}$ is a circularly symmetric Gaussian complex random vector with mean μ and covariance matrix $\boldsymbol{C}$

Acronyms

A/D	analog to digital
A&F	amplify and forward
AP	access point
AWGN	additive white gaussian noise
BER	bit error rate
BLAST	Bell laboratories layered space-time
BS	base station
BSC	binary symmetric channel
CBPS	coded bits per symbol
CBPSC	coded bits per subcarrier
CC	convolutional coding
CDF	cumulative distribution function
CDMA	code division multiple access
COFDM	coded-OFDM
CP	cyclic prefix
CSMA	carrier sense multiple access
CT	continuous-time
D/A	digital to analog
DAB	digital audio broadcasting
dB	decibel
DBPS	data bits per symbol
DC	diagonal coding
D&F	decode and forward
D-FRFD	distributed full-rate/full-diversity
DFT	discrete Fourier transform
Di-BLAST	distributed BLAST
D-OSTC	distributed orthogonal STC
DPC	dirty-paper coding

DPCCH	dedicated physical control channel
DPDCH	dedicated physical data channel
DS	direct-sequence
DSC	diagonally strict concave
DS-SS	direct-sequence spread spectrum
DSTC	distributed space-time coding
DTTB	digital terrestrial television broadcasting
DVB	digital video broadcasting
FDD	frequency division duplex
FDMA	frequency division multiple access
FEC	forward error correction
FFT	fast Fourier transform
FIR	finite impulse response
FRFD	full-rate full-diversity
FSK	frequency shift keying
HC	horizontal coding
Hz	Hertz
IBI	inter block interference
IFFT	inverse fast Fourier transform
iid	independent and identically distributed
IIR	infinite impulse response
I/O	input-output
ISI	inter symbol interference
IWFA	iterative water filling algorithm
LAN	local area network
LOS	line of sight
LPF	lowpass filter
LTI	linear time invariant
LTV	linear time varying
MA	multiple access
MAN	metropolitan area network
MC	multicarrier
MC-CDMA	multicarrier code division multiple access
MIMO	multiple input multiple output
MISO	multiple input single output
ML	maximum likelihood
MMSE	minimum mean square error
MRC	maximal ratio combining
MSE	mean square error
MT	mobile terminal
MU	multiuser

MUI	multiuser interference
NE	Nash equilibrium
OFDM	orthogonal frequency division multiplexing
OFDMA	orthogonal frequency division multiple access
1D	one dimensional
OSTBC	orthogonal space-time block coding
PAM	pulse amplitude modulation
pdf	probability density function
PEP	pairwise error probability
PO	Pareto optimum
PHY	physical
PSK	phase shift keying
P2P	peer to peer
QAM	quadrature amplitude modulation
QOD	quasi orthogonal design
QoS	quality of service
QRD	QR decomposition
RA	resource allocation
RD	resource discovery
RRM	radio resource management
RX	receiver
S/D&F	selective decode and forward
SDMA	space division multiple access
SER	symbol error rate
SIMO	single input multiple output
SIR	signal to interference ratio
SNIR	signal to noise plus interference ratio
SISO	single input single output
SNR	signal to noise ratio
SS	spread-spectrum
STC	space-time coding
SR2D	source-relay to destination
S2R	source to relay
SVD	singular value decomposition
TDD	time division duplex
TDMA	time division multiple access
3G	third generation
TOD	trace orthogonal design
TST	threaded space-time coding
TX	transmitter
2D	two dimensional

UMTS	universal mobile telecommunications system
UST	universal space-time
U-TOD	unitary TOD
VC	vertical coding
VoIP	voice-over-Internet protocol
WF	water filling
WH	Walsh-Hadamard
WLAN	wireless local area network
ZF	zero-forcing
ZFDP	zero-forcing dirty-paper

About the Author

Sergio Barbarossa received his B.Sc./M.Sc. in electrical engineering in 1984 and his PhD in in electrical engineering in 1988, from the University of Rome "La Sapienza," Italy. He started his research activity on synthetic aperture radar (SAR), in 1984, at Selenia, Rome. In 1987, he worked at the Environmental Research Institute of Michigan (ERIM), Ann Arbor, MI. From 1988 to 1991, he was an assistant professor at the University of Perugia. In November 1991, he joined the University of Rome "La Sapienza," where he is currently a full professor. From 1998 to 2004, he has been a member of the IEEE Signal Processing for Communications Technical Committee. He served as an associate editor for the IEEE Transactions on Signal Processing. He organized, as the general chairman, the Fourth IEEE Workshop on Signal Processing Advances in Wireless Communications, Rome, June 2003. He has held positions as visiting scientist and visiting professor at the University of Virginia (1995, 1997), the University of Minnesota (1999), and the Polytechnic University of Catalonia, Barcelona, Spain (2001). He received the 2000 IEEE Best Paper Award from the IEEE Signal Processing Society, in the area of Signal Processing for Communications, as the coauthor of a paper on optimal precoding and blind channel equalization. He has been one of the principal investigators in the international projects *SATURN*, on space-time coding, and *ROMANTIK*, on multihop wireless networks, funded by the European Union. He is currently a member of the editorial board of the IEEE Transactions on Signal Processing, focusing on sensor networks. His current research interests lie in the areas of cooperative communications, random graphs, self-organizing networks, sensor networks, and space-time coding.

Index

Recent Titles in the Artech House Mobile Communications Series

John Walker, Series Editor

3G CDMA2000 Wireless System Engineering, Samuel C. Yang

3G Multimedia Network Services, Accounting, and User Profiles, Freddy Ghys, Marcel Mampaey, Michel Smouts, and Arto Vaaraniemi

802.11 WLANs and IP Networking: Security, QoS, and Mobility, Anand R. Prasad, Neeli R. Prasad

Advances in 3G Enhanced Technologies for Wireless Communications, Jiangzhou Wang and Tung-Sang Ng, editors

Advances in Mobile Information Systems, John Walker, editor

Advances in Mobile Radio Access Networks, Y. Jay Guo

CDMA for Wireless Personal Communications, Ramjee Prasad

CDMA Mobile Radio Design, John B. Groe and Lawrence E. Larson

CDMA RF System Engineering, Samuel C. Yang

CDMA Systems Capacity Engineering, Kiseon Kim and Insoo Koo

CDMA Systems Engineering Handbook, Jhong S. Lee and Leonard E. Miller

Cell Planning for Wireless Communications, Manuel F. Cátedra and Jesús Pérez-Arriaga

Cellular Communications: Worldwide Market Development, Garry A. Garrard

Cellular Mobile Systems Engineering, Saleh Faruque

The Complete Wireless Communications Professional: A Guide for Engineers and Managers, William Webb

EDGE for Mobile Internet, Emmanuel Seurre, Patrick Savelli, and Pierre-Jean Pietri

Emerging Public Safety Wireless Communication Systems, Robert I. Desourdis, Jr., et al.

The Future of Wireless Communications, William Webb

GPRS for Mobile Internet, Emmanuel Seurre, Patrick Savelli, and Pierre-Jean Pietri

GPRS: Gateway to Third Generation Mobile Networks, Gunnar Heine and Holger Sagkob

GSM and Personal Communications Handbook, Siegmund M. Redl, Matthias K. Weber, and Malcolm W. Oliphant

GSM Networks: Protocols, Terminology, and Implementation, Gunnar Heine

GSM System Engineering, Asha Mehrotra

Handbook of Land-Mobile Radio System Coverage, Garry C. Hess

Handbook of Mobile Radio Networks, Sami Tabbane

High-Speed Wireless ATM and LANs, Benny Bing

Interference Analysis and Reduction for Wireless Systems, Peter Stavroulakis

Introduction to 3G Mobile Communications, Second Edition, Juha Korhonen

Introduction to Digital Professional Mobile Radio, Hans-Peter A. Ketterling

Introduction to GPS: The Global Positioning System, Ahmed El-Rabbany

An Introduction to GSM, Siegmund M. Redl, Matthias K. Weber, and Malcolm W. Oliphant

Introduction to Mobile Communications Engineering, José M. Hernando and F. Pérez-Fontán

Introduction to Radio Propagation for Fixed and Mobile Communications, John Doble

Introduction to Wireless Local Loop, Second Edition: Broadband and Narrowband Systems, William Webb

IS-136 TDMA Technology, Economics, and Services, Lawrence Harte, Adrian Smith, and Charles A. Jacobs

Location Management and Routing in Mobile Wireless Networks, Amitava Mukherjee, Somprakash Bandyopadhyay, and Debashis Saha

Mobile Data Communications Systems, Peter Wong and David Britland

Mobile IP Technology for M-Business, Mark Norris

Mobile Satellite Communications, Shingo Ohmori, Hiromitsu Wakana, and Seiichiro Kawase

Mobile Telecommunications Standards: GSM, UMTS, TETRA, and ERMES, Rudi Bekkers

Mobile Telecommunications: Standards, Regulation, and Applications, Rudi Bekkers and Jan Smits

Multiantenna Digital Radio Transmission, Massimiliano "Max" Martone

Multiantenna Wireless Communications Systems, Sergio Barbarossa

Multipath Phenomena in Cellular Networks, Nathan Blaunstein and Jørgen Bach Andersen

Multiuser Detection in CDMA Mobile Terminals, Piero Castoldi

Personal Wireless Communication with DECT and PWT, John Phillips and Gerard Mac Namee

Practical Wireless Data Modem Design, Jonathon Y. C. Cheah

Prime Codes with Applications to CDMA Optical and Wireless Networks, Guu-Chang Yang and Wing C. Kwong

QoS in Integrated 3G Networks, Robert Lloyd-Evans

Radio Engineering for Wireless Communication and Sensor Applications, Antti V. Räisänen and Arto Lehto

Radio Propagation in Cellular Networks, Nathan Blaunstein

Radio Resource Management for Wireless Networks, Jens Zander and Seong-Lyun Kim

RDS: The Radio Data System, Dietmar Kopitz and Bev Marks

Resource Allocation in Hierarchical Cellular Systems, Lauro Ortigoza-Guerrero and A. Hamid Aghvami

RF and Microwave Circuit Design for Wireless Communications, Lawrence E. Larson, editor

Sample Rate Conversion in Software Configurable Radios, Tim Hentschel

Signal Processing Applications in CDMA Communications, Hui Liu

Software Defined Radio for 3G, Paul Burns

Spread Spectrum CDMA Systems for Wireless Communications, Savo G. Glisic and Branka Vucetic

Third Generation Wireless Systems, Volume 1: Post-Shannon Signal Architectures, George M. Calhoun

Traffic Analysis and Design of Wireless IP Networks, Toni Janevski

Transmission Systems Design Handbook for Wireless Networks, Harvey Lehpamer

UMTS and Mobile Computing, Alexander Joseph Huber and Josef Franz Huber

Understanding Cellular Radio, William Webb

Understanding Digital PCS: The TDMA Standard, Cameron Kelly Coursey

Understanding GPS: Principles and Applications, Elliott D. Kaplan, editor

Understanding WAP: Wireless Applications, Devices, and Services, Marcel van der Heijden and Marcus Taylor, editors

Universal Wireless Personal Communications, Ramjee Prasad

WCDMA: Towards IP Mobility and Mobile Internet, Tero Ojanperä and Ramjee Prasad, editors

Wireless Communications in Developing Countries: Cellular and Satellite Systems, Rachael E. Schwartz

Wireless Intelligent Networking, Gerry Christensen, Paul G. Florack, and Robert Duncan

Wireless LAN Standards and Applications, Asunción Santamaría and Francisco J. López-Hernández, editors

Wireless Technician's Handbook, Second Edition, Andrew Miceli